Biotechnology for Alternative Fuels

Biotechnology for Alternative Fuels

Edited by **Craig Zodikoff**

SYRAWOOD
PUBLISHING HOUSE
New York

Published by Syrawood Publishing House,
750 Third Avenue, 9th Floor,
New York, NY 10017, USA
www.syrawoodpublishinghouse.com

Biotechnology for Alternative Fuels
Edited by Craig Zodikoff

International Standard Book Number: 978-1-68286-249-0 (Hardback)

Printed in China.

Contents

Preface

This book aims to highlight the current researches and provides a platform to further the scope of innovations in this area. This book is a product of the combined efforts of many researchers and scientists from different parts of the world. The objective of this book is to provide the readers with the latest information in the field.

There has been an increasing demand for alternative fuels because of the alarming fuel crisis. Biofuels like bioethanol, biodiesel, etc. have become prominent fuel sources. Biotechnology and its techniques play a major role in production of these fuels. Some of the crucial topics that provide a comprehensive understanding of this subject are included in this book like microbial biotechnology, metabolic engineering, waste management, industrial fermentation, etc. As this field is emerging at a fast pace, this book will help the readers to better understand the applications of biotechnology for production of alternative fuels.

I would like to express my sincere thanks to the authors for their dedicated efforts in the completion of this book. I acknowledge the efforts of the publisher for providing constant support. Lastly, I would like to thank my family for their support in all academic endeavors.

Editor

Comparative metagenomics of biogas-producing microbial communities from production-scale biogas plants operating under wet or dry fermentation conditions

Yvonne Stolze[1†], Martha Zakrzewski[2†], Irena Maus[1], Felix Eikmeyer[1], Sebastian Jaenicke[3], Nils Rottmann[4], Clemens Siebner[1], Alfred Pühler[1] and Andreas Schlüter[1*]

Abstract

Background: Decomposition of biomass for biogas production can be practiced under wet and dry fermentation conditions. In contrast to the dry fermentation technology, wet fermentation is characterized by a high liquid content and a relatively low total solid content. In this study, the composition and functional potential of a biogas-producing microbial community in an agricultural biogas reactor operating under wet fermentation conditions was analyzed by a metagenomic approach applying 454-pyrosequencing. The obtained metagenomic dataset and corresponding 16S rRNA gene amplicon sequences were compared to the previously sequenced comparable metagenome from a dry fermentation process, meeting explicitly identical boundary conditions regarding sample and community DNA preparation, sequencing technology, processing of sequence reads and data analyses by bioinformatics tools.

Results: High-throughput metagenome sequencing of community DNA from the wet fermentation process applying the pyrosequencing approach resulted in 1,532,780 reads, with an average read length of 397 bp, accounting for approximately 594 million bases of sequence information in total. Taxonomic comparison of the communities from wet and dry fermentation revealed similar microbial profiles with *Bacteria* being the predominant superkingdom, while the superkingdom *Archaea* was less abundant. In both biogas plants, the bacterial phyla *Firmicutes*, *Bacteroidetes*, *Spirochaetes* and *Proteobacteria* were identified with descending frequencies. Within the archaeal superkingdom, the phylum *Euryarchaeota* was most abundant with the dominant class *Methanomicrobia*. Functional profiles of the communities revealed that environmental gene tags representing methanogenesis enzymes were present in both biogas plants in comparable frequencies. 16S rRNA gene amplicon high-throughput sequencing disclosed differences in the sub-communities comprising methanogenic *Archaea* between both processes. Fragment recruitments of metagenomic reads to the reference genome of the archaeon *Methanoculleus bourgensis* MS2^T revealed that dominant methanogens within the dry fermentation process were highly related to the reference.

Conclusions: Although process parameters, substrates and technology differ between the wet and dry biogas fermentations analyzed in this study, community profiles are very similar at least at higher taxonomic ranks, illustrating that core community taxa perform key functions in biomass decomposition and methane synthesis. Regarding methanogenesis, *Archaea* highly related to the type strain *M. bourgensis* MS2^T dominate the dry fermentation process, suggesting the adaptation of members belonging to this species to specific fermentation process parameters.

Keywords: Biogas, Dry fermentation, Wet fermentation, Microbial communities, Methanogenesis, Metagenomics, 16S rRNA, *Methanoculleus bourgensis*, Fragment recruitment

* Correspondence: aschluet@cebitec.uni-bielefeld.de
†Equal contributors
[1]Institute for Genome Research and Systems Biology, CeBiTec, Bielefeld University, Universitätsstraße 25, D-33615 Bielefeld, Germany
Full list of author information is available at the end of the article

Background

Rising energy costs and considerations on long term environmental sustainability have placed renewable energy sources in the focus of debate. The development of renewable energy resources offers the chance to replace traditional fossil fuels and can help to reduce carbon dioxide emissions [1,2]. An economically attractive technology to generate bioenergy is the production of biogas that is a mixture of methane (CH_4) and carbon dioxide (CO_2) as the main components, with small amounts of hydrogen sulfide (H_2S), nitrogen (N_2), hydrogen (H_2), ammonia (NH_3) and carbon monoxide (CO) [3]. The most common and widespread utilization of biogas is the production of electricity and heat by its combustion in combined heat and power units.

The process of biogas production takes place under anaerobic conditions and involves microbial decomposition of organic matter, yielding methane as the main final product of underlying metabolic pathways. In Germany, mostly maize silage combined with liquid manure is utilized as the substrate for biogas production [1,4]. Complex consortia of microorganisms are responsible for biomass decomposition and biogas production involving the stages substrate hydrolysis, acidogenesis, acetogenesis and methanogenesis. However, most of these microbes, as well as their roles in biogas production, are currently unknown. Recently, the analysis of the structure, composition and activity of microbial communities in relation to input substrates and fermentation parameters in biogas plants have become the focus of research [5-7]. It is generally accepted that a better understanding of the composition and activity of the multifarious microbial community is crucial for further optimization of reactor performance and fermentation process technologies. Moreover, to increase the yield of biogas, a detailed insight into relevant microbial metabolic pathways involved in methane synthesis and syntrophy is necessary.

Previous studies analyzed the taxonomic structure and enzymatic potential of biogas communities residing in agricultural biogas reactors. A classical microbiological approach for the analysis of the communities' taxonomy is the generation of 16S rRNA gene clone libraries, followed by Sanger sequencing of the 16S rRNA gene fragments [8-12]. Sequencing of 16S rRNA gene clone libraries is limited since coverage of the microbial complexity frequently is laborious, costly and time-consuming. Moreover, sequence information on community 16S rRNA marker genes does not provide direct insights into functions of microorganisms. To achieve deeper insights into community structure and function, metagenome analyses applying high-throughput sequencing technologies were carried out [13-17]. Elaborate bioinformatics methods and analysis platforms facilitated metagenome sequence data interpretation and comparison [13,14,16,18]. Another approach for comparison of metagenome datasets is fragment recruitment of metagenomic sequences related to selected genomes of reference species. This approach provides insights into the degree of relatedness of indigenous species within a given habitat to known reference species. Recently, fragment recruitment has been applied for marine and silage microbial communities [19,20].

In principle, decomposition of biomass for biogas production can be practiced under wet or dry fermentation conditions. Wet fermentation is characterized by a high liquid content and a low total solid content, which usually is below 10%. In contrast to this, in dry fermentation the total solids content is between 15 and 35%. Biogas plants operating under dry fermentation conditions apply mostly maize silage, green rye (and similar biomass), dung (cow dung, poultry dry excrement and so forth) or municipal solid wastes as substrates without any continuous supplementation of liquid manure, which consequently leads to a low liquid content [1,21]. To control the water content, recirculation of digestate or liquid is applied, which may have a great impact on the activity of the underlying community. Recirculation may influence the pH, salt and organic loads, which could inhibit the microbial activity. Dry fermentation proved to be a convenient technology for the fermentation of substrates possessing relatively high dry matter contents. According to the German Renewable Energy Law (EEG), a technology bonus was granted for dry fermentation biogas plants built before the year 2008. Previously, a dry fermentation process of a production-scale biogas plant was characterized at the metagenomic level [13,14,16]. On the other hand, wet fermentation utilizing maize silage and liquid manure from cattle or swine is performed in most mid-sized, agricultural biogas plants in Germany. In this study, the microbial community of an agricultural biogas plant performing wet fermentation was analyzed by applying a metagenomic approach. Obtained results were compared to taxonomic community profiles deduced from a dry fermentation biogas plant analyzed previously. The present study adopted exactly the same methodology for processing samples, preparing total community DNA and metagenome sequence data analysis as for metagenome analysis of the dry fermentation process in the study mentioned above. It is hypothesized that biogas-producing microbial communities comprise a 'core' microbiome and variable sub-communities that respond to specific conditions and process parameters prevailing in particular reactor environments. The objective of this study was to differentiate biogas communities from biogas plants performing wet and dry fermentation, with respect to their taxonomic profiles, and to deduce correlations between these profiles and process parameters collected for both fermentation types. Another aim of this study was to identify key species specifically adapted to one process or the other and their predicted functions, focusing on methanogenic species.

Results and discussion

Analyzed biogas production plants differ in substrate input and chemical parameters

To compare taxonomic and functional profiles of the biogas-producing microbial communities from production-scale biogas plants operating under dry or wet fermentation conditions, samples from the primary digesters of two agricultural biogas plants differing in these fermentation types were analyzed. The biogas plant operating under dry fermentation conditions (BGP_DF) was sampled previously [16], whereas samples from the biogas plant operating under wet fermentation conditions (BGP_WF) were taken in March 2011. One of the major differences between BGP_WF and BGP_DF is their dry matter content (BGP_DF: 14% ± 2% and BGP_WF: 9% ± 1%). BGP_DF was fed with high amounts of plant silages and low amounts of chicken manure [16], whereas the substrate composition of BGP_WF mainly consisted of maize silage and a relatively high amount of liquid pig manure (Table 1). Moreover, both biogas plants can be distinguished according to their process parameters, such as volatile organic acids, total inorganic carbon, acetic and propionic acid concentrations and ammonium contents (see Table 1). BGP_DF is characterized by higher acetic and propionic acid concentrations as compared to BGP_WF, suggesting that consumption of these compounds is limited in BGP_DF. The biogas and methane yields (698.2 l/kg and 350.5 l/kg organic dry matter (oDM), see Table 1) are in the normal range of production in mesophilic production-scale biogas plants.

In a recent study, a metagenome approach was carried out to study the taxonomic composition and functional potential of the microbial community in the biogas plant BGP_DF [13,16]. In total, 1,347,644 sequencing reads were generated with an average read length of 367.7 bases providing approximately 496 million bases sequence information (Table 2). Most biogas plants in Germany practice wet fermentation utilizing liquid manure and maize silage for the production of methane. To obtain insights into the microbial community composition of this process, a metagenome sequencing approach for the biogas plant applying wet fermentation (BGP_WF) was carried out. The same sample preparation, DNA-extraction method and sequencing technique were applied as previously described for BGP_DF to ensure comparability of the metagenome datasets. Sequencing of the samples originating from BGP_WF resulted in 1,532,780 sequencing reads, with an average read length of 387.3 bases, accounting for approximately 594 million bases sequence information (Table 2). To include only high quality sequences, both datasets were filtered for GC (G: Guanine, C: Cytosine bias and duplicates as previously described [22]. After this filtering step, 1,019,333 sequences from the BGP_DF and 1,097,549 sequences from the BGP_WF remained and were used for downstream taxonomic and functional analyses (see Table 2).

Comparative analyses of taxonomic profiles obtained from wet and dry fermentation communities revealed high similarities

The community structures in the biogas plants operating under wet (BGP_WF) or dry fermentation (BGP_DF) conditions were studied using CARMA3 [23] and MetaSAMS [24]. The software CARMA3 was applied to calculate the taxonomic (microbial composition based on phylogenetic analyses) and functional profile (predicted Pfam protein families based on similarity searches) of

Table 1 Characteristics of the studied biogas plants performing wet or dry fermentation technology

	Biogas plant operating dry fermentation (Sampling date: 14 August 2007)	Biogas plant operating wet fermentation (Sampling date: 1 March 2011)
pH	7.7 ± 0.01	7.8 ± 0.01
Conductivity (mS/cm)	17.1 ± 1	21.6 ± 1
VOA (mg/l)	7,739 ± 60	3,987 ± 31
TIC	15,159 ± 120	14,517 ± 115
VOA/TIC	0.51	0.27
NH_4-N (g NH_4-N/l)	2.25 ± 0.02	2.85 ± 0.02
Acetic acid (mg/l)	2,628 ± 50	344 ± 7
Propionic acid (mg/l)	179 ± 3.6	14 ± 0.3
Fed substrates	Maize silage (63%), green rye (35%), chicken manure (2%)	Maize silage (approximately 72%), pig manure (approximately 28%)
Biogas yield (l/kg oDM)	698.2	810.5
Methane yield (l/kg oDM)	350.5	417.8

oDM: Organic dry matter; TIC: Total inorganic carbon; VOA: Volatile organic acids.

Table 2 Metagenome sequencing statistics of the DNA samples from the wet (BGP_WF) and dry (BGP_DF) fermentation biogas plants

	Unfiltered sequences[a]		Filtered sequences[a]	
	BGP_DF	BGP_WF	BGP_DF	BGP_WF
Reads (bp)	1,347,644	1,532,780	1,019,333	1,097,549
Average read length (bp)	363	387	366.0	387.5
Sequence information (Mbp)	495.5	593.7	373.1	424.3

[a]Sequencing data summary is shown before (unfiltered sequences) and after (filtered sequences) the filtering step for duplicates and GC bias. Mbp: mega base pairs.

the microbial community. CARMA3 is implemented in MetaSAMS (Metagenome Sequence Analysis and Management System), a software suited for the analysis of metagenome datasets. For the interpretation only taxonomic assignments with an E-value threshold of 10^{-5} were considered. In total, 711,293 sequences of the BGP_WF were assigned to a superkingdom representing 64.8% of the total number of analyzed sequences. In the BGP_DF dataset, 604,243 sequences (59.3% of the total sequences) were similar to known reference sequences at the rank superkingdom. Rarefaction analyses on the mean taxonomic richness on the taxonomic family rank showed a saturation from approximately 800,000 reads, which indicates that the majority of the microbial community has been captured, while the rarefaction analysis on the rank genus was nearly saturated (Additional file 1: Figure S1A and D).

At higher taxonomic ranks, the community compositions in BGP_WF and BGP_DF are very similar (Figure 1). Both communities are mainly composed of bacterial (59% in BGP_WF and 52.5% in BGP_DF of the total analyzed reads) and archaeal microorganisms (5.5% in BGP_WF and 6.6% in BGP_DF) (Figure 1). The bacterial superkingdom mainly comprises the phyla *Firmicutes* (32.5% in BGP_WF and 25.4% in BGP_DF), *Bacteroidetes* (8.4% in BGP_WF and 6.2% in BGP_DF), *Spirochaetes* (1.7% in BGP_WF and 0.5% in BGP_DF) and *Proteobacteria* (1.7% in BGP_WF and 1.2% in BGP_DF). These phyla were also common in other biogas plant microbiota [9,17,18]. Bacterial groups belonging to the taxa *Firmicutes, Bacteroidetes* and *Spirochaetes* are assumed to be involved in cellulolytic degradation, proteolysis, acidogenesis and homoacetogenesis [10]. Among the archaeal community, *Euryarchaeota* (5.2% in BGP_WF and 6.2% in BGP_DF of all analyzed sequences) is the most abundant phylum with *Methanomicrobia* as the largest class (Figure 1). Comparing both profiles on the phylum level, minor differences were observed in the relative abundances of *Firmicutes.* However, there are some noteworthy changes in the relative abundances of taxonomic groups between wet and dry fermentation digesters on the class level, in particular of classes belonging to the phylum *Firmicutes.*

Among the phylum *Firmicutes* in BGP_WF, *Clostridia* (19.5% of all analyzed sequences in BGP_WF) forms the largest class, followed by *Bacilli* (1.8%) and *Erysipelotrichi* (1.2%) (Figure 1). Likewise, *Clostridia* and *Bacilli* are also the most abundant classes of *Firmicutes* in the microbiome of the BGP_DF, with 14.3% and 1.4% of the total number of analyzed reads, respectively, whereas *Erysipelotrichi* is barely present (0.1%) in this digester. While *Clostridia* and *Bacilli* species are well described in the anaerobic digestion process in biogas plants, the evidence for *Erysipelotrichi* species is sparse [248]. Little is known about the family *Erysipelotrichaceae.* However, it was also identified in the microbial community of the gut [25]. An increase in members of this family was associated with a diet high in fat, increased body weight and decreased fecal short-chain fatty acid concentrations in mice [25]. Concurrently, occurrence of *Erysipelotrichi* members correlates with lower short-chain fatty acid concentrations in BGP_WF. Whether this observation is really due to the metabolic capabilities of this group of microorganisms remains to be determined.

In both biogas plants, the most abundant families belonging to the class *Clostridia* are *Clostridiaceae* (3.8% in BGP_WF and 1.8% in BGP_DF of all analyzed sequences), *Ruminococcaceae* (0.5% in BGP_WF and 0.3% in BGP_DF) and *Lachnospiraceae* (0.3% in BGP_WF and 0.1% in BGP_DF) (Figures 1 and 2). *Clostridium,* the prevalent genus within *Clostridiaceae,* seems to belong to the core set of organisms, as it is dominant in both biogas plants studied (Figure 1). Species of this genus, such as *Clostridium thermocellum* [26] and *Clostridium clariflavum* [27], produce cellulosomes, an extracellular multi-enzyme complex which is important for the degradation of complex carbohydrates such as cellulose. Indeed, environmental gene tags (EGTs) classified to the genus *Clostridium* encode enzymes relevant in the hydrolysis process of glycoside bonds (PF00150 and PF00759). In both biogas plants, *Alkaliphilus* is the second largest genus within the family *Clostridiaceae.* It has also been detected in high amounts in a biogas plant fed with plant biomass and pig manure slurry [17]. The species *Alkaliphilus peptidofermentans,* isolated from a soda lake, is described to ferment peptides to acetate and formate [28]. The functional profile of EGTs assigned to the genus *Alkaliphilus* in the wet fermentation process includes various peptidase families (PF00768 and PF05343).

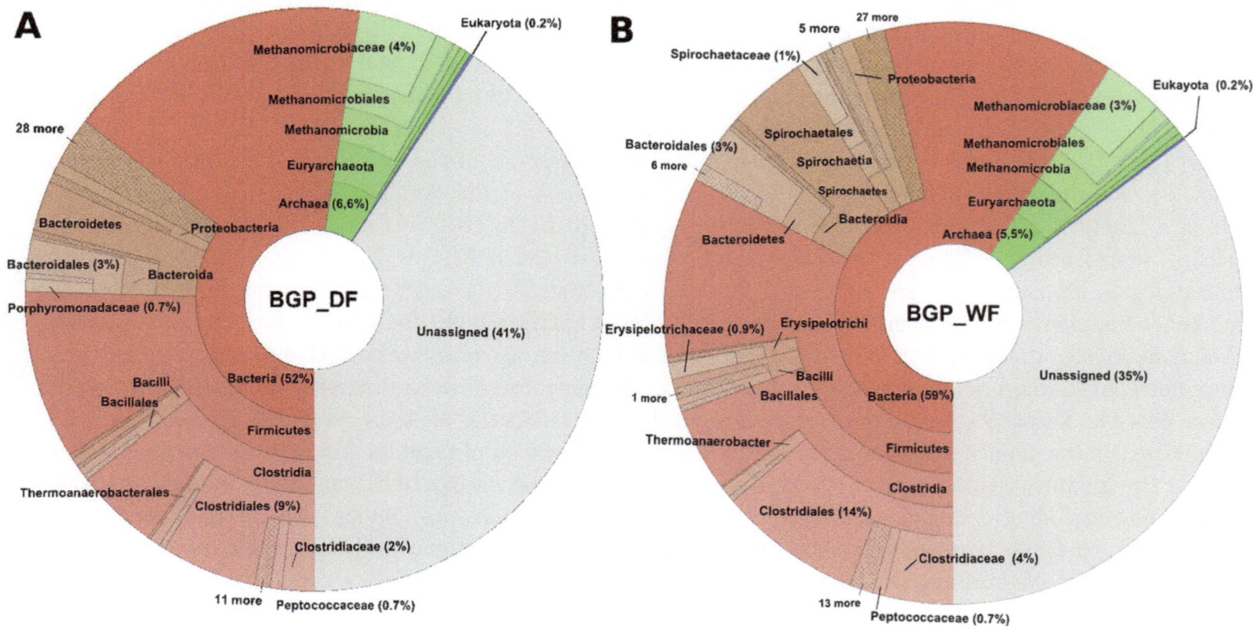

Figure 1 Taxonomic composition of microbial communities from the wet and dry fermentation processes. Taxonomic structure of the microbial community associated with a **(A)** dry (BGP_DF) or **(B)** wet (BGP_WF) fermentation process: The taxonomic composition of the microbial communities residing in the biogas plants analyzed was determined using CARMA3 and visualized by Krona plots. Only the ranks from superkingdom to family were considered.

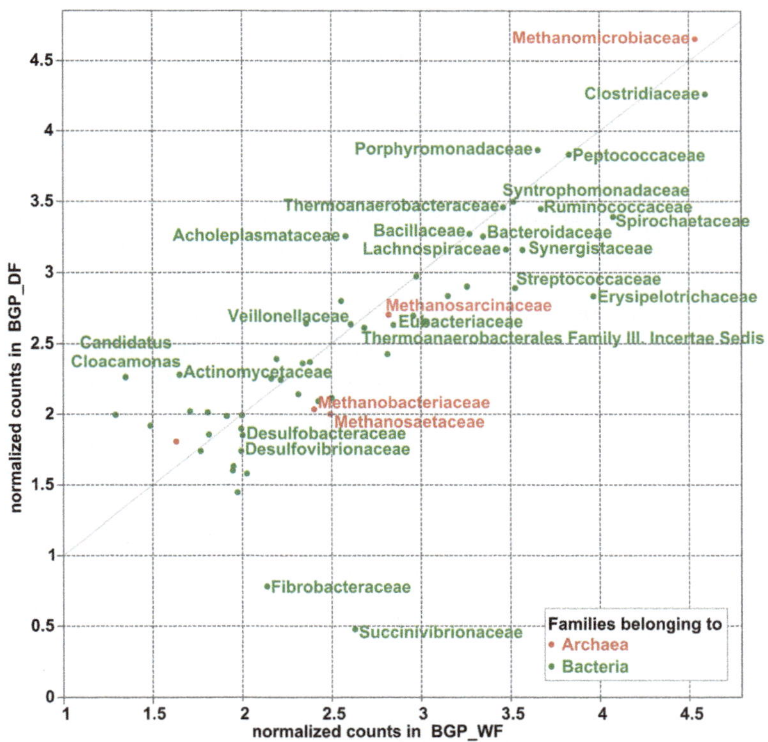

Figure 2 Scatterplot of sequence counts assigned to the taxonomic rank family for microbial communities from the wet and dry fermentation processes. For each taxonomic family, sequence abundances, normalized on the smallest dataset, with pseudocounts (for an explanation see Methods section) are plotted as a logarithm to the base 10. Red symbols indicate archaeal and green symbols indicate bacterial families. Only families for which the sum of the relative read abundances measured was at least 0.01% were considered.

Members of the genus *Ruminococcus* belonging to the family *Ruminococcaceae* (order *Clostridiales*) are cellulolytic organisms, which occur in the rumen community [29] and in biogas plants [17]. They play an important role in the digestion of plant cell wall material and produce acetate [29,30]. Among the functional profile of EGTs assigned to *Ruminococcus*, protein families for glycosyl hydrolase family 9, 26 and 48 (PF00759, PF02156 and PF04616) representing cellulose-, cellobiose- and hemicellulose-degrading enzymes were observed. Furthermore, gene fragments for dockerin and cohesin (PF00404 and PF00963), the scaffoldin units of cellulosomes, were detected in the functional profile of the wet fermentation process. Recently, the genome of *Acetivibrio cellulolyticus* of the family *Ruminococcus* was sequenced [31]. The genome carries genes for a complex cellulosome system including endoglucanases and cellobiohydrolases. Glycosyl family 8 and 9, both representing these enzymes, were detected among the EGTs assigned to the genus *Acetivibrio*.

The family *Spirochaetaceae* dominates the phylum *Spirochaetes* with 68% of the reads assigned to this phylum in BGP_DF (Figure 1). It is the third largest family, with 1.2% of all reads in the wet fermentation sample, whereas it belongs to the minor groups in the dry fermentation process (0.2%). Microorganisms related to *Treponema* have already been described in a mesophilic biogas digester treating pig manure using 16S rRNA clone libraries [10]. Genomes of the *Treponema* species encode proteolytic enzymes [32] and glycoside hydrolases [33]. EGTs classified to the genus *Treponema* in BGP_WF were also assigned to carbohydrate phosphorylase (PF00343), alpha amylase (PF00128) and 4-alpha-glucanotransferase (PF02446), which participate in starch and sucrose metabolism, carbohydrate phosphorylase (PF00343) and alphaamylase (PF00128). The family *Succinivibrionaceae* belonging to the *Gammaproteobacteria* is only predicted in the biogas plant operating wet fermentation (0.04% of all analyzed sequenced reads) and was sparsely detected in the dry fermentation process (Figures 1 and 2). This family is noted to use glucose and other carbohydrates as an energy source and to produce succinate and acetate [34]. Likewise, the family *Fibrobacteraceae* is more frequently present in the wet fermentation process. The known species *Fibrobacter succinogenes* is described to play a key role in the degradation process of cellulose [35].

Prevotellaceae belonging to the phylum *Bacteroidetes* is more abundant in the BGP_WF, with 0.18% of all analyzed sequences in the corresponding dataset as compared to 0.8% of all sequences in the BGP_DF (Figures 1 and 2). Species of the genus *Prevotella* are described to decompose hemicellulose, starch and pectin [36,37]. Various Pfam families were discovered in the wet fermentation digester that were assigned to *Prevotella* and are predicted to be involved in the degradation of hemicellulose (such as glycosyl hydrolase family 3 (PF00933, PF01915), glycosyl hydrolase 92 (PF07971) and alpha-L-arabinofuranosidase (PF06964)) or starch (such as glycosyl hydrolase family 31 (PF01055)).

Synergistaceae of the phylum *Synergistetes* belongs to the core-set of the families represented in both biogas plants (0.4% in BGP_WF and 0.1% in BGP_DF of all analyzed reads) (Figures 1 and 2). *Anaerobaculum* and *Aminobacterium* are genera within the phylum *Synergistetes*, which are both predicted in the biogas plants. Recently, *Anaerobaculum* was detected in anaerobic digestion of slaughterhouse waste mixture [38] and thermophilic sludge [11]. Species of *Anaerobaculum* are associated with the fermentation of peptides and produce short-chain fatty acids. Likewise, species of the genus *Aminobacterium* were isolated from anaerobic sludge and are described to ferment a range of amino acids to acetate, propionate and hydrogen [39,40].

Community profiles of the dry and wet fermentation process differ at lower taxonomic ranks

There are also a number of taxa that are slightly increased in the biogas plant operating the dry fermentation technology (Figures 1 and 2). For example, the two biogas plants differ in the proportion of sequences belonging to the family *Acholeplasmataceae*, of the phylum *Tenericutes*. The species *Acholeplasma laidlawii* was isolated from wastewater [15] and also has been identified in other biogas plants [1]. As a source for carbon, *A. laidlawii* utilizes glucose, fructose and galactose [15]. The genome of *A. laidlawii* harbors genes for enzymes that degrade starch, amino sugars and other sugars. In the functional profile of EGTs assigned to this family are glycoproteases (PF00814) and peptidases (PF01546) (data not shown).

In the dry fermentation process, *Candidatus Cloacamonas* is more prevalently present (Figure 2). The species *Candidatus Cloacamonas acidaminovorans* was recently detected to be highly abundant in other anaerobic digesters [17,41]. Previously, the genus was also identified in a 16S rRNA clone library of the same biogas plant [42]. Proteome analysis indicated that *C. acidaminovorans* might attain energy from sugars in the Embden-Meyerhof pathway and from the fermentation of amino acids, and thereby produces hydrogen and carbon dioxide [41].

Overall, the taxonomic profiles are similar for the wet and dry anaerobic digestion, especially at higher taxonomic ranks. In the BGP_WF, the families *Erysipelotrichaceae*, *Fibrobacteraceae*, *Succinivibrionaceae* and *Clostridiaceae* were found to be more abundant, whereas more sequences were assigned to *Acholeplasmataceae* and *Candidatus Cloacamonas* in the BGP_DF.

Hydrogenotrophic methanogens are dominant in the dry and wet fermentation processes

A detailed taxonomic analysis was performed for the archaeal sequences of both fermentation processes. In the wet fermentation process, less EGTs were assigned to *Methanomicrobiaceae* (3.4%) as compared to the dry fermentation digester (4.4%) (Figures 1 and 2). In both fermentation processes, *Methanomicrobiaceae* is the most abundant methanogenic family with *Methanoculleus* being the prevalent genus. *Methanoculleus* species conduct the hydrogenotrophic methanogenesis pathway, synthesizing methane from carbon dioxide and hydrogen. Further identified families that are capable of performing methanogenesis are *Methanosarcinaceae, Methanosaetaceae* and *Methanobacteriaceae* [43]. These families are only detected in low abundance in the microbial communities. In the BGP_WF, the acetoclastic methanogen *Methanosaeta* was measured in slightly higher frequencies as compared to the dry fermentation process (Figure 2). This genus was noted to be more dominant in biogas plants with low acetate concentrations [44], as it has a high affinity to acetate. The observation is supported by this study as the abundance of *Methanosaeta* correlates with the lower concentration of acetic acid in BGP_WF compared to BGP_DF (see Table 1).

In addition to the taxonomic comparison of the biogas plants, enzymes involved in the acetoclastic and hydrogenotrophic methanogenesis were categorized according to Pfam families and were searched in the functionally characterized metagenomes obtained from BGP_WF and BGP_DF. Reads assigned to the selected Pfam families and classified to the superkingdom *Archaea* were used for the subsequent analysis (see Figure 3). No EGTs were assigned to acetate kinase and phosphotransacetylase, which are key enzymes in the initial step of acetoclastic methanogenesis in *Methanosarcina* [45]. However, EGTs for all enzymes relevant in the hydrogenotrophic methanogenesis pathway were identified with a slightly higher amount in the dry fermenter (data not shown). Rarefaction analyses on the mean EGT richness showed a saturation at approximately 800,000 reads, which indicates that the majority of the microbial community has been captured (see Additional file 1: Figure S1B). Moreover, we emphasized the EGTs that are unique in each of the biogas plants' metagenomes, which means that they are absent in the one plant while they have an abundance of at least

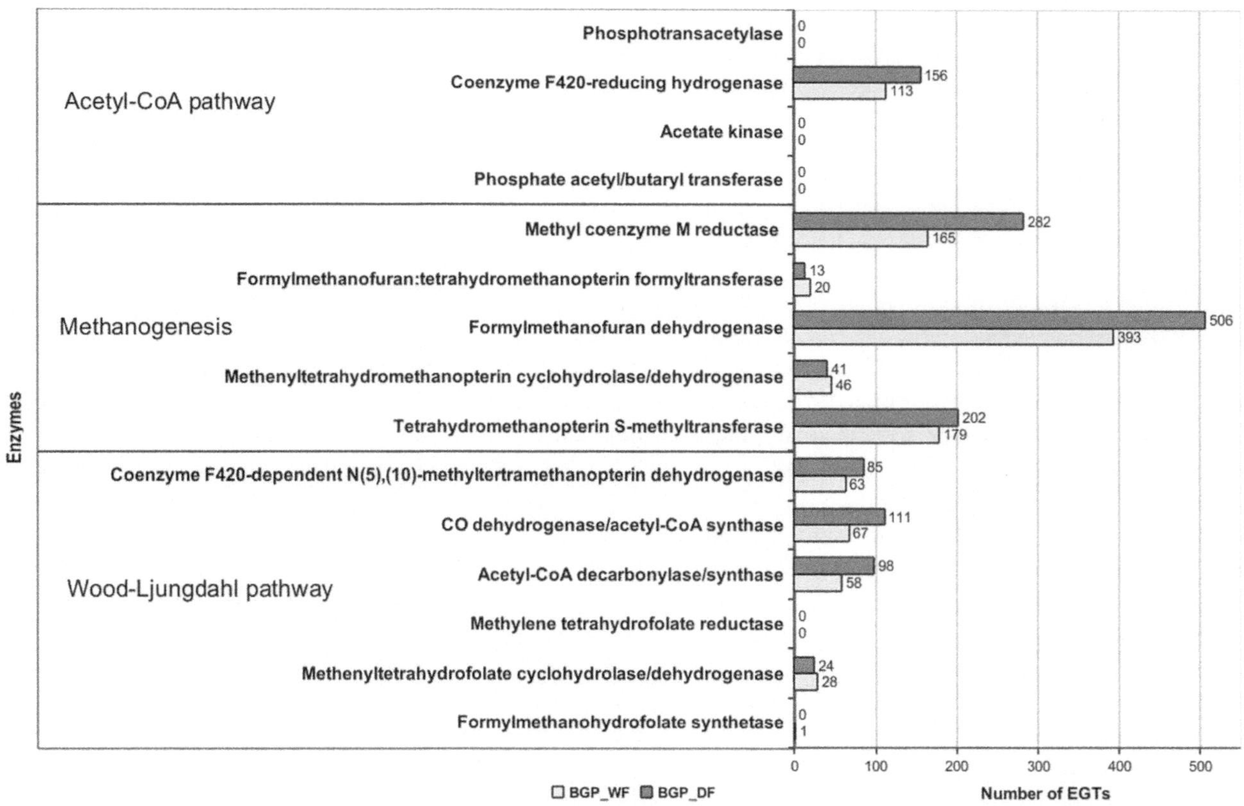

Figure 3 Comparative analysis of archaeal environmental gene tags (EGTs) representing methanogenesis-related enzymes in metagenome datasets for the wet or dry fermentation process. Comparison of the relative abundances of EGTs representing enzymes involved in the acetyl-CoA pathway, methanogenesis and Wood-Ljungdahl pathway. Only EGTs assigned to *Archaea* were considered and relative abundances were normalized based on the smallest dataset.

five in the other plant (normalized on the smallest dataset). This analysis showed that there are more unique EGTs in the wet fermentation process metagenome. However, none of the identified unique EGTs can be correlated to functions directly associated with the fermentation processes leading to methane production (Additional file 2: Table S1).

Differentiation of the methanogenic sub-community within the wet and dry fermentation process

To analyze the archaeal sub-community at a higher resolution, high-throughput 16S rRNA gene amplicon sequencing was carried out. The sequencing procedure for BGP_WF yielded 83,719 sequencing reads. A previous rRNA gene amplicon sequencing run of the sample obtained from BGP_DF resulted in 170,941 sequencing reads [46]. In total, 37 operational taxonomic units (OTUs) were of archaeal origin, representing 2,753 sequences of the BGP_WF dataset (8%). After manual inspection for chimeric sequences and excluding those sequences with a length below 100 bp, 15 OTUs remained comprising 2,691 reads. In the dry fermentation process, 22 OTUs representing 1,118 sequences (2%) were assigned to *Archaea*. After removing manually detected chimeric and short (<100 bp) sequences, nine OTUs remained representing 1,095 sequences. The number of sequences included in the OTUs was normalized to the number of the smallest filtered 16S rRNA gene amplicon dataset. Phylogenetic analysis of representative OTU sequences that were assigned to the superkingdom *Archaea* was carried out using FastTree. Rarefaction analyses on the mean OTU richness showed saturation at approximately 30,000 reads, which indicates that the majority of the microbial community has been captured (see Additional file 1: Figure S1C).

The largest archaeal OTUs of the wet fermentation and dry fermentation plant are similar to *Methanoculleus bourgensis* with 2,184 and 680 sequences, respectively (Figure 4). The presence of a phylogenetic cluster related *to M. bourgensis* is in agreement with previous studies based on 16S rRNA clone library analyses obtained from the biogas plant BGP_DF [42]. The representative sequences of the second largest OTU of each biogas plant cluster together. The representative amplicon sequences comprise 359 and 51, respectively, and are located in a large cluster formed by known *Methanoculleus* species. However, no sequence of a described *Methanoculleus* species could be identified in close proximity of these representative sequences, indicating a so far unknown archaeal species related to *Methanoculleus* (unknown *Methanoculleus* cluster II).

A further unknown *Methanoculleus* species is highly represented in the wet fermentation plant (unknown *Methanoculleus* cluster I). The corresponding OTU contains 98 sequence reads. A representative sequence of the dry fermentation biogas plant clusters close to the latter OTU, comprising only four reads. A so far unknown *Methanoculleus* species has also been described in the same biogas plant based on 16S rRNA library clones [42].

Moreover, representative sequences are located outside the phylogenetic cluster formed by the described *Methanoculleus* species. Corresponding archaeal species also belong to the class *Methanomicrobia* (cluster I and cluster II). The phylogenetic tree distantly affiliated a sequence from the dry fermentation plant to the recently identified species *Methanomassiliicoccus luminyensis* B10 from a human gut microbiome sample [47]. The representative sequence also clusters close to a so far unknown archaeal clone [48], which originates from an agricultural biogas plant supplied with water, maize silage and barley grains [49]. The distribution of similar sequences in various habitats suggests a wide occurrence of species related to *M. luminyensis*.

Besides *Methanomicrobia*, *Methanobacteria* were identified in the biogas plants based on phylogenetic analysis. Corresponding species are related to *Methanobrevibacter* and *Methanobacterium*. Both were identified in the 16S rRNA gene amplicon dataset, as well as in the whole metagenome approach, in the two biogas plants with minor frequencies. Some representative sequences (BGP_DF_8, BGP_WF_7 and BGP_WF_14) form a phylogenetic cluster with the 16S rRNA gene sequence of *Methanobacterium kanagiense* [50], a hydrogenotrophic archaeon isolated from an anaerobic propionate-oxidizing soil sample. Another representative sequence (BGP_WF_12) is 96% identical to the 16S rRNA gene sequence of *Methanobrevibacter olleyae* [51]. This species is present in sheep and bovine rumen and was described to use hydrogen and formate for methane formation. Further representative sequences (BGP_DF_9 and BGP_WF_15) are associated with *Methanobrevibacter woesei* and feature an identity of 97 to 98% to this species. *M. woesei* was isolated from goose gut and is a hydrogenotrophic archaeon that mainly uses hydrogen and carbon dioxide for methane production.

Finally, no representative sequences related to acetoclastic methanogens were identified in the biogas plants, showing a dominance of hydrogenotrophic methanogens in the biogas production process of these plants.

Overall, the constructed archaeal phylogenetic tree illustrates differences in the composition of the methanogenic sub-communities of both biogas plants. Besides the dominating *M. bourgensis* further *Methanoculleus* species are also present in the dry fermentation process (unknown *Methanoculleus* cluster I and II). In BGP_WF other methanogens comprising species related to unknown *Methanomicrobiaceae* species (cluster II and cluster III (see Figure 4) are prominent.

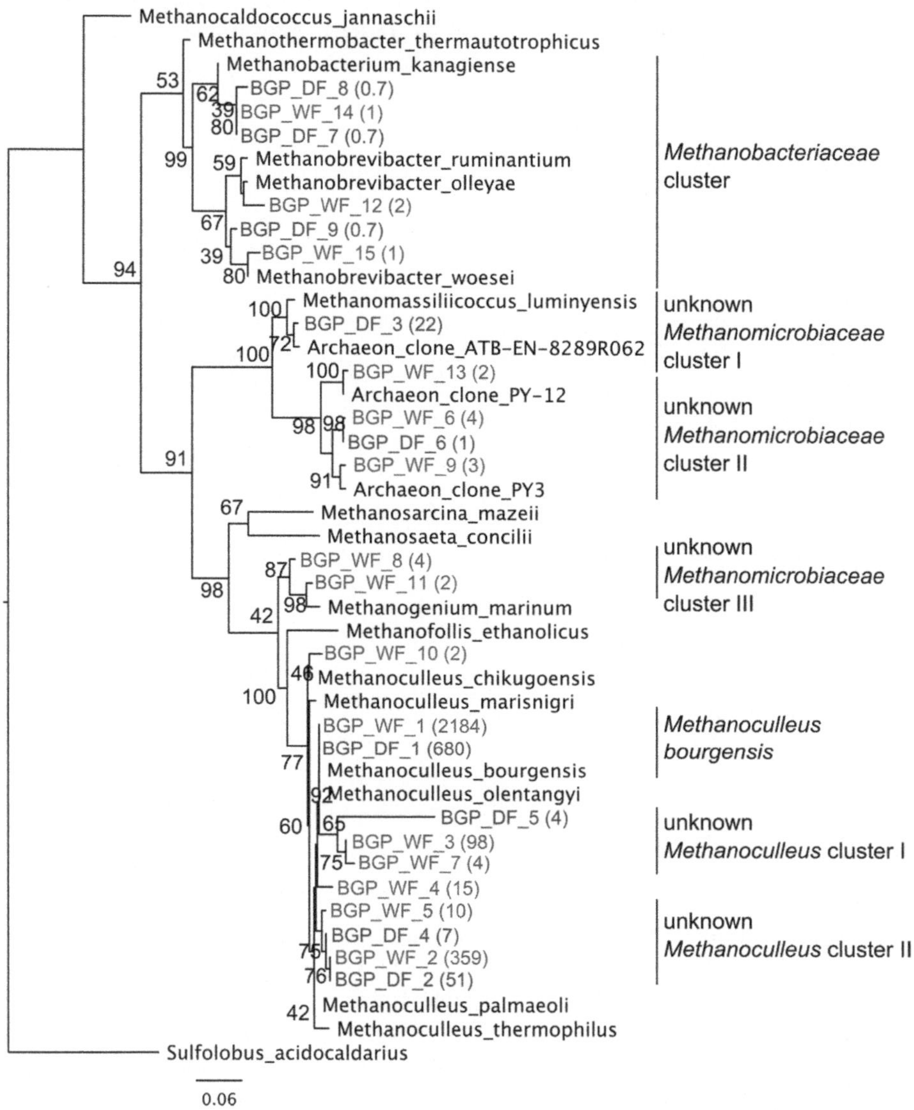

Figure 4 Archaeal phylogenetic tree of representative operational taxonomic units (OTUs) sequences from the wet and dry fermentation processes. Representative OTU sequences and corresponding counts assigned from the wet (BGP_WF, red) or dry fermentation (BGP_DF, green) process are labeled with an OTU identifier in brackets, and the nodes are labeled with the respective boot strap values. Representative sequences were assigned to one known Methanobacteriaceae and three unknown Methanomicrobiaceae clusters (I to III), and to one known and two unknown *Methanoculleus bourgensis* clusters (I and II), respectively. Sequence counts were normalized according to the smallest dataset.

Fragment recruitments revealed that methanogens of the dry fermentation plant are closely related to the type strain *Methanoculleus bourgensis* MS2^T at the genomic level

To analyze the degree of relatedness of biogas-producing community members to completely sequenced reference microorganisms, and to differentiate the metagenome datasets of both biogas plants referring to this, fragment recruitments were conducted as described previously [19]. Another objective of this analysis was to determine similarities of methanogens of both biogas plants to the genome of the type strain *M. bourgensis* MS2^T [43]. For this purpose, metagenome sequence reads were searched for matches to completely sequenced microbial genomes stored in the National Center for Biotechnology Information (NCBI) database applying the BLASTn algorithm [Basic Local Alignment Search Tool on nucleotide level]. Counts of metagenome reads that match to reference genomes with more than 90% sequence similarity are listed for the top 20 genomes in Table 3. Reference organisms appearing in this analysis taxonomically represent the orders *Methanomicrobiales*, *Clostridiales*, *Lactobacillales*, *Thermoanaerobacterales* and *Synergistales* and could be classified to belong to one of the following functional groups: cellulolytic organisms, secondary fermenters (acidogenic), syntropic organisms (acetogenic) and methanogens (see Table 4). Also, for BGP_DF, the orders *Thermotogales*

Table 3 List of 20 reference genomes showing the highest similarities to the metagenome datasets from biogas plants operating under wet or dry fermentation conditions as analyzed by fragment recruitment analysis

Reference sequence	Wet fermentation[a]	Dry fermentation[a]
Methanoculleus bourgensis MS2^T	15,992 (1.19%)	59,969 (4.45%)
Clostridium clariflavum DSM 19732	3,840 (0.28%)	3,282 (0.24%)
Clostridium thermocellum ATCC 27405	2,464 (0.18%)	2,201 (0.16%)
Clostridium kluyveri DSM 555	1,423 (0.11%)	807 (0.06%)
Streptococcus infantarius subsp. infantarius CJ18	1,360 (0.1%)	367 (0.03%)
Thermoanaerobacterium thermosaccharolyticum DSM 571	1,234 (0.09%)	1,454 (0.11%)
Mahella australiensis 50–1 BON	1,179 (0.09%)	1,319 (0.1%)
Methanoculleus marisnigri JR1	1,014 (0.08%)	3,298 (0.24%)
Desulfotomaculum carboxydivorans CO-1-SRB	944 (0.07%)	682 (0.05%)
Clostridium difficile M120	776 (0.06%)	386 (0.03%)
Thermoanaerovibrio acidaminovorans DSM 6589	688 (0.05%)	387 (0.03%)
Syntrophomonas wolfei subsp. wolfei str. Goettingen	622 (0.05%)	358 (0.03%)
Streptococcus gallolyticus UCN34	621 (0.05%)	191 (0.01%)
Streptococcus suis GZ1	575 (0.04%)	444 (0.06%)
Streptococcus pasteurianus ATCC 43144	556 (0.04%)	176 (0.01%)
Streptococcus macedonicus ACA-DC 198	523 (0.04%)	153 (0.01%)
Thermoanaerobacter sp. X514	502 (0.04%)	310 (0.02%)
Pelotomaculum thermopropionicum SI	452 (0.03%)	533 (0.04%)
Clostridium cellulolyticum H10	437 (0.03%)	321 (0.02%)
Clostridium cellulovorans 743B	386 (0.03%)	210 (0.01%)

[a]Number and percentage of reads recruited for each strain determined by BLASTn analyses (reads featuring at least 90% sequence similarity to the reference genome were counted).

and *Bacillales* were identified among the top 20 recruitments.

Additionally, BLASTn results were graphically evaluated in fragment recruitment plots, in which the degree of similarity of a single hit is plotted against the position of this hit on the reference genome sequence. This was done for those three microorganisms showing the highest sequence similarities to matching metagenome sequence reads (see Table 3), namely *M. bourgensis* MS2^T, *C. clariflavum* DSM 19732 and *C. thermocellum* ATCC 27505 (see Figure 5). Hit distribution plots show the numbers of hits with a similar degree of similarity to the reference genome sequence. Fragment recruitments revealed that methanogens within fermentation samples of both biogas plants are related to the reference type strain *M. bourgensis* MS2^T. It was observed that dominant methanogens of BGP_DF are more closely related to the reference genome as compared to those of BGP_WF (see Figure 5A). It was supposed that *M. bourgensis* may be very well adapted to adverse conditions prevailing in BGP_DF, such as high ammonium concentrations and osmolarity [12,77,78], due to the fed substrate, which was chicken dry excrement.

Surprisingly, *M. bourgensis* MS2^T represents the only organism that can be regarded as reasonable reference for members of the biogas-producing communities analyzed in this study. All other microorganisms listed in Table 3 feature a lesser degree of conservation to biogas community members and hence are more distantly related to them (exemplarily see Figure 5B,C). Moreover, differentiation of the microbial communities of both biogas plants is not apparent from these fragment recruitment analyses, since results obtained for both metagenome datasets uncovered nearly the same set of prominent reference organisms, with only slight differences in rankings (see Tables 3 and 4). This result confirms deduced taxonomic profiles for both biogas plants at higher taxonomic ranks which did not uncover pronounced differences between both microbial communities (see previous sections). However, it was observed that the diversity within the genus *Streptococcus* is greater for BGP_WF which was fed with swine manure. It is known that particular *Streptococcus* species such as *Streptococcus suis* and *Streptococcus pasteurianus* belong to the common microbiome of the pig's intestinal tract [52,79,80]. Moreover, it should be noted here that *Desulfotomaculum carboxydivorans* was identified at rank nine for BGP_WF and at rank eight for BGP_DF in fragment recruitments. This bacterium is able to catabolize low-molecular weight compounds

Table 4 Species identified during a fragment recruitment analysis using metagenome sequence data of wet and dry fermentation biogas plants

(x)[a] Species (y)[b]	Functional role	Taxonomy[c] (phylum, class, order)	Origin; attributes; metabolic features	Reference
(1) *Methanoculleus bourgensis* (1)	*Methanogenic*	*Euryarchaeota, Methanomicrobia, Methanomicrobiales*	Isolated from activated sludge; methanogen; hydrogenotroph	[43]
(2) *Clostridium clariflavum* (3)	*Cellulolytic*	*Firmicutes, Clostridia, Clostridiales*	Isolated from thermophilic anaerobic sludge; Cluster III *Clostridium*; cellulolytic; cellulosome	[52]
(3) *Clostridium thermocellum* (4)	*Cellulolytic*	*Firmicutes, Clostridia, Clostridiales*	Isolated from hot spring (Yellowstone), cotton bales, farm soil and other habitats; thermophilic; cellulolytic; cellulosome	[26,53]
(4) *Clostridium kluyveri* (7)	*Secondary fermenters*	*Firmicutes, Clostridia, Clostridiales*	Isolated from canal mud; fermentation of ethanol and acetate to butyrate, caproate and H_2	[54]
(5) *Streptococcus infantarius* (13)	*Secondary fermenters*	*Firmicutes, Bacilli, Lactobacillales*	Isolated from fermented dairy and plant products; associated with different human and animal infections; fermentative metabolism	[27]
(6) *Thermoanaerobacterium thermosaccharolyticum* (5)	*Syntrophic*	*Firmicutes, Clostridia, Thermoanaerobacterales*	Isolated from geothermal sites (Yellowstone); class V *Clostridia*; saccharolytic; fermentation of a wide range of carbohydrates to ethanol, acetic acid, lactic acid, H_2 and CO_2	[55,56]
(7) *Mahella australiensis* (6)	*Secondary fermenters*	*Firmicutes, Clostridia, Thermoanaerobacterales*	Isolated from oil field (Queensland, Australia); predicted to utilize pentoses; xylose metabolism	[57]
(8) *Methanoculleus marisnigri* (2)	*Methanogenic*	*Euryarchaeota, Methanomicrobia, Methanomicrobiales*	Isolated from marine sediment; methanogen; hydrogenotroph	[58,59]
(9) *Desulfotomaculum carboxydivorans* (8)	*Secondary fermenters*	*Firmicutes, Clostridia, Clostridiales*	Isolated from anaerobic bioreactor sludge; moderately thermophilic; fermentation of pyruvate, lactate, glucose and fructose; chemolithoheterotrophic; sulfate reduction	[60]
(10) *Costridium difficile* (12)	*Secondary fermenters*	*Firmicutes, Clostridia, Clostridiales*	Human isolate; pathogenic for humans and animals; causes diarrhea and colitis; mesophilic; chemoorganotroph	[61]
(11) *Thermoanaerovibrio acidaminovorans* (11)	*Syntrophic*	*Synergistetes, Synergistia, Synergistales*	Isolated from anaerobic reactor of a sugar refinery; fermentation of amino acids to acetate and propionate; metabolism enhanced by hydrogen scavenger	[62]
(12) *Syntrophomonas wolfei* subsp. wolfei (15)	*Syntrophic*	*Firmicutes, Clostridia, Clostridiales*	Isolated from anaerobic digester sludge; syntrophic fatty acid metabolism, syntrophic association with methanogenic archaeon	[63]
(13) *Streptococcus gallolyticus* (-)	*Secondary fermenters*	*Firmicutes, Bacilli, Lactobacillales*	Isolated from endocarditis patient; part of the rumen flora; pathogenic for ruminants, birds and humans; fermentation of carbohydrates of plant origin	[18,64-67]
(14) *Streptococcus suis* (10)	*Secondary fermenters*	*Firmicutes, Bacilli, Lactobacillales*	Clinical origin; zoonotic pathogen for pigs and humans; fermentation of carbohydrates	[2,32,67]
(15) *Streptococcus pasteurianus* (-)	*Secondary fermenters*	*Firmicutes, Bacilli, Lactobacillales*	Isolated from human blood; pathogenic; fermentation of carbohydrates	[52,67]
(16) *Streptococcus macedonicus* (-)	*Secondary fermenters*	*Firmicutes, Bacilli, Lactobacillales*	Isolated from fermented (dairy) foods; pathogenic; fermentation of carbohydrates	[67,68]
(17) *Thermoanaerobacter* sp. (18)	*Secondary fermenters*	*Firmicutes, Clostridia, Thermoanaerobacterales*	Isolated from deep sub-surface sample; thermophilic; fermentation of monomeric and polymeric carbohydrates to ethanol	[69,70]

Table 4 Species identified during a fragment recruitment analysis using metagenome sequence data of wet and dry fermentation biogas plants *(Continued)*

(18) *Pelotomaculum thermopropionicum* (9)	*Syntrophic*	*Firmicutes, Clostridia, Clostridiales*	Isolated from granular sludge of an upflow blanket reactor; thermophilic; fermentation of volatile fatty acids (propionate) in syntrophic association with methanogen	[71,72]
(19) *Clostridium cellulolyticum* (16)	*Cellulolytic*	*Firmicutes, Clostridia, Clostridiales*	Isolated from decayed grass compost; cellulolytic; cellulosome	[55]
(20) *Clostridium cellulovorans*	*Cellulolytic*	*Firmicutes, Clostridia, Clostridiales*	Isolated from methanogenic fermentation of hybrid poplar wood; mesophilic; cellulolytic; cellulosome	[73]
(-) *Petrotoga mobilis* (14)	*Secondary fermenters*	*Thermotogae, Thermotogae, Thermotogales*	Isolated from hot oil-field water from oil reservoir; heterotrophic; fermentation of different carbohydrates including xylan	[74]
(-) *Bacillus coagulans* (17)	*Secondary fermenters*	*Firmicutes, Bacilli, Bacillales*	Isolated from spoiled canned milk; thermotolerant; slightly acidophilic; carbohydrate utilization; production of lactic acid	[75]
(-) *Geobacillus* sp. (19)	*Secondary fermenters*	*Firmicutes, Bacilli, Bacillales*	thermophilic; chemoorganotrophic	CP001638 (GenBank Accession No., unpublished)
(-) *Syntrophothermus lipocaldicus* (20)	*Syntrophic*	*Firmicutes, Clostridia, Clostridiales*	Isolated from thermophilic upflow anaerobic sludge blanket; utilization of fatty acids (butyrate); syntrophic association with hydrogenotrophic organisms	[76]

[a]Ranking of fragment recruitments within the BGP_WF dataset.
[b]Ranking of fragment recruitments within the BGP_DF dataset.
[c]Taxonomic classification of the reference microorganism identified by fragment recruitment. Identified microorganisms were classified according to their functional role.

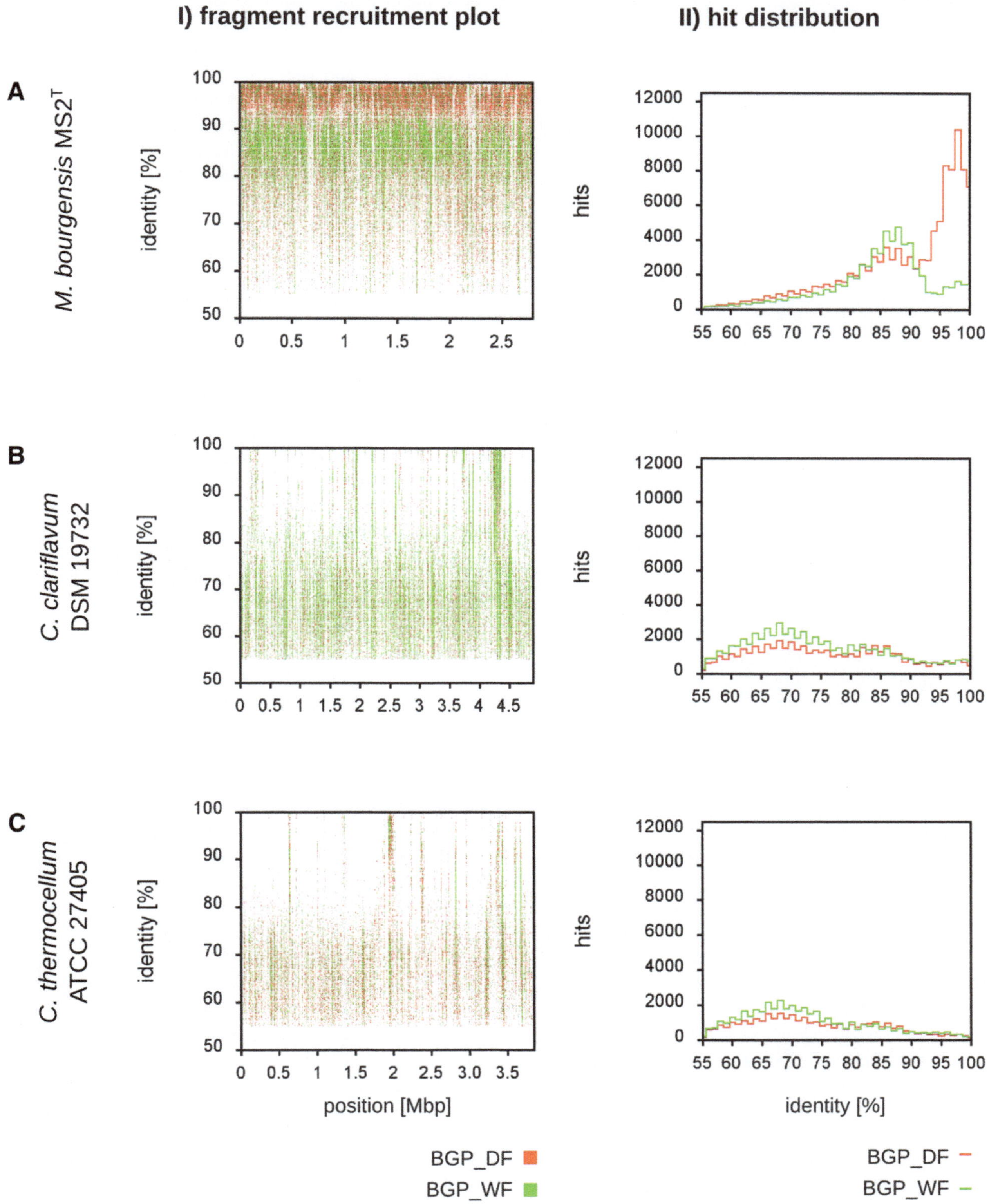

Figure 5 Fragment recruitment plots and hit distribution for metagenome sequence reads from the wet and dry fermentation processes on three reference genomes. Visualization of BLASTn analyses of metagenomic sequence reads to the reference genomes of *Methanoculleus bourgensis* MS2^T **(A)**, *Clostridium clariflavum* DSM 19732 **(B)** and *Clostridium thermocellum* ATCC 27405 **(C)**. Within the fragment recruitment plot (I) the sequence identities (>55%) between each hit of a metagenomic sequence read and the chromosomal reference sequence are plotted against the position of the alignment. In the hit distribution plot (II) the normalized numbers of reads featuring hits to the reference genome were plotted in intervals of 1% for metagenome reads showing 55 to 100% sequence identity to the reference.

(pyruvate, lactate, glucose etc.) and reduces sulfate as terminal electron acceptor to sulfide (hydrogen sulfide), which is deleterious in biogas since it leads to corrosion of combustion units and in pipelines.

In summary, fragment recruitments revealed that representation of biogas community members in databases is currently insufficient. The only exception is *M. bourgensis* $MS2^T$ which is very similar to dominant members of the methanogenic sub-community, at least in BGP_DF.

Conclusions

The methane-producing community of an agricultural biogas plant applying wet fermentation technology was characterized by means of a metagenomic approach and taxonomically compared with the microbial community of a biogas plant applying dry fermentation that was previously characterized [16]. This is the first study in which metagenomes of different biogas plants are compared using the same DNA preparation technique, sequencing technology and bioinformatics methods. Both communities' metagenomes were sequenced using 454-pyrosequencing, and the sequence data were analyzed based on their taxonomic composition and functional profile, with focus on the methanogenic sub-community and genes involved in methanogenesis, using the same bioinformatics tools and pipelines. Despite differences in the process parameters, such as acetic acid concentration and pH value and the biogas plant's substrate compositions, the microbial communities of BGP_WF and BGP_DF are similar in their composition on higher taxonomic ranks. Only a minor number of taxa on lower ranks differed between BGP_WF and BGP_DF. Accordingly, the majority of the taxa belonged to the core set of microorganisms residing in both biogas plants. Even though in both fermenters, the family of *Methanomicrobiaceae* is most dominant and *Methanoculleus* is the prevalent genus within this family, the composition of this genus differs between the different plants. As fragment recruitments revealed, within the genus *Methanoculleus, M. bourgensis* was the most dominant species in dry fermentation while this was not the case for the wet fermentation process.

The presented results clearly indicate that the hypothesis of a core community being present in biogas fermenters, operating either under wet or dry fermentation conditions, can be confirmed. This is in accordance with several studies examining the microbial taxonomic composition in biogas fermenters. Identified differences in taxonomic profiles can be associated with different process parameters of the fermenters, such as fed substrate, pH, acid and ammonium concentrations, as especially methanogens are sensitive regarding changes in acetate and propionate concentrations.

To further verify the hypothesis of a core microbiome and an adaptation of specific species to particular process parameters, further comparative studies with more biogas fermenters are required. Analyzing microbiota in biogas plants with altered physiochemical characteristics (temperature, pH and concentration of relevant metabolites) and substrate supply would aid in identifying the core species essential for the anaerobic digestion process, which in turn would provide information to control the biogas production process and prevent unstable conditions. Moreover, the metatranscriptome of the microbial biogas communities should be studied and compared to corresponding metagenomes to enable characterization not only of the taxonomic composition, but also to deduce the actual metabolic activity within the biogas fermenters.

Methods

Total community DNA isolation, purification and sequencing

The biogas plant featuring the mesophilic continuous dry fermentation technology (hereafter noted as BGP_DF) was designed for a capacity of 530 kW_{el} (combined heat and power (CHP)) and a daily input of maize silage (63%), green rye (35%) and chicken manure (2%), divided into 24 feedings per day. The process comprises two digesters; the primary digester (BIOGAS NORD GmbH, Bielefeld, Germany) (1,557 $m^3{}_{net\ volume}$, height of 6 m, diameter of 19 m) has an organic load of 4.8 kg oDM m^{-3} d^{-1}, a theoretical retention time of 59 days and a temperature of 40°C. At the end, the digestate is stored in a closed non-heated final storage reactor (BIOGAS NORD GmbH, Bielefeld, Germany) (2,987 $m^3{}_{net\ volume}$, height of 6 m, diameter of 26 m). The biogas and methane yields at the time of sampling were at 698.2 and 350.5 l/kg oDM, respectively.

The biogas plant applying the mesophilic wet fermentation technology (hereafter BGP_WF) was designed for a capacity of 537 kW_{el} CHP. The daily input of maize silage (approximately 72%) and liquid pig manure (approximately 28%) was divided into 24 feedings per day. The biogas plant is composed of two digesters and the storage tank (BIOGAS NORD GmbH, Bielefeld, Germany). The digester (2,041 $m^3{}_{net\ volume}$, height of 6.4 m, diameter of 21 m) has an organic load of 4.0 kg oDM m^{-3} d^{-1}, a theoretical retention time of 55 days and a temperature of 40°C. The digestate is stored in a closed non-heated final storage tank (4,742 $m^3{}_{net\ volume}$, height of 6 m, diameter of 32 m). The biogas and methane yields at the time of sampling were at 810.5 and 417.8 l/kg oDM, respectively.

Samples were taken from the primary digester of BGP_WF and total community DNA was extracted in triplicates applying the same procedure as described previously for the biogas plant featuring dry fermentation (BGP_DF). The triplicates were pooled prior to sequencing applying high-throughput sequencing using the Genome Sequencer FLX system (Roche Diagnostics GmbH,

Mannheim, Germany) [22]. The BGP_WF metagenomic DNA was sequenced omitting the additive for high-GC DNA in order to ensure comparability between the metagenomes of the BGP_DF and BGP_WF. Subsequently, both sequence datasets were processed to remove emulsion-PCR duplicates and sequences affected by GC-biases in the course of sequencing, as described previously [22].

Taxonomic and functional analysis of metagenome sequence data

The filtered metagenome sequences of BGP_DF and BGP_WF were imported into MetaSAMS [81] and taxonomically characterized using CARMA3 [23]. CARMA3 computes taxonomic and functional assignments for EGTs derived from a microbial community. The filtered metagenome sequences of BGP_DF and BGP_WF were imported into MetaSAMS to apply CARMA3, which is implemented in MetaSAMS. Moreover, MetaSAMS enables exploration, analysis, management and visualization of calculated observations for metagenome sequences. Using MetaSAMS, rarefaction curves were calculated based on the levels 'family' and 'genus'. Rarefaction analysis addresses the assessment of 'taxon' richness from different sub-sample sizes regarding metagenome sequence reads. In a rarefaction curve, the number of assigned taxa (on a specified level) is plotted as a function of the number of sequences within a selected sub-sample. The functional profile was calculated by searching for Pfam families matching with an E-value threshold of 10^{-5}. Subsequently, the functional profile for selected taxonomic units was evaluated regarding their functional roles in the anaerobic digestion process.

To determine whether the relative abundance of specific taxa changed in the biogas fermenters, abundances were visualized in a scatter plot. First, the absolute counts of reads assigned to a specific family were normalized according to the smallest dataset. Then, the logarithm was used to decrease the influence of more dominant families. To allow the logarithm of zero, a pseudo-count of one sequence was added for each family prior to the logarithm. Moreover, differences in the methanogenesis step were studied. Briefly, EGTs which were taxonomically characterized as originating from archaeal species and functionally assigned to genes involved in the acetyl-CoA, methanogenesis or Wood-Ljungdahl pathway were counted and normalized according to the smallest sample [19]. The metagenomic sequence data of BGP_WF can be found at the European Molecular Biology Laboratory - European Bioinformatics Institute (EMBL-EBI) database under the accession number [EMBL: PRJEB5813].

16S rDNA amplicon generation, sequencing and analysis

16S rRNA gene amplicons were generated and sequenced as described recently [24], applying high-throughput sequencing using the Genome Sequencer FLX system (Roche Diagnostics GmbH, Mannheim, Germany). First, a PCR was performed to amplify a region covering V3 and V4 using universal 16S rDNA primers. Next, gel electrophoresis and gel extraction were applied to obtain only amplicons with the correct length. The PCR was repeated in order to attach barcode tags as well as adaptors to the amplicons. Finally, the amplicons were sequenced on a 454 Genome Sequencer (GS) device using FLX Titanium chemistry.

16S rRNA amplicon sequences of the BGP_DF and BGP_WF were simultaneously processed using the QIIME (Quantitative Insights Into Microbial Ecology) pipeline [82]. First, barcode and primer sequences were removed allowing 0 and 2 mismatches, respectively, and sequences with ambiguous bases were discarded. The option 'truncate_only' was used meaning that reverse primer sequences will be trimmed only if they are identified at the end of the amplicon sequences. To obtain high quality sequences for phylogenetic analyses, strict quality processing was carried out (window size of 25 bases, average quality score 25). Subsequently, the software package USEARCH version 6.0 was applied for denoising, chimera detection (*de novo* mode) and clustering into OTUs based on a 97% sequence identity [83,84]. Afterwards, representative sequences were selected for each cluster and assigned to taxonomic groups using the RDP Classifier 2.5 [85]. Only assignments with a confidence value of at least 0.8 were considered. Rarefaction curves based on OTUs were calculated to determine the coverage of the microbial community by the sequenced metagenome reads. For phylogenetic analysis, representative sequences assigned to *Archaea* with a confidence value of at least 0.8 were selected and aligned with the Infernal 1.1 software [86] using the *Archaea* SSU rRNA model (RF01959) from Rfam [48]. Finally, the alignment was used as a basis for tree reconstruction using FastTree [87]. The tree was rooted with the sequence of the *Crenarchaeota Sulfolobus acidocaldarius* covering the V3-V4 region.

Fragment recruitments

Fragment recruitments were performed as described previously [19,20] to compare the relatedness of metagenomic reads to the genomes of reference microorganisms. BLASTn analyses of metagenomic reads against a database containing all genomes of completely sequenced microorganisms were accomplished. Identity values of hits were computed by dividing the number of identical bases by the sequence length of the read. Hits with an identity value below 55% were discarded. Results of the BLASTn analysis were then visualized by plotting the calculated identity of each sequence read against the alignment position on the reference sequence. Moreover, a histogram was generated to display the distribution of hit identities. Numbers of hits displayed in fragment recruitments and in histograms were normalized based on the smallest sample size.

Additional files

Additional file 1: Figure S1. Rarefaction analyses of sequenced metagenomes and 16S rRNA gene amplicons originating from dry (BGP_DF)
and wet fermentation biogas plants (BGP_WF) microbial communities. Rarefaction analysis plots on (A) taxonomic mean richness at the family rank derived from metagenome data, (B) environmental gene tags (EGT) derived from metagenome data and (C) operational taxonomic unit (OTU) richness derived from 16S rRNA gene amplicons in correlation with the sampled reads.

Additional file 2: Table S1. Unique environmental gene tags (EGTs) encoding different proteins found in metagenome datasets for the wet and dry fermentation processes. Only EGTs featuring an abundance of 0 in the one and at least five in the other metagenome dataset were taken into account. Relative abundances were normalized based on the smallest dataset.

Abbreviations

BGP_DF: Biogas plant operating under dry fermentation conditions; BGP_WF: Biogas plant operating under wet fermentation conditions; EGT: Environmental gene tag; oDM: Organic dry matter; OTU: Operational taxonomic unit.

Competing interests

The authors declare that they have no competing interests.

Authors' contributions

YS participated in functional sequence data analysis and coordinated and drafted the manuscript. MZ participated in sequence data processing, carried out the taxonomic analysis and participated in the functional analysis of the sequencing data. IM participated in manuscript preparation. FE did the laboratory work and participated in fragment recruitment analysis and presentation. SJ participated in sequence data processing and analysis. NR provided biogas plant and metadata information. AP supervised the study and contributed to the draft of the manuscript. AS conceived the study, participated in its design and the fragment recruitment analysis. All authors read and approved the final manuscript.

Acknowledgements

This research was supported by the Bundenministerium für Ernährung, Landwirtschaft und Verbraucherschutz (BMEL) and Fachagentur für nachwachsenende Rohstoffe (FNR) projects, grant numbers 22006712 and 22028711. FE and IM acknowledge the receipt of a scholarship from the CLIB Graduate Industrial Biotechnology. The Bioinformatics Resource Facility (BRF) is thanked for their support on running the bioinformatics software at the Center for Biotechnology (CeBiTec), Bielefeld, Germany. Many thanks to the sequencing department at the Center of Biotechnology (CeBiTec).

Author details

[1]Institute for Genome Research and Systems Biology, CeBiTec, Bielefeld University, Universitätsstraße 25, D-33615 Bielefeld, Germany. [2]QIMR Berghofer Medical Research Institute Herston, 300 Herston Road, Brisbane QLD 4006, Australia. [3]Bioinformatics Resource Facility, CeBiTec, Bielefeld University, Universitätsstraße 25, D-33615 Bielefeld, Germany. [4]NORTH-TEC Maschinenbau GmbH, Oldenhörn 1, 25821 Bredstedt, Germany.

References

1. Weiland P. Biogas production: current state and perspectives. Appl Microbiol Biotechnol. 2010;85:849–60.
2. Yadvika S, Sreekrishnan TR, Kohli S, Rana V. Enhancement of biogas production from solid substrates using different techniques - a review. Bioresour Technol. 2004;95:1–10.
3. Ohmiya K, Sakka K, Kimura T. Anaerobic bacterial degradation for the effective utilization of biomass. Biotechnol Bioprocess Eng. 2005;10:482–93.
4. Braun R. Biogas from Energy Crop Digestion. In: IEA Task 37 Brochure. Paris, France: International Energy Agency; 2009.
5. Ács N, Kovács E, Wirth R, Bagi Z, Strang O, Herbel Z, et al. Changes in the *Archaea* microbial community when the biogas fermenters are fed with protein-rich substrates. Bioresour Technol. 2012;131:121–7.
6. Kampmann K, Ratering S, Kramer I, Schmidt M, Zerr W, Schnell S. Unexpected stability of *Bacteroidetes* and *Firmicutes* communities in laboratory biogas reactors fed with different defined substrates. Appl Environ Microbiol. 2012;78(7):2106–19.
7. Wei H, Tucker MP, Baker JO, Harris M, Luo Y, Xu Q, et al. Tracking dynamics of plant biomass composting by changes in substrate structure, microbial community, and enzyme activity. Biotechnol Biofuels. 2012;5:20.
8. Bengelsdorf FR, Gerischer U, Langer S, Zak M, Kazda M. Stability of a biogas-producing bacterial, archaeal and fungal community degrading food residues. FEMS Microbiol Ecol. 2012;84(1):201–12.
9. Klocke M, Mähnert P, Mundt K, Souidi K, Linke B. Microbial community analysis of a biogas-producing completely stirred tank reactor fed continuously with fodder beet silage as mono-substrate. Syst Appl Microbiol. 2007;30:139–51.
10. Liu FH, Wang SB, Zhang JS, Zhang J, Yan X, Zhou HK, et al. The structure of the bacterial and archaeal community in a biogas digester as revealed by denaturing gradient gel electrophoresis and 16S rDNA sequencing analysis. J Appl Microbiol. 2009;106(3):952–66.
11. Weiss A, Jerome V, Freitag R, Mayer HK. Diversity of the resident microbiota in a thermophilic municipal biogas plant. Appl Microbiol Biotechnol. 2008;81(1):163–73.
12. Weiss A, Jérôme V, Burghardt D, Likke L, Peiffer S, Hofstetter EM, et al. Investigation of factors influencing biogas production in a large-scale thermophilic municipal biogas plant. Appl Microbiol Biotechnol. 2009;84:987–1001.
13. Jaenicke S, Ander C, Bekel T, Bisdorf R, Dröge M, Gartemann K-H, et al. Comparative and joint analysis of two metagenomic datasets from a biogas fermenter obtained by 454-pyrosequencing. PLoS One. 2011;6:e14519.
14. Krause L, Diaz NN, Edwards RA, Gartemann K-H, Krömeke H, Neuweger H, et al. Taxonomic composition and gene content of a methane-producing microbial community isolated from a biogas reactor. J Biotechnol. 2008;136(1–2):91–101.
15. Li A, Chu Y, Wang X, Ren L, Yu J, Liu X, et al. A pyrosequencing-based metagenomic study of methane-producing microbial community in solid-state biogas reactor. Biotechnol Biofuels. 2013;6:3.
16. Schlüter A, Bekel T, Diaz NN, Dondrup M, Eichenlaub R, Gartemann K-H, et al. The metagenome of a biogas-producing microbial community of a production-scale biogas plant fermenter analysed by the 454-pyrosequencing technology. J Biotechnol. 2008;136(1–2):77–90.
17. Wirth R, Kovacs E, Maróti G, Bagi Z, Rákhely G, Kovács KL. Characterization of a biogas-producing microbial community by short-read next generation DNA sequencing. Biotechnol Biofuels. 2012;5:41.
18. Rademacher A, Zakrzewski M, Schlüter A, Schönberg M, Szczepanowski R, Goesmann A, et al. Characterization of microbial biofilms in a thermophilic biogas system by high-throughput metagenome sequencing. FEMS Microbiol Ecol. 2012;79(3):785–99.
19. Eikmeyer F, Köfinger P, Poschenel A, Jünemann S, Zakrzewski M, Heinl S, et al. Metagenome analyses reveal the influence of the inoculant *Lactobacillus buchneri* CD034 on the microbial community involved in grass silaging. J Biotechnol. 2013;167(3):334–43.
20. Rusch DB, Halpern AL, Sutton G, Heidelberg KB, Williamson S, Yooseph S, et al. The Sorcerer II global ocean sampling expedition: northwest Atlantic through eastern tropical Pacific. PLoS Biol. 2007;5(3):e77.
21. Weiland P. Biomass digestion in agriculture: a successful pathway for the energy production and waste treatment in Germany. Eng Life Sci. 2006;6(3):302–9.
22. Jans C, Gerber A, Bugnard J, Njage PM, Lacroix C, Meile L. Novel Streptococcus infantarius subsp. infantarius variants harboring lactose

metabolism genes homologous to *Streptococcus thermophilus*. Food Microbiol. 2012;31(1):33–42.

23. Gerlach W, Stoye J. Taxonomic classification of metagenomic shotgun sequences with CARMA3. Nucleic Acids Res. 2011;39:e91.
24. Zakrzewski M, Bekel T, Ander C, Pühler A, Rupp O, Stoye J, et al. MetaSAMS-A novel software platform for taxonomic classification, functional annotation and comparative analysis of metagenome datasets. J Biotechnol. 2012;167(2):156–65.
25. Fleissner CK, Huebel N, Abd El-Bary MM, Loh G, Klaus S, Blaut M. Absence of intestinal microbiota does not protect mice from diet-induced obesity. Br J Nutr. 2010;104(6):919–29.
26. Brown SD, Lamed R, Morag E, Borovok I, Shoham Y, Klingeman DM, et al. Draft genome sequences for *Clostridium thermocellum* wild-type strain YS and derived cellulose adhesion-defective mutant strain AD2. J Bacteriol. 2012;194(12):3290–1.
27. Izquierdo JA, Goodwin L, Davenport KW, Teshima H, Bruce D, Detter C, et al. Complete genome sequence of *Clostridium clariflavum* DSM 19732. Stand Genomic Sci. 2012;6(1):104–15.
28. Zhilina TN, Zavarzina DG, Kolganova TV, Lysenko AM, Turova TP. New alkaliphilic peptide fermenting and Fe(III) regenerating Bacteria *Alkaliphilus peptidofermentans* sp.nov. from soda lake. Mikrobiologiia. 2012;78(4):496–505.
29. Berg Miller ME, Antonopoulos DA, Rincon MT, Band M, Bari A, Akraiko T, et al. Diversity and strain specificity of plant cell wall degrading enzymes revealed by the draft genome of *Ruminococcus flavefaciens* FD-1. PLoS One. 2009;4(8):e6650.
30. Miller TL, Wolin MJ. Bioconversion of cellulose to acetate with pure cultures of *Ruminococcus albus* and a hydrogen-using acetogen. Appl Environ Microbiol. 2012;61(11):3832–5.
31. Dassa B, Borovok I, Lamed R, Henrissat B, Coutinho P, Hemme CL, et al. Genome-wide analysis of *Acetivibrio cellulolyticus* provides a blueprint of an elaborate cellulosome system. BMC Genomics. 2012;13:210.
32. Hu P, Yang M, Zhang A, Wu J, Chen B, Hua Y, et al. Complete genome sequence of *Streptococcus suis* serotype 14 strain JS14. J Bacteriol. 2012;193(9):2375–6.
33. Kasperowicz A, Stan-Glasek K, Guczynska W, Piknová M, Pristas P, Nigutová K, et al. Fructanolytic and saccharolytic enzymes of the rumen bacterium *Pseudobutyrivibrio ruminis* strain 3–preliminary study. Folia Microbiol. 2010;55(4):329–31.
34. Hippe H, Hagelstein A, Kramer I, Swiderski J, Stackebrandt E. Phylogenetic analysis of *Formivibrio citricus*, *Propionivibrio dicarboxylicus*, *Anaerobiospirillum thomasii. Succinimonas amylolytica* and *Succinivibrio dextrinosolvens* and proposal of *Succinivibrionaceae* fam. nov. Int J Syst Bacteriol. 1999;49(Pt 2):779–82.
35. Suen G, Weimer PJ, Stevenson DM, Aylward FO, Boyum J, Deneke J, et al. The complete genome sequence of *Fibrobacter succinogenes* S85 reveals a cellulolytic and metabolic specialist. PLoS One. 2011;6(4):e18814.
36. Deublein D, Steinhauser A. Biogas from Waste And Renewable Resources: An Introduction. Hoboken, New Jersey, USA: John Wiley & Sons; 2010.
37. Purushe J, Fouts DE, Morrison M, White BA, Mackie RI, North American Consortium for Rumen Bacteria, et al. Comparative genome analysis of *Prevotella ruminicola* and *Prevotella bryantii*: insights into their environmental niche. Microb Ecol. 2010;60(4):721–9.
38. Palatsi J, Viñas M, Guivernau M, Fernandez B, Flotats X. Anaerobic digestion of slaughterhouse waste: main process limitations and microbial community interactions. Bioresour Technol. 2011;102(3):2219–27.
39. Baena S, Fardeau ML, Labat M, Ollivier B, Thomas P, Garcia JL, et al. *Aminobacterium colombiensegen* nov. sp. nov., an amino acid-degrading anaerobe isolated from anaerobic sludge. Anaerobe. 1998;4(5):241–50.
40. Chertkov O, Sikorski J, Brambilla E, Lapidus A, Copeland A, Glavina Del Rio T, et al. Complete genome sequence of *Aminobacterium colombiense* type strain (ALA-1). Stand Genomic Sci. 2010;2(3):280–9.
41. Pelletier E, Kreimeyer A, Bocs S, Rouy Z, Gyapay G, Chouari R, et al. "*Candidatus Cloacamonas acidaminovorans*": genome sequence reconstruction provides a first glimpse of a new bacterial division. J Bacteriol. 2008;190(7):2572–9.
42. Kröber M, Bekel T, Diaz NN, Goesmann A, Jaenicke S, Krause L, et al. Phylogenetic characterization of a biogas plant microbial community integrating clone library 16S-rDNA sequences and metagenome sequence data obtained by 454-pyrosequencing. J Biotechnol. 2009;142(1):38–49.
43. Maus I, Wibberg D, Stantscheff R, Eikmeyer FG, Seffner A, Boelter J, et al. Complete genome sequence of the hydrogenotrophic, methanogenic archaeon *Methanoculleus bourgensis* strain MS2(T), isolated from a sewage sludge digester. J Bacteriol. 2012;194(19):5487–8.
44. Walter A, Knapp BA, Farbmacher T, Ebner C, Insam H, Franke-Whittle IH. Searching for links in the biotic characteristics and abiotic parameters of nine different biogas plants. Microb Biotechnol. 2012;5(6):717–30.
45. Smith KS, Ingram-Smith C. *Methanosaeta*, the forgotten methanogen? Trends Microbiol. 2007;15(4):150–5.
46. Zakrzewski M, Goesmann A, Jaenicke S, Jünemann S, Eikmeyer FG, Szczepanowski R, et al. Profiling of the metabolically active community from a production-scale biogas plant by means of high-throughput metatranscriptome sequencing. J Biotechnol. 2012;158(4):248–58.
47. Dridi B, Fardeau ML, Ollivier B, Raoult D, Drancourt M. *Methanomassiliicoccus luminyensis* gen. nov., sp. nov., a methanogenic archaeon isolated from human faeces. Int J Syst Evol Microbiol. 2012;62(Pt 8):1902–7.
48. Griffiths-Jones S, Bateman A, Marshall M, Khanna A, Eddy SR. Rfam: an RNA family database. Nucleic Acids Res. 2003;31(1):439–41.
49. Nettmann E, Bergmann I, Pramschüfer S, Mundt K, Plogsties V, Herrmann C, et al. Polyphasic analyses of methanogenic archaeal communities in agricultural biogas plants. Appl Environ Microbiol. 2010;76(8):2540–8.
50. Kitamura K, Fujita T, Akada S, Tonouchi A. *Methanobacterium kanagiense* sp. nov., a hydrogenotrophic methanogen, isolated from rice-field soil. Int J Syst Evol Microbiol. 2011;61(Pt 6):1246–52.
51. Rea S, Bowman JP, Popovski S, Pimm C, Wright AD. *Methanobrevibacter millerae* sp. nov. and *Methanobrevibacter olleyae* sp. nov., methanogens from the ovine and bovine rumen that can utilize formate for growth. Int J Syst Evol Microbiol. 2007;57(Pt 3):450–6.
52. Hu P, Yang M, Zhang A, Wu J, Chen B, Hua Y, et al. Complete genome sequence of *Streptococcus suis* serotype 3 strain ST3. J Bacteriol. 2011;193(13):3428–9.
53. Feinberg L, Foden J, Barrett T, Davenport KW, Bruce D, Detter C, et al. Complete genome sequence of the cellulolytic thermophile *Clostridium thermocellum* DSM1313. J Bacteriol. 2011;193(11):2906–7.
54. Rusniok C, Couvé E, Da Cunha V, El Gana R, Zidane N, Bouchier C, et al. Genome sequence of *Streptococcus gallolyticus*: insights into its adaptation to the bovine rumen and its ability to cause endocarditis. J Bacteriol. 2010;192(8):2266–76.
55. Heinl S, Wibberg D, Eikmeyer F, Szczepanowski R, Blom J, Linke B, et al. Insights into the completely annotated genome of *Lactobacillus buchneri* CD034, a strain isolated from stable grass silage. J Biotechnol. 2012;161(2):153–66.
56. Lazarev VN, Levitskii SA, Basovskii YI, Chukin MM, Akopian TA, Vereshchagin VV, et al. Complete genome and proteome of *Acholeplasma laidlawii*. J Bacteriol. 2011;193(18):4943–53.
57. Sikorski J, Teshima H, Nolan M, Lucas S, Hammon N, Deshpande S, et al. Complete genome sequence of *Mahella australiensis* type strain (50–1 BON). Stand Genomic Sci. 2011;4(3):331–41.
58. Anderson I, Ulrich LE, Lupa B, Susanti D, Porat I, Hooper SD, et al. Genomic characterization of *Methanomicrobiales* reveals three classes of methanogens. PLoS One. 2009;4(6):e5797.
59. Anderson IJ, Sieprawska-Lupa M, Lapidus A, Nolan M, Copeland A, Glavina Del Rio T, et al. Complete genome sequence of *Methanoculleus marisnigri* Romesser *et al.* 1981 type strain JR1. Stand Genomic Sci. 2009;1(2):189–96.
60. Parshina SN, Sipma J, Nakashimada Y, Henstra AM, Smidt H, Lysenko AM, et al. *Desulfotomaculum carboxydivorans* sp. nov., a novel sulfate-reducing bacterium capable of growth at 100% CO. Int J Syst Evol Microbiol. 2005;55(Pt 5):2159–65.
61. He M, Sebaihia M, Lawley TD, Stabler RA, Dawson LF, Martin MJ, et al. Evolutionary dynamics of *Clostridium difficile* over short and long time scales. Proc Natl Acad Sci U S A. 2010;107(16):7527–32.
62. Chovatia M, Sikorski J, Schröder M, Lapidus A, Nolan M, Tice H, et al. Complete genome sequence of *Thermanaerovibrio acidaminovorans* type strain (Su883). Stand Genomic Sci. 2009;1(3):254–61.
63. Sieber JR, Sims DR, Han C, Kim E, Lykidis A, Lapidus AL, et al. The genome of *Syntrophomonas wolfei*: new insights into syntrophic metabolism and biohydrogen production. Environ Microbiol. 2010;12(8):2289–301.
64. Seedorf H, Fricke WF, Veith B, Brüggemann H, Liesegang H, Strittmatter A, et al. The genome of *Clostridium kluyveri*, a strict anaerobe with unique metabolic features. Proc Natl Acad Sci U S A. 2008;105(6):2128–33.
65. Hemme CL, Mouttaki H, Lee YJ, Zhang G, Goodwin L, Lucas S, et al. Sequencing of multiple clostridial genomes related to biomass conversion and biofuel production. J Bacteriol. 2010;192(24):6494–6.

66. Hinse D, Vollmer T, Rückert C, Blom J, Kalinowski J, Knabbe C, et al. Complete genome and comparative analysis of *Streptococcus gallolyticus* subsp. *gallolyticus*, an emerging pathogen of infective endocarditis. BMC Genomics. 2011;12:400.
67. Schlegel L, Grimont F, Ageron E, Grimont PA, Bouvet A. Reappraisal of the taxonomy of the *Streptococcus bovis/Streptococcus equinus* complex and related species: description of *Streptococcus gallolyticus* subsp. *gallolyticus* subsp. nov., *S. gallolyticus* subsp. *macedonicus* subsp. nov. and *S. gallolyticus* subsp. *pasteurianus* subsp. nov. Int J Syst Evol Microbiol. 2003;53(Pt 3):631–45.
68. Papadimitriou K, Ferreira S, Papandreou NC, Mavrogonatou E, Supply P, Pot B, et al. Complete genome sequence of the dairy isolate *Streptococcus macedonicus* ACA-DC 198. J Bacteriol. 2012;194(7):1838–9.
69. Feng X, Xu Y, Chen Y, Tang YJ. MicrobesFlux: a web platform for drafting metabolic models from the KEGG database. BMC Syst Biol. 2012;6:94.
70. Lee YE, Jain MK, Lee C, Zeikus JG. Taxonomic distinction of saccharolytic thermophilic anaerobes: description of *Thermoanaerobacterium xylanolyticum* gen. nov., sp. nov., and *Thermoanaerobacterium saccharolyticum* gen. nov., sp. nov.; reclassification of *Thermoanaerobium brockii, Clostridium thermosulfurogenes*, and *Clostridium thermohydrosulfuricum* E100-69 as *Thermoanaerobacter brockii* comb. nov., *Thermoanaerobacterium thermosulfurigenes* comb. nov., and *Thermoanaerobacter thermohydrosulfuricus* comb. nov., respectively; and transfer of *Clostridium thermohydrosulfuricum* 39E to *Thermoanaerobacter ethanolicus*. Int J Syst Bacteriol. 1993;43:41–51.
71. Kosaka T, Uchiyama T, Ishii S, Enoki M, Imachi H, Kamagata Y, et al. Reconstruction and regulation of the central catabolic pathway in the thermophilic propionate-oxidizing syntroph *Pelotomaculum thermopropionicum*. J Bacteriol. 2006;188(1):202–10.
72. Kosaka T, Kato S, Shimoyama T, Ishii S, Abe T, Watanabe K. The genome of *Pelotomaculum thermopropionicum* reveals niche-associated evolution in anaerobic microbiota. Genome Res. 2008;18(3):442–8.
73. Tamaru Y, Miyake H, Kuroda K, Nakanishi A, Matsushima C, Doi RH, et al. Comparison of the mesophilic cellulosome-producing *Clostridium cellulovorans* genome with other cellulosome-related clostridial genomes. Microb Biotechnol. 2011;4(1):64–73.
74. Lien T, Madsen M, Rainey FA, Birkeland NK. *Petrotoga mobilis* sp. nov., from a North Sea oil-production well. Int J Syst Bacteriol. 1998;48(Pt 3):1007–13.
75. Su F, Tao F, Tang H, Xu P. Genome sequence of the thermophile *Bacillus coagulans* Hammer, the type strain of the species. J Bacteriol. 2012;194(22):6294–5.
76. Djao OD, Zhang X, Lucas S, Lapidus A, Del Rio TG, Nolan M, et al. Complete genome sequence of *Syntrophothermus lipocalidus* type strain (TGB-C1). Stand Genomic Sci. 2010;3(3):268–75.
77. Goberna M, Insam H, Franke-Whittle IH. Effect of biowaste sludge maturation on the diversity of thermophilic *Bacteria* and *Archaea* in an anaerobic reactor. Appl Environ Microbiol. 2009;75(8):2566–72.
78. Schnürer J, Olsson J, Börjesson T. Fungal volatiles as indicators of food and feeds spoilage. Fungal Genet Biol. 1999;27(2–3):209–17.
79. Lin IH, Liu TT, Teng YT, Wu HL, Liu YM, Wu KM, et al. Sequencing and comparative genome analysis of two pathogenic *Streptococcus gallolyticus* subspecies: genome plasticity, adaptation and virulence. PLoS One. 2011;6(5):e20519.
80. Zheng X, Zheng H, Lan R, Ye C, Wang Y, Zhang J, et al. Identification of genes and genomic islands correlated with high pathogenicity in *Streptococcus suis* using whole genome tiling microarrays. PLoS One. 2011;6(3):e17987.
81. Wibberg D, Blom J, Jaenicke S, Kollin F, Rupp O, Scharf B, et al. Complete genome sequencing of *Agrobacterium sp.* H13-3, the former *Rhizobium lupini* H13-3, reveals a tripartite genome consisting of a circular and a linear chromosome and an accessory plasmid but lacking a tumor-inducing Ti-plasmid. J Biotechnol. 2011;155(1):50–62.
82. Caporaso JG, Kuczynski J, Stombaugh J, Bittinger K, Bushman FD, Costello EK, et al. QIIME allows analysis of high-throughput community sequencing data. Nat Methods. 2010;7(5):335–6.
83. Edgar RC. Search and clustering orders of magnitude faster than BLAST. Bioinformatics. 2010;26(19):2460–1.
84. Edgar RC, Haas BJ, Clemente JC, Quince C, Knight R. UCHIME improves sensitivity and speed of chimera detection. Bioinformatics. 2010;27(16):2194–200.
85. Wang Q, Garrity GM, Tiedje JM, Cole JR. Naive Bayesian classifier for rapid assignment of rRNA sequences into the new bacterial taxonomy. Appl Environ Microbiol. 2007;73(16):5261–7.
86. Nawrocki EP, Kolbe DL, Eddy SR. Infernal 1.0: inference of RNA alignments. Bioinformatics. 2009;25(10):1335–7.
87. Price MN, Dehal PS, Arkin AP. FastTree: computing large minimum evolution trees with profiles instead of a distance matrix. Mol Biol Evol. 2009;26(7):1641–50.

Improved glycerol utilization by a triacylglycerol-producing *Rhodococcus opacus* strain for renewable fuels

Kazuhiko Kurosawa[1], Andreas Radek[1,3], Jens K Plassmeier[1] and Anthony J Sinskey[1,2*]

Abstract

Background: Glycerol generated during renewable fuel production processes is potentially an attractive substrate for the production of value-added materials by fermentation. An engineered strain MITXM-61 of the oleaginous bacterium *Rhodococcus opacus* produces large amounts of intracellular triacylglycerols (TAGs) for lipid-based biofuels on high concentrations of glucose and xylose. However, on glycerol medium, MITXM-61 does not produce TAGs and grows poorly. The aim of the present work was to construct a TAG-producing *R. opacus* strain capable of high-cell-density cultivation at high glycerol concentrations.

Results: An adaptive evolution strategy was applied to improve the conversion of glycerol to TAGs in *R. opacus* MITXM-61. An evolved strain, MITGM-173, grown on a defined medium with 16 g L^{-1} glycerol, produced 2.3 g L^{-1} of TAGs, corresponding to 40.4% of the cell dry weight (CDW) and 0.144 g g^{-1} of TAG yield per glycerol consumed. MITGM-173 was able to grow on high concentrations (greater than 150 g L^{-1}) of glycerol. Cultivated in a medium containing an initial concentration of 20 g L^{-1} glycerol, 40 g L^{-1} glucose, and 40 g L^{-1} xylose, MITGM-173 was capable of simultaneously consuming the mixed substrates and yielding 13.6 g L^{-1} of TAGs, representing 51.2% of the CDM. In addition, when 20 g L^{-1} glycerol was pulse-loaded into the culture with 40 g L^{-1} glucose and 40 g L^{-1} xylose at the stationary growth phase, MITGM-173 produced 14.3 g L^{-1} of TAGs corresponding to 51.1% of the CDW although residual glycerol in the culture was observed. The addition of 20 g L^{-1} glycerol in the glucose/xylose mix resulted in a TAG yield per glycerol consumed of 0.170 g g^{-1} on the initial addition and 0.279 g g^{-1} on the pulse addition of glycerol.

Conclusion: We have generated a TAG-producing *R. opacus* MITGM-173 strain that shows significantly improved glycerol utilization in comparison to the parental strain. The present study demonstrates that the evolved *R. opacus* strain shows significant promise for developing a cost-effective bioprocess to generate advanced renewable fuels from mixed sugar feedstocks supplemented with glycerol.

Keywords: Triacylglycerol, Renewable fuels, *Rhodococcus opacus*, Adaptive evolution, Glycerol utilization, Co-fermentation

Background

The development of upgrading technologies capable of producing substitutes for petroleum-based fuels and chemicals has gained momentum [1-3]. Glycerol is an inevitable by-product generated during biodiesel and bioethanol production processes [4-7]. In fact, in the biodiesel production process, glycerol is the main by-product—approximately 10% (*w*/*w*) by weight of the total product—of the transesterification of triacylglycerols from animal fats and plant oils with an alcohol [6]. Bioethanol produced through the fermentation of sugars by yeasts is also accompanied by the generation of glycerol as a fermentation by-product, comprising up to 2% by volume of the liquid fraction in the whole stillage [8]. The utilization of glycerol as a carbon source for microbial production processes has not been studied extensively, because glycerol was more expensive than other carbon sources used in the conventional fermentation

* Correspondence: asinskey@mit.edu
[1]Department of Biology, Massachusetts Institute of Technology, 77 Massachusetts Avenue, Cambridge, MA 02139, USA
[2]Engineering Systems Division, Massachusetts Institute of Technology, 77 Massachusetts Avenue, Cambridge, MA 02139, USA
Full list of author information is available at the end of the article

industry. On the contrary, methods for producing glycerol by fermentation had been studied [9]. However, due to the dramatic increase in renewable fuel productions worldwide over the last few years, glycerol production has increased as a waste product, which has led to a decrease in its price, thereby allowing it to become an attractive feedstock for production of value-added materials by fermentation [4,6].

In microorganisms, glycerol metabolism has been better understood in *Escherichia coli* [10]. Although *E. coli* was long thought to require the presence of external electron acceptors (respiratory metabolism) for glycerol utilization, it has recently been shown that the bacterium can metabolize glycerol in a fermentative manner (in the absence of electron acceptors) [11]. Glycerol dissimilation in *E. coli* can proceed through three different routes to produce the glycolytic intermediate dihydroxyacetone phosphate (DHAP): the aerobic GlpK (glycerol kinase encoded by *glpK*)-GlpD (aerobic glycerol-3-phosphate dehydrogenase encoded by *glpD*) and anaerobic GlpK (glycerol kinase encoded by *glpK*)-GlpABC (anaerobic glycerol-3-phosphate dehydrogenase encoded by *glpABC*) respiratory routes, or the GldA (glycerol dehydrogenase encoded by *gldA*)-DhaKLM (dihydroxyacetone kinase encoded by *dhaKLM*) fermentative route [12-14]. Of these routes, the GldA-DhaKLM fermentive route has been shown to enable efficient utilization of glycerol under both anaerobic and microaerobic conditions. The glycerol metabolism is regulated in various ways. In particular, in the presence of glycerol with glucose, wild-type *E. coli* has been known to exhibit diauxic growth, showing that glucose is preferentially consumed before glycerol [15,16]. While many microorganisms have been shown to ferment glycerol, the fermentative metabolism of glycerol has been reported only in species of the genera *Anaerobiospirillum* [17], *Bacillus* [18], *Citrobacter* [18], *Clostridium* [19], *Enterobacter* [20], *Escherichia* [11], *Klebsiella* [21], *Lactobacillus* [18], and *Propionibacterium* [22]. More recently, there have been extensive studies for the development of microbiological processes to convert glycerol into various value-added materials, aimed at the production of arabitol [23], 2,3-butanediol [24], butanol [25], citric acid [26], ethanol [27,28], hydrogen [29,30], lactic acid [31], polyhydroxybutyrate [32], 1,2-propanediol [33], 1,3-propanediol [34], propionic acid [35], succinate [14], and triacylglycerols (TAGs) [36,37].

TAGs are esters in which three molecules of fatty acids are linked to glycerol and exploited as the precursor to produce lipid-based biofuels such as biodiesel and hydrocarbon fuels [38,39]. TAGs are found extensively as the carbon storage molecule in animals, plants, algae, and microorganisms [40,41]. Many sources of TAGs, with the exception of those with very short chain fatty acids, are converted to hydrocarbon-based biofuels that are identical in virtually every respect to commercially available petroleum-derived fuels [42,43]. Practically, renewable jet fuel, termed hydroprocessed esters and fatty acids (HEFA), made from vegetable-based TAG-containing feedstocks, has been successfully tested in military and commercial aircrafts [44,45]. However, technologies for cost-effectively converting renewable natural resources to biofuel molecules, in order to minimize the conflict between food and fuel use, have not yet been developed [46]. Oleaginous microorganisms that utilize a great variety of substrates offer benefits for TAG production from biological resources such as waste glycerol and lignocellulosic biomass [47,48].

Rhodococcus opacus PD630 produces TAGs consisting primarily of C16 and C18 series of long chain fatty acids, which are quite similar to those of vegetable-derived TAGs [49]. Steinbüchel and coworkers [50] demonstrated that *R. opacus* PD630 grown on gluconate is capable of accumulating up to 76% of the cell dry weight (CDW) as TAGs. We have reported that *R. opacus* PD630 has a rare ability to produce large amounts of TAGs when grown in the presence of high concentrations of glucose [51,52]. In addition, we recently engineered xylose fermenting strains that are capable of completely and simultaneously utilizing both xylose and glucose to produce large amounts of TAGs in the presence of high sugar concentrations [53,54]. Strain MITXM-61 grown in corn stover hydrolysate containing 118 g L^{-1} of initial sugars was capable of completely utilizing both xylose and glucose in the genuine lignocellulosic feedstock and yielded 15.9 g L^{-1} of TAGs with a productivity of 0.133 g L^{-1} h^{-1}, corresponding to 54% of the CDW [54]. However, the strain does not produce TAGs on glycerol, and the growth is poor. Here, we aimed to improve glycerol utilization in *R. opacus* to ensure the effective use of glycerol and constructed a TAG-producing *R. opacus* strain capable of high-cell-density cultivation at high concentrations of glycerol or mixtures of glucose/xylose/glycerol.

Results

Construction of a robust glycerol-fermenting *R. opacus* strain

The cell growth of *R. opacus* PD630 was extremely poor in glycerol fermentations. In order to generate a TAG-producing strain with improved growth on glycerol, we sought to apply an adaptive evolution approach. Competent cells of the engineered xylose-fermenting *R. opacus* MITXM-61 strain, a PD630 derivative strain, were treated by electroporation and spread on a defined agar medium containing 16 g L^{-1} glycerol as the sole carbon source. Several colonies appeared on the plates, and the clones were isolated after 10 days of cultivation (Additional file 1: Figure S1). The growth of five isolates, termed MITGM-71, −72, −73, −74, and −81, was tested in a defined medium with 16 g L^{-1} glycerol in flasks. The cell growth of those strains started after a long lag phase that lasted from 3 to 4 days, as shown in Additional file 1: Figure S2. Strain MITGM-73 exhibited robust cell growth, reaching an optical density (OD_{660}) of approximately 10 after 6 days of cultivation, but exhibited very slow growth on glycerol compared to other carbon sources. To further improve its glycerol utilization, MITGM-73 was subjected to an adaptive evolution procedure to select fast-growing glycerol-fermenting variants. Sequential transfers of MITGM-73 cells in batch flask cultivations with defined medium, supplemented 100 g L^{-1} glycerol, were carried out. Following inoculation, the culture was allowed to grow until the early stationary phase and then inoculated into the fresh medium. This transfer procedure was repeated for four iterations. The cell growth during the adaptive evolution is presented in Additional file 1: Figure S3. The initial culture grew slowly with a lag phase of 3 days, but after one transfer, it eventually grew faster with a short lag phase of 1 day. After five batch cultures, the culture was streaked on the plates with glycerol for isolated colonies. One of the fastest growing isolates was selected and designated strain MITGM-173 after comparing 20 isolates for their growth in the glycerol medium.

Growth of *R. opacus* MITGM-173 with high glycerol concentrations

Our previous studies [51,54] showed that *R. opacus* PD630 derivatives were capable of growing on glucose and xylose at high initial concentrations of greater than 200 g L^{-1} as the sole carbon source. We examined the growth of strain MITGM-173 on defined media with initial glycerol concentrations of 16, 40, 80, 120, 160, or 200 g L^{-1} in flask cultures (Figure 1). Growth of the strain began after 2 days of cultivation in media containing up to 160 g L^{-1} of glycerol, and the culture reached stationary phase after 5 days of cultivation on concentrations of 16, 40, 80, and 120 g L^{-1} glycerol.

TAG production of *R. opacus* MITGM-173 on glycerol and/or glucose

To elucidate the glycerol-assimilation profile by strain MITGM-173, we carried out flask cultivations in defined media containing either 16 g L^{-1} glycerol, a mixture of 8 g L^{-1} glycerol and 8 g L^{-1} glucose, or 16 g L^{-1} glucose. In these cultures, the kinetics of TAG production as fatty acids, CDW, fatty acid content as percent of CDW, and residual sugars and $(NH_4)_2SO_4$ present in culture supernatants were determined (Figure 2a–c). When MITGM-173 was cultivated on glycerol alone (Figure 2a), the growth began after 1 day of cultivation and TAG accumulation increased after $(NH_4)_2SO_4$ was depleted. In these cultures, maximum TAG production of 2.3 (±0.3) g L^{-1}, representing 40.4 (±3.1) % of CDW occurred after 6 days of cultivation, at which point the residual glycerol was completely consumed. Growing on a glycerol/glucose mix (Figure 2b) and glucose alone (Figure 2c), resulted in TAG production of 2.7 (±0.3) g L^{-1} and 2.9 (±0.2) g L^{-1}, respectively, corresponding to yields of 44.4 (±3.9) % and 47.8 (±3.6) % of CDW, respectively, during the stationary phase, which was 5 and 3 days post-inoculation, respectively. MITGM-173 grown on a glycerol/glucose mix had a short lag phase and a fast generation time in comparison with that on glycerol alone. The concentrations of glycerol and glucose in the medium simultaneously decreased over time although utilization of glycerol was delayed in the first day, and the complete consumption of glycerol and glucose occurred after 4 and 2 days of cultivation, respectively. During 6 days of cultivation, the maximum yield of TAG per gram of carbon source consumed was 0.144 (±0.015) g g^{-1} on glycerol alone, 0.169 (±0.018) g g^{-1} on a glycerol/glucose mix, and 0.181 (±0.002) g g^{-1} on glucose alone, respectively. When a mixture of glycerol and glucose was used as the carbon substrates in the medium, both the maximum TAG production and TAG yield per carbon source consumed were lower than those attained with glucose alone but higher than those with glycerol alone. The identity of the lipids and the fatty acid composition profiles of MITGM-173 cells grown under these conditions were quite similar to one another. A thin-layer chromatography (TLC) analysis of the crude organic extract from the cells showed that the TAG fraction amounted to approximately 90% (*w/w*) of the extractable lipids (Figure 2d). The fatty acid composition of the lipids was composed mainly of palmitic acid (22% to 28%), *cis*-10-heptadecenoic acid (16% to 20%), and oleic acid (17% to 24%), as revealed by gas chromatography (GC) analysis (Figure 2e).

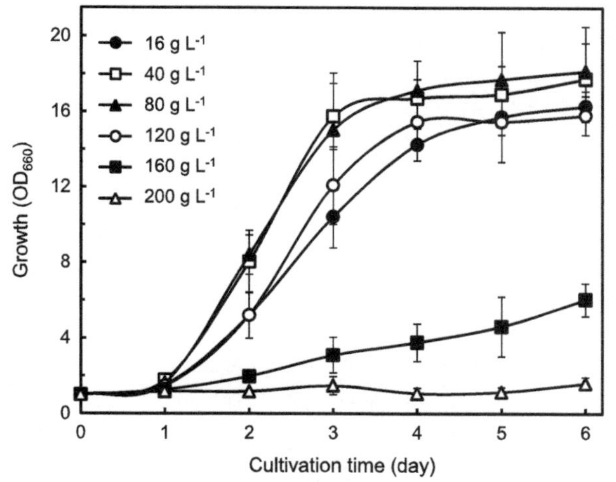

Figure 1 Growth of *R. opacus* MITGM-173 on varying concentrations of glycerol. Glycerol concentrations of defined media were 16, 40, 80, 120, 160, and 200 g L^{-1} in shake flasks. Values and error bars represent the mean and s.d. of triplicate experiments.

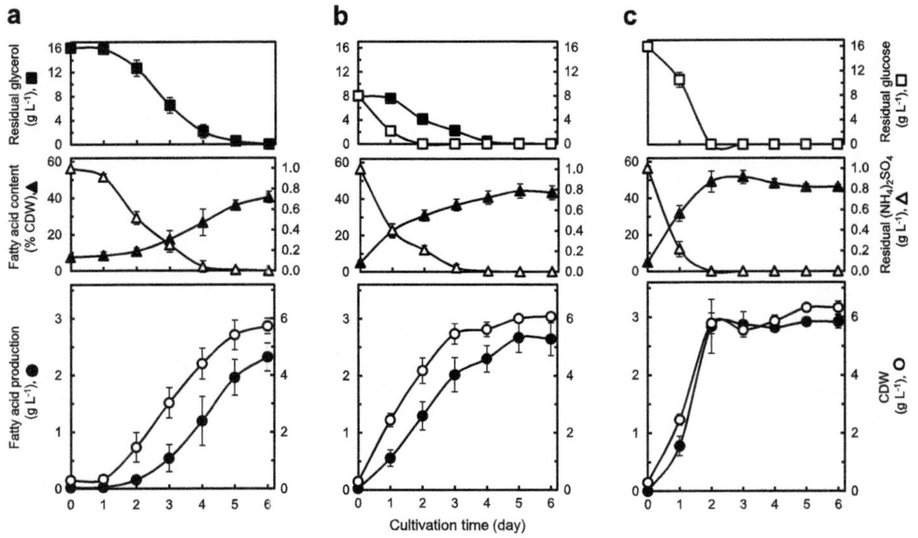

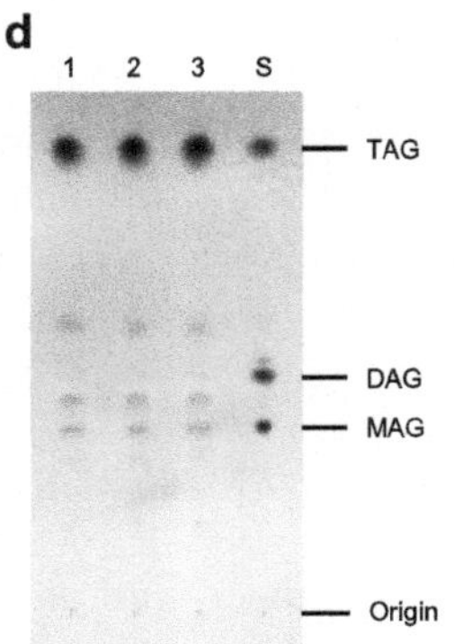

e

Fatty acid species	Carbon source		
	Glycerol	Glycerol/glucose	Glucose
Myristic acid (C14:0)	2.7 (±0.6)	2.7 (±0.6)	2.0 (±0.0)
Pentadecanoic acid (C15:0)	8.9 (±2.7)	7.2 (±0.6)	5.2 (±0.6)
Palmitic acid (C16:0)	24.8 (±1.0)	22.8 (±1.0)	28.0 (±2.1)
Palmitoleic acid (C16:1)	16.1 (±0.6)	13.6 (±1.2)	10.6 (±0.6)
Heptadecanoic acid (C17:0)	7.6 (±2.3)	9.6 (±0.6)	9.2 (±0.6)
cis-10-Heptadecenoic acid (C17:1)	19.5 (±0.6)	20.0 (±1.5)	16.5 (±0.6)
Stearic acid (C18:0)	2.0 (±0.0)	2.0 (±0.0)	4.3 (±0.6)
Oleic acid (C18:1)	17.5 (±4.6)	21.1 (±1.2)	23.1 (±1.2)
Linoleic acid (C18:2)	0.9 (±0.0)	1.0 (±0.0)	1.1 (±0.0)

Figure 2 TAG production from glycerol and/or glucose by *R. opacus* MITGM-173. (a-c) Time course kinetics of TAG production as fatty acids. The strain was grown in defined media containing 16 g L^{-1} glycerol **(a)**, a mixture of 8 g L^{-1} glycerol and 8 g L^{-1} glucose **(b)**, and 16 g L^{-1} glucose **(c)** in shake flasks. Values and error bars represent the mean and s.d. of triplicate experiments. **(d)** Thin-layer chromatography analysis of the crude organic extracts obtained from the cells grown on glycerol **(a)**, glycerol/glucose **(b)**, and glucose **(c)** for 6 days. Lipids were extracted and separated on a silica gel plate as described in the "Methods" section. Lipid standards of TAG (1,2-dioleoyl-3-stearoyl-*rac*-glycerol), DAG (1,2-dipalmitoyl-*rac*-glycerol), and MAG (DL-α-palmitin) were used to identify the Rf value for TAG under the conditions used. Lanes: 1, crude lipid extract (10 μg) on glycerol; 2, crude lipid extract (10 μg) on glycerol/glucose; 3, crude lipid extract (10 μg) on glucose; S, TAG (3 μg)/DAG (3 μg)/MAG (3 μg) mixtures. **(e)** Fatty acid composition as percentage of total fatty acids (g g^{-1}) of lipids from the cells growing in the defined medium containing glycerol **(a)**, glycerol/glucose **(b)**, or glucose **(c)** for 6 days. Data are results of triplicate experiments, ±s.d.

Meanwhile, cultivating in defined media containing either 16 g L^{-1}glycerol, a mixture of 8 g L^{-1} glycerol and 8 g L^{-1} glucose, or 16 g L^{-1} glucose, the parental strain MITXM-61 was hardly able to grow on glycerol alone (Additional file 1: Figure S4a). In the case of culturing on glucose alone (Additional file 1: Figure S4c) and mixed glycerol/glucose substrates (Additional file 1: Figure S4b), the TAG production in the stationary phase 3 days post-inoculation, at which point the residual glucose was completely consumed, was 2.9 (±0.2) g L^{-1} and 1.0 (±0.2) g L^{-1}, respectively, corresponding to 48.9 (±1.9) % CDW and 30.4 (±2.9) % CDW, respectively. MITXM-61 was also unable to utilize glycerol in the mixed substrates when glycerol was included in the medium together with glucose. The results demonstrated that the evolved strain MITGM-173 harbored significantly improved glycerol-utilizing capacity, compared to the parental strain MITXM-61.

Optimization of TAG production from glycerol by *R. opacus* MITGM-173

High-cell-density cultivation is essential to maximize volumetric productivity and to reduce production costs, and the media used should be composed of highly concentrated carbon sources [55,56]. We have demonstrated that TAG production in *R. opacus* is greatly affected by the ratio of carbon to nitrogen (C/N) in the medium [51-54]. The operational C/N ratio of glycerol and $(NH_4)_2SO_4$ in a defined medium for maximum production of TAGs by strain MITGM-173 was optimized using a response surface methodology. The experimental design model assigned nine combinations of glycerol and $(NH_4)_2SO_4$ concentrations including three repetitions at a central point (100 g L^{-1} glycerol and 6.23 g L^{-1} $(NH_4)_2SO_4$) for a total of 11 bioreactors in batch fermentations. The design matrix of the variables in coded units and actual concentrations along with the experimental response is presented in Table 1. Strain MITGM-173 grown in the medium containing 100 g L^{-1} glycerol and 6.23 g L^{-1} $(NH_4)_2SO_4$ yielded maximum TAGs of 13.8 (±0.6) g L^{-1}, representing 45.2 (±0.7) % of the CDW, at which point the residual glycerol was almost completely consumed. Strain MITGM-173 had a critical feature capable of performing high-cell-density cultivation at high glycerol concentrations and producing large amounts of TAGs. We performed analysis on the experimental data using the software StatGraphics. The coefficient of determination (R^2) was 0.9314, suggesting a relatively high correlation between predicted and experimental values. The estimated parameters from simulation with the model equation are as follows:

$$Y = -8.66238 + 0.246034X_1 + 3.26793X_2 - 0.0013664{X_1}^2 + 0.00250372X_1X_2 - 0.267951{X_2}^2$$

where Y is the predicted response (TAG production, g L^{-1} as fatty acids) and X_1 and X_2 are coded values of glycerol concentration (g L^{-1}) and $(NH_4)_2SO_4$ concentration (g L^{-1}), respectively. The surface plots illustrated by the equation are presented (Figure 3). The experimental design predicted that growing MITGM-173 cells in a defined medium with a C/N ratio of 14.7 containing 96.0 g L^{-1} glycerol and 6.55 g L^{-1} $(NH_4)_2SO_4$ would result in a maximal TAG production of 13.8 g L^{-1} as fatty acids. The C/N ratio for maximum TAG production of MITGM-173 on glycerol was slightly lower than that (C/N of 17.8) of PD630 on glucose [51] and that (C/N of 16.5) of MITXM-61 on xylose [54]. The predicted yield was validated by batch-culture fermentations with the optimized conditions. As predicted, the maximum TAG production of 13.4 (±1.5) g L^{-1} as fatty acids corresponding to 44.0 (±2.6) % of the CDW occurred after 10 days of cultivation, which was close to the predicted yield (Figure 4). Under these cultural conditions, the yield of total fatty acids per gram of glycerol consumed was 0.147 (±0.007) g g^{-1}.

Table 1 Central composite experimental design matrix defining glycerol and $(NH_4)_2SO_4$ concentrations

Run	Real values		Coded values		Fatty acid production	
	X_1	X_2	X_1	X_2	% CDW	g L^{-1} of culture
1	100	6.23	0	0	45.9	13.9
2	30	1.83	−1	−1	44.3	5.1
3	100	6.23	0	0	45.2	14.3
4	200	6.23	1.41	0	6.0	0.2
5	30	10.63	−1	1	18.2	2.1
6	100	6.23	0	0	44.5	13.2
7	170	10.63	1	1	10.8	0.5
8	170	1.83	1	−1	14.6	0.4
9	0	6.23	−1.41	0	0.0	0.0
10	100	12.45	0	1.41	23.0	6.5
11	100	0	0	−1.41	0.0	0.0

R. opacus MITGM-173 was inoculated in the modified defined media in bioreactors. The data for TAG production as fatty acids represent the maximum values during 10 days of cultivation.
X_1 glycerol concentration (g L^{-1}), X_2 $(NH_4)_2SO_4$ concentration (g L^{-1}).

Batch fermentation of *R. opacus* MITGM-173 on mixtures of glucose and xylose with glycerol

We next explored the impact of TAG production by *R. opacus* MITGM-173 on mixtures of xylose and glucose, supplemented with glycerol, in the bioreactor system. Lignocellulosic hydrolysates are mixtures of hexoses and pentoses, mostly glucose and xylose with a typical mass ration around 2:1 [57]. Glycerol is produced as a by-product at levels of approximately 10% (*w*/*w*) of the total biodiesel generated [4-7]. Consequently, we investigated TAG production of strain MITGM-173 on mixed carbon sources in a 2:2:1 ratio of glucose/xylose/glycerol with an initial total substrate concentration of 100 g L^{-1}. When cells of MITGM-173 were inoculated into a modified defined medium containing initial concentrations of 40 g L^{-1} xylose and 40 g L^{-1} glucose (Figure 5a), the cell growth increased rapidly after 12 h of cultivation, and nitrogen depletion occurred between 2 and 3 days. Glucose was completely depleted after 2 days and xylose after 3 days as a result of the concomitant consumption of xylose and glucose in the medium. The TAG production as fatty acids after 3 and 7 days of cultivation was 9.5 (±1.1) g L^{-1} and 10.2 (±1.5) g L^{-1}, respectively, corresponding to 42.3 (±0.8) % and 44.0 (±3.9) %, respectively, of the CDW. The maximum yield of TAGs as fatty acids per gram of carbon source consumed was 0.128 (±0.009) g g^{-1} during 7 days of cultivation. In contrast, when MITGM-173 was cultivated in media containing initial concentrations of 40 g L^{-1} xylose, 40 g L^{-1} glucose, and 20 g L^{-1} glycerol (Figure 5b), and initial concentrations of 40 g L^{-1} xylose and 40 g L^{-1} glucose with pulse loading of 20 g L^{-1} glycerol after 2 days of cultivation (Figure 5c) and the TAG production after 3 and 7 days of cultivation were 10.0 (±0.1) g L^{-1} and 13.6 (±1.4) g L^{-1}, respectively, corresponding to 44.4 (±1.4) % CDW and 51.2 (±2.1) % CDW, respectively, on the xylose/glucose/glycerol mix, and 11.1 (±1.4) g L^{-1} and 14.3 (±1.2) g L^{-1}, respectively, corresponding to 39.8 (±0.6) % CDW

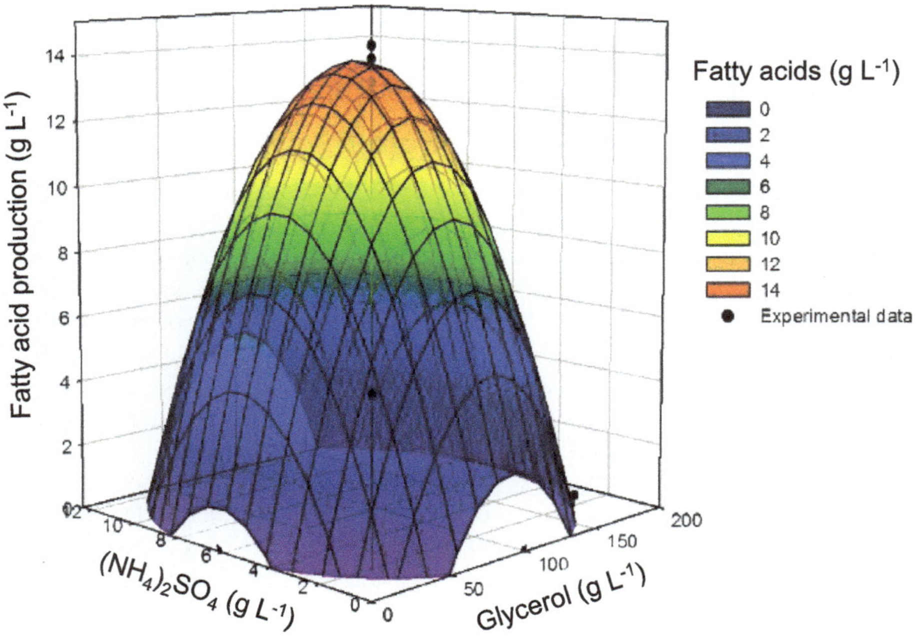

Figure 3 Response surface plot of the effect of glycerol and $(NH_4)_2SO_4$ concentrations on TAG production. As fatty acids by *R. opacus* MITGM-173. Curves and points represent predicted values and experimental data, respectively.

and 51.1 (±3.1) % CDW, respectively, on the xylose/glucose mix with the pulse loading glycerol. The concentrations of all three carbon sources in the media simultaneously decreased over time, although the consumption of glycerol was delayed in the first day after the addition of glycerol as compared to that of glucose or xylose. The consumption of xylose, glucose and glycerol in the xylose/glucose/glycerol mix was complete after 4, 2 and 7 days, respectively, of cultivation (Figure 5b). When the glycerol pulse was applied to the mixed sugars, xylose and glucose were consumed after 3 and 2 days, respectively, of cultivation and a residual glycerol amount of 5.3 g L^{-1} was observed after 7 days of cultivation (Figure 5c). The maximum yield of TAGs per gram of carbon source consumed during 7 days of cultivation was 0.136 (±0.006) g g^{-1} on the xylose/glucose/glycerol mix, and 0.151 (±0.004) g g^{-1} on the xylose/glucose mix with pulse loading glycerol. When glycerol in the xylose/glucose mix medium was added, the maximum TAG yield per carbon source consumed was higher than that with the xylose/glucose mix alone. The identity of the lipids and the fatty acid composition of MITGX-173 grown under those conditions were similar to one another. The major component of the intracellular lipids was TAGs, and the fatty acids consisted primarily of palmitic acid (26 to 29%), *cis*-10-heptadecenoic acid (18 to 21%) and oleic acid (13 to 15%) (Figure 5d, e).

Discussion

The capability to utilize glycerol efficiently is advantageous for microbial conversion process to develop cost-effective, streamlined bioprocesses in renewable fuel production. In addition, development of microbial technology for efficiently converting lignocellulosic biomass to biofuels has been considered as a potential solution for reducing current petroleum consumption and carbon emissions. Previously, we have developed a TAG-producing *R. opacus* MITXM-61 strain, which can efficiently coferment with glucose and xylose present predominantly in hydrolysates of lignocellulosic biomass [54]. Strain MITXM-61 was able to grow faintly on glycerol. The use of evolutionary engineering has proven to be instrumental in obtaining phenotypes of microorganisms with improved properties [58-60]. In present work, therefore, we applied an evolutionary approach to improve glycerol utilization in the *R. opacus*.

As a result, we constructed a TAG-producing *R. opacus* MITGM-173 strain capable of growing on glycerol concentrations greater than 150 g L^{-1} (Figure 1), as well as xylose. The strain grown on a mixture of 40 g L^{-1} glucose, 40 g L^{-1} xylose and 20 g L^{-1} glycerol utilized the substrates at maximum consumption rates of 0.833 g L^{-1} h^{-1}, 0.721 g L^{-1} h^{-1} and 0.119 g L^{-1} h^{-1}, respectively, and yielded 13.6 g L^{-1} of TAGs after 7 days of cultivation (Figure 5b). Strain MITGM-173 was capable of completely utilizing carbon sources in a 2:2:1 ratio of glucose/xylose/glycerol with an initial total substrate concentration of 100 g L^{-1}. The maximum TAG production of MITGM-173 grown on a mixture of 40 g L^{-1} glucose and 40 g L^{-1} xylose was 10.2 g L^{-1} with maximum consumption rates of 0.923 g L^{-1} h^{-1} by glucose, and 0.843 g L^{-1} h^{-1} by xylose (Figure 5a). The TAG yield per substrate consumed

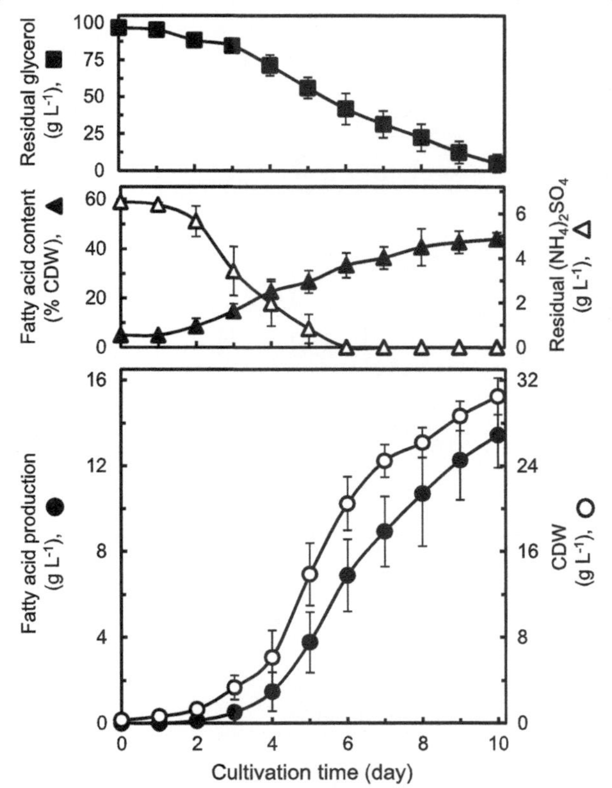

Figure 4 Time course of TAG production as fatty acids from glycerol by *R. opacus* MITGM-173. Performed under the optimized conditions. The strain was grown in a modified defined medium containing 96 g L^{-1} glucose and 6.55 g L^{-1} $(NH_4)_2SO_4$ in bioreactors. Values and error bars represent the mean and s.d. of triplicate experiments.

(0.136 g g^{-1}) on a 2:2:1 ratio of glucose/xylose/glycerol was slightly higher than that (0.128 g g^{-1}) observed on the 2:2 glucose/xylose without glycerol. The consumption curve of three substrates on the 2:2:1 glucose/xylose/glycerol culture sequentially exhibited a simultaneous pattern in contrast to a diauxie. Most microorganisms preferentially consume glucose in mixed substrates due to carbon catabolite repression or allosteric competition of the sugar in sugar transport [61,62]. The diauxic metabolism has been a major bottleneck to complete conversion and efficient utilization of multiple substrates [63]. Strain MITGM-173 was capable of utilizing multiple substrates independent of each other, although the consumption rate on glycerol was remarkably slower than that on glucose or xylose.

In addition, cultivating in a mixture of 40 g L^{-1} xylose and 40 g L^{-1} glucose with pulse loading of 20 g L^{-1} glycerol after 2 days of cultivation, MITGM-173 consumed the pulsed glycerol at the maximum consumption rate of 0.123 g L^{-1} h^{-1} and produced 14.3 g L^{-1} of TAGs corresponding to 0.151 g g^{-1} yield per carbon sources consumed after 7 days of cultivation, at which point the observed residual glycerol concentration was 5.3 g L^{-1} (Figure 5c). Interestingly, the maximum TAG production and the TAG yield per carbon sources consumed in the mixture with the pulse loading of glycerol were higher than those with the initial addition of glycerol. Based on the results obtained in mixtures containing 40 g L^{-1} glucose and 40 g L^{-1} xylose (Figure 5a), excessive addition of 20 g L^{-1} glycerol in the glucose/xylose resulted in an increase of 3.4 g L^{-1} TAGs (no residual glycerol) on the initial addition, and 4.1 g L^{-1} (residual glycerol of 5.3 g/L) on the pulse loading, corresponding to 0.170 g g^{-1} and 0.279 g g^{-1}, respectively, of the TAG yield per glycerol consumed (Figure 5b, c). The results showed that the pulse loading is a better approach compared to initial addition in regard to improvement of TAG productivity on glycerol in the *R. opacus*. The pulsed glycerol might serve as the direct intracellular precursor for synthesis of TAGs. More importantly, elucidation of this unique glycerol-assimilation mechanism should be of great importance in developing a superior strain, which results in a high yield of TAGs. The maximum theoretical conversion of glucose to TAGs in microorganisms is approximately 0.316 g g^{-1} (TAG yield per sugar consumed) [64]. One of the major challenges for developing the cost-effective streamlined bioprocess is to achieve yields that are close to theoretical limits [65].

Thus, the fermentation performance of strain MITGM-173 on glycerol may provide important clues to the construction of an industrial strain with high TAG yield. While the simultaneous and complete utilization of multiple substrates of strain MITGM-173 is ideal for production from renewable resources, the very slow consumption of glycerol would seem to be a barrier to its industrial application. The relatively shorter consumption by *R. opacus* on multiple substrates simultaneously could make this organism a practical bioconverter for TAG production. Glycerol metabolism by *Rhodococcus* species has not been studied thoroughly. Further studies need to be carried out to elucidate the genetic changes involved in the improvement of glycerol utilization and the underlying glycerol assimilation mechanism as well as the mechanism and regulation of simultaneous metabolism of multiple carbon substrates by *R. opacus*. The present study demonstrated that an evolved strain, MITGM-173, has potential towards production of advanced biofuels from low-cost feedstocks of glycerol and lignocellulosic biomass.

Conclusions

The adaptive evolution approach has proved useful for optimization of the inefficient glycerol-utilizing *R. opacus* strain. The evolved strain, MITGM-173, shows significantly improved glycerol metabolizing capacity in comparison to the parental strain, MITXM-61. Strain MITGM-173 was capable of utilizing high concentrations of glycerol or

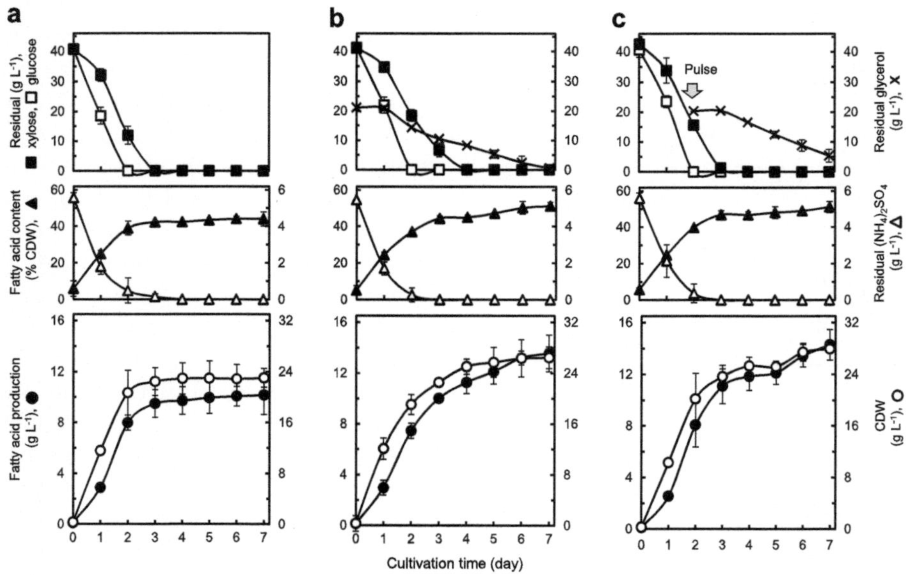

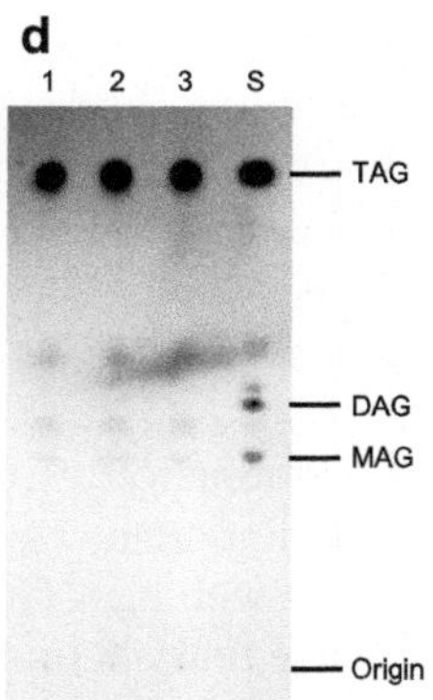

e

Fatty acid species	Carbon source		
	Xylose/glucose	Xylose/glucose /glycerol	Xylose/glucose with pulse loading of glycerol
Myristic acid (C14:0)	2.7 (±0.1)	2.7 (±0.1)	2.7 (±0.1)
Pentadecanoic acid (C15:0)	9.1 (±0.2)	9.8 (±0.2)	10.4 (±0.9)
Palmitic acid (C16:0)	28.3 (±0.2)	26.0 (±0.4)	25.7 (±1.3)
Palmitoleic acid (C16:1)	14.7 (±0.7)	14.9 (±0.4)	15.8 (±0.9)
Heptadecanoic acid (C17:0)	9.0 (±0.5)	9.6 (±0.2)	8.8 (±0.6)
cis-10-Heptadecenoic acid (C17:1)	18.6 (±0.5)	19.5 (±0.6)	20.1 (±1.2)
Stearic acid (C18:0)	2.2 (±0.2)	2.1 (±0.2)	1.8 (±0.1)
Oleic acid (C18:1)	14.6 (±0.8)	14.5 (±0.5)	13.8 (±1.1)
Linoleic acid (C18:2)	0.8 (±0.0)	0.9 (±0.0)	0.9 (±0.0)

Figure 5 TAG production from mixed substrates of glucose, xylose, and glycerol by *R. opacus* MITGM-173. (a-c) Time course kinetics of TAG production as fatty acids. The strain was grown in modified defined media supplemented with 5.56 g L^{-1} $(NH_4)_2SO_4$ containing a mixture of 40 g L^{-1} xylose and 40 g L^{-1} glucose **(a)**, a mixture of 40 g L^{-1} xylose, 40 g L^{-1} glucose and 20 g L^{-1} glycerol **(b)**, and a mixture of 40 g L^{-1} xylose and 40 g L^{-1} glucose with pulse loading of 20 g L^{-1} glycerol after 2 days of cultivation **(c)** in bioreactors. Values and error bars represent the mean and s.d. of triplicate experiments. **(d)** Thin-layer chromatography analysis of the crude organic extracts from the cells growing on xylose/glucose **(a)**, xylose/glucose/glycerol **(b)**, and xylose/glucose with pulse loading of glycerol **(c)** for 7 days. Lanes: 1, crude lipid extract (10 μg) on xylose/glucose; 2, crude lipid extract (10 μg) on xylose/glucose/glycerol; 3, crude lipid extract (10 μg) on xylose/glucose with pulse loading of glycerol; S, TAG (6 μg)/DAG (3 μg)/MAG (3 μg) mixtures. **(e)** Fatty acid composition as percentage of total fatty acids (g g^{-1}) of lipids from the cells growing in the defined medium containing xylose/glucose **(a)**, xylose/glucose/glycerol **(b)**, or xylose/glucose with pulse loading of glycerol **(c)** for 7 days. Data are results of triplicate experiments, ±s.d.

mixed glycerol/glucose/xylose simultaneously and producing large amounts of TAGs. The pulse addition of glycerol into the glucose/xylose mixture resulted in conspicuously increased TAG yield per glycerol consumed. Elucidation of the underlying glycerol-assimilation mechanism will provide great potential in constructing an industrial strain with high TAG yield. This study revealed that the evolved *R. opacus* strain has critical features of the biocatalyst for developing a cost-effective manufacturing paradigm to generate advanced renewable fuels. To our knowledge, this is the first report of any *Rhodococcus* strains capable of cofermenting glycerol, glucose and xylose.

Methods

Bacterial strains and media

A list of strains used in this study is given in Additional file 2: Table S1. *R. opacus* MITXM-61 was constructed in our previous study [54]. Glycerol-fermenting MITXM-61-derivatives were constructed in this study. The culture media used were LB broth (BD Diagnostic Systems, Sparks, MD) and a defined medium containing the following composition per liter: 16 g glycerol, 1.0 g $(NH_4)_2SO_4$, and mineral components consisted of 1.0 g $MgSO_4 \cdot 7H_2O$, 0.015 g $CaCl_2 \cdot 2H_2O$, 1.0 ml of a trace element solution, 1.0 ml stock A solution, and 35.2 ml 1.0 M phosphate buffer as described [66]. Modifications of the defined medium are stated in table and figure legends. Solid media were supplemented with 2% (*w*/*v*) agar. The strains were routinely maintained on LB agar medium and preserved in 20% (*v*/*v*) glycerol at −80 °C. All chemicals were obtained from Sigma-Aldrich (St. Louis, MO) unless otherwise noted.

Strain construction

Competent cells of *R. opacus* MITXM-61 were prepared, as previously described [54], and treated with electroporation (Bio-Rad gene pulser, Hercules, CA) at 2.5 kV, 25 μF, and 200 Ω in a 2-mm electroporation cuvette (VWR, Radnor, PA). The pulsed cells were diluted with LB broth, regenerated for 3 h with gentle agitation, plated onto a defined agar medium containing 16 g L^{-1} glycerol, and incubated to harbor glycerol utilizing strains. After 10 days of cultivation, spontaneous mutants that exhibited robust growth on glycerol were isolated.

Strain MITGM-73, one of the glycerol-utilizing isolates, was used for serial transfers of cells using repetitive cultures in flasks. The cells grown on LB agar medium for 3 days were inoculated into a 250-mL baffled flask with 50 mL of the defined medium containing 100 g L^{-1} glycerol to an initial OD of 1.0. When the cells were grown to early stationary phase, 5 mL of culture broth from the preceding flask was transferred to a new flask culture with the same medium composition. This procedure was repeated for four iterations. From the culture broth after a total of 4 generations (22 days), 20 colonies were randomly isolated by plating for single clones on a defined agar medium with 16 g L^{-1} glycerol and tested again for growth in flasks with a defined medium containing 100 g L^{-1} glycerol. One of the fastest growing strains was named MITGM-173 and used for further experiments.

Fermentation conditions

All cultures were grown at 30°C. Cell growth was monitored by determining the optical density (OD) at 660 nm (Thermo Scientific GENESYS 20, Waltham, MA). *R. opacus* seed cultures were prepared in a modified defined medium supplemented with 16 g L^{-1} glucose. Cells from colonies grown on LB agar medium for 3 days were inoculated into the modified medium in a flask. The culture was cultivated for 2 days until the late exponential phase. Unless otherwise stated, cultures for flask and bioreactor experiments were inoculated with the seed culture to an initial OD of 1.0 (2.5×10^8 cfu mL^{-1}). Shake flask experiments were carried out using 250-mL baffled flasks with a working volume of 50 mL and incubated on a rotary shaker at 200 rpm (Multitron, Infors, Bottmingen, Switzerland). Bioreactor experiments were conducted in a 2-L fermentor (Bioengineering bioreactor, R'ALF, Wald, Switzerland) with a working volume of 1 L. The pH of the medium was kept constant at 6.9 ± 0.1 by the automatic addition of 2 M NaOH. The dissolved oxygen level was monitored using an Ingold polarographic probe (Mettler-Toledo Ingold Inc., Bedford, MA). The dissolved oxygen tension was maintained above 60% by using an adjusted stirrer profile for increasing the agitation speed from 300 rpm up to 1,000 rpm and automatically sparging with a mixture of air and pure oxygen at a constant gas flow rate of 1.0 vvm. When necessary, polypropylene glycol P 2,000 was manually added to each vessel to prevent foam formation.

Response surface methodology for optimization of TAG production

The ratio of carbon to nitrogen (C/N) in the medium is the most important for increased TAG production in *R. opacus* [51,52]. The statistical experimental approach using Box-Wilson central composite design [67] with five settings for each of two factors was applied in the optimization of TAG production from glycerol. Glycerol concentration (g L^{-1}) and $(NH_4)_2SO_4$ concentration (g L^{-1}) were chosen as independent variables, and TAG production (g L^{-1}) was used as a dependent output variable. A set of 11 runs was conducted with nine combinations of glycerol and $(NH_4)_2SO_4$ concentrations including three repetitions at the central point in submerged batch fermentations (Table 1). The software StatGraphics (StatPoint Inc., USA) was used for regression and graphical analysis of the data. The optimum values of the selected variables were obtained by solving the regression equation. Three-dimensional surface plots were drawn by SigmaPlot 11 (Systat Software, Inc., San Jose, CA) to display the interaction among various variables.

Analytical methods

CDW was determined by lyophilizing cell pellet after centrifuging 10 mL of culture broth at 8,000 g for 15 min and washing the cell pellet twice in deionized water. The lyophilized cell pellet was used to analyze the identity of lipids and the fatty acid composition. For the identification of lipids, TLC experiments were carried out using a two-step resolution method as previously described [51]. Lyophilized cell pellets were extracted with methanol and chloroform (1:1, *v*/*v*) and incubated at room temperature for 1 h with gentle agitation. Ten micrograms of crude lipid extract was spotted onto silica gel 60 plates (EMD Chemicals Inc., Gibbstown, NJ). Samples were resolved using an initial polar solvent system consisting of 60:35:5 chloroform/methanol/water, followed by a second solvent system containing 70:30:1 hexane/diethyl ether/acetic acid. Resolved lipids were visualized by charring. Plates were sprayed with a 3% cupric acetate-8% aqueous phosphoric acid solution followed by baking in a 200°C oven for 5 min. To determine the fatty acid content of the cells and the composition of lipids, the whole cells were subjected to methanolysis and the resulting fatty acid methyl esters (FAMEs) were analyzed by GC as described in detail recently [51,53]. GC analysis of FAMEs was performed by using an Agilent 6850 series II network GC system equipped with an Agilent DB-Wax column (30 m by 0.32 mm, 0.5 μm film) (Agilent Technologies, Santa Clara, CA) with hydrogen as the carrier gas. A 2-μL portion of the sample was injected with a 30:1 split ratio. The inlet was maintained at 250°C. The oven was held at 80°C for 5 min, heated to 220°C at 20°C min^{-1}, and then held at 220°C for 5 min. Peak detection was carried out by a flame ionization detector, which was maintained at 300°C. The fatty acids were identified and quantified by comparison to standard FAMEs. Fatty acid content was defined as the percentage of the ratio of fatty acids to cell dry weight (% CDW). Total lipid content was calculated as the sum of total fatty acid contents for nine FAMEs: methyl myristate (C14:0), methyl pentadecanoate (C15:0), methyl palmitate (C16:0), methyl palmitoleate (C16:1), methyl heptadecanoate (C17:0), methyl *cis*-10-heptadecenoate (C17:1), methyl stearate (C18:0), methyl oleate (C18:1), and methy linoleate (C18:2). The supernatants of the culture broth were used for analyses of residual glycerol, glucose, xylose, and $(NH_4)_2SO_4$ after filtration through 0.2-μm syringe filters. Glycerol, glucose, and xylose concentrations in the culture were measured by high-performance liquid chromatography (HPLC; Agilent 1100 system) fitted with an Aminex HPX-87H column (300 × 7.8 mm, Bio-Rad) coupled to a refractive index (RI) detector as previously described [51,53]. Ammonia concentration in the culture was measured using a Sigma Ammonia Assay Kit according to the manufacturer's instructions.

Additional files

Additional file 1: Figure S1. Glycerol-utilizing colonies of *R. opacus* on plates. Cells of a xylose-fermenting MITXM-61 strain after electroporation were spread on a defined agar medium containing 16 g L^{-1} glycerol and incubated for 10 days. **Figure S2.** Growth of *R. opacus* MITXM-61 derivatives on glycerol. Each strain was grown in a defined medium containing 16 g L^{-1} glycerol in shake flasks. Values and error bars represent the mean and s.d. of triplicate experiments. **Figure S3.** Adaptive evolution of *R. opacus* MITGM-73 for improved glycerol utilization. The strain was grown in a modified defined medium containing 100 g L^{-1} glycerol in a flask. Five milliliters of the culture were sequentially transferred into a flask containing 50 mL of the fresh modified medium after 6, 10, 14, and 18 days of cultivation as indicated by the arrow. **Figure S4.** Time course kinetics of TAG production as fatty acids from glycerol and/or glucose by *R. opacus* MITXM-61. The strain was grown in defined media containing 16 g L^{-1} glycerol (a), a mixture of 8 g L^{-1} glycerol and 8 g L^{-1} glucose (b), and 16 g L^{-1} glucose (c) in shake flasks. Values and error bars represent the mean and s.d. of triplicate experiments.

Additional file 2: Table S1. Bacterial strains used in this study.

Abbreviations

TAG: Triacylglycerol; CDW: Cell dry weight; OD: Optical density: C/N: Carbon to nitrogen ratio; HPLC: High-performance liquid chromatography; GC: Gas chromatography; FAME: Fatty acid methyl ester; TLC: Thin-layer chromatography.

Competing interests

The authors declare that they have no competing interests.

Authors' contributions

KK conducted all the experiments, carried out the strain construction and the flask cultivations, analyzed the results, and wrote the manuscript. AR performed the fermentor experiments and analyzed the results. JKP and AJS helped the study and reviewed the manuscript drafting. All authors read and approved the final manuscript.

Acknowledgements

We thank the Sweetwater Energy, Inc. and the MIT Energy Initiative for financial support. The authors would like to thank Christopher J. Brigham for a critical review of the manuscript.

Author details

[1]Department of Biology, Massachusetts Institute of Technology, 77 Massachusetts Avenue, Cambridge, MA 02139, USA. [2]Engineering Systems Division, Massachusetts Institute of Technology, 77 Massachusetts Avenue, Cambridge, MA 02139, USA. [3]Present address: Institute of Bio- and Geosciences, IBG-1: Biotechnology, Systems Biotechnology, Forschungszentrum Juelich, 52425 Juelich, Germany.

References

1. Ragauskas AJ, Williams CK, Davison BH, Britovsek G, Cairney J, Eckert CA, et al. The path forward for biofuels and biomaterials. Science. 2006;311:484–9.
2. Robertson GP, Dale VH, Doering OC, Hamburg SP, Melillo JM, Wander MM, et al. Agriculture. Sustainable biofuels redux. Science. 2008;322:49–50.
3. Chu S, Majumdar A. Opportunities and challenges for a sustainable energy future. Nature. 2012;488:294–303.
4. da Silva GP, Mack M, Contiero J. Glycerol: a promising and abundant carbon source for industrial microbiology. Biotechnol Adv. 2009;27:30–9.
5. Li C, Lesnik KL, Liu H. Microbial conversion of waste glycerol from biodiesel production into value-added products. Energies. 2013;6:4739–68.
6. Yang F, Hanna MA, Sun R. Value-added uses for crude glycerol—a byproduct of biodiesel production. Biotechnol Biofuels. 2012;5:13.
7. Clomburg JM, Gonzalez R. Anaerobic fermentation of glycerol: a platform for renewable fuels and chemicals. Trends Biotechnol. 2013;31:20–8.
8. Rausch KD, Belyea RL. The future of coproducts from corn processing. Appl Biochem Biotechnol. 2006;128:47–86.
9. Wang ZX, Zhuge J, Fang H, Prior BA. Glycerol production by microbial fermentation: a review. Biotechnol Adv. 2001;19:201–23.
10. Murarka A, Dharmadi Y, Yazdani SS, Gonzalez R. Fermentative utilization of glycerol by Escherichia coli and its implications for the production of fuels and chemicals. Appl Environ Microbiol. 2008;74:1124–35.
11. Dharmadi Y, Murarka A, Gonzalez R. Anaerobic fermentation of glycerol by Escherichia coli: a new platform for metabolic engineering. Biotechnol Bioeng. 2006;94:821–9.
12. Gonzalez R, Murarka A, Dharmadi Y, Yazdani SS. A new model for the anaerobic fermentation of glycerol in enteric bacteria: trunk and auxiliary pathways in Escherichia coli. Metab Eng. 2008;10:234–45.
13. Durnin G, Clomburg J, Yeates Z, Alvarez PJ, Zygourakis K, Campbell P, et al. Understanding and harnessing the microaerobic metabolism of glycerol in Escherichia coli. Biotechnol Bioeng. 2009;103:148–61.
14. Blankschien MD, Clomburg JM, Gonzalez R. Metabolic engineering of Escherichia coli for the production of succinate from glycerol. Metab Eng. 2010;12:409–19.
15. Lin ECC. Glycerol dissimilation and its regulation in bacteria. Annu Rev Microbiol. 1976;30:535–78.
16. Martinez-Gomez K, Flores N, Castaneda HM, Martinez-Batallar G, Hernandez-Chavez G, Ramirez OT, et al. New insights into Escherichia coli metabolism: carbon scavenging, acetate metabolism and carbon recycling responses during growth on glycerol. Microb Cell Fact. 2012;11:46.
17. Lee PC, Lee WG, Lee SY, Chang HN. Succinic acid production with reduced by-product formation in the fermentation of Anaerobiospirillum succiniciproducens using glycerol as a carbon source. Biotechnol Bioeng. 2001;72:41–8.

18. Yazdani SS, Gonzalez R. Anaerobic fermentation of glycerol: a path to economic viability for the biofuels industry. Curr Opin Biotechnol. 2007;18:213–9.
19. Biebl H. Fermentation of glycerol by Clostridium pasteurianum—batch and continuous culture studies. J Ind Microbiol Biotechnol. 2001;27:18–26.
20. Ito T, Nakashimada Y, Senba K, Matsui T, Nishio N. Hydrogen and ethanol production from glycerol-containing wastes discharged after biodiesel manufacturing process. J Biosci Bioeng. 2005;100:260–5.
21. Jarvis GN, Moore ER, Thiele JH. Formate and ethanol are the major products of glycerol fermentation produced by a Klebsiella planticola strain isolated from red deer. J Appl Microbiol. 1997;83:166–74.
22. Bories A, Himmi E, Jauregui JJA, Pelayo-Ortiz C, Gonzales VA. Glycerol fermentation with Propionibacteria and optimisation of the production of propionic acid. Sci Des Aliments. 2004;24:121–35.
23. Yoshikawa J, Habe H, Morita T, Fukuoka T, Imura T, Iwabuchi H, et al. Production of D-arabitol from raw glycerol by Candida quercitrusa. Appl Microbiol Biotechnol. 2014;98:2947–53.
24. Petrov K, Petrova P. High production of 2,3-butanediol from glycerol by Klebsiella pneumoniae G31. Appl Microbiol Biotechnol. 2009;84:659–65.
25. Jensen TO, Kvist T, Mikkelsen MJ, Christensen PV, Westermann P. Fermentation of crude glycerol from biodiesel production by Clostridium pasteurianum. J Industrial Microbiol Biotechnol. 2012;39:709–17.
26. Rymowicz W, Fatykhova AR, Kamzolova SV, Rywinska A, Morgunov IG. Citric acid production from glycerol-containing waste of biodiesel industry by Yarrowia lipolytica in batch, repeated batch, and cell recycle regimes. Appl Microbiol Biotechnol. 2010;87:971–9.
27. Liu XY, Jensen PR, Workman M. Bioconversion of crude glycerol feedstocks into ethanol by Pachysolen tannophilus. Bioresour Technol. 2012;104:579–86.
28. Yu KO, Kim SW, Han SO. Engineering of glycerol utilization pathway for ethanol production by Saccharomyces cerevisiae. Bioresour Technol. 2010;101:4157–61.
29. Pott RWM, Howe CJ, Dennis JS. The purification of crude glycerol derived from biodiesel manufacture and its use as a substrate by Rhodopseudomonas palustris to produce hydrogen. Bioresour Technol. 2014;152:464–70.
30. Tran KT, Maeda T, Wood TK. Metabolic engineering of Escherichia coli to enhance hydrogen production from glycerol. Appl Microbiol Biotechnol. 2014;98:4757–70.
31. Mazumdar S, Blankschien MD, Clomburg JM, Gonzalez R. Efficient synthesis of L-lactic acid from glycerol by metabolically engineered Escherichia coli. Microb Cell Fact. 2013;12:7.
32. Ibrahim MH, Steinbüchel A. Poly(3-hydroxybutyrate) production from glycerol by Zobellella denitrificans MW1 via high-cell-density fed-batch fermentation and simplified solvent extraction. Appl Environ Microbiol. 2009;75:6222–31.
33. Clomburg JM, Gonzalez R. Metabolic engineering of Escherichia coli for the production of 1,2-propanediol from glycerol. Biotechnol Bioeng. 2011;108:867–79.
34. Tang XM, Tan YS, Zhu H, Zhao K, Shen W. Microbial conversion of glycerol to 1,3-propanediol by an engineered strain of Escherichia coli. Appl Environ Microbiol. 2009;75:1628–34.
35. Zhang A, Yang ST. Propionic acid production from glycerol by metabolically engineered Propionibacterium acidipropionici. Process Biochem. 2009;44:1346–51.
36. Xu JY, Zhao XB, Wang WC, Du W, Liu DH. Microbial conversion of biodiesel byproduct glycerol to triacylglycerols by oleaginous yeast Rhodosporidium toruloides and the individual effect of some impurities on lipid production. Biochem Eng J. 2012;65:30–6.
37. Yu KO, Jung J, Ramzi AB, Choe SH, Kim SW, Park C, et al. Development of a Saccharomyces cerevisiae strain for increasing the accumulation of triacylglycerol as a microbial oil feedstock for biodiesel production using glycerol as a substrate. Biotechnol Bioeng. 2013;110:343–7.
38. Vasudevan PT, Briggs M. Biodiesel production—current state of the art and challenges. J Ind Microbiol Biotechnol. 2008;35:421–30.
39. Lestari S, Maki-Arvela P, Beltramini J, Lu GQ, Murzin DY. Transforming triglycerides and fatty acids into biofuels. ChemSusChem. 2009;2:1109–19.
40. Murphy DJ. The biogenesis and functions of lipid bodies in animals, plants and microorganisms. Prog Lipid Res. 2001;40:325–438.
41. Hu Q, Sommerfeld M, Jarvis E, Ghirardi M, Posewitz M, Seibert M, et al. Microalgal triacylglycerols as feedstocks for biofuel production: perspectives and advances. Plant J. 2008;54:621–39.
42. Santillan-Jimenez E, Crocker M. Catalytic deoxygenation of fatty acids and their derivatives to hydrocarbon fuels via decarboxylation/decarbonylation. J Chem Technol Biotechnol. 2012;87:1041–50.
43. Furimsky E. Hydroprocessing challenges in biofuels production. Catal Today. 2013;217:13–56.
44. Fairley P. Introduction: next generation biofuels. Nature. 2011;474:S2–5.
45. Pearlson M, Wollersheim C, Hileman J. A techno-economic review of hydroprocessed renewable esters and fatty acids for jet fuel production. Biofuels Bioprod Bioref. 2013;7:89–96.
46. Peralta-Yahya PP, Zhang F, del Cardayre SB, Keasling JD. Microbial engineering for the production of advanced biofuels. Nature. 2012;488:320–8.
47. Kosa M, Ragauskas AJ. Lipids from heterotrophic microbes: advances in metabolism research. Trends Biotechnol. 2011;29:53–61.
48. Dobson R, Gray V, Rumbold K. Microbial utilization of crude glycerol for the production of value-added products. J Ind Microbiol Biotechnol. 2012;39:217–26.
49. Alvarez HM, Mayer F, Fabritius D, Steinbüchel A. Formation of intracytoplasmic lipid inclusions by Rhodococcus opacus strain PD630. Arch Microbiol. 1996;165:377–86.
50. Waltermann M, Luftmann H, Baumeister D, Kalscheuer R, Steinbüchel A. Rhodococcus opacus strain PD630 as a new source of high-value single-cell oil? Isolation and characterization of triacylglycerols and other storage lipids. Microbiology. 2000;146:1143–9.
51. Kurosawa K, Boccazzi P, de Almeida NM, Sinskey AJ. High-cell-density batch fermentation of Rhodococcus opacus PD630 using a high glucose concentration for triacylglycerol production. J Biotechnol. 2010;147:212–8.
52. Brigham CJ, Kurosawa K, Rha C, Sinskey AJ. Bacterial carbon storage to value added products. J Microbial Biochem Technol. 2011;S3:002. doi:10.4172/1948-5948.
53. Kurosawa K, Wewetzer SJ, Sinskey AJ. Engineering xylose metabolism in triacylglycerol-producing Rhodococcus opacus for lignocellulosic fuel production. Biotechnol Biofuels. 2013;6:134.
54. Kurosawa K, Wewetzer SJ, Sinskey AJ. Triacylglycerol production from corn stover using a xylose-fermenting Rhodococcus opacus strain for lignocellulosic biofuels. J Microbial Biochem Technol. 2014;6:254–9.
55. Riesenberg D, Guthke R. High-cell-density cultivation of microorganisms. Appl Microbiol Biotechnol. 1999;51:422–30.
56. Park EY. Recent progress in microbial cultivation techniques. Adv Biochem Eng Biotechnol. 2004;90:1–33.
57. Carroll A, Somerville C. Cellulosic biofuels. Annu Rev Plant Biol. 2009;60:165–82.
58. Sauer U. Evolutionary engineering of industrially important microbial phenotypes. Adv Biochem Eng Biotechnol. 2001;73:129–69.
59. Wisselink HW, Toirkens MJ, Wu Q, Pronk JT, van Maris AJ. Novel evolutionary engineering approach for accelerated utilization of glucose, xylose, and arabinose mixtures by engineered Saccharomyces cerevisiae strains. Appl Environ Microbiol. 2009;75:907–14.
60. Cheng KK, Lee BS, Masuda T, Ito T, Ikeda K, Hirayama A, et al. Global metabolic network reorganization by adaptive mutations allows fast growth of Escherichia coli on glycerol. Nat Commun. 2014;5:3233.
61. Dien BS, Cotta MA, Jeffries TW. Bacteria engineered for fuel ethanol production: current status. Appl Microbiol Biotechnol. 2003;63:258–66.
62. Hahn-Hagerdal B, Karhumaa K, Fonseca C, Spencer-Martins I, Gorwa-Grauslund MF. Towards industrial pentose-fermenting yeast strains. Appl Microbiol Biotechnol. 2007;74:937–53.
63. Kim JH, Block DE, Mills DA. Simultaneous consumption of pentose and hexose sugars: an optimal microbial phenotype for efficient fermentation of lignocellulosic biomass. Appl Microbiol Biotechnol. 2010;88:1077–85.
64. Ratledge C. The role of malic enzyme as the provider of NADPH in oleaginous microorganisms: a reappraisal and unsolved problems. Biotechnol Lett. 2014;36:1557–68.
65. Janssen HJ, Steinbüchel A. Fatty acid synthesis in Escherichia coli and its applications towards the production of fatty acid based biofuels. Biotechnol Biofuels. 2014;7:7.
66. Chartrain M, Jackey B, Taylor C, Sandford V, Gbewonyo K, Lister L, et al. Bioconversion of indene to cis (1S,2R) indandiol and trans (1R,2R) indandiol by Rhodococcus species. J Ferment Bioeng. 1998;86:550–8.
67. Sudarjanto G, Keller-Lehmann B, Keller J. Optimization of integrated chemical-biological degradation of a reactive azo dye using response surface methodology. J Hazard Mater. 2006;138:160–8.

3

Large-scale robot-assisted genome shuffling yields industrial *Saccharomyces cerevisiae* yeasts with increased ethanol tolerance

Tim Snoek[1,2†], Martina Picca Nicolino[1,2†], Stefanie Van den Bremt[3], Stijn Mertens[1,2], Veerle Saels[1,2], Alex Verplaetse[3], Jan Steensels[1,2] and Kevin J Verstrepen[1,2*]

Abstract

Background: During the final phases of bioethanol fermentation, yeast cells face high ethanol concentrations. This stress results in slower or arrested fermentations and limits ethanol production. Novel *Saccharomyces cerevisiae* strains with superior ethanol tolerance may therefore allow increased yield and efficiency. Genome shuffling has emerged as a powerful approach to rapidly enhance complex traits including ethanol tolerance, yet previous efforts have mostly relied on a mutagenized pool of a single strain, which can potentially limit the effectiveness. Here, we explore novel robot-assisted strategies that allow to shuffle the genomes of multiple parental yeasts on an unprecedented scale.

Results: Screening of 318 different yeasts for ethanol accumulation, sporulation efficiency, and genetic relatedness yielded eight heterothallic strains that served as parents for genome shuffling. In a first approach, the parental strains were subjected to multiple consecutive rounds of random genome shuffling with different selection methods, yielding several hybrids that showed increased ethanol tolerance. Interestingly, on average, hybrids from the first generation (F1) showed higher ethanol production than hybrids from the third generation (F3). In a second approach, we applied several successive rounds of robot-assisted targeted genome shuffling, yielding more than 3,000 targeted crosses. Hybrids selected for ethanol tolerance showed increased ethanol tolerance and production as compared to unselected hybrids, and F1 hybrids were on average superior to F3 hybrids. In total, 135 individual F1 and F3 hybrids were tested in small-scale very high gravity fermentations. Eight hybrids demonstrated superior fermentation performance over the commercial biofuel strain Ethanol Red, showing a 2 to 7% increase in maximal ethanol accumulation. In an 8-l pilot-scale test, the best-performing hybrid fermented medium containing 32% (*w/v*) glucose to dryness, yielding 18.7% (*v/v*) ethanol with a productivity of 0.90 g ethanol/l/h and a yield of 0.45 g ethanol/g glucose.

Conclusions: We report the use of several different large-scale genome shuffling strategies to obtain novel hybrids with increased ethanol tolerance and fermentation capacity. Several of the novel hybrids show best-parent heterosis and outperform the commonly used bioethanol strain Ethanol Red, making them interesting candidate strains for industrial production.

Keywords: Genome shuffling, Bioethanol, Ethanol tolerance, Breeding, Heterosis, Hybridization, Complex phenotype, Evolutionary engineering, Very high gravity, Fermentation

* Correspondence: kevin.verstrepen@biw.vib-kuleuven.be
[†]Equal contributors
[1]Laboratory for Genetics and Genomics, Centre of Microbial and Plant Genetics (CMPG), KU Leuven, Kasteelpark Arenberg 22, 3001 Leuven, Belgium
[2]Laboratory for Systems Biology, VIB, Bio-Incubator, Gaston Geenslaan 1, 3001 Leuven, Belgium
Full list of author information is available at the end of the article

Background

With an annual production of over 50 billion liters per year, bioethanol has become a valuable alternative for non-renewable fossil fuels, yet increases in productivity can aid in further establishing this biofuel. The yeast *Saccharomyces cerevisiae* is often the preferred micro-organism in ethanolic fermentations, due to its high natural ethanol tolerance and excellent fermentation capacity [1]. Very high gravity (VHG) fermentations, in which high (>270 g/l) initial sugar concentrations are used, can in theory yield much improved efficiency because higher final ethanol concentrations are reached [2]. However, stress factors encountered during the fermentation, especially ethanol stress, negatively impact the viability and fermentative performance of the yeast cells, which leads to sluggish or stuck fermentations [3]. Therefore, novel strains with improved ethanol tolerance may further increase productivity by allowing higher initial sugar concentrations that yield increased final ethanol concentrations.

Improving the ethanol tolerance of yeast is not straightforward. First, ethanol tolerance is a complex phenotype, which is underscored by transcriptome and single-gene knockout studies of ethanol-exposed strains that identified hundreds of genes with very different functions to be associated with ethanol tolerance (reviewed by Ma and Liu [4]). Although these studies have shed some light on the mechanisms of ethanol tolerance, detailed mechanistic insight explaining how and why cells can resist high ethanol concentrations is still lacking. Moreover, it is not clear if and how results obtained in laboratory strains, which are in general more ethanol sensitive, can be extrapolated to inherently more robust industrial strains [5]. In addition, there is no strict definition for ethanol tolerance and it is unclear how the different methods used to measure ethanol tolerance relate to the stress encountered in industrial fermentations [6].

Because of the complexity of ethanol tolerance, rational methods for strain improvement, such as genetic engineering, have only yielded limited success to improve the ethanol tolerance of industrial yeasts. By contrast, approaches that generate artificial variation in a non-targeted fashion, for instance by means of evolutionary engineering or mutagenesis, have proven more successful to improve this complex phenotype [7]. In the last decade, genome shuffling was added to the arsenal of techniques to enhance complex phenotypes in microbes without the need for insight into the molecular mechanisms governing the trait of interest (reviewed by Biot-Pelletier and Martin [8]). Genome shuffling consists of three steps [9]. First, a genetically diverse population is generated; mostly by induction of mutations in a single strain, for instance by using chemicals or UV radiation. Second, this diverse population is subjected to screening or selection in order to identify or enrich for the best-performing mutants. Third, the genomes of these superior cells are shuffled by means of asexual (protoplast fusion) or sexual (sporulation and mating) hybridization. In this way, novel combinations of beneficial mutations can be created and/or deleterious mutations can be removed. This whole procedure is typically repeated several times in order to further improve the phenotype, and thereby, genome shuffling allows for much larger leaps in the fitness landscape as compared to classical strain improvement by means of mutagenesis and screening [9]. Another advantage of genome shuffling is that this method is not considered to be genetic modification, in contrast to methods such as protoplast fusion. Therefore, novel strains generated by genome shuffling can be readily applied anywhere in the world without complex procedures to obtain approval to use the strains, even for the production of ethanol that would be used in foods and beverages. Multiple industrially relevant phenotypes in various micro-organisms, including tolerance to industrial stress factors as well as production titers, were successfully improved using genome shuffling [8].

Genome shuffling has also yielded yeast strains with improved ethanol tolerance and fermentation performance (see for an overview [7]). Whereas all these studies report an improvement of the parental strains, the industrial relevance of the novel strains is not always clear. Some researchers used laboratory strains [10,11], which likely do not perform well under industrial conditions. In other studies, the novel hybrids were not tested on a pilot scale [12-15]. Moreover, genome shuffling is often combined with genetic engineering [16,17], making some industrial applications problematic. In addition, most studies used only a limited number of different strains or relied on artificial diversity created by mutagenesis of a single strain. In this way, the vast potential of the rich natural genetic diversity of *Saccharomyces* yeasts is largely ignored. Moreover, mutagenesis may yield mutations that increase certain selected parameters, but negatively affect others. Even though such problems may also arise when combining naturally occurring mutations, in theory, these mutations have withstood the test of natural selection, increasing the chance that they do not negatively affect fitness. However, using naturally occurring mutations is more complex because it requires finding suitable parental strains that combine different positive mutations and also show sufficient sporulation efficiency and spore viability to allow genome shuffling, which is often a problem for industrial strains.

Here, we set out to characterize a large collection of *Saccharomyces* yeasts to identify eight parental strains that could be used for genome shuffling aimed at obtaining novel yeast hybrids that show increased ethanol tolerance compared to the currently used industrial yeasts, such as Ethanol Red. We explored different genome shuffling strategies and selection methods, including repeated rounds of random mating and robot-assisted targeted mating with selection for outcrossing. Our efforts resulted in several hybrids showing best-parent heterosis for ethanol production. One of those hybrids was tested in pilot-scale fermentations and retained its superior performance in these semi-industrial conditions, making it an interesting candidate for commercial bioethanol production.

Results

Large-scale screening of *Saccharomyces* yeasts for ethanol production and sporulation

Because we aimed to exploit the natural variation among *Saccharomyces* yeasts to create novel hybrids with increased ethanol tolerance, we screened a collection consisting of 318 *Saccharomyces* strains isolated from different natural niches and industries such as wine, ale, or biofuel production (see 'Methods'). We first determined the ethanol accumulation capacity under lab-scale VHG conditions (fermentation of 150 ml YP + 35% (*w*/*v*) glucose). The average ethanol production varied widely among strains from different industries, with strains used for bioethanol production showing the highest average production, followed by wine yeasts, wild isolates, sake yeasts, spirits yeasts, and beer and bakery strains (see Additional file 1: Figure S1A). The strain showing the highest ethanol titer was Ethanol Red (Fermentis, S. I. Lesaffre) (Additional file 1: Figure S1B), a strain commonly used in first-generation bioethanol production.

Since many industrial *S. cerevisiae* strains are known to show poor sexual reproduction, we tested the capacity of all 318 yeasts to generate viable spores. We found that 172 (54%) of the strains formed spores, yet the sporulation efficiency and spore viability varied both within and among strains from different origins (Additional file 1: Figure S2A). Most of the lager, sake, and ale strains did not show sporulation, confirming previous studies reporting low sporulation efficiency among industrial strains [18,19]. Next, we determined the spore viability of each sporulating strain by dissecting four tetrads and observed that 144 strains yielded at least one viable spore (Additional file 1: Figure S2B). The majority (74/144) of the strains were found to be homothallic, whereas 46 strains gave rise to both haploid and diploid colonies after tetrad dissection. Twenty-three strains were found to be heterothallic and only produced colonies with a single mating-type. Only 12 of these heterothallic strains consistently showed 2a:2α segregation of the mating-type locus (Additional file 1: Figure S2C).

We hypothesized that shuffling the genomes of genetically divergent strains increases the chance of obtaining hybrids that show best-parent heterosis for our traits of interest (because each strain may contribute different beneficial genetic variation). To estimate the genetic relatedness of the different strains, we performed interdelta PCR fingerprinting on all 318 yeasts [20,21]. Figure 1 shows a heat map that represents the results of the phenotypic and genotypic screen.

Eight genetically divergent heterothallic strains showing high spore viability were selected as parental strains for genome shuffling. This set included Ethanol Red (P7), the best-performing strain from the screening for VHG fermentation capacity described above. Given its widespread commercial application in bioethanol fermentations, Ethanol Red is used as a reference strain throughout this paper. The other seven strains included a sake strain (P4) that showed a very high maximal ethanol accumulation, but fermented a bit slower than Ethanol Red, three strains (P1, P6, P8) that showed efficient fermentation during the first phase of the fermentation but reached lower final ethanol titers, two strains (P2, P3) that did not show exceptional fermentation efficiency, but showed efficient growth in the presence of ethanol (see later), and one strain (P5) that reached a lower final ethanol titer (Additional file 1: Table S1). This last strain was chosen because previous studies found that weak strains often do contain some beneficial genetic variation that is absent from superior strains [5,22,23].

Growth in the presence of ethanol is correlated with fermentation performance

Genome shuffling typically produces a large pool of hybrids, many of which are inferior to the parental strains. Selecting the few superior hybrids from this pool is therefore not always straightforward, especially for traits like ethanol production, which cannot be tested in bulk. Therefore, we explored the correlation between ethanol production and the ability to grow in the presence of high concentrations of ethanol, a trait that is easier to select for. We measured the growth capacity of each strain on solid YPD supplemented with different ethanol concentrations (5 to 14% (*v*/*v*)) using a robot-assisted screen (see 'Methods'). We found a significant positive correlation between production and growth in the presence of ethanol, with the strongest correlation in medium supplemented with 11% (*v*/*v*) ethanol (Pearson $r = 0.51$, $P < 0.0001$, Additional file 1: Figure S3). Hence, we decided to pre-select hybrids generated by genome shuffling for their capacity to grow in the presence of ethanol, and then test individual hybrids for their capacity to produce high concentrations of ethanol (see further). In addition, we used growth in the presence of ethanol as a selection method between different rounds of genome shuffling in some approaches.

Genome shuffling based on random mating generates improved F3 hybrids

Starting from the eight selected parental strains, we created several pools of genome-shuffled hybrids using two different strategies. First, we used random genome shuffling, where the eight parental strains were allowed to

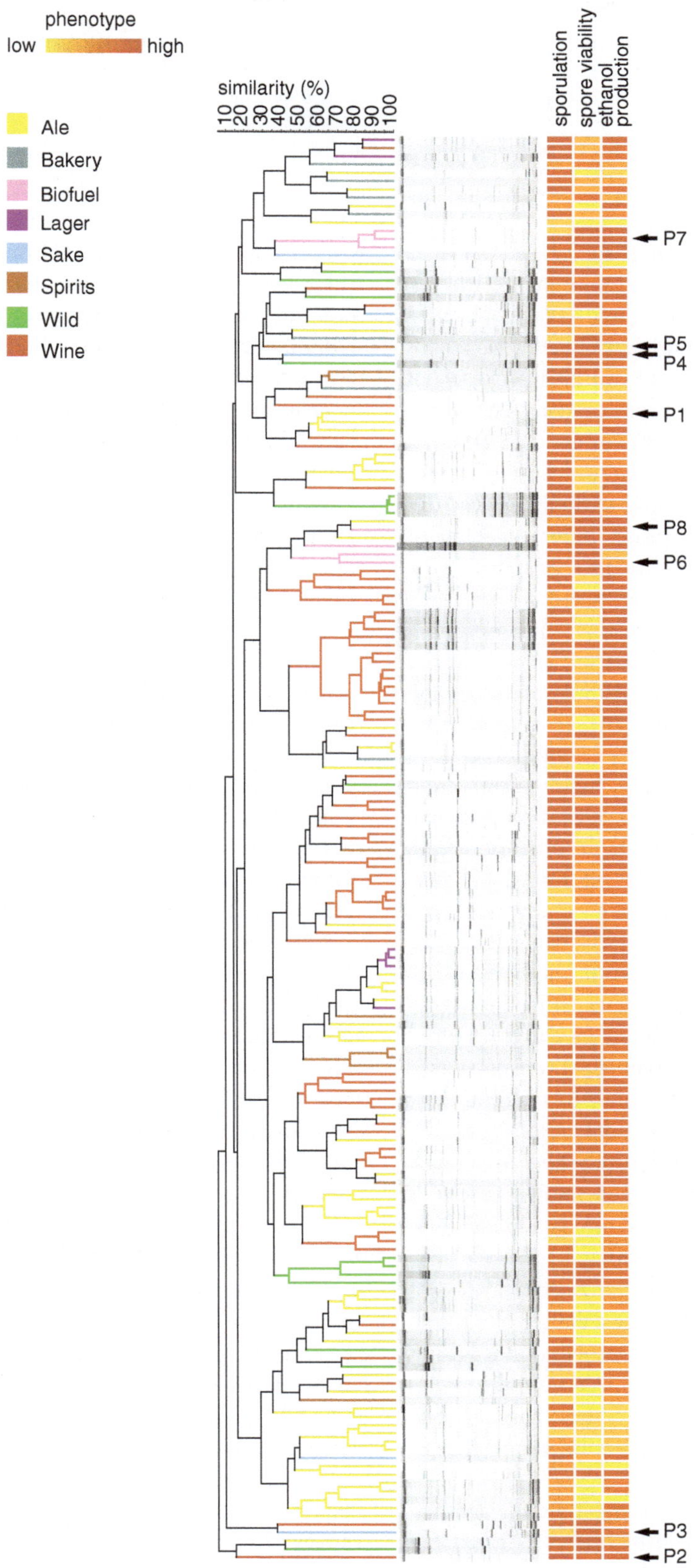

Figure 1 (See legend on next page.)

(See figure on previous page.)

Figure 1 Large-scale genotypic and phenotypic screening of *Saccharomyces* yeasts to select parental strains for genome shuffling. Strains that show good sporulation were clustered according to their genetic relatedness as estimated by interdelta fingerprinting (see 'Methods'). Phenotypes shown include sporulation capacity, spore viability, and ethanol production from a VHG substrate. Each of these phenotypes was measured for all the strains, and the measurements were then normalized and displayed using a color scale (heat map) as indicated in the figure. Eight genetically divergent heterothallic strains were selected that could sporulate and displayed high spore viability (indicated with arrows and codes P1-P8). This strain set included the commonly-used bioethanol strain Ethanol Red (P7); the best-fermenting strain out of the collection.

interbreed randomly. In a second approach, we used large-scale robot-assisted targeted genome shuffling, where specific pairs of strains were mated (see 'Genome shuffling based on targeted mating yields hybrids with increased ethanol tolerance' and Figure 2A). For each of these strategies, we performed three rounds of shuffling, thereby generating three generations of genome-shuffled pools of hybrids (referred to as F1, F2, and F3).

For the random genome shuffling strategy, spores derived from the eight parental strains were subjected to mass mating (see 'Methods') to create a heterogeneous F1 mixture. We used three different selection methods to obtain F3 hybrids starting from this F1 generation. First, we selected the pool of hybrids for growth in the presence of ethanol after each round of shuffling, thereby exploiting the correlation between ethanol production and growth in high ethanol conditions (see above). In a second approach, the hybrids generated after each round of shuffling were selected for survival in medium containing extremely high ethanol levels. Third, we also performed the different rounds of random genome shuffling in the absence of any selection. Figure 2B shows a schematic overview of these different genome shuffling strategies and selection methods.

Random genome shuffling using selection for growth in the presence of ethanol was carried out for three consecutive rounds of shuffling in three biological replicates (see 'Methods' and Additional file 1: Table S2). To directly compare the ethanol tolerance of populations obtained after different rounds of shuffling, we tested from each biological replicate of the F2 and F3 generations as well as from the unselected F1 pool 24 random isolates for their capacity to grow on medium containing 10 to 12% (*v*/*v*) ethanol (Figure 3A and Additional file 1: Table S2). We found that the ethanol tolerance increased after one round of shuffling and selection, but that another round did not increase the tolerance further (Figure 3A). Multiple clones (29/72 from F2, 21/72 from F3) showed best-parent heterosis (that is, they showed higher ethanol tolerance than all parental strains) (Figure 3B).

The second selection method during random genome shuffling was based on survival in high ethanol levels (18 to 22% (*v*/*v*)). Although the ethanol tolerance of microbes is often measured by the capability to survive exposure to lethal ethanol levels, this is the first study using this phenotype as a selection method in genome shuffling experiments. After each round of shuffling, pools of hybrids were grown to the late log-phase, and from each pool, 10^9 cells were harvested and exposed to rich medium supplemented with ethanol. After incubation for 16 h, cells were plated on solid glycerol medium to select for cells that survived exposure to ethanol and that were still able to respire (since ethanol is a potent inducer of so-called petite mutations that lead to a loss of mitochondrial function [24]). We carried out this selection method in three biological replicates. After the first round of shuffling, we selected F1 hybrids that survived 18% (*v*/*v*) ethanol, and after the second round of shuffling, F2 hybrids were selected that survived exposure to 19% (*v*/*v*) ethanol (see 'Methods' and Additional file 1: Table S3). This selection strategy was highly selective: whereas 10^9 cells were exposed to ethanol, only about 30,000 cells (approximately 0.003% of the population) were recovered for the F1 generation after exposure to 18% (*v*/*v*) ethanol (Additional file 1: Table S3). For each generation, we also tested the capacity of the hybrids to survive a range of ethanol concentrations and found that the survival capability increased with each round of shuffling, indicating that we were successfully selecting cells with improved tolerance for high ethanol concentrations (Figure 3C and Additional file 1: Table S3). Moreover, subjecting unselected and growth-selected F3 populations to high ethanol levels gave results similar to or worse than those obtained for F1 populations, indicating that the increased survival capacity resulted from the selection and not merely from repeated rounds of sporulation and mating (see Additional file 1: Table S4).

Different selection methods yield hybrids with different fermentation performance

To investigate how the different selection methods used in our random genome shuffling experiments (selection for growth, survival, or no selection) affected the performance of the final F3 populations in VHG fermentations, we inoculated subpopulations (that is, samples of the pools of hybrids and not pre-selected pure individual hybrids) of the different F3 populations into YP medium supplemented with 35% (*w*/*v*) glucose. None of the hybrid populations fermented faster or more completely than the control strain Ethanol Red. Interestingly, among hybrid populations, unselected populations showed the fastest fermentation, whereas growth-selected populations fermented to a same

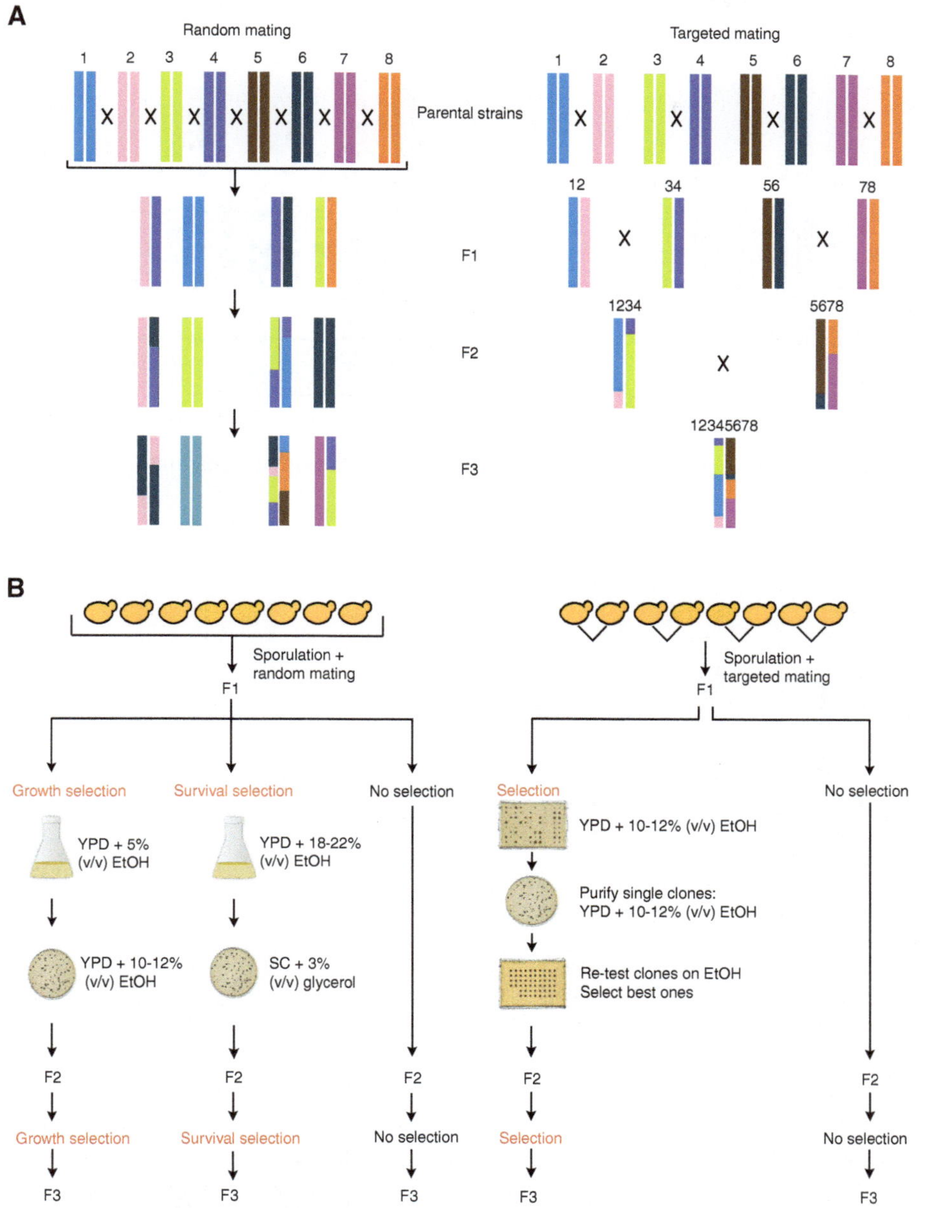

Figure 2 Two different genome shuffling strategies used to generate hybrids starting from the eight parental strains. (A) Conceptual outline of targeted and random genome shuffling. Genome shuffling based on random mating (left) allows random mating between spores form the eight parental strains (F1) and between spores derived from genome-shuffled hybrid populations (F2 and F3). Genome shuffling based on targeted mating (right) exploits the selection of true outcrossed hybrids using plasmid-based markers at each stage, ensuring the presence of the eight initial genomes in the final F3 hybrids. For simplicity, homologous chromosomes of the parental strains have been given the same color, although parental strains were heterozygous diploids. **(B)** Details of the experimental procedures used for genome shuffling strategies. Genome shuffling based on random mating (left) was performed using two different types of selection after each round of mating. First, hybrids were selected for their capacity to grow in the presence of ethanol by first inoculating them into medium containing 5% (*v/v*) ethanol followed by growth in the presence of 10 to 12% (*v/v*) ethanol (referred to as 'growth selection,' similar to the selection applied for targeted mating). Alternatively, the hybrids were selected for survival in medium containing very high (18 to 22% (*v/v*)) ethanol levels ('survival selection'). In parallel to these two approaches, we also carried out shuffling without any selection in between the different rounds of hybridization ('no selection'). For targeted mating (right), a robot is used to perform specific crosses between the eight parental strains in all pairwise combinations, followed by screening for ethanol tolerance (growth capacity in the presence of ethanol). The best-performing hybrids were used as parental strains for the next round of robot-based targeted mating, and in parallel similar breeding schemes were carried out without applying any selection after each round of shuffling. See 'Methods' and Additional file 1: Figure S5 for more details about these procedures.

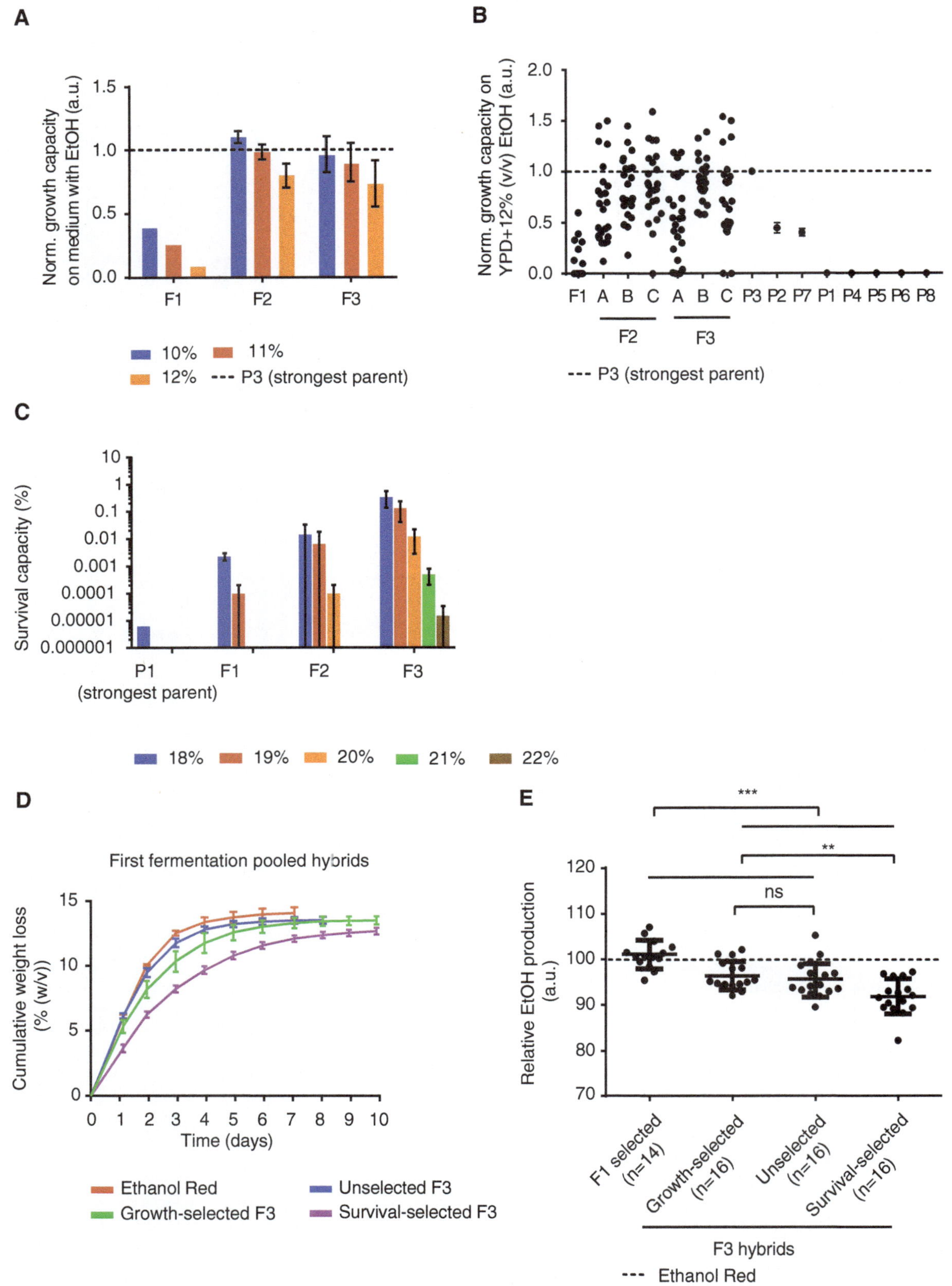

Figure 3 (See legend on next page.)

(See figure on previous page.)

Figure 3 Genome shuffling based on random mating generates hybrids with increased ethanol tolerance. (A) Growth capacity in the presence of ethanol of clones obtained before selection (F1), after one round (F2) and two rounds of shuffling and selection (F3). Whereas F2 hybrids show a clear increase in growth capacity, F3 hybrids do not show further increases. **(B)** Same data as Figure 3A for the individual isolates at 12% (*v/v*) ethanol and the parental strains ($n = 2$ for each parent). F2 and F3 isolates on average performed better than F1 isolates, but did not show statistical differences between different replicates and generations (unpaired *t*-test $P > 0.05$). A, B, and C are biological replicates. **(C)** Selection for survival in high ethanol yields hybrids with increased survival capacity which further increases after each round of shuffling. Each bar represents the average and standard deviation of three biological replicates, except for F1 19% (*v/v*) where $n = 2$. Survival after exposure to 21% (*v/v*) and 22% (*v/v*) was only measured for F3 populations. The data for the strongest parental strain (P1) is displayed for 18% (*v/v*); this strain completely lost its viability after exposure to 19% (*v/v*). **(D)** Fermentation performance of F3 populations in YP + 35% (*w/v*) glucose. The cumulative weight loss, a proxy for CO_2 production, during the fermentations is shown. Each line represents the average of two replicate fermentations for the reference strain (Ethanol Red), six replicates for growth-selected and survival-selected populations, or twelve replicates for unselected populations. Error bars represent standard deviations. **(E)** Selected isolates from several pools of hybrids were tested for their maximal ethanol accumulation in YP + 35% (*w/v*) glucose. On average, F1 hybrids showed the highest average ethanol production levels. Unselected and growth-selected F3 hybrids on average performed similar, but better than survival-selected F3 hybrids. The dotted line indicates the ethanol production of Ethanol Red. Unpaired *t*-test: ns, not significant; $^{**}P \leq 0.01$; $^{***}P \leq 0.001$.

degree, but more slowly (Figure 3D). On the other hand, survival-selected populations showed less efficient fermentation. Re-inoculating the cells for two subsequent fermentations confirmed these differences (Additional file 1: Figure S4). After three subsequent fermentations, we selected 16 colonies from populations derived from each genome shuffling strategy. In addition, we also selected 14 different F1 clones that showed a high growth capacity in the presence of ethanol. We evaluated the fermentation performance of all these isolates by measuring the maximal ethanol accumulation in VHG fermentations, and normalized this value to the average ethanol level produced by Ethanol Red in the same batch. Interestingly, on average, individual hybrids selected from the F1 population showed superior fermentation performance compared to the hybrids isolated from the F3 pools, with 10 out of 14 F1 hybrids reaching a higher final ethanol level than Ethanol Red (Figure 3E). There was no statistical difference in the final ethanol titer between growth-selected and unselected F3 hybrids, whereas all categories performed statistically better than survival-selected hybrids (Figure 3E and Additional file 2). In this screening, three growth-selected F3 hybrids and two unselected F3 hybrids showed a higher maximal ethanol accumulation than Ethanol Red (relative ethanol production >100).

Genome shuffling based on targeted mating yields hybrids with increased ethanol tolerance

In addition to the random genome shuffling strategy described above, we mated the same eight parental strains using a targeted funnel-breeding scheme based on non-stable plasmid-based dominant markers (see Figure 2A and 'Methods'). This strategy involved three rounds of shuffling; F1 hybrids incorporated two of the initial genomes; F2 hybrids incorporated four parental genomes; and finally, F3 hybrids were a mosaic of the eight parental genomes (Figures 2A and 4). We carried out two parallel experiments: one in which we performed selection for growth in the presence of ethanol after each round of shuffling and one experiment that started from unselected F1 hybrids and in which no intermediate selection was carried out (Figure 2B).

To generate a large number of F1 hybrids with known ancestry, we crossed the eight parental strains in all pairwise combinations in multiple replicates by placing spore suspensions of strains carrying different antibiotic resistance markers in close proximity on solid rich medium using a robot (see Additional file 1: Figure S5 and 'Methods' for details). We obtained 500 outcrossed and 72 inbred F1 pools of hybrids, covering all possible crossings at least once. We assayed the growth of these pools of outcrossed and inbred F1 hybrids as well as the parental strains in the presence of different concentrations of ethanol as well as on medium without ethanol using a robot-based spotting assay (Additional file 1: Figure S6). Next, the ethanol tolerance values of all pools derived from a certain cross were averaged per tested ethanol concentration, and ranked based on the highest ethanol concentration measured (13% (*v/v*); see Figure 5A). A striking result was that the average ethanol tolerance of outcrossed hybrids was significantly higher than that of inbreds (Figure 5B and Additional file 1: Figure S7).

As a next step, we purified and re-tested single isolates from eight different F1 populations that showed high ethanol tolerance as well as from inbred populations from each parental strain. We found that isolated inbreds performed worse than their respective parents except for strain P3, whereas seven outcrossed F1 strains showed best-parent heterosis and one F1 hybrid performed in between its respective parents, underscoring the large phenotypic improvement reached by outcrossing (Figure 5C). Eight highly ethanol-tolerant F1 clones were used to create three funnel breeding schemes for the next round of genome shuffling based on targeted mating (see Figure 4 for details). In total, we assayed the ethanol tolerance of 720 outcrossed and 236 inbred pools of hybrids for the F2 generation, and we found that, similar to the first round of targeted mating, the

F1	1	2	3	4	5	6	7	8
1	11	12	13	14	15	16	17	18
2	21	22	23	24	25	26	27	28
3	31	32	33	34	35	36	37	38
4	41	42	43	44	45	46	47	48
5	51	52	53	54	55	56	57	58
6	61	62	63	64	65	66	67	68
7	71	72	73	74	75	76	77	78
8	81	82	83	84	85	86	87	88

FUNNEL A

F2	12	36	45	78
12	1212	1236	1245	1278
36	3612	3636	3645	3678
45	4512	4536	4545	4578
78	7812	7836	7845	7878

F3	4578	3678	3645	1236	1245	1278
1236	12364578					
1245		12453678				
1278			12783645			
4578				45781236		
3678					36781245	
3645						36451278

FUNNEL B

F2	36	45	17	28
36	3636	3645	3617	3628
45	4536	4545	4517	4528
17	1736	1745	1717	1728
28	2836	2845	2817	2828

F3	3645	3617	2836	1728	2845	1745
1728	17283645					
2845		28453617				
1745			17452836			
3645				36451728		
3617					36172845	
2836						28361745

FUNNEL C

F2	12	36	48	75
12	1212	1236	1248	1275
36	3612	3636	3648	3675
48	4812	4836	4848	4875
75	7512	7536	7548	7575

F3	3657	4857	3648	1248	1236	1275
1248	12483657					
1236		12364857				
1275			12753648			
3657				36571248		
4857					48571236	
3648						36481275

Figure 4 Schematic overview of targeted genome shuffling. For the first round of targeted genome shuffling (creating pools of F1 hybrids), the eight parental strains (numbered 1 to 8) were crossed in all pairwise combinations, including inbreeding. After screening and selection (see main text), isolated F1 hybrids were used to create three funnels (A, B, C). In each funnel, the F1 hybrids were crossed in all pairwise combinations to make F2 hybrids, followed by screening and selection for ethanol tolerance. For the last round of genome shuffling, F2 hybrids were crossed in such a way that F3 hybrids incorporated genetic material from all eight initial parental strains. Note that these funnel schemes were also carried out for F1 hybrids from the same parental strains that were not selected for growth in the presence of ethanol, and for which no selection was performed in between the different rounds of hybridization.

ethanol tolerance of outcrossed F2 hybrids was higher than the ethanol tolerance of inbreds (Additional file 1: Figure S8). Per funnel, we isolated six F2 isolates that showed high ethanol tolerance. For the final round of mating, we crossed the selected F2 hybrids of each funnel in such a way that F3 hybrids incorporated the eight parental genomes. In parallel, these exact procedures, that is, the same targeted mating schemes consisting of multiple rounds of shuffling, were carried out using unselected F1 hybrids without intermediate selection for ethanol tolerance.

In order to compare the ethanol tolerance of selected and unselected F3 hybrids, we assayed the ethanol tolerance for 784 selected pools of F3 crosses and 596 pools of unselected F3 crosses. Next, we isolated 310 clones (10 single colonies from 31 pools) for the funnels where selection was applied after each round of mating and another 310 clones for funnels where no selection was

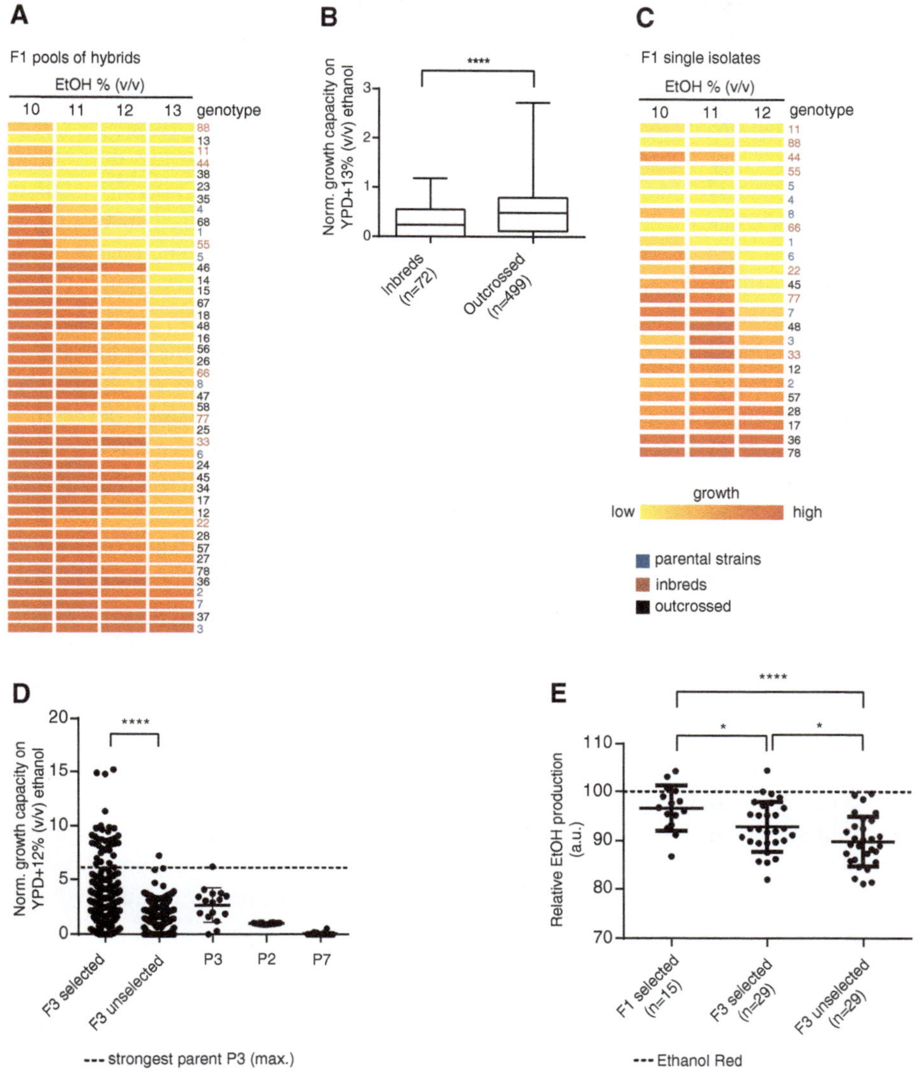

Figure 5 Genome shuffling based on targeted mating yields hybrids with increased growth and fermentation capacity. (A) The eight parental strains were crossed in all pairwise combinations. Each horizontal line represents the average growth in medium with different concentrations of ethanol for all hybrid populations, as well as the parental strains. The data was normalized and converted to a heat map, and strains were ranked from low to high based on their growth in 13% (*v/v*) ethanol. **(B)** Outcrossed F1 hybrids display higher ethanol tolerance than F1 inbreds (Mann-Whitney test: ****$P \leq 0.0001$). **(C)** Most isolated F1 outcrossed hybrids show heterosis for ethanol tolerance, whereas most inbreds perform poorer than their parents ('inbreeding depression'). Each horizontal line represents the average performance of all single clones from a certain cross, inbred or parental strain on different concentrations of ethanol. Strains were ranked from low to high based on 12% (*v/v*) ethanol data. **(D)** F3 hybrids subjected to selection for ethanol tolerance (measured by the capacity to grow in the presence of ethanol) after each round of genome shuffling show higher ethanol tolerance than unselected F3 hybrids (Mann-Whitney test: ****$P \leq 0.0001$). The dot plots show the relative growth of all F3 isolates, obtained with and without selection after each round of shuffling, respectively, on 12% (*v/v*) ethanol. The data from the strongest parental strains (P3, P2, and P7) are shown for comparison. **(E)** Individual hybrids show different maximal ethanol accumulation in VHG fermentations. F1 hybrids selected for their capacity to grow in the presence of high ethanol levels on average show the highest ethanol production. Next, F3 hybrids subjected to selection after each round of shuffling on average show higher ethanol production that unselected F3 hybrids. The dotted line indicates the ethanol production level of Ethanol Red. Unpaired *t*-test: *$P \leq 0.05$; ****$P \leq 0.0001$.

applied in between each round mating, and tested the growth capacity of these pure isolates in the presence of ethanol. As expected, we found that selected hybrid isolates grew better than unselected hybrids (Figure 5D) on ethanol, and some showed best-parent heterosis and performed better than the strongest parental strain.

On average, targeted F1 hybrids show superior VHG fermentation over targeted F3 hybrids

To find out whether increased ethanol tolerance leads to increased maximal ethanol accumulation in VHG fermentations, we tested the best-performing clones of selected and unselected targeted F3 hybrids in lab-scale

VHG fermentations. In addition, we included 15 selected targeted F1 hybrids (the eight F1 isolates used in the funnels plus seven additional isolates). Interestingly, F1 hybrids on average showed higher ethanol production than both selected and unselected F3 hybrids (Figure 5E). This may be partly explained by the presence of a larger fraction of the Ethanol Red genome in these F1 strains (Additional file 1: Figure S9). In addition, selection for ethanol tolerance after each round of shuffling yielded F3 hybrids that show a higher ethanol production than those obtained from crosses where no selection was applied until after the last round of genome shuffling, indicating that selection for ethanol tolerance can in some cases increase the chances of selecting hybrids with increased ethanol production. Overall, in this screening, we identified four F1 hybrids and one F3 hybrid that showed a higher ethanol production than Ethanol Red (Additional file 3).

Confirmation of the increased fermentation capacity of random and targeted hybrids

To confirm that some hybrids obtained in our different genome shuffling experiments outperform the commonly used bioethanol strain Ethanol Red, we re-tested thirteen hybrids (four F1 and four F3 hybrids from the 'random' approach and four F1 and one F3 hybrid from the 'targeted' approach; see Additional files 2 and 3) in triplicate in small-scale VHG fermentations. In total, eight hybrids (H1-H8) showed a statistically significant higher maximal ethanol accumulation (unpaired *t*-test: $P < 0.05$) than Ethanol Red (Additional file 4). Since it took these hybrids longer to complete the fermentation than Ethanol Red, we calculated an estimated ethanol productivity after 7 days (when Ethanol Red finished the fermentation). Seven hybrids (H1-H7) showed a significantly higher ethanol production rate, whereas the eighth hybrid (H8) fermented more slowly (Table 1). The average ethanol production for four hybrids (H9-H12) was statistically identical to Ethanol Red, yet one of those hybrids (H9) completed the fermentation 1 day faster than Ethanol Red (see Additional file 4 and Table 1). One hybrid (H13) showed a lower maximal ethanol accumulation than Ethanol Red. For the improved hybrids, the ethanol yield (grams of ethanol produced per grams of sugar consumed) was not statistically different from the yield of Ethanol Red (0.49 g/g), indicating that these hybrids ferment with an efficiency similar to Ethanol Red (Additional file 4).

Pilot-scale testing of one superior hybrid

We set out to examine whether the performance of one of the best hybrids, H1, could be confirmed in semi-industrial conditions. First, we screened the fermentation capacity of H1 and 26 natural strains from the collection in static lab-scale fermentations (100 ml fermentation medium containing 320 g/l glucose) and measured the final ethanol production. Hybrid H1 ranked second in this screening, yet among the top strains, no statistical differences in ethanol titer could be detected (Additional file 5). Next, we tested the performance of H1 and the three other best-performing strains in 8 l VHG medium in a bioreactor. Notably, Ethanol Red was not tested on this scale since it ranked only ninth in the screening. H1 and Y145 showed superior fermentation performance over the other two strains (Figure 6); both strains completely utilized all the initial glucose, yet H1 showed the highest final ethanol titer, yield, and ethanol productivity (Table 2).

Discussion

In this study we developed multiple large-scale genome shuffling strategies, which yielded several new hybrid yeasts that show improved maximal ethanol accumulation and ethanol tolerance compared to the commonly used

Table 1 Hybrids with improved fermentation performance identified in this study

			Ethanol production		Estimated ethanol productivity $d = 7$			Yield
Code	Genotype	Generation	Relative	(% (v/v))	(g/l)	Relative	(g/l/h)	(g/g)
H1	47	F1	107.3 ± 1.16	20.0 ± 0.22	158.1 ± 1.70	105.7 ± 1.27	0.94 ± 0.01	0.49 ± 0.00
H2	Unknown	F1	107.3 ± 0.21	20.0 ± 0.04	158.2 ± 0.32	106.7 ± 0.19	0.95 ± 0.00	0.49 ± 0.00
H3	Unknown	F3	106.2 ± 1.16	19.8 ± 0.22	156.6 ± 1.71	103.6 ± 1.32	0.93 ± 0.01	0.49 ± 0.00
H4	Unknown	F1	105.4 ± 2.35	19.4 ± 0.44	153.2 ± 0.66	104.8 ± 1.89	0.94 ± 0.02	0.49 ± 0.00
H5	12364578	F3	105.0 ± 0.41	19.6 ± 0.08	154.8 ± 0.60	104.5 ± 0.31	0.93 ± 0.00	0.48 ± 0.00
H6	Unknown	F1	104.9 ± 1.01	19.6 ± 0.19	154.6 ± 1.49	104.7 ± 0.51	0.94 ± 0.00	0.49 ± 0.01
H7	Unknown	F1	103.9 ± 0.45	19.4 ± 0.08	153.2 ± 0.66	102.3 ± 0.45	0.91 ± 0.00	0.50 ± 0.01
H8	Unknown	F3	102.0 ± 0.70	19.0 ± 0.13	150.3 ± 1.03	98.3 ± 0.59	0.88 ± 0.01	0.49 ± 0.00
H9	78	F1	100.0 ± 1.05	19.1 ± 0.20	151.0 ± 1.58	NA	NA	0.48 ± 0.01
P7	EtOH Red	-	100 ± 0.83	19.0 ± 0.40	149.7 ± 2.94	100 ± 0.84	0.90 ± 0.03	0.49 ± 0.01

Genotype indicates the predicted genotype; for hybrids generated by random mating the genotypes are unknown, whereas for targeted hybrids the parental strains are indicated. Hybrids H1 up to H7 show both increased maximal ethanol accumulation and estimated ethanol productivity relative to Ethanol Red. Data for Ethanol Red is based on 30 biological replicates. See Additional file 4 for the complete dataset.

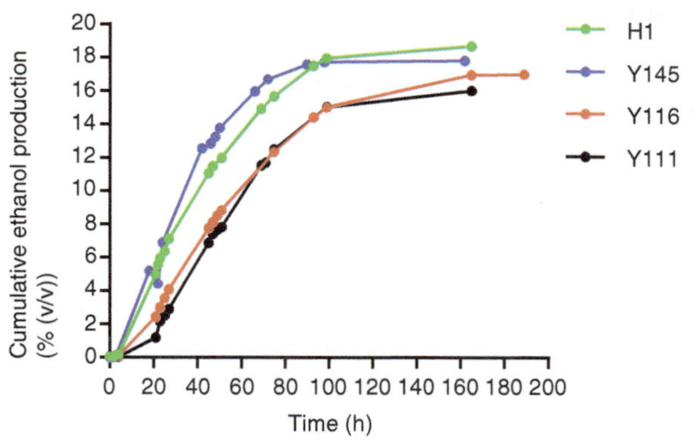

Figure 6 A targeted hybrid (H1) generated by genome shuffling excels in pilot scale fermentations. H1 and three high-ethanol producing industrial strains (Y145, Y116 and Y111) were each inoculated into 8 l VHG medium with 32% (*w/v*) glucose in a bioreactor. On regular time points the ethanol concentration was determined.

biofuel strain Ethanol Red. Yeast strains with increased fermentation performance can help to increase the productivity and economic viability of biofuel production. A potential problem of genome shuffling is that many industrial strains show poor sexual reproduction and therefore are not suitable for strain improvement by means of mating. However, after screening more than 300 strains for their ability to form spores, we found the total number of strains that could potentially be used for genome shuffling to be sufficient to cover a large proportion of the existing natural genetic variation in *S. cerevisiae* (Figure 1).

The use of genome shuffling has already been applied to multiple phenotypes in *S. cerevisiae*, including tolerance to various stressors such as acetic acid [25], spent sulfite liquor [26], heat [27,28], ethanol, and fermentation capacity (as discussed by Steensels et al. [7]). In contrast to most previous genome shuffling experiments, we exploited the large natural variation among *Saccharomyces* yeasts. Instead of starting from one mutagenized strain, we selected eight parental strains, all shown to be genetically divergent heterozygous diploids (ongoing work in our laboratory by Brigida Gallone et al.). In addition, we used robotized assays that allowed us to generate and test many more hybrids than previous studies, thereby maximizing the chances of generating and finding superior hybrids.

Since we generated a huge number of different hybrids, testing each hybrid individually in a small-scale fermentation was impossible. Therefore, we used a robot-assisted screen for growth in the presence of a range of different ethanol concentrations (5 to 14% (*v*/*v*)), a phenotype for which we found significant correlations with ethanol production. Our results demonstrate that selecting for growth capacity in the presence of ethanol offers an efficient way to perform a first screening to select variants that are more likely to show increased fermentation capacity. Previously, Zheng et al. [12] explored the correlation between different industrially relevant phenotypes and did not find a significant correlation between growth in the presence of ethanol and ethanol production. However, we tested a much higher number of different yeast strains (308 versus 15). In addition, we found the highest correlation with growth in the presence of 11% (*v*/*v*) ethanol, a concentration not tested by Zheng et al. [12], suggesting that higher ethanol stress is more efficient for pre-screening or selecting strains for VHG fermentation. Therefore, we used selection for growth in the presence of ethanol as one of the selection methods between each round of genome shuffling. It is important to note that growth and production are at best weakly correlated, as also reported by Pais and colleagues [29], suggesting that both phenotypes have a partly different genetic basis.

Apart from yielding superior novel hybrid yeasts, our experiments also generated insight into the dynamics and efficiency of different genome shuffling strategies to obtain yeasts with improved ethanol tolerance and maximal ethanol production. We tested two different genome shuffling strategies, one based on random mating and one based on targeted mating. Moreover, for these two strategies, we compared the effects of selecting a subpopulation of superior hybrids in terms of ethanol tolerance after each round of shuffling, versus a parallel experiment in which we did not apply any selection until after the last round of shuffling (see further).

The first selection method encompassed selection for growth in the presence of ethanol after each round of shuffling. Interestingly, we found that, on average, F1 hybrids selected for their growth capacity in the presence of ethanol showed superior fermentation performance compared to F3 hybrids, both for genome shuffling with random mating and targeted mating. This can partly be explained by the finding that out of the eight selected parental strains, some specific combinations of parents provide a superior genetic basis, so that further crossing

Table 2 The performance of hybrid H1 and three high ethanol-producing strains on a pilot scale

Strain	Ethanol production (% (*v*/*v*))	Yield (%)	Yield (g/g)	Ethanol productivity (g/l/h)
H1	18.7	88.8	0.45	0.90
Y145	17.9	86.5	0.44	0.87
Y116	17.0	88.8	0.45	0.71
Y111	16.0	88.0	0.45	0.77

Each strain was inoculated into 8 l of VHG medium. After 7 days, the fermentations were terminated, and the ethanol and glucose content were determined.

with other strains often results in weaker hybrids (see further). Another factor that might contribute is sign epistasis, which occurs when two alleles that each are beneficial in a particular genetic background result in a detrimental effect when combined in the same strain. Since F3 hybrids are expected to possess more complex mosaic genetic backgrounds, where alleles from different strains are combined, the effect of sign epistasis might be more pronounced. On the other hand, at least in theory, if sufficient F2 and F3 hybrids are tested, it should be possible to isolate hybrids that are superior to the best F1 hybrids, even if these superior F2 and F3 hybrids are very rare. Indeed, while superior F3 hybrids were more rare than improved F1 hybrids, we were able to isolate several superior F3 strains.

Since the targeted mating strategy allowed us to keep track of the ancestry of the hybrids and examine the contribution of the different parental genomes, we found that the targeted F1 hybrids that we tested in fermentations were considerably enriched in the Ethanol Red (P7) genome, and to a lesser degree in the P4 background, whereas the P1, P5, and P6 genomes seem under-represented (Additional file 1: Figure S9). The depletion of these genomes was perhaps not unexpected since we used growth in the presence of ethanol as a pre-selection method; P1, P5, and P6 grow relatively poorly in the presence of ethanol (Additional file 1: Figure S6). Indeed, all four targeted F1 hybrids that reached a higher final ethanol titer than Ethanol Red share Ethanol Red as one of their parents (Additional file 3).

Interestingly, we recovered only 1 (out of 29) targeted F3 hybrids selected for growth in the presence of ethanol that reached a higher ethanol titer than Ethanol Red. This seems to suggest that it may not be efficient to repeat the genome shuffling for several rounds and that it is instead better to stop after one round of shuffling. There are multiple possible explanations for the superior fermentation performance of (targeted) F1 compared to F3 hybrids. Firstly, some strains, such as Ethanol red and P4, already show excellent ethanol accumulation in VHG fermentation. Hence, it seems likely that these parents possess interesting combinations of alleles. It seems probable that at least some F1 hybrids inherit the majority of these beneficial alleles, but this chance decreases with each round of crossing. In theory, further rounds of crossing may result in the combination of beneficial alleles from more strains, but in reality, this would require generating and testing a very large number of hybrids. Second, the scatterplot showing the correlation between growth and ethanol production suggests that strains that grow weakly in the presence of ethanol are unlikely to produce high ethanol levels (Additional file 1: Figure S3B). Nevertheless, strains that grow well in the presence of ethanol are not always the best ethanol-producing strains, which was also illustrated by our parental strains: the best-fermenting strain out of the collection, Ethanol Red, grows well in the presence of ethanol, but not as good as P3, which produces considerably lower ethanol levels. The repeated selection for ethanol tolerance in consecutive rounds of genome shuffling may therefore exclude some of the high ethanol-producing strains.

In general, F1 hybrids generated by random mating also performed better than F3 hybrids. Specifically, 10 (out of 14) F1 hybrids from the random shuffling approach that were screened in VHG fermentations reached a higher ethanol titer than Ethanol Red, as compared to only 3 (out of 16) F3 hybrids. Based on these frequencies, genome shuffling using random mating followed by pre-selecting the F1 hybrids for their growth capacity in the presence of ethanol appears to be the best strategy to identify novel strains with increased fermentation performance. Nevertheless, we also confirmed the increased performance of a number of F3 hybrids from both strategies. For reasons described above, we hypothesize that although F3 hybrids with increased VHG performance were less frequent than superior F1 hybrids, these F3 pools might in theory contain hybrids that outperform some of the best F1 hybrids because the theoretical genetic diversity is larger in the F3 pool compared to the F1 pool.

For both genome shuffling based on random mating and on targeted mating, we also investigated the dynamics of not applying any selection in between the different rounds of genome shuffling. The idea behind not selecting the hybrids after each round of genome shuffling is that, in theory, it is possible that some F1 or F2 hybrids that show weak ethanol tolerance harbor beneficial (recessive) alleles that can aid in bringing forth superior hybrids. By selecting after each round of shuffling, such F1 and F2 hybrids would not be selected for the next round of crossing. However, our results demonstrate that on average, F3 hybrids that were obtained from genome shuffling schemes that involved selection for growth in the presence of ethanol after each round of crossing showed better ethanol tolerance and ethanol production than F3 hybrids that did not undergo selection after each round of shuffling.

In addition to selection for increased growth in the presence of high ethanol concentrations, we also explored a selection strategy based on survival in medium containing even higher ethanol levels. Ethanol tolerance has multiple definitions, including the ability to survive high ethanol concentrations [6,30], yet until this report, no study has used this phenotype as a selection method in genome shuffling experiments. The advantage of selecting for survival is that extremely large numbers of hybrids can be subjected to selection in bulk, which in theory should increase the chance of selecting the very best hybrids present in the population. This selection strategy proved to be quite powerful and efficient, leading to a stepwise improvement of the capacity to survive high ethanol levels with each round of shuffling and selection. In a control experiment, we found that genome-shuffled F3 populations that were subjected to selection for growth in the presence of ethanol after each round of shuffling as well as F3 populations resulting from genome shuffling schemes without selection did not show increased survival capacity when subjected to high ethanol levels (Additional file 1: Table S4). Actually, these populations performed similar to F1 populations, suggesting that growth capacity in the presence of ethanol and the capacity to survive high ethanol levels as measured in our experimental setups are not correlated. Even though the selection strategy for survival in high ethanol indeed generated hybrids showing superior survival in high ethanol concentrations, the method did not yield any hybrids that showed superior ethanol production, suggesting that selecting for survival in ethanol is not a suitable strategy to isolate variants that show increased fermentation performance.

We chose to solely rely on natural variation for our genome shuffling strategies and not create extra genetic variation by means of mutagenesis. This choice implies that we only work with natural alleles that have been subjected to natural selection, which might decrease the chances of introducing alleles that are beneficial for the phenotype of interest (ethanol production) but detrimental for other vital phenotypes. However, the strategy also comes with limitations. By only using natural strains, the number of beneficial mutations present in the pool might be smaller than when an extensive mutagenesis treatment is applied on the starting strains.

The results obtained with targeted crosses demonstrate that most selected outcrossed F1 hybrids showed best-parent heterosis and outperformed both of their parental strains. The superior performance of hybrids compared to their parents (best-parent heterosis) has been observed in yeast before. Systematic analyses of pairwise crosses of strains described by Liti and colleagues [31] showed that heterosis does occur, although the extent varies from study to study [32-34]. Heterosis was also shown for ethanol production [35] and ethanol tolerance [36]. Nevertheless, the exact mechanisms underlying heterosis remain largely unknown. For future work, it would be interesting to characterize the novel strains we generated with increased fermentation capacity at the genetic and phenotypic level, for example by using a QTL mapping approach as described by Duitama et al. [37].

Previous genome shuffling studies also created yeast hybrids with increased ethanol production. Many of these studies were based on the similar principles than the ones we used, that is, selecting for stress resistance after each round of shuffling and testing the obtained hybrids (often F3 hybrids) for their fermentation capacity [10,17,28,38]. Alternatively, individual fermentation trials were performed after each round of shuffling and only hybrids with increased fermentation performance advanced to the next round of genome shuffling [12,14-16,25,27]. However, most of these studies started from a mutagenized single strain, of which the initial performance is often not compared to a larger set of industrial strains. This makes it difficult to compare the results of these previous studies to those reported in this study and to estimate whether these previous studies yielded strains that could be interesting for industrial applications.

Our study yielded eight hybrids that showed a statistically higher maximal ethanol accumulation compared to Ethanol Red, which was unable to completely ferment all the available sugar in the VHG (350 g glucose per liter) medium that we used in our small-scale trials. One of these hybrids was further tested and shown to retain its superior fermentation performance on an 8-l pilot scale, making it an attractive candidate strain for commercial bioethanol production. We can imagine that our best hybrids could be further improved, using for instance directed evolution or mutagenesis and screening, to fit to the specific conditions of a particular fermentation process.

Conclusions

In this study, we have shown that genome shuffling of natural, genetically divergent *Saccharomyces* strains can generate novel hybrids with increased ethanol tolerance and fermentation capacity. In addition, we found that selecting for growth in the presence of ethanol increases the efficiency of finding those (rare) variants that show increased fermentation performance. Although additional rounds of genome shuffling yielded some hybrids with increased fermentation capacity, on average, F1 hybrids showed higher ethanol production than F3 hybrids. The two best-performing hybrids, H1 and H2, produced approximately 1.0% (*v*/*v*) more ethanol under VHG conditions than Ethanol Red, the best-fermenting strain out of the collection and one of the parental strains. Hybrid H1 also showed an excellent fermentation capacity on a pilot scale and is therefore an interesting candidate strain for industrial applications.

Methods

Strains and media

In this study, we screened 318 industrial and wild *Saccharomyces* yeasts, including 26 representative (homozygous diploid) *S. cerevisiae* strains described in a previous paper [31]. The strains can be divided according to their origin as shown in Table 3.

The interdelta genetic fingerprinting assay was carried out as described in previous studies by Legras and Karst [20] and Steensels and colleagues [21].

The eight parental strains used in the different genome shuffling strategies were all diploid heterothallic *S. cerevisiae* strains and are listed in Table 4.

Table 3 Overview of strain collection used in this study

Origin	Number of strains
Ale beer	132
Lager beer	46 (*S. pastorianus*)
Wine	79
Sake	14
Spirits	11
Bakery	10
Biofuels	7
Wild	19

Yeast strains were routinely grown in YPD medium (1% (*w*/*v*) yeast extract (LabM, Heywood, UK), 2% (*w*/*v*) peptone (BD Biosciences, San Jose, CA, USA), 2% (*w*/*v*) D-glucose (Sigma-Aldrich, St. Louis, MO, USA), solidified with 2% (*w*/*v*) agar (Invitrogen, Carlsbad, CA, USA) to make plates. For long-time storage at −80°C, YPD + 25% (*v*/*v*) glycerol was used. For growth on glucose in the presence of ethanol, we used YP + 4%(*w*/*v*) glucose supplemented with ethanol to enhance the Crabtree effect. To select for antibiotic markers, solid YPD medium was supplemented with 200 μg/ml G418 (ForMedium, Hunstanton, UK), 200 μg/ml hygromycin B (Invitrogen), or both for dual selection. Sporulation was induced on minimal sporulation medium (1% (*w*/*v*) KAc, 0.05% (*w*/*v*) amino acids, 2% (*w*/*v*) agar) at 23°C after pre-growth in YPD (see also further). Tetrad dissection was carried out using a Singer micromanipulator as described [39]; mating-type determination of germinated spores was carried out by mating-type PCR [40].

Mass sporulation, random spore isolation and mass mating

To create an F1 pool for genome shuffling with random mating, each parental strain was sporulated and subjected to random spore isolation individually (see further). Equal amounts of spores (around 6.25×10^6 spores per strain, that is, 5×10^7 spores in total) were mixed in 50 ml rich medium and allowed to mate randomly for 16 h. We verified that each parental strain yielded a viable spore suspension (50 to 90% viability by randomly picking 20 spores for each strain). After this period of germination, mating and proliferation (approximately six population doublings), the whole culture (F1) was split into aliquots which were frozen at −80°C. Each genome shuffling with random mating experiment was started from these F1 aliquots. To start up sporulation for populations of hybrids, cells were thawed, added to 50 ml YPD and grown for three to four population doublings after which approximately 3×10^8 cells were collected and put on solid minimal sporulation medium. After 5 to 10 days of incubation, random spore isolation was carried out using a modified version of a previously described protocol [41]; all biomass from a square sporulation plate was scraped/washed off, digested overnight (35°C, 80 rpm) in 8 ml 0.1 mg/ml zymolyase (100 T, Amsbio), transferred to a falcon tube, and vortexed in the presence of approximately 400 μl sterile glass beads (150 to 212 μm, Sigma) for 3 min, and subjected to sonication (Branson Digital Sonifier, amplitude = 50%) with cooling in between as described. In the end we determined the density and purity (typically >99% pure) of spore suspensions microscopically. Mass mating was carried out in 50 ml of rich liquid medium (1% (*w*/*v*) yeast extract, 3% (*w*/*v*) peptone, 5% (*w*/*v*) glucose) by inoculating purified single spores to a starting density of 10^6 spores/ml, followed by 16 h of incubation at 30°C, 80 rpm. After mating, cells were frozen down until further processing (ethanol tolerance assay for genome shuffling with selection or pre-growth before sporulation for genome shuffling without selection, see also Additional file 1: Supplemental Text).

Table 4 Overview of strains selected for genome shuffling

Code paper	Original code (origin)
P1	Y244 (ale)
P2	Y310 (wine)
P3	Y142 (sake)
P4	Y317 (sake)
P5	Y148 (spirits)
P6	Y114 (biofuels)
P7	Y115 (Ethanol Red)
P8	Y242 (biofuels)

Genome shuffling with random mating and selection strategies

For random genome shuffling with growth selection, one F1 aliquot was grown overnight in 50 ml YPD, subcultured into YPD + 5% (*v*/*v*) ethanol and grown for 8 h. The OD_{600} was measured and serial dilutions were made and spread onto YPD agar plates supplemented with ethanol (10 to 12% (*v*/*v*)). The plates were sealed with parafilm and put in a plastic bag to minimize ethanol evaporation, and incubated at 30°C. The appearance of colony-forming units was monitored for up to 3 weeks. Next, we selected a plate from which yeasts were recuperated for further breeding, based on the following criteria: 1) an ethanol percentage as high as possible, 2) substantial growth on the plate (>150 CFUs), and 3) variation in colony size within the plate, indicating enrichment for the best-performing hybrids (see also Additional file 1: Supplemental Text). For replicate A, we harvested F1 hybrids from plates containing 10% (*v*/*v*) ethanol and F2 hybrids from 11% (*v*/*v*). For replicates B and C, F1 hybrids were selected from 11% (*v*/*v*) and F2 hybrids from plates containing 11.3% (*v*/*v*) ethanol. Biomass was washed off, mixed, and frozen at −80°C. For the next round of genome shuffling, an entire vial was thawed and subjected to pre-growth and mass sporulation as described. To compare ethanol tolerance between different rounds of shuffling, for each replicate and the unselected F1 pool 24, random isolates were obtained after streaking frozen populations, which were obtained after washing off the biomass after selection, on YPD agar. Each of these colonies was grown individually in liquid and had its ethanol tolerance measured using a robot-based spotting assay (see further). In this assay, values were normalized to parental strain P3, the strongest parental strain.

For random genome shuffling with selection for survival in high ethanol levels, five F1 aliquots were combined and grown in 50 ml YPD to an OD_{600} of approximately 5.0 (approximately 10^8 cells/ml). Per survival assay, 10^9 cells were collected, spun down, resuspended in 10 ml of selective medium (YP + 4% glucose + ethanol) and subjected to ethanol shock for 16 h (18 to 22% (*v*/*v*)). After this period, all biomass was spun down, washed with PBS, and resuspended in 1 ml PBS. Ninety percent of this biomass was plated on nine plates containing synthetic complete (SC) medium (6.7 g/l Difco™ Yeast Nitrogen Base (BD), 2 g/l SC amino acids (MP Biomedicals, Santa Ana, CA, USA)) +3% (*v*/*v*) glycerol + 2% (w/v) agar to specifically select for surviving cells that were still able to respire (non-petites). The remaining biomass was used to prepare serial dilutions which were also plated on SC + 3% (*v*/*v*) glycerol + 2% (*w*/*v*) agar. Plates were incubated for five days at 30°C after which the number of CFUs was determined. Per round of shuffling, we tested the survival capacity in a range of ethanol concentrations. By selecting biomass from a certain concentration of ethanol for the next round of shuffling, we used three selection criteria: 1) an ethanol percentage as high as possible, 2) substantial number of genetically unique hybrids (>150 unique hybrids which corresponds to >5400 CFUs; see also Additional file 1: Supplemental Text), 3) stringent selection (<1% of population should survive). We harvested F1 hybrids that were exposed to 18% (*v*/*v*) ethanol for all three replicates and F2 hybrids that survived 19% (*v*/*v*) (only for replicate C-F2 we harvested less biomass, that is, 2492 CFUs, in order to not select from the same ethanol concentration (18% (*v*/*v*)) for two subsequent rounds of shuffling). All biomass was washed off, mixed, and frozen at −80°C. For the next round of genome shuffling, an entire vial was thawed and subjected to pre-growth and mass sporulation as described.

Genome shuffling with targeted mating and selection strategies

For targeted genome shuffling, we used a similar approach to Zheng et al. [25] to select for outcrossed hybrids. For the first round of mating, for each parental strain, we generated a pRS41H-transformed and a pRS41K-transformed strain (plasmids are described by Taxis and Knop [42]) using a standard yeast transformation protocol [43]. pRS41H and pRS41K are ARS/CEN yeast shuttle vectors harboring a resistance gene against hygromycin B (*hphNT1*) or Geneticin (*kanMX4*), respectively. We confirmed that these markers were largely retained after sporulation and germination (data not shown). For the first round of mating, the parental strains were sporulated individually, and tetrads were semi-digested using the random spore isolation protocol described above (omitting the sonication steps). These suspensions were divided over 96-well plates to generate masterplates to use for pairwise combinations of spore suspensions using the Singer ROTOR HDA©. Masterplates were designed to allow all pairwise matings between parental strains to take place in multiple biological replicates. Mating occurred on Singer Plus Plates filled with YPD agar, by combining approximately 10^5 spores of each suspension on the same position of the plate using sterile RePad 96LW pins using the Singer ROTOR HDA©. After 24 h of mating at room temperature, we replica-plated to YPD plates supplemented with both G418 and hygromycin B, and repeated this the next day. As a negative control, we always included a mating spot in which strains with the same antibiotic resistance marker were mated. For the next round of shuffling, hybrids that were selected for ethanol tolerance (see further) or unselected hybrids were re-streaked on YPD until both markers were lost, plasmids were retransformed, and sporulation and pairwise matings were carried out similarly as described above. Isolates that were used for a next round of shuffling were always tested for sporulation capacity and spore viability. Unselected hybrids were always taken from a random mating event without carrying out a test for ethanol tolerance.

After mating and replica-plating (see before), we subcultured F1 hybrids from solid medium into liquid YPD (96-well plate filled with 150 μl medium/well) using the ROTOR, incubated the cells overnight (horizontal shaker 900 rpm, 30°C), subcultured them the next day into fresh YPD + 5% (*v*/*v*) ethanol (selection), and froze the cells down (no selection). Then, a robot-assisted screen for ethanol tolerance was carried out (see further). After quantification, hybrid populations were selected from which single isolates were to be obtained. In order to obtain these single clones, selected populations were entirely scraped off, cultured overnight in 5 ml YPD, subcultured into 5 ml YPD + 8% (*v*/*v*) to a starting OD_{600} = 0.5 and after overnight growth serially diluted and plated for single colonies on YPD plates with 10 to 11% (*v*/*v*) ethanol. The fastest-growing colonies were retested for their ethanol tolerance as described before (pre-growth in YPD + 5% (*v*/*v*) ethanol followed by spotting on different ethanol concentrations). The best clone was used as a parental strain for the next round of shuffling.

Robot-assisted screen for ethanol tolerance

To measure the ethanol tolerance of a large number of strains or hybrids in a high-throughput fashion, we used a robot-assisted spotting assay. First, strains or hybrids were freshly grown on YPD agar (approximately 50 ml Singer Plus Plate) in a 96-well format. Then, using the ROTOR, these stains were subcultured into liquid YPD (96-well plate filled with 150 μl medium/well); next, the cells were incubated overnight (horizontal shaker 900 rpm, 30°C) and subcultured the next day into fresh YPD + 5% (*v*/*v*) ethanol. Cells were grown for 2 days to saturation, after which the plate served as a source to spot cells on multiple solid Singer Plus Plates containing YPD supplemented with various levels (10 to 13% (*v*/*v*)) of ethanol using the ROTOR equipped with RePad 96LW pins. After spotting, plates were incubated at 30°C (sealed with parafilm and put in a plastic bag) and scanned using an Epson Perfection V700 PHOTO at 300 dpi at regular intervals, typically for up to 2 to 3 weeks. YPD plates (control) were scanned after two days. The scans were quantified in ImageJ using the ScreenMill software suite operated in summation mode [44]. For screening natural strains, ethanol tolerance was expressed as the end point area on medium with ethanol normalized by the area on YPD. For screening hybrids during genome shuffling with targeted mating, this YPD-normalized area was normalized by the area of a strong parental strain on the same plate (P2).

Lab-scale VHG fermentations

Lab-scale fermentations under VHG conditions were started with an overnight pre-growth of a single colony into 3 ml YPD, followed by a transfer of the entire culture to 30 ml YP + 4% (*w*/*v*) glucose and additional growth for 48 h to the stationary phase (200 rpm, 30°C); 250-ml Schott bottles each filled with 150 ml YP + 35% (*w*/*v*) glucose and a magnetic rod (35 × 5 mm) were inoculated to a starting $OD_{600} = 1.0$ (approximately 2.0×10^7 cells/ml). These bottles were sealed with a waterlock and stirred continuously at 150 rpm on a magnetic stirring platform (IKA® RO 15) at 30°C. The bottles were weighed on a daily base to determine the cumulative weight loss, a proxy for CO_2 production. Each fermentation was terminated when its weight loss dropped below 0.20 g/day. In each fermentation batch, we included three biological replicates of Ethanol Red as a control strain; the ethanol production of each tested strain was subsequently expressed relative to the average ethanol production of these three control fermentations. The screening of the yeast collection for VHG fermentation capacity was carried out similarly; however, fermentations were kept static and were terminated after 14 days. We found a good correlation (Spearman $r = 0.85$, $P \leq 0.001$) between the ethanol level reached in static and stirred fermentations (Additional file 1: Table S1).

Fermentations of pools of hybrids were carried out under stirred conditions as described above, but started by thawing a vial of hybrids and adding it completely to 30 ml YP + 4% (*w*/*v*) glucose for pre-growth. Re-inoculation took place by harvesting two thirds of all the biomass and inoculating it directly into fresh VHG medium. The ethanol production was determined using Anton Paar Alcolyzer and the end glucose concentration using a commercial GOD-PAP assay (Dialab).

VHG fermentations under semi-industrial conditions

To pre-screen the strains under semi-industrial VHG conditions, each strain was grown overnight in 5 ml YPD, subcultured by transferring 1 ml of this culture to 10 ml YP + 5% (*w*/*v*) glucose and grown for 3 days, after which 10 ml of this culture was added to 100 ml YP + 10% (*w*/*v*) glucose, which was grown overnight. Then, after washing one time with Ringer solution, a biomass suspension containing 50 g wet biomass in 100 ml Ringer solution was created. Next, bottles containing 100 ml VHG medium (2% (*w*/*v*) peptone, 2% (*w*/*v*) malt extract, 0.12% (*w*/*v*) Fermaid, 109.9 mg/ml $ZnSO_4.7H_2O$, 320 g/l glucose, antifoam) were inoculated with 1 ml of this biomass suspension (approximately 0.5 g wet biomass). The fermentations were kept static at 30°C and terminated after 7 days after which the ethanol concentration was determined using Anton Paar Alcolyzer and HPLC.

To determine the performance of the best strains of the pre-screening on a pilot scale (8 l), strains were grown overnight in 5 ml YPD, subcultured by transferring the entire culture to 150 ml YP + 5% (*w*/*v*) glucose followed by 1 day of growth. Then, this entire culture was added to 2 l YP + 10% (*w*/*v*) glucose which was grown for three days. Biomass was washed one time with Ringer solution, after which a biomass suspension containing 50 g wet biomass in 100 ml Ringer solution was created. Sartorius Biostat B plus fermentors filled with 8 l VHG medium were inoculated with 80 ml biomass suspension. This medium was aerated for 30 min pre-fermentation; the pH was kept at 4.5 as soon as it reached this value by addition of 12% ammonia. Fermentations were stirred at 300 rpm for the first 4 h after which the stirring speed was adjusted to 50 rpm for the remaining time. Fermentations were terminated after 7 days.

Data analysis

Genetic fingerprints were analyzed using BioNumerics (Applied Maths, Belgium) and clustered according to the dice correlation coefficient to build a similarity matrix and a Unweighted Pair Group Method with Arithmetic Mean (UPGMA) algorithm. Heat maps were created by converting VHG ethanol production data to *Z*-scores; sporulation, spore viability, and ethanol tolerance data were normalized to values between 0 and 1.

Statistical analyses were carried out using Graphpad Prism.

Additional files

Additional file 1: Supplemental Figures, Tables and Text.

Additional file 2: Fermentation data hybrids generated with genome shuffling with random mating.

Additional file 3: Fermentation data hybrids generated with genome shuffling with targeted mating.

Additional file 4: Fermentation data hybrids that were tested in biological triplicates. Promising hybrids (see Additional files 2 and 3) were subjected to VHG fermentations. For each parameter the mean and standard deviation are shown. Yield was calculated based on the maximal amount of ethanol that could be formed based on glucose consumption. It is also indicated whether the ethanol production is significantly different from the ethanol production by Ethanol Red in the same batch (t-test).

Additional file 5: Performance of hybrid H1 and 26 additional strains in small-scale VHG fermentations under static conditions. Indicated is the average ethanol production with standard deviation and the number of biological replicates this is based on. It is also indicated whether the ethanol production significantly differs from the ethanol production of hybrid H1 (t-test).

Abbreviations

EtOH: ethanol; OD: optical density; SC: Synthetic complete; VHG: Very high gravity; YPD: Yeast extract peptone dextrose; QTL: Quantitative trait locus.

Competing interests

The authors declare that they have no competing interests.

Authors' contributions

TS, MPN, SvdB, SM, JS, and VS carried out experimental work. TS, SvdB, JS, AV, and KJV designed the experiments and analyzed the results. TS and KJV wrote the paper. All authors read and approved the final manuscript.

Acknowledgements

The authors thank all Verstrepen lab members for their help and suggestions. Research in the lab of K.J.V. is supported by Barry Callebaut, ERC Starting Grant 241426, VIB, EMBO YIP program, FWO, Human Frontier Science program, IWT and KU Leuven NATAR Program Financing.

Author details

[1]Laboratory for Genetics and Genomics, Centre of Microbial and Plant Genetics (CMPG), KU Leuven, Kasteelpark Arenberg 22, 3001 Leuven, Belgium. [2]Laboratory for Systems Biology, VIB, Bio-Incubator, Gaston Geenslaan 1, 3001 Leuven, Belgium. [3]Laboratory of Enzyme, Fermentation and Brewing Technology, KU Leuven technologiecampus Ghent, Gebroeders De Smetstraat 1, 9000 Ghent, Belgium.

References

1. Bai FW, Anderson WA, Moo-Young M. Ethanol fermentation technologies from sugar and starch feedstocks. Biotechnol Adv. 2008;26:89–105.
2. Puligundla P, Smogrovicova D, Obulam VSR, Ko S. Very high gravity (VHG) ethanolic brewing and fermentation: a research update. J Ind Microbiol Biotechnol. 2011;38:1133–44.
3. Gibson BR, Lawrence SJ, Leclaire JP, Powell CD, Smart KA. Yeast responses to stresses associated with industrial brewery handling. FEMS Microbiol Rev. 2007;31:535–69.
4. Ma M, Liu ZL. Mechanisms of ethanol tolerance in *Saccharomyces cerevisiae*. Appl Microbiol Biotechnol. 2010;87:829–45.
5. Swinnen S, Schaerlaekens K, Pais T, Claesen J, Hubmann G, Yang Y, et al. Identification of novel causative genes determining the complex trait of high ethanol tolerance in yeast using pooled-segregant whole-genome sequence analysis. Genome Res. 2012;22:975–84.
6. Casey GP, Ingledew WM. Ethanol tolerance in yeasts. Crit Rev Microbiol. 1986;13:219–80.
7. Steensels J, Snoek T, Meersman E, Picca Nicolino M, Voordeckers K, Verstrepen KJ. Improving industrial yeast strains: exploiting natural and artificial diversity. FEMS Microbiol Rev. 2014;38:947–95.
8. Biot-Pelletier D, Martin VJ. Evolutionary engineering by genome shuffling. Appl Microbiol Biotechnol. 2014;98:3877–87.
9. Zhang YX, Perry K, Vinci VA, Powell K, Stemmer WP, del Cardayre SB. Genome shuffling leads to rapid phenotypic improvement in bacteria. Nature. 2002;415:644–6.
10. Hou L. Novel methods of genome shuffling in *Saccharomyces cerevisiae*. Biotechnol Lett. 2009;31:671–7.
11. Liu JJ, Ding WT, Zhang GC, Wang JY. Improving ethanol fermentation performance of *Saccharomyces cerevisiae* in very high-gravity fermentation through chemical mutagenesis and meiotic recombination. Appl Microbiol Biotechnol. 2011;91:1239–46.
12. Zheng DQ, Wu XC, Tao XL, Wang PM, Li P, Chi XQ, et al. Screening and construction of *Saccharomyces cerevisiae* strains with improved multi-tolerance and bioethanol fermentation performance. Bioresour Technol. 2011;102:3020–7.
13. Hou L. Improved production of ethanol by novel genome shuffling in *Saccharomyces cerevisiae*. Appl Biochem Biotechnol. 2010;160:1084–93.
14. Zheng D, Zhang K, Gao K, Liu Z, Zhang X, Li O, et al. Construction of novel *Saccharomyces cerevisiae* strains for bioethanol active dry yeast (ADY) production. PLoS One. 2013;8:e85022.
15. Zheng DQ, Chen J, Zhang K, Gao KH, Li O, Wang PM, et al. Genomic structural variations contribute to trait improvement during whole-genome shuffling of yeast. Appl Microbiol Biotechnol. 2014;98:3059–70.
16. Tao X, Zheng D, Liu T, Wang P, Zhao W, Zhu M, et al. A novel strategy to construct yeast *Saccharomyces cerevisiae* strains for very high gravity fermentation. PLoS One. 2012;7:e31235.
17. Wang PM, Zheng DQ, Liu TZ, Tao XL, Feng MG, Min H, et al. The combination of glycerol metabolic engineering and drug resistance marker-aided genome shuffling to improve very-high-gravity fermentation performances of industrial *Saccharomyces cerevisiae*. Bioresour Technol. 2012;108:203–10.
18. Mortimer RK, Romano P, Suzzi G, Polsinelli M. Genome renewal: a new phenomenon revealed from a genetic study of 43 strains of *Saccharomyces cerevisiae* derived from natural fermentation of grape musts. Yeast. 1994;10:1543–52.
19. Codón AC, Gasent-Ramírez JM, Benítez T. Factors which affect the frequency of sporulation and tetrad formation in *Saccharomyces cerevisiae* baker's yeasts. Appl Environ Microbiol. 1995;61:1677.
20. Legras JL, Karst F. Optimisation of interdelta analysis for *Saccharomyces cerevisiae* strain characterisation. FEMS Microbiol Lett. 2003;221:249–55.
21. Steensels J, Meersman E, Snoek T, Saels V, Verstrepen KJ. Large-scale selection and breeding to generate industrial yeasts with superior aroma production. Appl Environ Microbiol. 2014;80:6965–75.
22. Hu XH, Wang MH, Tan T, Li JR, Yang H, Leach L, et al. Genetic dissection of ethanol tolerance in the budding yeast *Saccharomyces cerevisiae*. Genetics. 2007;175:1479–87.
23. Yang Y, Foulquié-Moreno MR, Clement L, Erdei E, Tanghe A, Schaerlaekens K, et al. QTL analysis of high thermotolerance with superior and downgraded parental yeast strains reveals new minor QTLs and converges on novel causative alleles involved in RNA processing. PLoS Genet. 2013;9:e1003693.
24. Ibeas JI, Jimenez J. Mitochondrial DNA loss caused by ethanol in Saccharomyces flor yeasts. Appl Environ Microbiol. 1997;63:7–12.
25. Zheng DQ, Wu XC, Wang PM, Chi XQ, Tao XL, Li P, et al. Drug resistance marker-aided genome shuffling to improve acetic acid tolerance in *Saccharomyces cerevisiae*. J Ind Microbiol Biotechnol. 2011;38:415–22.
26. Pinel D, D'Aoust F, del Cardayre SB, Bajwa PK, Lee H, Martin VJ. *Saccharomyces cerevisiae* genome shuffling through recursive population mating leads to improved tolerance to spent sulfite liquor. Appl Environ Microbiol. 2011;77:4736–43.
27. Shi DJ, Wang CL, Wang KM. Genome shuffling to improve thermotolerance, ethanol tolerance and ethanol productivity of *Saccharomyces cerevisiae*. J Ind Microbiol Biotechnol. 2009;36:139–47.
28. Lu Y, Cheng YF, He XP, Guo XN, Zhang BR. Improvement of robustness and ethanol production of ethanologenic *Saccharomyces cerevisiae* under co-stress of heat and inhibitors. J Ind Microbiol Biotechnol. 2012;39:73–80.

29. Pais TM, Foulquié-Moreno MR, Hubmann G, Duitama J, Swinnen S, Goovaerts A, et al. Comparative polygenic analysis of maximal ethanol accumulation capacity and tolerance to high ethanol levels of cell proliferation in yeast. PLoS Genet. 2013;9:e1003548.
30. Alper H, Moxley J, Nevoigt E, Fink GR, Stephanopoulos G. Engineering yeast transcription machinery for improved ethanol tolerance and production. Science. 2006;314:1565–8.
31. Liti G, Carter DM, Moses AM, Warringer J, Parts L, James SA, et al. Population genomics of domestic and wild yeasts. Nature. 2009;458:337–41.
32. Zörgö E, Gjuvsland A, Cubillos FA, Louis EJ, Liti G, Blomberg A, et al. Life history shapes trait heredity by accumulation of loss-of-function alleles in yeast. Mol Biol Evol. 2012;29:1781–9.
33. Plech M, de Visser JA, Korona R. Heterosis is prevalent among domesticated but not wild strains of *Saccharomyces cerevisiae*. G3: Genes| Genomes|. Genetics. 2014;4:315–23.
34. Shapira R, Levy T, Shaked S, Fridman E, David L. Extensive heterosis in growth of yeast hybrids is explained by a combination of genetic models. Heredity. 2014;113:316–26.
35. Marullo P, Bely M, Masneuf-Pomarède I, Pons M, Aigle M, Dubourdieu D. Breeding strategies for combining fermentative qualities and reducing off-flavor production in a wine yeast model. FEMS Yeast Res. 2006;6:268–79.
36. Timberlake WE, Frizzell MA, Richards KD, Gardner RC. A new yeast genetic resource for analysis and breeding. Yeast. 2011;28:63–80.
37. Duitama J, Sánchez-Rodríguez A, Goovaerts A, Pulido-Tamayo S, Hubmann G, Foulquié-Moreno MR, et al. Improved linkage analysis of Quantitative Trait Loci using bulk segregants unveils a novel determinant of high ethanol tolerance in yeast. BMC Genomics. 2014;15:207.
38. Wang H, Hou L. Genome shuffling to improve fermentation properties of top-fermenting yeast by the improvement of stress tolerance. Food Sci Biotechnol. 2010;19:145–50.
39. Sherman F, Hicks J. Micromanipulation and dissection of asci. Methods Enzymol. 1990;194:21–37.
40. Huxley C, Green ED, Dunham I. Rapid assessment of *S. cerevisiae* mating type by PCR. Trends Genet. 1990; 6:236.
41. Treco DA, Winston F. Growth and manipulation of yeast. Current Protocols in Molecular Biology. 2008:13.12. 11–13.12. 12.
42. Taxis C, Knop M. System of centromeric, episomal, and integrative vectors based on drug resistance markers for *Saccharomyces cerevisiae*. Biotechniques. 2006;40:73.
43. Gietz RD, Schiestl RH, Willems AR, Woods RA. Studies on the transformation of intact yeast cells by the LiAc/ss-DNA/PEG procedure. Yeast. 1995;11:355–60.
44. Dittmar JC, Reid RJ, Rothstein R. ScreenMill: a freely available software suite for growth measurement, analysis and visualization of high-throughput screen data. BMC Bioinformatics. 2010;11:353.

4

Characterization of cell wall components of wheat bran following hydrothermal pretreatment and fractionation

Zara Merali[1], Samuel R A Collins[1], Adam Elliston[1], David R Wilson[1], Andres Käsper[2] and Keith W Waldron[1*]

Abstract

Background: Pretreatments are a prerequisite for enzymatic hydrolysis of biomass and production of ethanol. They are considered to open up the plant cell wall structure by altering, moving or solubilizing lignin and hydrolyzing a proportion of hemicellulosic moieties. However, there is little information concerning pretreatment-induced changes on wheat bran cell wall polymers and indeed on changes in cell wall phenolic esters in bran or other lignocellulosic biomass. Here, we evaluate polymeric changes (chemical and physical) as a result of selected hydrothermal pretreatment conditions on destarched wheat bran using controlled polymer extraction methods. Quantification of cell wall components together with soluble oligosaccharides, the insoluble residues and ease of extractability and fractionation of biomass residues were conducted.

Results: Pretreatment solubilized selected arabinoxylans and associated cross-linking ferulic and diferulic acids with a concomitant increase in lignin and cellulosic glucose. The remaining insoluble arabinoxylans were more readily extractable in alkali and showed considerable depolymerization. The degree of arabinose substitution was less in xylans released by higher concentrations of alkali. The recalcitrant biomass which remained after pretreatment and alkali extraction contained mostly cellulosic glucose and Klason lignin. Pretreatment generated small but insignificant amounts of yeast-inhibiting compounds such as furfural and hydroxymethyl furfural. As such, simultaneous saccharification and fermentation of the hydrothermally pretreated bran resulted in increased ethanol yields compared to that of the control (97.5% compared to 63% theoretical).

Conclusion: Hydrothermal pretreatment of destarched wheat bran resulted in degradation and depolymerization of the hemicellulosic arabinoxylans together with some breakdown of cellulosic glucose. This was accompanied by a significant reduction in the cross-linking phenolic acids such as ferulic and diferulic acids. The results suggest that hydrothermal pretreatment enhances enzymatic digestibility of the cellulose not only by depolymerization and solubilization of the hemicelluloses but by breakdown of interpolymeric phenolic cross-links between the remaining insoluble polymers. This allows easier access of hydrolytic enzymes by opening or loosening of the cell wall thus resulting in enhanced saccharification of cellulose and subsequent fermentation to ethanol. The reduction in cinnamic acids by selected breeding or biotechnological approaches could provide a useful basis for improved saccharification and fractionation of wheat bran polysaccharides.

Keywords: Arabinoxylans, Phenolics, Hydrothermal pretreatment, Simultaneous saccharification and fermentation, Wheat bran

* Correspondence: keith.waldron@ifr.ac.uk
[1]The Biorefinery Centre, Institute of Food Research, Norwich Research Park, Colney, Norwich NR4 7UA, UK
Full list of author information is available at the end of the article

Background

Wheat bran consists of the outer coat (pericarp, testa and aleuron layers) of the wheat grain (*Triticum aestivum* L). Wheat bran is separated from the other parts of the wheat kernel by milling, and the chemical composition of wheat bran predominantly comprises non-starch polysaccharides (approximately 38%), starch (approximately 19%), protein (approximately 18%) and lignin (approximately 6%), with the non-starch polysaccharides being approximately 70% arabinoxylans, approximately 19% cellulose and approximately 6% β-(1,3)/β-(1,4)-glucan [1]. It also contains significant amounts of phenolic acids such as ferulic and *p*-coumaric acids [2] which are esterified to arabinofuranosyl residues [3]. The structural and physical properties of these arabinoxylans are largely determined by the degree, type and distribution pattern of the substitutions along the xylan backbone [4]. Furthermore, the substitution pattern and the interpolymeric cross-linkages such as those mediated through phenolics acids will affect their digestibility [5] and enzymatic degradation [6]. Wheat bran is an important by-product of the cereal industry produced in vast quantities worldwide, and it is estimated that more than a 100 million tons of bran are available in the European Union [7]. Global demand for sustainable resources means that there is an increasing socio-economic pressure to ensure complete utilization of such feedstocks [8]. Industrial wheat bran represents 14%–19% of the wheat grain and has the potential to serve as a low-cost feedstock for the production of renewable energy or chemicals. It is envisaged that utilization of both the starch and hemicellulosic/cellulosic part of wheat bran would greatly facilitate potential applications in a biorefinery concept. However, hydrolysis of bran using hemicellulolytic and cellulolytic enzymes is not adequate to liberate sufficient simple sugars; therefore, a range of physical or chemical pretreatments are required [9].

Compared with other pretreatments, the advantages of hydrothermal pretreatment (HT) include a significantly lower environmental impact, lower capital investment, the avoidance of chemicals and low by-product generation [9]. It is now clear that hot water and related thermophysical pretreatments enhance saccharification of lignocellulose by disrupting the cell wall matrix [10-12], thereby improving the accessibility of cellulases to the cellulosic microfibrils.

Recent research has indicated that the pretreatment-induced changes in the cell walls are highly complex and differ from species to species and tissue to tissue reflecting the variation in wall chemistries [13,14]. More detailed information on chemical changes during pretreatment and enzymolysis has resulted from the use of antibodies to cell-wall polymer epitopes in wheat straw [15] and poplar [16] providing further information on the protective roles of different polymers throughout the process. It is becoming evident that the mechanisms of pretreatment are far from being fully resolved and differ between different lignocellulosic sources.

Most research on the impacts of pretreatment on cell walls involves studies on plant stem and leaf tissues [13-16]. There is relatively little information available on how pretreatments affect the polymeric nature of cell walls of wheat bran which surrounds the seed and has much less of a structural role in the plant [17]. This is of significance if the non-cellulosic polysaccharides are to be additionally exploited e.g. in food production or materials. Alkaline extraction/fractionation of wheat bran arabinoxylans have been reported using hydrogen peroxide [18,19] and potassium hydroxide [20,21]. Some enzymatic hydrolysis of arabinoxylan fractions from wheat bran have also been described [22]. However, although pretreatment of wheat bran for bioethanol production has been studied (e.g. [23]) and pretreatment conditions have been used for extraction of arabinoxylans of varying quality [24], there is limited knowledge of pretreatment-induced changes on wheat bran cell walls or alterations at a polymeric level relevant to fractionation. Since arabinoxylans in wheat bran are physically and chemically associated with each other and with other cell wall components such as lignin, an alteration in any of these components is likely to affect the accessibility of cellulases during saccharification as shown in cell walls of other biomass sources [25]. Previously, we reported the effects of hydrothermal pretreatment on the degradation and sequential alkali extractability of wheat straw cell wall polymers [17]. We demonstrated that under conditions that augment enzymatic saccharification, significant changes occur in the extractability/fractionation and chemistry of cell wall hemicelluloses, lignin and cross-linking phenolics. Of particular note was the reduction in the molecular weight (Mw) of the remaining arabinoxylans and a considerable reduction in ferulic and diferulic acids. In the present paper, the polymeric changes (chemical and physical) are reported as a result of hydrothermal pretreatment (180°C and 200°C) on wheat bran. Using controlled polymer-extraction/fractionation methods [26], we have evaluated pretreatment-induced changes in the constituent polymers and associated phenolic cross-linkages. In order to account for the fate of cellulose and hemicellulose through the solid and liquid streams, mass balance calculations were conducted together with the quantification of yeast-inhibiting compounds such as furfural and hydroxymethylfurfural (HMF). The impact of the pretreatments on the production of ethanol from bran was evaluated by simultaneous saccharification and fermentation (SSF) process using a commercial enzyme (CTec2®) and Ethanol Red® yeast.

Results and discussion

Physical effect of hydrothermal pretreatment on wheat bran

Wheat bran was destarched and pretreated as shown in Figure 1 after which it was extracted as an alcohol insoluble residue (AIR). Low magnification imaging of the unstained AIRs exhibited layers of wheat bran such as the aleurone (the outer layer of the endosperm) and the outer pericarp, together with the starchy endosperm (Figure 2A-C). Microscopy depicted the aleurone as the most abundant layer present in wheat bran and in agreement with previous work on separated bran layers which showed that the largest fraction consisted of the aleurone (62%) followed by the pericarp layer (27.1%) [2]. The most noticeable visual effect of hydrothermal pretreatment (HT) on the aleurone layer compared to the untreated control (Figure 2D) was the brown coloration (Figure 2E,F) which intensified with the severity of treatment. A closer examination depicted the presence of lipids which (colourless in the untreated aleurone layer; Figure 2G) had darkened probably as a result of oxidation occurring during the pretreatment (Figure 2H,I). Under UV irradiation, lignin-rich material has been shown to autofluoresce blue at neutral pH and the presence of phenolics (such as ferulic acid) is indicated by turquoise/green autofluorescence at pH 11 [5]. A strong positive result was observed for the untreated, control aleurone (Figure 2J,M) indicating the presence of phenolics and related moieties esterified to these cell walls. However, the HT aleurone layers showed no such autofluorescence suggesting a degradation and reduction in the levels of fluorescing moieties (Figure 2K,L,N,O).

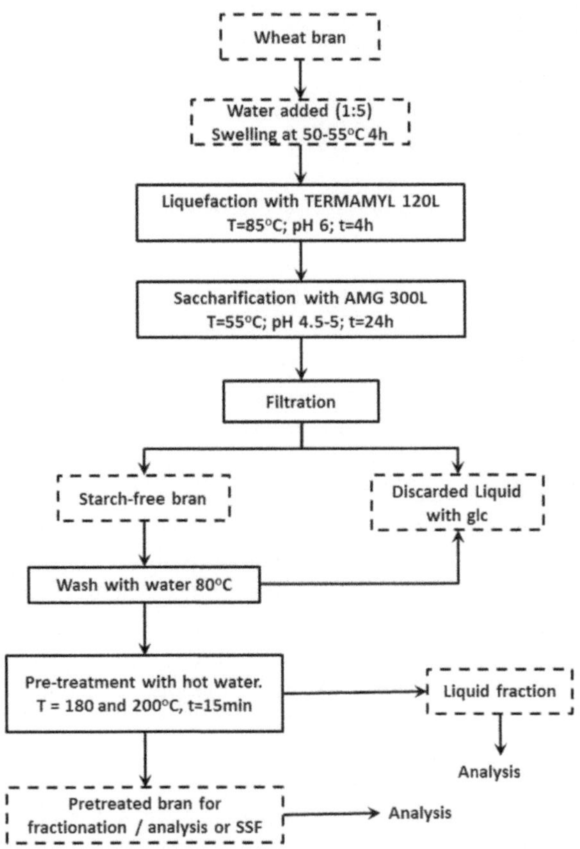

Figure 1 Pretreatment and destarching procedure developed for wheat bran.

Effect of hydrothermal pretreatment on cell wall composition of wheat bran

Table 1 shows the effect of hydothermal pretreatment (HT) on the gross compositions of alcohol insoluble residues (AIRs) with detailed sugar mass balance and compositions shown in Table 2 and Figure 3A. Untreated wheat bran AIR consisted of 77% carbohydrate. Most of this carbohydrate comprised of glucose, xylose and arabinose, reflecting 18.5% (cellulosic) glucan and 51.8% (xylan) hemicellulose and is in agreement with previous reports [18,20]. A large proportion of the insoluble hemicelluloses were solubilized as a result of HT (40% and 51% at 180°C and 200°C, respectively; Table 2) and a smaller but not inconsequential amount of (cellulosic) glucose (Figure 3A). This is caused by acid hydrolysis consistent with the high temperature and mildly acidic conditions as observed previously for instance in wheat straw [27,17] and arabinoxylan extraction from destarched wheat bran [24].

The untreated bran AIR contained 10.8% lignin (w/w) in agreement with previously reported data of 10% [28]. The pretreatment-induced reduction in insoluble carbohydrate resulted in a concomitant increase in the level of Klason lignin to nearly 15% (w/w; Table 1) at 200°C. Ferulic and *p*-coumaric acid were the most abundant phenolic acids present in wheat bran (details in Table 3). The amount of phenolic acids recovered in the untreated control was ca 1.5%. These were reduced during hydrothermal pretreatment, consistent with previous studies for instance in steam exploded barley bran where the effect of high temperatures (above 220°C) resulted in decreasing levels of ferulic acid compared to *p*-coumaric acid [29]. However, a significant proportion (0.6% and 0.4% at 180°C and 200°C, respectively) remained esterified to the arabinoxylans. A considerable amount of ferulic acid was present in the dimer form (19% of the total; Table 3), indicating a high degree of cell wall cross-linking [30]. The presence of ferulic acid dehydrodimers is likely to have a profound effect on the functional characteristics of the cell walls as indicated by previous studies on wheat bran [2] and Chinese water chestnut [5,31]. The reduction in interpolymeric cross-linking such as those mediated via diferulic acid would be expected to affect the ease of extraction of

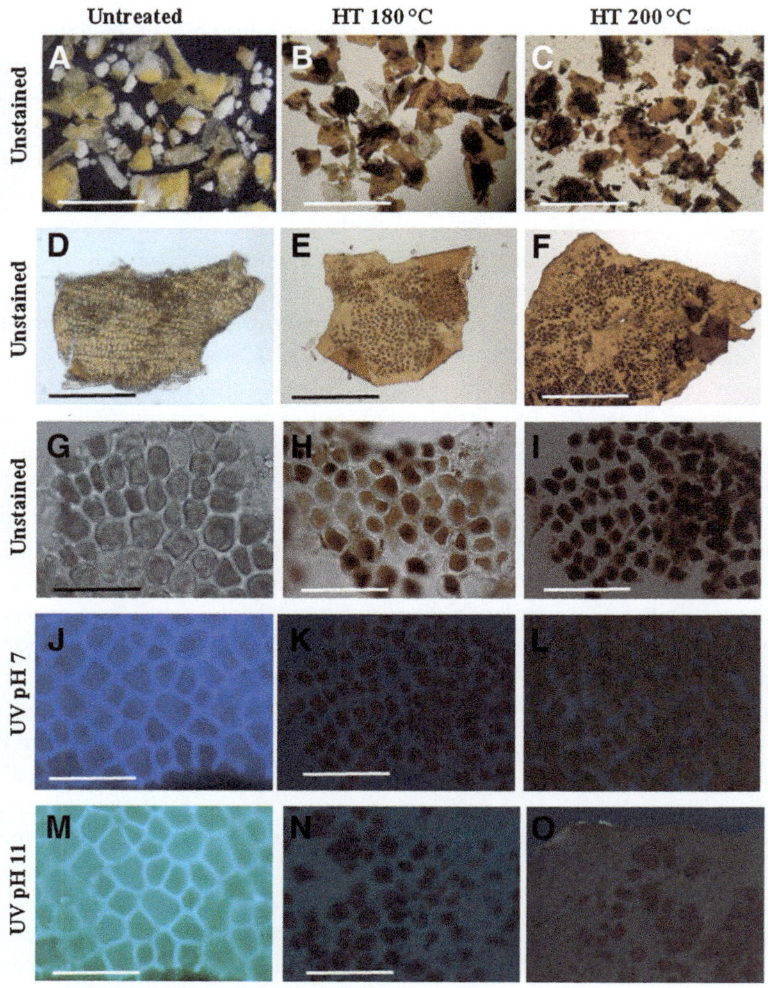

Figure 2 Photomicrographs of control and pretreated wheat bran. Images of unstained bran **(A-C)** and at higher magnification **(D-I)** showing colouration due to Maillard reactions in the pretreated materials. Under UV, lignified native bran tissues autofluoresce blue at neutral pH **(J)** whilst at alkaline pH, cinnamic acid and related moieties autofluoresce green **(M)**. However in the HT samples, this autofluorescence is visibly reduced. Bars: 1 mm **(A-C)**; 0.5 mm **(D-L)**; 100 μm **(G-O)**.

arabinoxylans and enhance the accessibility of cell wall degrading cellulases during saccharification. Interestingly, whilst the pretreatment-related reduction in the measured levels of ferulic and diferulic acid corresponded to the decrease in alkaline autofluorescence

Table 1 Compositional data of the (destarched) control and hydrothermally pretreated (HT) wheat bran (analysed in alcohol insoluble residue (AIR))

Component	Content (% w/w original starting material)		
	Untreated	HT-180°C	HT-200°C
DM	89.0 ± 0.4	19.1 ± 0.7	18.3 ± 1.7
Cellulose	18.5 ± 2.1	21.9 ± 2.0	24.2 ± 2.6
Hemicellulose (total):	51.8 ± 4.2	31.2 ± 1.7	26.0 ± 2.1
Starch	4.1 ± 0.5	1.9 ± 0.1	0.8 ± 0.0
Phenolic acids (esters)	1.5 ± 1.2	0.6 ± 0.3	0.4 ± 0.1
Klason lignin	10.8 ± 1.3	12.4 ± 2.4	14.7 ± 1.5

under UV (Figure 2M-O), this was not the case for lignin. Figure 2J-L shows a reduction in autofluorescence of lignin whilst the measured Klason lignin increased (Table 1). The loss of fluorescence may have resulted from the degradation of fluorescing moieties during pretreatment. It may also be the case that other wall components such as carbohydrate and protein breakdown products have interacted with the lignin resulting in chemical change or perhaps quenching autofluorescence. Indeed, previous studies have indicated that in some biomass, pretreatment results in the formation of "pseudo lignin" [32,33] which would be quantified by the Klason methodology. Such uncharacterized material may not exhibit fluorescence associated with native lignin. In addition, much lignin is known to be solubilized under pretreatment conditions [25] and it is possible that such solubilization involves the loss or relocation of autofluorescing components.

Table 2 Carbohydrate composition (μg/mg dry weight) of untreated control and hydrothermally pretreated wheat bran following sequential extraction

	[a]Yield (g)	Rhamnose	Fucose	Arabinose	Xylose	Mannose	Galactose	Glucose	GlcA	Total
Ctrl AIR	2.02	1.4 ± 0.3	0.7 ± 0.1	228 ± 3	291 ± 3	5.5 ± 0.1	15.2 ± 1.2	185 ± 2	42.8 ± 0.5	769 ± 3
Ctrl HW	0.02	nd	nd	139 ± 2	228 ± 3	8.3 ± 0.3	17.6 ± 1.1	140 ± 2	35.1 ± 0.4	568 ± 4
Ctrl 0.5 mol/L	0.15	nd	nd	167 ± 3	223 ± 4	1.7 ± 0.1	8.1 ± 0.3	162 ± 1	32.3 ± 0.3	594 ± 5
Ctrl 1 mol/L	0.18	nd	nd	190 ± 4	433 ± 6	2.9 ± 0.2	8.3 ± 0.4	177 ± 2	31.1 ± 0.3	841 ± 8
Ctrl 4 mol/L	0.20	nd	nd	196 ± 4	466 ± 5	1.1 ± 0.1	10.4 ± 1.6	173 ± 2	25.2 ± 0.2	871 ± 6
Ctrl residue	1.00	nd	nd	83 ± 2	103 ± 3	0.8 ± 0.1	3.3 ± 0.8	204 ± 2	30.6 ± 0.5	425 ± 6
180°C AIR	2.05	nd	nd	126 ± 2	188 ± 2	5.5 ± 0.6	14.1 ± 1.6	220 ± 4	40.6 ± 1.1	593 ± 7
180°C HW	0.05	nd	nd	119 ± 2	192 ± 3	6.1 ± 0.8	15.3 ± 1.2	185 ± 3	32.1 ± 0.5	550 ± 6
180°C 0.5 mol/L	0.16	nd	nd	228 ± 3	290 ± 2	1.0 ± 0.2	6.4 ± 1.1	196 ± 2	27.8 ± 0.3	749 ± 6
180°C 1 mol/L	0.17	nd	nd	190 ± 2	289 ± 3	1.8 ± 0.3	6.7 ± 1.1	85 ± 1	29.2 ± 0.3	601 ± 5
180°C 4 mol/L	0.24	nd	nd	75 ± 1	185 ± 2	1.2 ± 0.4	9.3 ± 2.0	86 ± 1	21.6 ± 0.4	379 ± 3
180°C residue	0.76	nd	nd	63 ± 1	97 ± 1	0.7 ± 0.1	2.6 ± 0.5	265 ± 2	28.3 ± 0.3	457 ± 5
200°C AIR	1.98	nd	nd	101 ± 1	152 ± 2	5.5 ± 0.5	13.5 ± 1.1	242 ± 3	37.8 ± 0.8	551 ± 6
200°C HW	0.07	nd	nd	92 ± 2	107 ± 2	5.9 ± 0.5	13.8 ± 1.2	184 ± 1	24.2 ± 0.5	427 ± 4
200°C 0.5 mol/L	0.19	nd	nd	207 ± 4	218 ± 3	1.4 ± 0.4	5.8 ± 0.3	153 ± 1	27.9 ± 0.7	613 ± 6
200°C 1 mol/L	0.19	nd	nd	102 ± 2	238 ± 4	2.2 ± 0.2	6.8 ± 0.2	75 ± 1	29.7 ± 0.6	453 ± 4
200°C 4 mol/L	0.25	nd	nd	76 ± 1	161 ± 3	1.3 ± 0.1	8.2 ± 0.7	88 ± 2	28.9 ± 0.8	364 ± 3
200°C residue	0.71	nd	nd	59 ± 1	95 ± 2	0.4 ± 0.1	1.6 ± 0.2	291 ± 3	28.0 ± 0.5	476 ± 5

Residue represents the insoluble material after fractionation. Values are expressed as means of duplicate analysis ± SD.
GlcA glucuronic acid, *Ctrl* untreated control, *nd* none detected.
[a]Yield is the dry weight of fraction recovered after sequential extraction of the AIR in hot water (HW) and progressively stronger alkali (0.5, 1 and 4 mol/L KOH).

Characterization of AIRs, sequentially extracted fractions and the insoluble residues

Sequential extraction in progressively strong alkali provides a basis for comparing changes in cell walls during physiological development as well as physicochemical/thermal processing [34]. AIRs were sequentially extracted in hot water followed by progressively stronger alkali (in degassed aqueous solution and in the presence of $NaBH_4$ to minimize alkaline peeling) which progressively extracts the non-cellulosic cell wall polysaccharides. Table 2 (and Figure 3B) shows the recoveries and yields of total carbohydrate whereas the total phenolic compositions are shown in Table 3 (and Figure 3C). Sequential extraction of the control AIR solubilized 49% of the material (Table 2). In contrast, alkaline extraction of the HT AIR samples resulted in the solubilization of larger quantities of material (63% and 64% at 180°C and 200°C, respectively; Table 2).

Most solubilization of AIR carbohydrate was effected by 1 and 4 mol/L alkali treatments, particularly in the 200°C HT material. This resulted in relatively less material remaining in the insoluble residues from HT biomass (Figure 3B). The main carbohydrate released from the AIR during alkali fractionation comprised arabinoxylan hemicelluloses (Table 2). The extracted polymer also retained phenolic acid groups which had resisted HT and saponification (Table 3, Figure 3C). Very little soluble material was recovered in the hot water (HW) fraction of the non-pretreated wheat bran, and what was solubilized consisted mostly of arabinose; xylose and non-cellulosic glucose (Table 2). Yields of approximately 9% have been obtained for water-extractable material from purified wheat bran using more aggressive treatment at 100°C, 30 min [18]. Nevertheless, the results indicate that most wheat bran arabinoxylans are not extractable in water probably because of their structural features and/or association with each other and other cell wall components. After pretreatments, slightly larger quantities of hot water soluble material was released from the AIRs (<4%; in the 200°C HT). The relative degree of xylan branching could be inferred from the ratio of arabinose to xylose and in the HT polymers this decreased (0.78 in untreated AIR compared to 0.66 in 200°C AIR, respectively), consistent with a decrease in water solubility. The level of arabinose substitution further decreased with increasing concentrations of KOH to 0.42 in untreated and 0.38 in 200°C HT in 4 mol/L extracts. This reflects a higher degree of xylan-cellulose hydrogen bonding in the latter polymers, thereby requiring more concentrated alkali for their solubilization. The decrease in water solubility is consistent with previous studies on wheat bran [35] and brewers' spent grain [20]. The isolated fractions contained uronic acid units probably derived from both glucuronoxylan and pectic

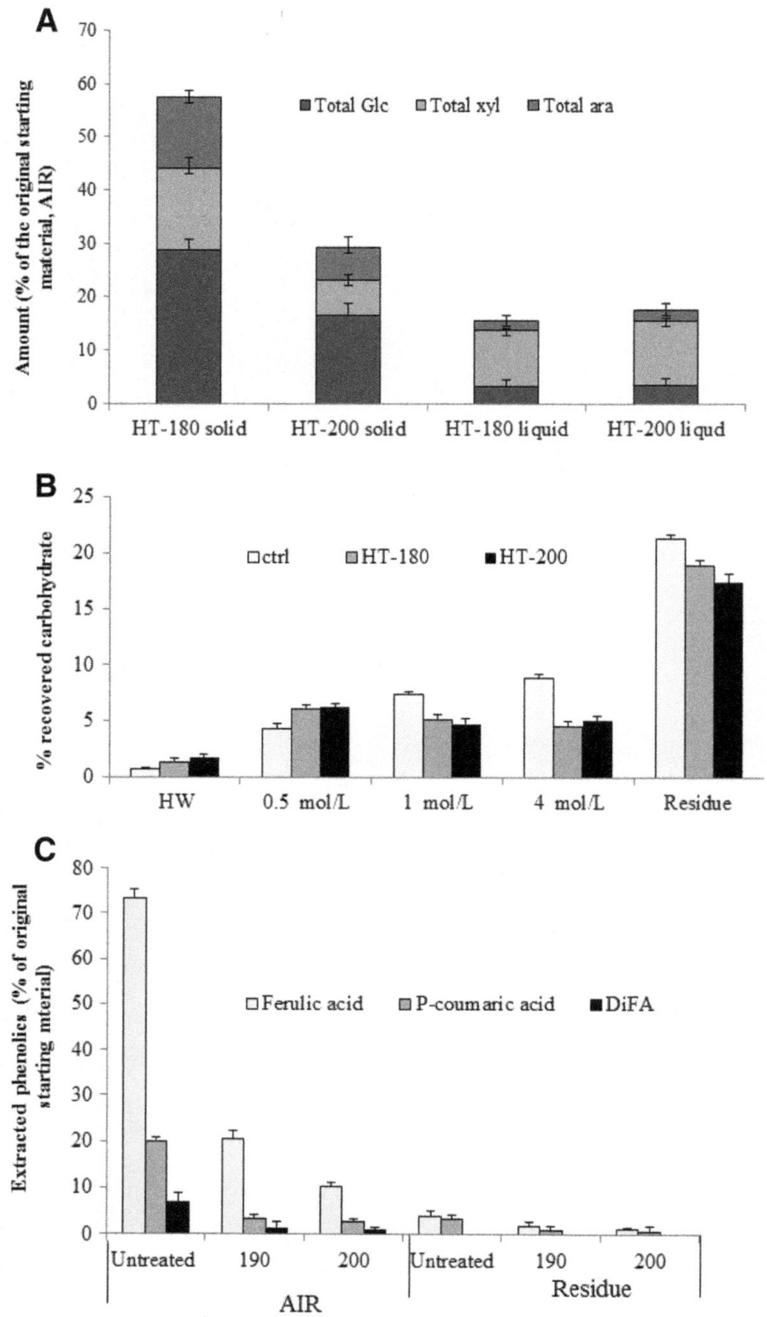

Figure 3 Cell wall composition of wheat straw (A) showing the total amount of key carbohydrates recovered in the solid biomass and pretreatment liquors as a function of original starting material and (B) total recovery of carbohydrates present in alkali fractions and (C) selected phenolic acids from the control and pretreated wheat bran. *PT* pretreated, *pCA p*-coumaric acid, *FA* ferulic acid, *DiFA* diferulic acids.

polymers (indicated by the presence of small quantities of rhamnose) which are thought to have originated from epidermal cells of the inner pericarp [2].

The insoluble residues consisted mainly of arabinoxylan and cellulosic material (Table 2). A significant level of xylose remained in the cellulose-rich residue after the 4 mol/L KOH extraction (approximately 20% of the total carbohydrate, Table 2), indicating that some xylan was cross-linked to other alkali-stable wall components. These residues were also analysed for esterified cinnamic acid derivatives (data in Table 3). The bulk of the phenolics, especially ferulic acid, was lost as a result of the pretreatment (Figure 3C) and further due to sequential alkaline extraction which hydrolyzes the labile ester bond between

Table 3 Phenolic acid composition (μg/mg) of wheat bran before (in AIRs) and after alkaline sequential extraction

	p-coumaric acid	Ferulic acid	8,5'-DiFA	5,5'-DiFA	8-*O*-4'-DiFA
Ctrl AIR	2.6 ± 0.5	9.7 ± 0.4	0.9 ± 0.1	0.9 ± 0.2	1.0 ± 0.4
190°C AIR	1.0 ± 0.2	3.6 ± 0.4	0.3 ± 0.1	0.5 ± 0.1	0.7 ± 0.0
200°C AIR	0.8 ± 0.2	2.7 ± 0.3	0.2 ± 0.4	0.2 ± 0.0	0.4 ± 0.0
Ctrl residue	0.4 ± 0.1	0.5 ± 0.1	nd	nd	nd
190°C residue	0.2 ± 0.0	0.5 ± 0.1	nd	nd	nd
200°C residue	0.1 ± 0.0	0.2 ± 0.0	nd	nd	nd

Values are expressed as means of duplicate analysis ± SD.
nd none detected.

the phenolics and arabinose. However, a significant amount (approximately 7% of the total in untreated, Table 3) remained esterified to the isolated arabinoxylan and recalcitrant residue. The most prominent phenolics were ferulic acid, followed by *p*-coumaric acid and dehydrodimers of ferulic acid. In the HT AIRs, the reduction in phenolics correlated with increasing severity of temperature; and in terms of the remaining phenolics present in the HT residues, only ferulic acid (thought to be involved in cell-to-cell adhesion) and *p*-coumaric acid (which is thought to be ether linked with the matrix polymers and/or with lignin) were detected.

Molecular weight (Mw) distribution of solubilized xylan following fractionation

Molecular masses of xylans solubilized during fractionation and following hydrothermal pretreatment were evaluated by high-performance size exclusion chromatography (HPSEC) in combination with RI detection. The representative spectra are presented in Figure 4A. Three dominant peaks were identified using elution profiles of standard controls (Figure 4A,B); high Mw range (1,660 to 48 kDa), medium (Mw range 48 to 5.8 kDa) and low (Mw range 5.8 to 0.18 kDa). The smaller but nonetheless significant peaks (approximately 6% of the total area) observed at high Mw range (1,660 to 48 kDa) may have originated from the adherent wheat endosperm (Mw 850 kDa of the alkaline extract) [36]. Generally, the relative amount of medium Mw xylans decreased in stronger alkali with a concomitant increase in the amount of low Mw xylans (Figure 4B). Thus, the xylan moieties remaining in the bran cell walls after hydrothermal pretreatments are of a lower molecular weight indicating that during milder conditions larger structures of xylan are solubilized and at extreme conditions these high Mw xylans are subsequently hydrolyzed to medium and low Mw xylans. This is in agreement with the xylan solubility of various feedstocks producing polymers of varying Mw sizes for instance during hot water pretreatment of wheat straw [27,17], eucalyptus wood [37] and corn cobs [22]. Furthermore, in all bran preparations (including the control), the xylans extracted by the higher alkali concentrations exhibit lower molecular weights than those extracted in less concentrated alkali. The ease of alkali extraction is likely to reflect the degree of xylan substitution, whether it be other sugars (e.g. arabinose) or acetyl groups. The latter have been highlighted during microarray polymer profiling of pretreated wheat straw polysaccharides [15].

There is varied information available on the Mw profiling of alkali-extracted wheat bran hemicelluloses [18,19,21]. Maes and Delcour [18] also observed two peaks at high and low Mw sizes, and the xylan components are of the order reported here for wheat bran. However, in contrast, Hollmann and Lindhauer [19] and Zhang et al. [21] only observed one peak at a Mw range of 100–110 kDa corresponding to high Mw range (116–48 kDa) in this study. Both studies used relatively aggressive extraction methods; peroxidation, which is highly degradative, and 24% KOH followed by HPSEC in DMSO could result in more destructive solubilization and include aggregates derived from lignin. In the present study, alkaline degradation was minimized by utilizing degassed water and the addition of 20 mmol/L $NaBH_4$ prior to extraction.

Evaluation of pretreatment liquor

The concentration of solubilized sugars together with fermentation inhibitors is shown in Figure 5. Mass balances (shown in Figure 3A) together with details of xylan solubility by HPSEC (above) provide information of the effect of HT on carbohydrate accessibility and its saccharification. The liquors contained mostly arabinoxylan-derived xylose (3.5 and 7.6 g/L; 180°C and 200°C, respectively) and glucose (2.4 and 5.3 g/L; 180°C and 200°C, respectively; Figure 5). Small quantities of xylose-associated fragments were also detected in the liquors using NMR and increased in line with the severity of pretreatment (data not shown). This is consistent with previous reports on HT wheat straw [27,17,15] reflecting their release from the solid residues during pretreatment, followed by further degradation into monomeric xylose or converted to furfural derivatives (Figure 5). The presence of these inhibitors is undesirable because

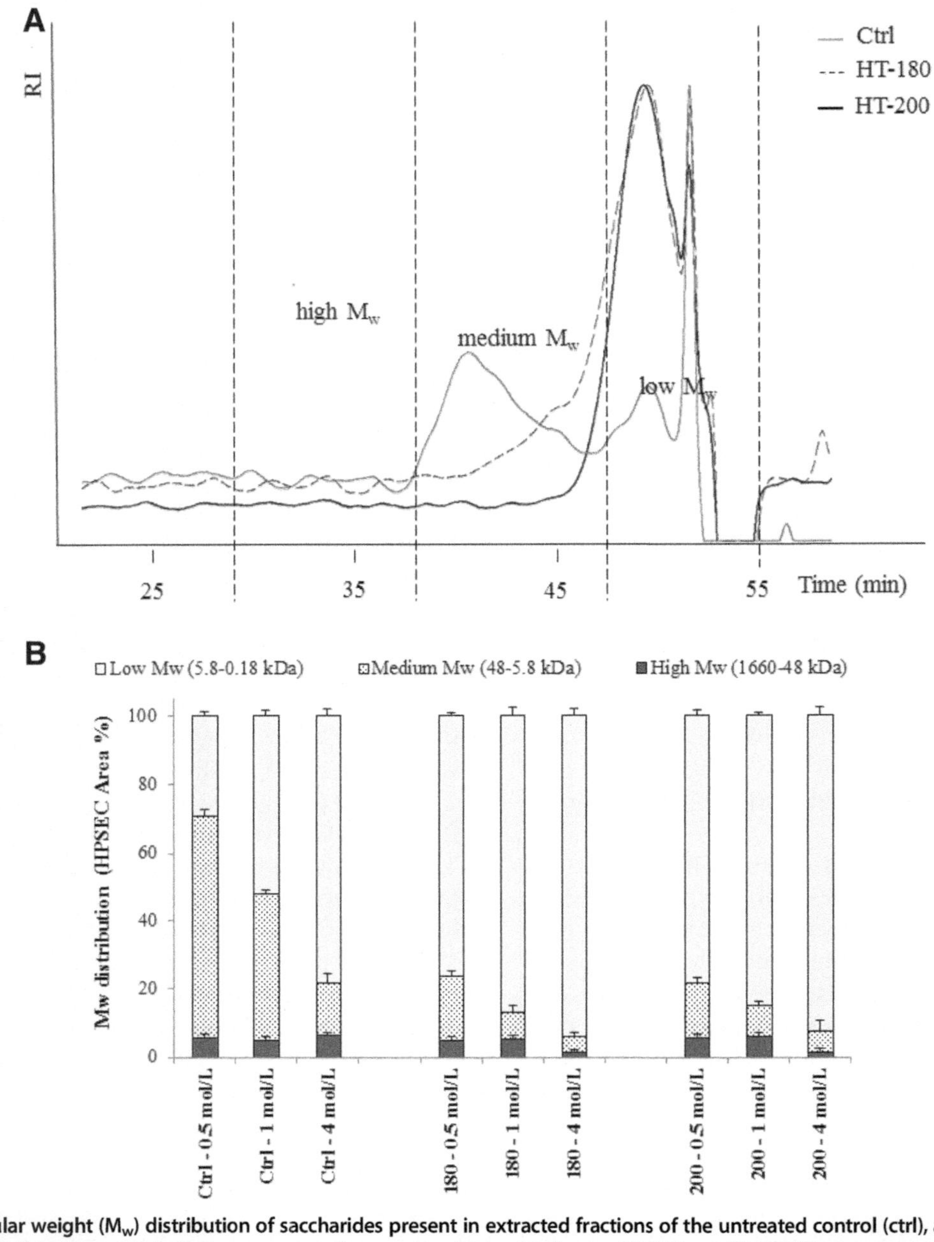

Figure 4 Molecular weight (M_w) distribution of saccharides present in extracted fractions of the untreated control (ctrl), and HT treated at 180°C and 200°C wheat bran (A) depicting representative spectra obtained following HPSEC and (B) HPSEC area percentages based on peak areas calculated using dextran standards.

they are toxic to yeast; of greater significance is the release of acetate from acetylated xylans which exhibits a higher level of toxicity [38]. Data presented here indicate that the amounts of degradation products are expected to have a limited effect on fermentation. The most sensitive yeast strains decrease fermentation rates when mixtures of inhibitors reach 1.2 g/L 5-HMF, 0.7 g/L furfural and 6 g/L for acetic acid [39]; and most strains are only inhibited at twice this concentration. The maximum concentration of furfural at 200°C was just below this limit at 1.31 g/L therefore not expected to have any deleterious effects; however, large batch fermentations will need to be conducted in order to verify this.

Simultaneous saccharification and fermentation (SSF) of HT wheat bran

SSF of raw and pretreated wheat bran was carried out in order to confirm the beneficial effect of HT on ethanol yields. SSF was performed with CTec2® and Ethanol Red® yeast with a substrate loading of 10%, a cellulase loading of 20 FPU/g dry cellulosic source, performed at 35°C for 96 h. Figure 6 depicts the amount of ethanol in

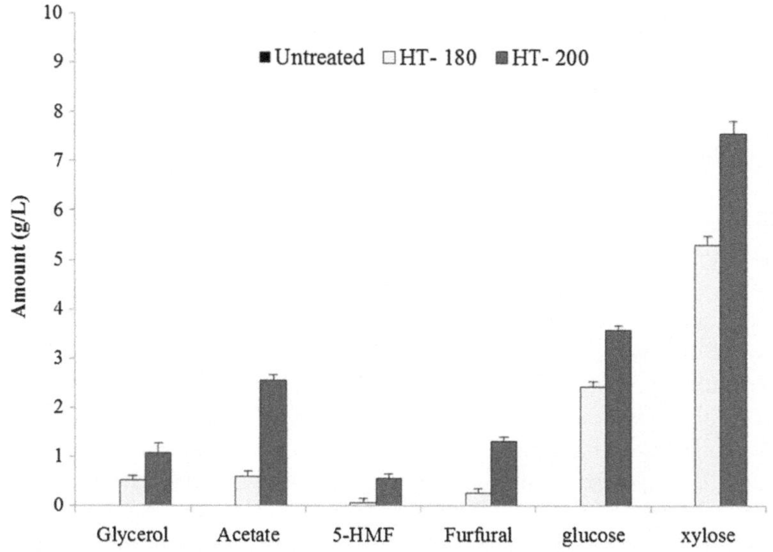

Figure 5 Characterization of the liquor obtained after HT of wheat bran depicting fermentation inhibitor compounds and solubilized sugars.

produced using SSF without added nutrients or aeration. The theoretical ethanol yield possible was 10.5, 12.5 and 13.7 g per 100 g dry bran for the control, HT at 180°C and HT at 200°C, respectively (calculated from the glucose composition in Table 2). A high ethanol conversion (of 97.5% and 96.5% of the theoretical) was obtained at both pretreatment temperatures (180°C and 200°C, respectively; Figure 6). The least amount of ethanol was produced in the untreated control bran (Figure 6) and represented 63% of the theoretical yield. Ethanol yield of 85% of theoretical has been obtained by acid hydrolysis (1% H_2SO_4) of wheat bran followed by enzymatic hydrolysis with Celluclast and Ultraflo, fermented using baker's yeast (1 g/L) [23]. However, in that study, only acid-pretreated filtrates were fermented and very high concentrations of enzyme were used. In the present study, a commercial preparation of Cellic CTec2 was used which contained enzyme cocktails specifically developed for the efficient conversion of cellulose and also contained some hemicellulosic activity (www.bioenergy.novozymes.com). As such, only half the enzyme dosage was required (equivalent to 1% w/w; g enzyme/g cellulose).

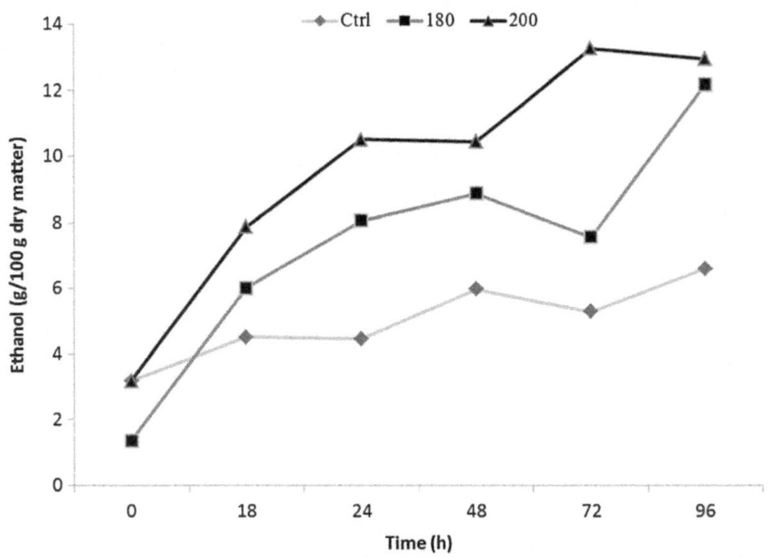

Figure 6 Shake flask SSFs with Ethanol Red ® culture, a substrate loading of 10% and cellulase load corresponding to 20 FPU g/L dry cellulosic source performed at 35°C for 96 h.

Similar findings were also reported for SSF of corn fibre, cobs and stover using Cellic CTec2 whereby the production of ethanol increased from 5% (v/v) to 7.2% (v/v) [22]. Fermentations were conducted using *Saccharomyces cerevisiae* (Ethanol Red®) which is already being used in distilleries for increased ethanol production. However, results from a comparative study [40] suggested that under optimum conditions the differences between ethanol yields from Ethanol Red and baker's yeast were negligible; although the levels of acetic acid produced by Ethanol Red yeast during fermentation were significantly lower. Data presented here indicate that wheat bran is a good feedstock for conversion to ethanol considering that SSF of the untreated bran resulted in 63% of theoretical yield from glucose (cf. wheat straw) [17], probably reflecting the relatively lower level of lignification. It also suggests that the conditions used for HT were more than adequate, even at 180°C to destabilize the cell wall polymer for the efficient enzyme conversion to ethanol and relatively low concentrations of inhibitors.

Conclusions

Hydrothermal pretreatment of destarched wheat bran released cell wall arabinoxylans, cinnamic acids and breakdown products including furfural and acetic acid. Up to 68% of the pretreated biomass could be further solubilized in alkali mostly as arabinoxylans. These exhibited significant thermochemically induced depolymerization; the level of arabinose substitution also decreased with increasing alkali concentrations. The residues contained mostly cellulosic glucose and lignin. Their simultaneous saccharification and fermentation resulted in ethanol yields of up to 97.5% theoretical. The depolymerization of the hemicelluloses would have been augmented by the considerable reduction in ferulic acid cross-links, thus contributing towards the accessibility of cellulases during enzymolysis resulting in enhanced saccharification. The reduction in cinnamic acids by selected breeding or biotechnological approaches could provide a useful basis for improved saccharification and fractionation of wheat bran polysaccharides.

Methods

Raw material

Milled wheat bran (WB; *Triticum aestivum* L. cv. Avalon) was obtained from G.R. Wright and Sons Ltd, Middlesex, UK and following dry matter determination stored in the freezer until use.

Starch hydrolysis

Hydrothermal pretreatment was performed on destarched and washed wheat bran (Figure 1). Wheat bran was treated in a two-step enzymatic hydrolysis consisting of liquefaction by thermostable α-amylases (Termamyl 120 L) and saccharification by amyloglucosidase (AMG 300 L) to remove the starch. The hydrolysis was performed at a bran/water ratio of 1:5. Approximately 6 kg of bran (four batches) were hydrolysed using a Buchi Rotavapor R-153 (Büchi Labortechnik AG, Flawil, Switzerland). Firstly, bran slurry supplemented with 4 mL Termamyl 120 L/kg slurry was treated at 85°C and pH 6 for 4 h. Secondly, AMG 300 L was added at a ratio of 12 mL/kg of slurry from the first step at pH 5 and the slurry treated at 55°C for 48 h to ensure total starch digestion. The slurry was filtered through a Munktell filter paper (quality 5), and the solid material was washed with its own volume of deionized water (×3).

Hydrothermal pretreatment (HT)

For each pretreatment, 400 g (dry weight) of destarched bran was loaded into the reactor (solid-to-water-ratio 1:10) and water was pumped through the packed material bed at a circulation flow rate of 4–5 L/min. The reactor (inner diameter 106 mm, material AISI 316) had a total volume of 6 L, with an electric heater and liquid phase circulation pump (Micropump Inc. Series 2200. The destarched bran was then pretreated at two holding temperatures (180°C/1.1 MPa and 200°C/1.55 MPa) for 10 and 15 min, respectively.

Analysis of raw and pretreated material

Dry matter determination

The dry matter (DM) of the samples was determined by weighing triplicate samples in a Mettler Toledo LP16 Infrared Dryer balance (Mettler Toledo Ltd., Beaumont Leys, Leicester, UK) as described in [17].

Preparation of alcohol insoluble residues (AIR)

AIRs were prepared from the control and pretreated wheat bran as described in [17].

Cell wall fractionation

Sequential extraction (in progressively stronger alkali) was conducted by a modification of the method in [26]. Pretreated AIRs together with the control (2 g) were suspended in hot water (60°C, 200 mL) and cell wall components extracted (shaking, 2 h, 25°C). The extracts were centrifuged (10,000 × g, 1 h) and the supernatant filtered through Whatman GF/C filter paper (Whatman, Maidstone, UK) and freeze dried. The insoluble residue was further extracted with 0.5 mol/L KOH with 20 mmol/L $NaBH_4$ for 2 h (shaking, 25°C) centrifuged (10,000 × g, 60 min) and the supernatant filtered. The filtrate was first neutralized with acetic acid, extensively dialyzed (tubing size 30/32″, Medicell International, London, UK, 7 days, changing three times daily) and then freeze dried. The residue was further extracted as above in 1 and 4 mol/L KOH containing 20 mmol/L $NaBH_4$, filtered and neutralized. All the filtrates were freeze dried as above following dialysis. The freeze-dried extracts, insoluble residues and AIRs were biochemically analysed in duplicate.

Carbohydrate analysis

Sugars were released from the fractions by hydrolysis with H_2SO_4 ($w = 72\%$) for 3 h, followed by dilution to 1 mol/L (Saemen hydrolysis). Hydrolyzed monosaccharides were analysed as their alditol acetates by GC as described in [17].

Starch analysis

Samples were analysed for starch using the Megazyme total starch kit (α-amylase K-TSTA 05/2008; Megazyme International Ireland Ltd, Ireland, UK) according to the manufacturer's instructions. The analyses were performed in triplicate.

Molecular weight (Mw) distributions

Freeze-dried alkali fractions (2 mg) were dissolved in ammonia solution ($w = 30\%$) and heated (60°C, 1 h) to solubilize the extracts. HPSEC was performed on a Dionex (Sunnyvale, USA) Ultimate 3000 UPLC system equipped with three TSKgel SuperAW columns (6.0 mm ID × 150 mm per column; 4 µm) in series (SuperAW4000, SuperAW3000, SuperAW2500; Tosoh Bioscience, Stuttgart, Germany), in combination with a guard column (Tosoh Bioscience, Stuttgart, Germany). Columns were eluted at 40°C with 0.2 mol/L sodium nitrate at 0.6 mL/min. The eluate was monitored with a Shodex RI-101 (Kawasaki, Japan) refractive index (RI) detector. Dextran standards (Sigma-Aldrich, Poole, Dorset, UK) were used for Mw calibration.

Lignin and phenolic acids

Klason lignin was quantified gravimetrically as described in [17]. The samples were analysed in triplicate. The total alkali-extractable hydroxycinnamate content of the AIRs and insoluble residues after fractionation was determined by saponification of 5 mg of sample in 4 mol/L sodium hydroxide as described in [17].

Acetic acid, furfural and hydroxymethylfurfural (HMF)

Degradation products were quantified in the pretreated liquids using a Shimadzu HPLC system (Hertogenbosch, The Netherlands) with column oven CTO-10A-vp and Autoinjector SIL-10 AD-vp equipped with a guard column (Bio-Rad H cartridge, Bio-Rad Laboratories Ltd, Hemel Hempstead, Herts., UK) and an Aminex HPX-87H column (300 × 7.8 mm; Bio-Rad Laboratories Ltd, Hemel Hempstead, Herts., UK). Elution took place at 80°C with 5 mmol/L H_2SO_4 at 0.6 mL/min. The eluate was monitored using a Refractive Index detector RID-10A (Shimadzu, Kyoto, Japan).

Light and autofluorescence microscopy

AIRs of samples were observed in their native state and photographed using a Wild M8 stereomicroscope. For higher magnification, samples were observed with an Olympus BX60 microscope (Olympus, Tokyo, Japan) interfaced with a ProRes© Capture Pro 2.1 camera and software (Jenoptik, Jena, Germany). The autofluorescence was recorded using the UV filter cube with a filter cube configuration of excitation filter band pass 330–385 nm, barrier filter 420 nm.

Simultaneous saccharification and fermentation (SSF)

Yeast cultivation

Ethanol Red® was prepared after pre-culturing aerobically at 30°C in yeast propagating medium (YP; Difco, Lawrence, Kansas, USA) containing yeast extract 1%, peptone 2% and glucose 2%. The yeast cells were transferred to YP medium supplemented with 5% glucose and cultured aerobically at 30°C overnight. The cells were harvested, washed with 0.9% NaCl solution to remove any residual nutrients, resuspended in ten times concentrated yeast nitrogen base (YNB; Formedium, Hunstanton, UK) which contained no glucose and inoculated (10% v/v) in fermentation flasks to give initial yeast concentration of 5 g/L (dry weight). Active cultures for inoculation were prepared by growing the organism in YM (growth medium Sigma-Aldrich, Poole, Dorset, UK) supplemented with: 0.3% yeast extract, 0.3% malt extract, 0.5% peptone and 1% dextrose (Difco, Lawrence, Kansas, USA) shaking (rotary shaker, 200 rpm, 18 h, 25°C).

Fermentation

Fermentations were performed on washed pretreated wheat bran slurry (10%) in closed universal bottles (20 mL, VWR International Ltd, Leicester, Leicestershire, UK) containing 9 mL of YNB without nutrients in sodium acetate buffer pH 5.0 (autoclaved, 121°C, 90 min), 1 mL of YNB in sodium acetate buffer pH 5.0 (autoclaved, 121°C/90 min) and 1 mL of yeast cell suspension (10%). A cellulase load (Cellic® CTec2, Novozymes, Bagsvaerd, Denmark) corresponding to 20 FPU/g dry cellulosic source was transferred to the system prior to the SSF process which was performed at 35°C for 96 h (anaerobically, shaking). After the reaction time, enzyme deactivation (100°C, 10 min) was followed by quantification using HPLC (Waters®, Waters Ltd, Hertfordshire, UK) equipped with a refractive index detector using a Bio-Rad Aminex HPX-87P column (Bio-Rad Laboratories Ltd, Hemel Hempstead, UK) at 65°C and MQ water (Millipore®, Millipore Ltd, Watford, UK) as mobile phase at a flow rate of 0.6 mL/min. Positive (copier paper) and negative (MQ water) controls were prepared using exactly the same procedure as the test samples, and the ethanol peaks in the negative controls were subtracted from the test peaks prior to tabulation. Ethanol values were expressed as theoretical yield (from carbohydrate data in Table 2).

Calculations

Calculations were performed using Microsoft Office Excel (2010; Microsoft, Berkshire, UK). The percentage carbohydrate yields were calculated from the monomeric amounts given in Table 1. Percentage changes as a result of the pretreatment have been calculated based on the original starting material (control AIR). The closing mass balance was generated from the dry biomass entering the hydrothermal pretreatment (HT) system, the amount of water added and the final mass of the solid residue obtained following pretreatment together with the liquid fractions. The key sugars were calculated as monomeric equivalents present in the control and pretreated AIRs, taking into account the dry matter (DM) content and back calculated to the original starting material entering the pretreatment stream. As a starting point, the amount given in Table 1 was multiplied by the DM content to determine the starting amount of each sugar. So for each pretreatment, key sugars were calculated as follows:

For example, for glucose yield

$$(g/kg) = \frac{\text{HT biomass } (g) * 1,000 * \text{DM} * \text{amount of glucose in HT AIR } (g)}{\text{Starting biomass } (g)}$$

The ethanol yields were calculated as follows:

Ethanol (g) = Glucose present (%) × Weight of material × 0.511 (factor for ethanol conversion) × 1.111 (water of hydrolysis).

Abbreviations

AIR: alcohol insoluble residue; DiFA: diferulic acid; DM: dry matter; HT: hydrothermal pretreatment; HPSEC: high-performance size exclusion chromatography; HW: hot water extraction; Mw: molecular weight; SSF: simultaneous saccharification and fermentation.

Competing interests

The authors declare that they have no competing interests.

Authors' contributions

All authors contributed intellectually via scientific discussions during the work. ZM designed the study, executed the experimental work and drafted the manuscript. SRAC and AE contributed towards the design of molecular mass analyses. DW conducted the fermentation and AK carried out the hydrothermal pretreatments at Biogold. KWW provided academic expertise in the interpretation of the data and refined the manuscript. All authors read and approved the final version.

Acknowledgements

This work has been part funded by the Biotechnology and Biological Sciences Research Council (BBSRC), UK, and the Commission of the European Communities, Framework 7, Grant Agreement 211863. This study does not necessarily reflect the views of the commission and its future policy in this area. We are grateful to Dr Mary Parker and Kathryn Cross for microscopy and to Novozymes for providing the enzymes.

Author details

[1]The Biorefinery Centre, Institute of Food Research, Norwich Research Park, Colney, Norwich NR4 7UA, UK. [2]Biogold OÜ, Lossi 19A, 12616 Tallinn, Estonia.

References

1. Bergmans MEF, Beldman G, Gruppen H, Voragen AGJ. Optimisation of the selective extraction of (glucurono) arabinoxylans from wheat bran: use of barium and calcium hydroxide solution at elevated temperatures. J Cereal Sci. 1996;23:235–45.
2. Parker ML, Ng A, Waldron KW. The phenolic acid and polysaccharide composition of cell walls of bran layers of mature wheat (*Triticum aestivum* L. cv. Avalon) grains. J Sci Food Agri. 2005;85:2539–47.
3. Bunzel M, Ralph J, Steinhart H. Phenolic compounds as cross-links of plant derived polysaccharides. Czech J Food Sci. 2004;22:64–7.
4. Dervilly G, Saulnier L, Roger P, Thibault JF. Isolation of homogeneous fractions from wheat water-soluble arabinoxylans. Influence of the structure on their macromolecular characteristics. J Agri Food Chem. 2000;48:270–8.
5. Parker ML, Waldron KW. Texture of Chinese water chestnut—involvement of cell wall phenolics. J Sci Food Agric. 1995;68:337–46.
6. Tejirian A, Xu F. Inhibition of enzymatic cellulolysis by phenolic compounds. Enz Micro Technol. 2011;48:239–47.
7. Martel F, Estrine B, Plantier-Royon R, Hoffmann N, Portella C. Development of agriculture left-overs: fine organic chemicals from wheat hemicellulose-derived pentoses. Top Curr Chem. 2010;294:79–115.
8. Waldron KW. Bioalcohol production: biochemical conversion of lignocellulosic biomass. Woodhead: Oxford; 2010.
9. Yang B, Wyman CE. Pretreatment: the key to unlocking low-cost cellulosic ethanol. Biofuel Bioprod Bior. 2008;2:26–40.
10. Kumar P, Barrett DM, Delwiche MJ, Stroeve P. Methods for pretreatment of lignocellulosic biomass for efficient hydrolysis and biofuel production. Ind Eng Chem Res. 2009;48:3713–29.
11. Brodeur G, Yau E, Badal K, Collier J, Ramachandran KB, Ramakrishnan S. Chemical and physicochemical pretreatment of lignocellulosic biomass: a review. Enzyme Res. 2011;2011:787532.
12. Sarkar N, Ghosh SK, Bannerjee S, Aikat K. Bioethanol production from agricultural wastes: An overview. Renew Energ. 2012;37:19–27.
13. Collins SRA, Wellner N, Martinez Bordonado I, Harper AL, Miller CN, Bancroft IB, et al. Variation in the chemical composition of wheat straw: the role of tissue ratio and composition. Biotechnol Biofuels. 2014;7:121.
14. Zhang H, Thygesen LG, Mortensen K, Kadar Z, Lindedam J, Jorgensen H, et al. Structure and enzymatic accessibility of leaf and stem from wheat straw before and after hydrothermal pretreatment. Biotechnol Biofuels. 2014;7:74.
15. Hansen MAT, Ahl LI, Pedersen HL, Bestereng B, Willats WGT, Jorgensen H, et al. Extractibility and digestibility of plant cell wall polysaccharides during hydrothermal and enzymatic degradation of wheat straw (Triticum aestivum L.). Indust Crops and Prod. 2014;55:63–9.
16. DeMartini JD, Pattathil S, Avci U, Szekalski K, Mazumder K, Hahn MG, et al. Application of monoclonal antibodies to investigate plant cell wall deconstruction for biofuels production. Energy and Environmen Sci. 2014;4:4332–9.
17. Merali Z, Ho JD, Collins SRA, Le Gall G, Elliston A, Käsper A, et al. Characterization of cell wall components of wheat straw following hydrothermal pretreatment and fractionation. Biores Technol. 2013;131:226–34.
18. Maes C, Delcour JA. Structural characterisation of water-extractable and water-unextractable arabinoxylans in wheat bran. J Cer Sci. 2002;35:316–26.
19. Hollmann J, Lindhauer MG. Pilot-scale isolation of glucuronoarabinoxylansfrom wheat bran. Carb Polymers. 2005;59:225–30.
20. Mandalari G, Faulds CB, Sancho AI, Saija A, Bisignano G, LoCurto R, et al. Fractionation and characterisation of arabinoxylans from brewers' spent grain and wheat bran. J Cer Sci. 2005;42:205–12.
21. Zhang Y, Pitkanenc L, Dougladed J, Tenkanenc M, Remonda C, Joly C. Wheat bran arabinoxylans: chemical structure and film properties of three isolated fractions. Carb Pol. 2011;86:852–9.
22. Van Eylen D, Van Dongen F, Kabel M, De Bont J. Corn fiber, cobs and stover: enzyme-aided saccharification and co-fermentation after dilute acid pretreatment. Bioresour Technol. 2011;102:5995–6004.

23. Palmarola-Adrados B, Choteborska P, Galbe M, Zacchi G. Ethanol production from non-starch carbohydrates of wheat bran. Bioresour Technol. 2005;96:843–50.
24. Aguedo M, Fougnies C, Dermience M, Richel A. Extraction by three processes of arabinoxylans from wheat bran and characterisation of the fractions obtained. Carbohydr Polym. 2014;105:317–24.
25. Pu Y, Hu F, Huang F, Davison BH, Ragauskas AJ. Assessing the molecular structure basis for biomass recalcitrance during dilute acid and hydrothermal pretreatments. Biotechnol Biofuels. 2013;6:15.
26. Redgwell RJ, Selvendran RR. Structural features of cell-wall polysaccharides of onion *Allium cepa*. Carbohydr Res. 1986;157:183–99.
27. Kabel MA, Bos G, Zeevalking J, Voragen AG, Schols HA. Effect of pretreatment severity on xylan solubility and enzymatic breakdown of the remaining cellulose from wheat straw. Bioresour Technol. 2007;98:2034–42.
28. Theander O, Aman P, Westerlund E, Andersson R, Petersson D. Total dietary fiber determined as neutral sugar residues, uronic acid residues, and Klason lignin (the Uppsala method): collaborative study. J AOAC Int. 1995;78:1030–44.
29. Gong L, Huang L, Zhang Y. Effect of steam explosion treatment on barley bran phenolic compounds and antioxidant capacity. J Agri Food Chem. 2012;60:7177–84.
30. Waldron KW, Smith AC, Parr AJ, Ng A, Parker ML. New approaches to understanding and controlling cell separation in relation to fruit and vegetable texture. Trends Food Sci Technol. 1997;8:213–21.
31. Parker CC, Parker ML, Smith AC, Waldron KW. Thermal stability of texture in Chinese water chestnut may be dependent on 8, 8-dehydrodiferulic acid (aryltetralyn form). J Agric Food Chem. 2003;51:2034–9.
32. Kumar R, Hu F, Sannigrahi P, Jung S, Ragauskas AJ, Wyman CE. Carbohydrate derived pseudo lignin can retard cellulose biological conversion. Biotechnol Bioeng. 2013;110:737–353.
33. Wood IP, Elliston A, Collins SRA, Wilson D, Bancroft I, Waldron KW. Steam explosion of oilseed rape straw: extablishing key determinants of saccharification efficiency. Bioresource Technol. 2014;162:175–83.
34. Brett C, Waldron KW. Cell wall structure and the skeletal functions of the wall. In: Brett C, Waldron KW, editors. Physiology and Biochemistry of plant cell walls. 2nd ed. London: Chapman and Hall London; 1996. p. 2–24.
35. Kabel MA, Carvalheiro F, Garrote G, Avgerinos E, Koukios E, Parajo JC, et al. Hydrothermally treated xylan rich byproducts yield different classes of xylo-oligosaccharides. Carbohydr Polym. 2002;50:47–56.
36. Gruppen H, Hamer RJ, Voragen AGJ. Water- unextractable cell wall material from wheat flour 2. Fractionation of alkali-extracted polymers and comparison with water-extractable arabinoxylans. J Cer Sci. 1992;16:53–67.
37. Garrote G, Kabel MA, Schols HA, Falque E, Dominguez H, Parajo JC. Effects of Eucalyptus globulus wood autohydrolysis conditions on the reaction products. J Agric Food Chem. 2007;55:9006–13.
38. Bellido C, Bolado S, Coca M, Lucas S, Gonzalez-Benito G, Garcia-Cubero MT. Effect of inhibitors formed during wheat straw pretreatment on ethanol fermentation by Pichia stipitis. Bioresour Technol. 2011;102:10868–74.
39. Martin C, Jönsson LJ. Comparison of the resistance of industrial and laboratory strains of saccharomyces and zygosaccharomyces to lignocellulose-derived fermentation inhibitors. Enzyme Microb Technol. 2003;32:386–95.
40. Mukhtar K, Asgher M, Afghan S, Hussain K, Zia-ul-Hussnain S. Comparative study on two commercial strains of saccharomyces cerevisiae for optimum ethanol production on industrial scale. J Biomed Biotechnol. 2010;Article ID 419586:1–5.

5

A microbial platform for renewable propane synthesis based on a fermentative butanol pathway

Navya Menon[1], András Pásztor[2], Binuraj RK Menon[1], Pauli Kallio[2], Karl Fisher[1], M Kalim Akhtar[2], David Leys[1], Patrik R Jones[2,3*] and Nigel S Scrutton[1*]

Abstract

Background: Propane (C_3H_8) is a volatile hydrocarbon with highly favourable physicochemical properties as a fuel, in addition to existing global markets and infrastructure for storage, distribution and utilization in a wide range of applications. Consequently, propane is an attractive target product in research aimed at developing new renewable alternatives to complement currently used petroleum-derived fuels. This study focuses on the construction and evaluation of alternative microbial biosynthetic pathways for the production of renewable propane. The new pathways utilize CoA intermediates that are derived from clostridial-like fermentative butanol pathways and are therefore distinct from the first microbial propane pathways recently engineered in *Escherichia coli*.

Results: We report the assembly and evaluation of four different synthetic pathways for the production of propane and butanol, designated a) *atoB-adhE2* route, b) *atoB-TPC7* route, c) *nphT7-adhE2* route and d) *nphT7-TPC7* route. The highest butanol titres were achieved with the *atoB-adhE2* (473 ± 3 mg/L) and *atoB-TPC7* (163 ± 2 mg/L) routes. When aldehyde deformylating oxygenase (ADO) was co-expressed with these pathways, the engineered hosts also produced propane. The *atoB-TPC7-ADO* pathway was the most effective in producing propane (220 ± 3 μg/L). By (i) deleting competing pathways, (ii) including a previously designed ADO_{A134F} variant with an enhanced specificity towards short-chain substrates and (iii) including a ferredoxin-based electron supply system, the propane titre was increased (3.40 ± 0.19 mg/L).

Conclusions: This study expands the metabolic toolbox for renewable propane production and provides new insight and understanding for the development of next-generation biofuel platforms. In developing an alternative CoA-dependent fermentative butanol pathway, which includes an engineered ADO variant (ADO_{A134F}), the study addresses known limitations, including the low bio-availability of butyraldehyde precursors and poor activity of ADO with butyraldehyde.

Keywords: Propane, Butanol, Microbial pathway engineering, Cyanobacteria, Aldehyde deformylating oxygenase, *Escherchia coli*

* Correspondence: p.jones@imperial.ac.uk; nigel.scrutton@manchester.ac.uk
[2]Molecular Plant Biology, Department of Biochemistry, Tykistökatu 6A 6krs, University of Turku, FI 20014 TURUN YLIOPISTO, Turku, Finland
[1]BBSRC/EPSRC Centre for Synthetic Biology of Fine and Speciality Chemicals, Manchester Institute of Biotechnology, Faculty of Life Sciences, 131 Princess Street, The University of Manchester, Manchester M1 7DN, UK
Full list of author information is available at the end of the article

Background

Propane, a major component of autogas or liquefied petroleum gas (LPG), is an emerging fuel for future energy supply and transportation [1]. Propane is the third most widely used motor fuel and about 20 million tonnes of propane gas are used per year to fuel motor vehicles [2]. It is estimated that propane provides heat and energy for more than 14 million homes worldwide annually [3,4]. Propane also has an existing global market for a wide number of other stationary and mobile applications, such as low emission vehicles, gas burners and refrigeration systems [5]. Easy separation from liquid biotechnological processes as a gas and less energy requirements for liquefaction and storage offers potential advantages to propane over other gaseous fuels [6].

Natural metabolic pathways for the renewable biosynthesis of propane do not exist. The discovery of an aldehyde deformylating oxygenase (ADO) from cyanobacteria, however, has paved the way for synthetic alkane pathways to be constructed [7-9]. A microbial platform for propane generation dependent on fatty acid biosynthesis was recently reported [10]. The authors concluded that the pathway was limited by total flux through fatty acid synthesis (FAS). The most obvious example of this limitation comes from the markedly enhanced rate of propane synthesis observed when fatty acids were supplied to the external media. In the present study, we sought to bypass this limitation by generating new synthetic pathways that are not dependent on FAS. In this study, we designed a series of modified butyraldehyde pathways based on the CoA-dependent butanol pathways commonly found in *Clostridium* spp. Propane biosynthesis was thereafter achieved by interrupting the route to alcohol by the addition of ADO (Figure 1).

The butanol pathway in *Clostridium* proceeds either via a keto acid route (Ehrlich pathway) or a CoA-dependent route. Higher yields of branched chain alcohols and aldehyde precursors (for example, isobutyraldehyde) from the decarboxylation of keto acids make the Ehrlich pathway less attractive because ADO has a strong preference for straight chain aldehyde substrates [11-13]. By contrast, butanol production by the CoA-dependent route initiates with the condensation of two molecules of acetyl CoA. Reduction in subsequent steps produces the end-product 1-butanol via a butyraldehyde intermediate. There are several reports of engineered CoA-dependent butanol pathways in *Escherichia coli* and other host organisms [14-20]. In the following work, we constructed and evaluated a series of CoA-dependent butyraldehyde pathways that eliminate the dependency on AdhE2, thereby allowing butyraldehyde to be re-routed towards propane instead of butanol.

The use of ADO (from *Prochlorococcus marinus* MIT9313) as a terminal decarbonylase has been used for the production of medium/long chain (C9-C17) as well as short chain-length alkanes (C3, C7) [8,10]. Variant forms of ADO have demonstrated improved activity with the shorter chain aldehydes that are not encountered in native cyanobacteria [7,12]. These variant forms of ADO are therefore attractive enzyme components for building new synthetic pathways with a greater productivity, as addressed in the current study.

Here, we report on the construction and evaluation of novel pathways for propane production in *E. coli* that are independent of the FAS pathway in a recent study, thereby opening up possibilities for further optimization of short chain-length alkane biosynthesis in future studies [10].

Results

Synthesis of butyraldehyde based on the CoA-dependent 1-butanol pathway

For propane production, the ADO enzyme requires butyraldehyde as a precursor. We therefore constructed two biosynthetic pathways for butyraldehyde synthesis based on the fermentation pathway of 1-butanol, as summarized in Figure 1. The initial step of the pathway consisted of i) either AtoB from *E. coli* (Figure 1; *atoB* route) or NphT7 from *Streptomyces* sp. (Figure 1; *nphT7* route) to convert the metabolic pathway intermediate, acetyl-CoA to acetoacetyl-CoA [17,21]. Although the latter step has not previously been evaluated in *E. coli*, it was recently shown to be superior to AtoB for butanol production in *Synechococcus elongatus* PCC 7942 [18]. For the second and third steps, clostridial 3-hydroxybutyryl-CoA dehydrogenase (Hbd) and 3-hydroxybutyryl-CoA dehydratase (crotonase or Crt) were used to convert acetoacetyl-CoA to crotonyl-CoA. For the fourth step, the oxygen-sensitive flavoenzyme butyryl-CoA dehydrogenase, present in the native clostridial pathway, was replaced with NADH-dependent trans-enoyl-CoA reductase (Ter) from *Treponema denticola* to reduce crotonyl-CoA to butyryl-CoA [21]. In the final step, butyraldehyde was synthesized by NAD (P)H-dependent reduction of butryl CoA catalysed by AdhE2. To verify pathway functionality for butyraldehyde synthesis, we used butanol as a molecular reporter since numerous studies have shown that it can be stably accumulated in *E. coli* and easily monitored [20]. Expression of all pathway components were confirmed by sodium dodecyl sulfate-polyacrylamide gel electrophoresis (SDS-PAGE) (Additional file 1: Figures S1 to S3), and pathway functionality was verified by mass spectrometry. Both routes led to the production of 1-butanol clearly indicating that both pathways were capable of butyraldehyde synthesis. The strain harbouring the *atoB* route led to a 6.2-fold higher production of 1-butanol production (473.3 ± 3.2 mg/L) compared to the nphT7 route (Figure 2).

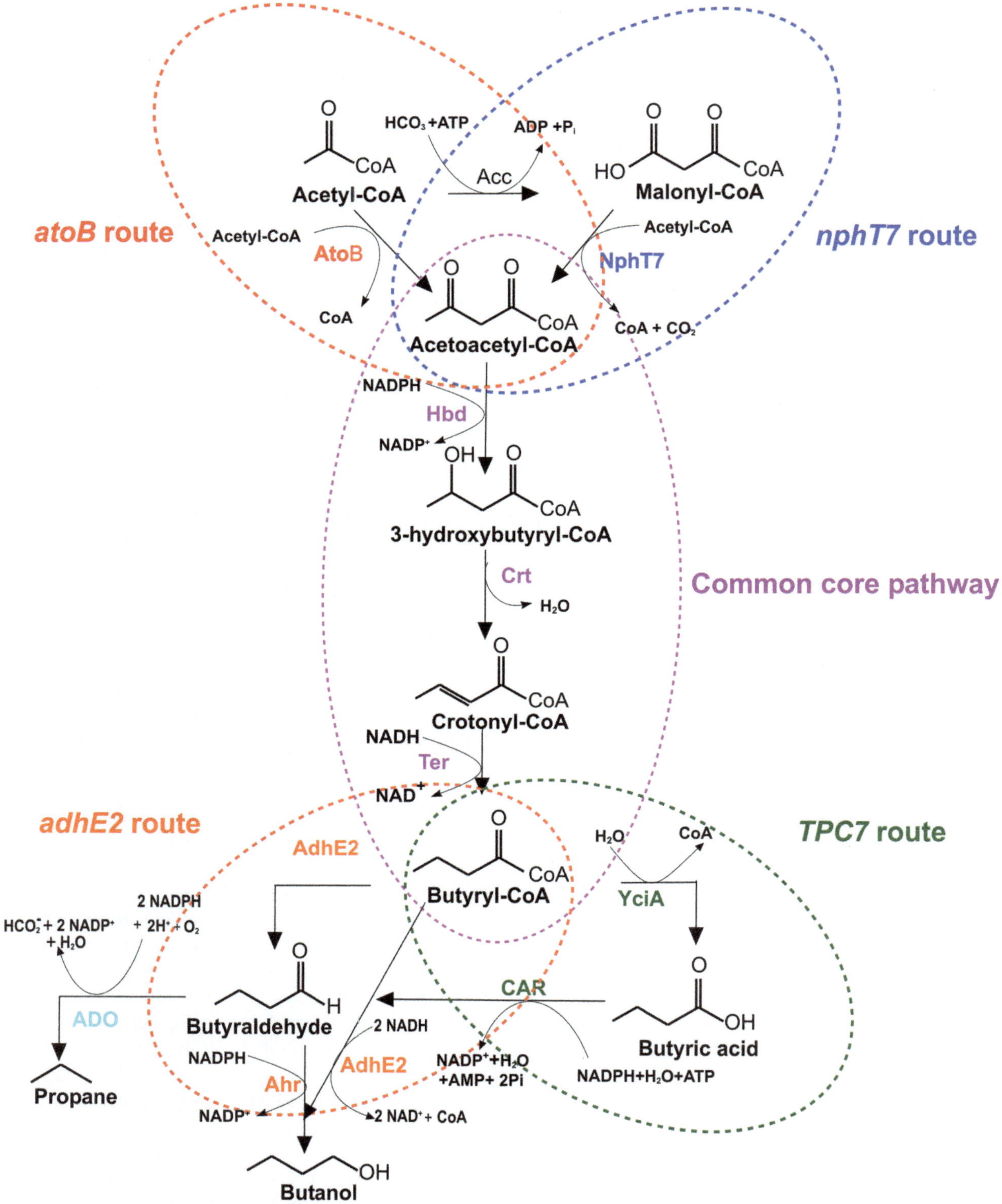

Figure 1 The CoA-dependent butanol pathways used for the production of propane in *E. coli*. The four CoA-dependent butanol producing synthetic routes (*atoB-adhE2* route, *atoB-TPC7* route, *nphT7-adhE2* route and *nphT7-TPC7* route) explored for butanol production in *E. coli* are shown. AdhE2, aldehyde-alcohol dehydrogenase; ADO, aldehyde deformylating oxygenase; Ahr, aldehyde reductase; AtoB, acetyl-CoA acetyltransferase; CAR, carboxylic acid reductase; Crt, 3-hydroxybutyryl-CoA dehydratase; Hbd, 3-hydroxybutyryl-CoA dehydrogenase; NphT7, acetoacetyl CoA synthase; Ter, trans-2-enoyl-CoA reductase; YciA, acyl-CoA thioester hydrolase.

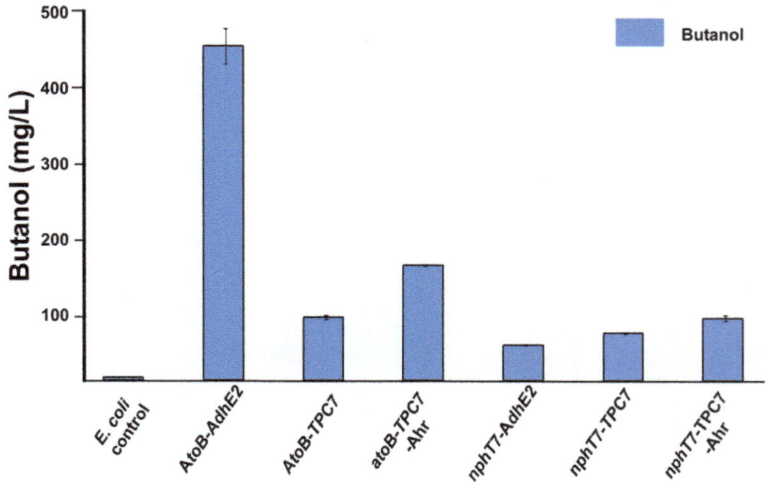

Figure 2 Total butanol produced by the engineered *E. coli* BL21 strains. Total butanol concentration obtained after 72 h of cultivation of *E. coli* wild-type cells harbouring engineered constructs is shown. The experimental protocol is described in the 'Materials and methods' section. Error bars are standard deviation ($n = 4$).

Modification of the butyraldehyde pathway by replacement of AdhE2

In the clostridial pathway, butyryl-CoA is converted to butanol in two successive catalytic steps by the bifunctional aldehyde/alcohol dehydrogenase AdhE2 (Figure 1; *AdhE2* route) [22]. From an engineering perspective, this is highly undesirable since the local presence of an aldehyde reductase component of AdhE2 is likely to compete for the ADO substrate. Given this possibility and to avoid creating internal metabolic competition, we therefore replaced AdhE2 with a i) thioesterase (YciA) from *Haemophilus influenzae* to cleave butyryl-CoA to butyric acid and ii) ATP/NADPH-dependent carboxylic acid reductase (CAR) from *Mycobacterium marinum* to convert butyric acid to butyraldehyde [8,23]. Two variant routes were thus created based on this modification, namely *atoB-TPC7*, *atoB-TPC7-ahr*, *nphT7-TPC7* and *nphT7-TPC7-ahr* (Figure 1; *TPC7* route). As before, 1-butanol was monitored to verify the functionality of the parts for butyraldehyde synthesis. In both cases, 1-butanol production was observed signifying yet again that butyraldehyde synthesis was achievable with both routes. 1-Butanol production was increased with the *atoB-TPC7* and *nphT7-TPC7* routes when Ahr was overexpressed (1.5 and 1.3 times for *atoB-TPC7* and *nphT7-TPC7* routes, respectively), indicating that the conversion of butraldehyde to butanol by the endogenous aldehyde reductases was limiting total pathway flux.

Evaluation of pathway routes for propane synthesis

The combined analysis clearly indicated that all pathway components for butyraldehyde synthesis were functional. In order to convert butyraldehyde to propane, the aldehyde reductases in the four butanol pathways were thereafter replaced or complemented by ADO. The TPC7-based strains produced relatively more propane (*atoB-TPC7-ADO* and *nphT7-TPC7-ADO*) compared to other strains (Figure 3; Additional file 1: Figures S4 and S5), and the majority of the pathway flux in all four routes was still directed towards 1-butanol. Interestingly, the lack of propane with the two *AdhE2* pathways clearly supported our initial hypothesis regarding internal metabolic competition with AdhE2.

The two functional *TPC7*-dependent propane pathways were subsequently modified by replacing ADO with the ADO$_{A134F}$ variant enzyme (Figure 4). This enzyme was initially designed to overcome some of the kinetic constraints in propane production resulting from very low activity of the enzyme towards short-chain substrates [12]. The A134F substitution alters the topology of the substrate-access channel and has been shown to result in improved activity towards low chain-length aldehydes butyraldehyde and pentaldehyde. The ADO$_{A134F}$ has also been shown to generate propane in both *in vitro* and *in vivo* biotransformations, and the *E. coli* strain expressing the variant enzyme produced approximately two times more propane (0.47 ± 0.04 mg/L) in comparison to the corresponding wild-type ADO system when fed with externally added butyraldehyde [12]. As expected, introduction of the variant ADO into the pathways engineered in this study also resulted in enhanced propane production in all cases (Figure 4). The best result was obtained with the *atoB-TPC7-ADO$_{A134F}$* combination, showing a 1.8-fold improvement, in comparison to the wild-type ADO system, and reaching a maximum propane titre of 0.29 ± 0.02 mg/L (Figure 4).

Increasing electron supply to ADO via ferredoxin

It has previously been shown that the supply of electrons to ADO via endogenous enzymes was inadequate for effective

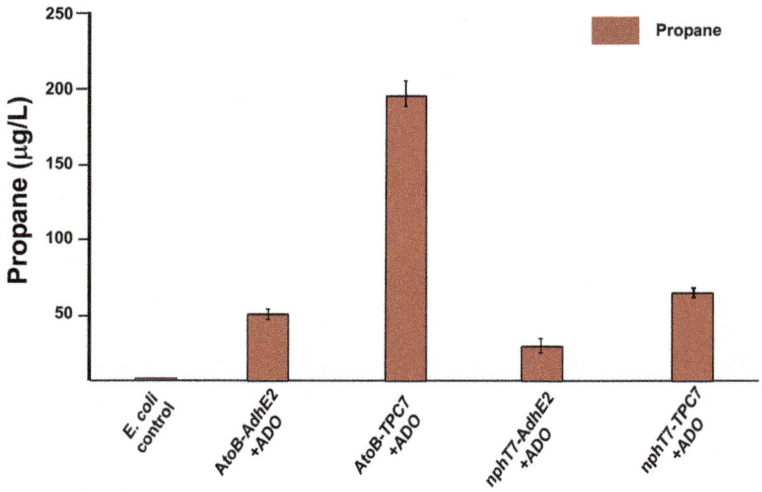

Figure 3 Total propane produced by the engineered *E. coli* BL21 strains. Total propane accumulated over 4 h of reaction under assay conditions performed in GC vials for the pathway-engineered *E. coli* cells, after overexpressing with ADO. The experimental protocol is described in the 'Materials and methods' section. Error bars are standard deviation ($n = 4$). ADO, aldehyde deformylating oxygenase; IPTG, isopropyl β-D-1-thiogalactopyranoside.

alkane biosynthesis in *E. coli* [10]. To alleviate this constraint, the heterologous ferredoxin PetF (*ssl0020*), the presumed natural electron acceptor/donor to ADO in *Synechocystis* sp. PCC 6803, was overexpressed as part of the engineered pathways. Consequently, co-expression of PetF improved propane production by roughly twofold for both the *atoB-TPC7* and *nphT7-TPC7* pathway in combination with both ADO and ADO_{A134F} (Figure 4). In order to verify the impact of the optimization efforts, residual butanol was also measured in selected strains (Additional file 1: Figures S4 and S5). If ADO was able to compete for butyraldehyde, the common precursor for both propane and butanol, one would expect to observe a lower butanol titre in strains with enhanced propane production. Indeed, residual butanol levels were decreased when ADO and ferredoxin were overexpressed. Similarly, when the ADO_{A134F} variant replaced wild-type ADO, the accumulation of butanol in the media was also lower. This is consistent with the

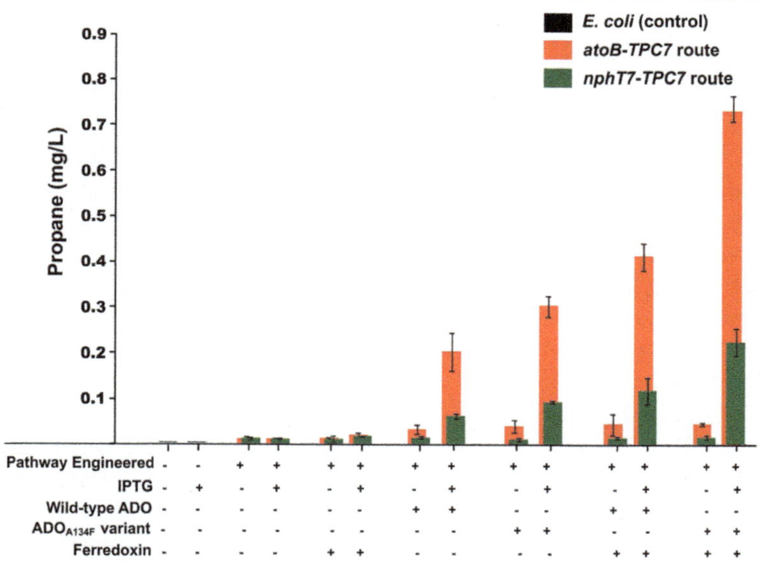

Figure 4 Propane production in pathway-engineered *E. coli* strains containing wild-type ADO or the ADO_{A134F} variant enzyme. *E. coli* cells were either pathway-engineered to include the *atoB-TPC7* route (indicated by red colour) or the nphT7-*TPC7* route (indicated by green colour) and contain wither wild-type ADO or the ADO_{A134F} variant enzyme in the absence/presence of ferredoxin (Fdx) from *Synechocystis* sp. PCC 6803. A detailed protocol for the pathway engineering and propane detection is included in the 'Materials and methods' section and with the supporting information (Additional file 1: Figure S6). Error bars are standard deviation ($n = 4$).

terminal ADO (and the ADO_{A134F} variant) acting on butyraldehyde produced by the engineered synthetic pathway (Additional file 1: Figures S4 and S5).

Removing competing pathways for propane synthesis

E. coli cells contain a wide range of aldehyde reductases and alcohol dehydrogenases which act to scavenge potentially toxic intracellular aldehydes [24]. Two of the native aldehyde reductases in *E. coli*, Ahr and YqhD, were previously shown to compete for butyraldehyde [10]. In order to further optimize propane production, *Δahr* and *ΔyqhD* single and double knockout *E. coli* strains were therefore tested (Figure 5A). Among the single knockout strains, *ΔyqhD* showed a greater increase in propane production compared to *Δahr*. The double knockout strain *Δahr/ΔyqhD* showed a cumulative effect of the gene knockouts, reaching a propane titre of 2.05 ± 0.12 mg/L for the *atoB-TPC7-ADO*$_{A134F}$ strain. The single knockout strains were also tested for propane production in the presence of *Synechocystis* sp. ferredoxin (Fdx), reaching a titre of up to 3.40 ± 0.19 mg/L in the *ΔyqhD* background (Figure 5B).

Potential for large-scale production of propane

As propane biosynthesis was only analysed in small-scale cultures (2.0 mL gas chromatography (GC) vials), larger scale productivity and longevity of propane production was also investigated with the best performing strains (that is, *atoB-TPC7* or *nphT7-TPC7* routes incorporated into *E. coli ΔyqhD* in the presence of the ADO_{A134F} variant and ferredoxin). The culture volume was scaled 400-fold to 200 mL, in 300-mL flasks sealed with airtight rubber septa, and propane production and accumulated residual butanol were monitored over 12 h (Figure 6 and Additional file 1: Figure S7). The highest propane accumulation was observed between 8 and 10 h. For the *atoB-TPC7* route, 1.38 ± 0.06 mg/L propane was accumulated within 12 h, while 0.4 ± 0.1 mg/L propane was accumulated for the *nphT7-TPC7* pathway (Figure 6). The ability to scale propane production is an important proof-of-concept that suggests large-scale propane production should be possible in a fermentor setup with appropriate optimisation of growth and feeding conditions. Controlled cultivation in a fermentor should bypass potential problems associated with flask cultivation, for example, non-optimal oxygen supply and the production of undesirable byproducts.

Discussion

Recent progress in exploring various heterologous microbial platforms towards metabolites such as fatty acids, alcohols and alkanes has improved the prospect of advanced biofuel production. Propane (C_3H_8) was proposed as a new microbial biofuel target as it would be able to act as a direct drop-in replacement for the corresponding non-renewable products currently in use (for example, autogas, LPG). Propane also has very good physicochemical properties which allow it to be stored and transported in a compressed liquid form, while under ambient conditions, it is a clean-burning gas used in various applications

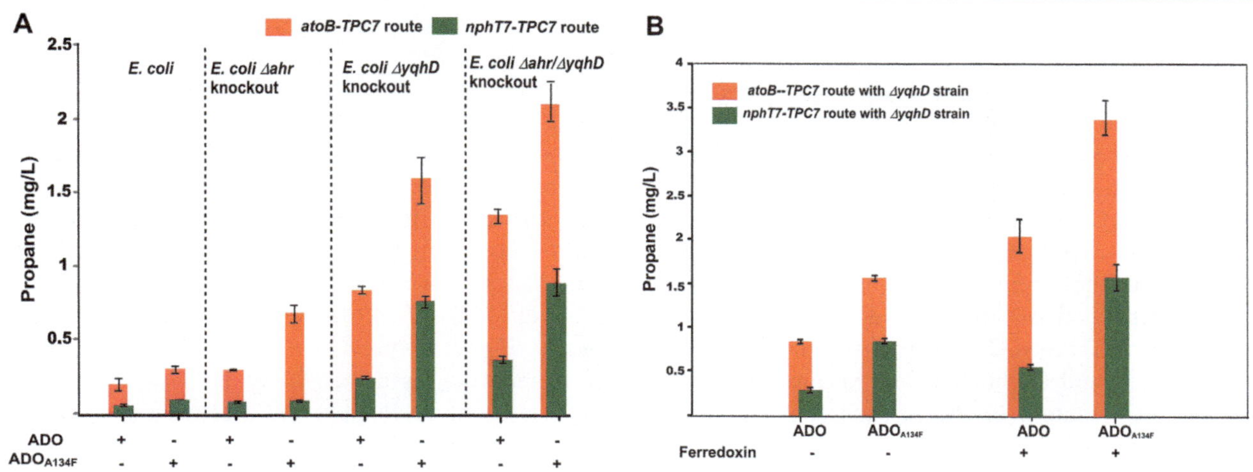

Figure 5 Propane produced in pathway-engineered Δahr/ΔyqhD single or double knockout *E. coli* strains and the effects of co-expressing a ferredoxin electron donating system. Propane production in the pathway-engineered *ΔyqhD* knockout cells with wild-type ADO or with the ADO_{A134F} variant enzyme is shown **(A)**. The *ΔyqhD* knockout strains were either engineered to contain the *atoB-TPC7* route (indicated by red colour) or the *nphT7-TPC7* route (indicated by green colour). Wild-type ADO or the ADO_{A134F} variant enzyme was co-expressed in the engineered cells either in combination with or without ferredoxin (Fdx) from *Synechocystis* sp. PCC 6803 **(B)**. A detailed protocol for the pathway engineering and propane detection is included in the 'Materials and methods' section and supporting information (Additional file 1: Figure S6). Error bars are standard deviation ($n = 4$). ADO, aldehyde deformylating oxygenase.

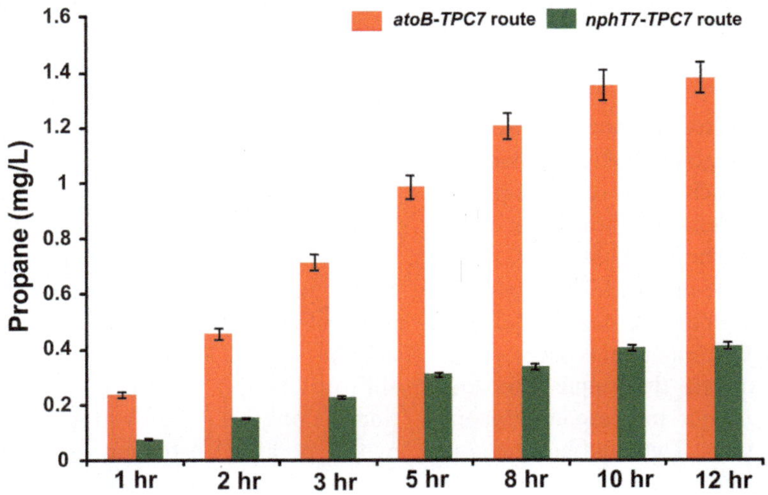

Figure 6 Larger scale cultures of the best performing propane-producing pathways. The *atoB-TPC7* route (indicated by red colour) and *nphT7-TPC7* (indicated by green colour) engineered in *ΔyqhD* knockout cells in the presence of the ADO_{A134F} variant and ferredoxin system were analysed at larger scale. The culture volume was scaled up to 400-fold to 200 mL, in a 300-mL flask sealed with airtight rubber septum. The propane accumulation for 12 h is shown. Error bars are standard deviation ($n = 3$).

ranging from heating to utilization as a transport fuel. The very first microbial biosynthetic pathway for producing propane was recently engineered in *E. coli* [10]. In that study, it was found that the biosynthesis of propane was restricted by the poor activity of the enzyme ADO towards short-chain substrates in combination with competing pathways that limits the availability of the ADO substrate, butyraldehyde. The study reported here therefore explored alternative biosynthetic approaches to provide a more comprehensive understanding of the limiting factors in microbial propane biosynthesis and to find possible ways to overcome metabolic bottlenecks.

The pathways assembled in the present study are not dependent on type II fatty acid biosynthesis unlike in the previous study and instead use CoA- rather than ACP-based intermediates. Importantly, this strategy allowed us to bypass the strict regulatory control on native fatty acid biosynthesis flux [10,25,26]. The pathways engineered in this study comprise four parallel variations of the Clostridial butanol pathway, differing from one another in the conversion of (1) acetyl-CoA to acetoacetyl-CoA (*atoB vs. nphT7* routes) and (2) butyryl-CoA to butyraldehyde (*adhE2 vs. TPC7* routes). In the first stage of pathway evaluation, production of butanol was used as a measure of metabolic flux towards butyraldehyde, the immediate substrate for propane production by ADO. The TPC7 variant routes were developed in order to avoid the bifunctional activity of AdhE2 which includes the reduction of butyraldehyde to 1-butanol. Such a reaction would be expected to compete with ADO for the butyraldehyde intermediate and lower propane production. Therefore, in order to separate these functional activities and allow greater pathway flexibility, AdhE2 was replaced with a 'TPC7' module in combination with Ahr.

For the first two variant pathways, the (co)substrate requirements for AtoB and NphT7 are quite distinct; AtoB requires two molecules of acetyl CoA whereas nphT7 requires acetyl-CoA along with malonyl-CoA. Both reactions result in the release of CoA, though in the latter case, there is also release of CO_2. The greater 1-butanol levels observed with the AtoB pathway suggest that it is a preferred route for acetoacetyl CoA production in *E. coli*. However, it is also important to note that AtoB has a higher protein expression level in *E. coli* compared to nphT7 (Additional file 1: Figure S1), and this may have limited 1-butanol production via the *nphT7* route. When coupled with ADO, butyraldehyde is converted to propane instead of butanol, and propane production was observed in the majority of the alternative strains. Consequently, the most productive propane pathway in the wild-type *E. coli* cells was observed with the *atoB-TPC7-ADO* route (220 ± 3 μg/L). The flux towards propane through these pathways was compromised by competing native aldehyde reductase activites, as demonstrated by a significant increase in propane production and a concomitant reduction in butanol production following the deletion of two native genes encoding for such activities. In contrast to the two *TPC7* pathways, the *adhE2*-dependent routes produced significantly less propane. This observation reinforced the view that AdhE2 converts butyraldehyde to butanol in successive steps without release of the butyraldehyde, thus preventing efficient interception of the pathway by ADO towards propane.

Quantitative comparisons revealed a difference of almost 3 orders of magnitude between the titres of butanol and propane (173 *vs.* 0.2 mg/L) in the wild-type strains of AtoB-TPC7-Ahr and AtoB-TPC7-ADO. This difference can be attributed partly to the poor efficiency of the last biosynthetic step (conversion of butyraldehyde to propane). This is consistent with the earlier reports that have demonstrated *in vitro* and *in vivo* in *E. coli* that ADO has very low activity even towards the most preferred native substrates (C10-C15) and especially towards C4 substrates [10,12]. To partially alleviate this constraint, the ADO_{A143F} variant was introduced into the engineered pathways to replace native ADO. This resulted in a 1.8-fold improvement in propane, most likely reflecting more efficient utilization of the intracellular pool of available butyraldehyde. However, the remaining large difference between 1-butanol and propane levels suggests that the optimization efforts employed in the present study are still not sufficient and/or that other metabolic constraints are limiting. One obvious example, as confirmed in this study as well as others, is the reduction of ADO in the presence of a 2Fe-2S ferredoxin [27]. Ferredoxin systems are known to improve the metabolic reactions by mediating electron transfer in partner enzymes as observed here [27-29]. Such an approach could be further aided by additional reducing components such as NAD(P)H:flavodoxin/ferredoxin oxidoreductase (for example, fpr) as demonstrated in the previous work [10].

The present work introduces an alternative and conceptually different pathway for propane production in comparison to the previous work [10] despite the fact that the propane titre remained slightly lower. However, a direct comparison between the FAS and CoA pathways is not suitable, particularly since the analysed experimental parameters differed between the two studies. Both the current, and previously reported, pathways will benefit from further optimisation strategies to improve carbon flux, involving detailed targeted metabolomics and mass balance calculation, pathway modelling and pathway refactoring along with modifications to the known limitations in the turnover rate constant of ADO.

Conclusions

We have reported here new pathways for propane production based on a fermentative butanol pathway. These are distinct from previously reported pathways that were based on fatty acid synthesis, where propane production is limited by the availability of butyraldehyde precursors and the poor activity of ADO with butyraldehyde. We have demonstrated that the new pathways are plausible alternatives for the construction of next-generation microbial propane production platforms. While the majority of the metabolic flux is still directed towards other products (that is, butanol), our work also demonstrates that there is strong potential for further optimization aimed at increasing propane production.

Materials and methods

All chemicals, solvents and standards were purchased from Sigma-Aldrich (St. Louis, MO, USA) and Fisher Scientific (Waltham, MA, USA) and were of analytical grade. Media components were obtained from Formedium (Norfolk, UK). Gene sequencing and oligonucleotide synthesis were performed by Eurofins MWG (Ebersberg, Germany). D-Glucose (GOPOD Format) assay kit was from Megazyme (Wicklow, Ireland).

Strains and plasmids

BL21 (DE3) (*fhuA2 [lon] ompT gal (λ DE3) [dcm] ΔhsdS*) cells from Novagen (Madison, WI, USA) were used for protein expression. The *ahr* (GenBank ID: ACT44688.1) and *yqhD* (GenBank ID: AAA97166.1) single and double knockout strains were generated in a previous study [10]. The structure of all plasmids is graphically depicted in Figure 7, and their preparation is described below.

pET-TPC7

The pET-TPC7 plasmid was constructed in a pETDuetT-1 vector (ColE1 replicon, Ampicillin^R, from Novagen) by replacing TE4 acyl-ACP thioesterase with a synthesized gene encoding YciA (*Haemophilus influenza*, GenBank ID: AAC22485.1) in pET-TPC4 backbone plasmid described in the previous study using restriction sites HindIII and NcoI [10]. The final plasmid contained synthesized genes encoding Sfp (maturation factor phosphopantetheinyl transferase from *Bacillus subtilis*, GenBank ID: X65610.1) and CAR (carboxylic acid reductase from *M. marinum*, GenBank ID: ACC40567.1) upstream to *yciA*.

pACYC-NHCT

The pACYC-NHCT plasmid was constructed in a pACYCDuet-1 vector (P15A replicon, Chloramphenicol^R, from Novagen) by subcloning a synthesized NcoI-BamHI gene fragment (GenScript, Piscataway, NJ, USA) carrying genes encoding NphT7 (acetoacetyl-CoA synthase from *Streptomyces* sp. CL190, GenBank ID: D7URV0.1) and Hbd (3-hydroxybutyryl-CoA dehydrogenase from *Clostridium acetobutylicum* ATCC 824, GenBank ID: P52041.2) into a synthesized pACYC construct harbouring *crt* (3-hydroxybutyryl-CoA dehydratase from *C. acetobutylicum* ATCC 824, GenBank ID: P52046.1) and *ter* (trans-2-enoyl-CoA reductase from *T. denticola* ATCC 35405, GenBank ID: Q73Q47.1).

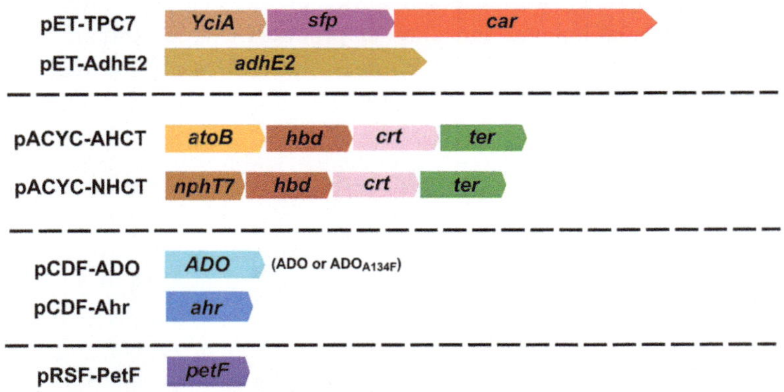

Figure 7 The plasmid design used to construct engineered propane producing pathways in *E. coli*. The structure of all plasmids used in this study for *E. coli* pathway engineering is shown. The preparation and design of these plasmids are described in the 'Materials and method' section.

pACYC-AHCT

pACYC-AHCT is a pACYCDuet-1 vector (P15A replicon, Chloramphenicol^R, from Novagen) with *atoB* (Acetyl-CoA acetyltransferase from *E. coli*, GenBank ID: P76461.1), *hbd* (3-hydroxybutyryl-CoA dehydrogenase from *C. acetobutylicum* ATCC 824, GenBank ID: P52041.2), *crt* (3-hydroxybutyryl-CoA dehydratase from *C. acetobutylicum* ATCC 824, GenBank ID: P52046.1) and *ter* (trans-2-enoyl-CoA reductase from *T. denticola* ATCC 35405, GenBank ID: Q73Q47.1) genes inserted in the order. Gene *atoB* (acetyl-CoA acetyltransferase from *E. coli*, GenBank ID: P76461.1) was PCR-amplified from *E. coli* K-12 purified genome, using primers 5′-attaggtaccAAAAATTGTGTCATCGTC AGTGCGGTAC and 5′-attaaagcTTAATTCAACCGTTC AATCACCATCGCAAT, showing complementary regions in capital letters. The *atoB* PCR fragment was then subcloned into pACYC-NHCT, thus replacing the KpnI-HindIII fragment carrying NphT7, resulting pACYC-AHCT plasmid.

pET-AdhE2

The pET-AdhE2 plasmid was constructed in a pET-Duet vector (f1 origin, Ampicillin^R, from Novagen) with *adhE2* by subcloning synthesized gene encoding AdhE2 (aldehyde-alcohol dehydrogenase, from *C. acetobutylicum* ATCC 824, GenBank ID: Q9ANR5) from a pUC57 parent vector provided by GenScript into pET-Duet vector, using restriction sites NcoI and AvrII.

pCDF-ADO and pCDF-ADO$_{A134F}$

The pCDF-ADO plasmid was constructed in a pCDFDuet-1 vector (CDF replicon, streptomycin/spectinomycin^R, from Novagen) with ADO (aldehyde deformylating oxygenase from *Prochlorococcus marinus* MIT9313, GenBank ID: Q7V6D4.1) cloned into the vector using NcoI and EcoRI restriction sites. The pCDF-ADO$_{A134F}$ vector was created by mutating the ADO insert, using A134F_forward (5′-GCA TTT GCG ATT TCT *TTT* TAT CAT ACG TAC-3′) and A134F_reverse primers (5′-GTA CGT ATG ATA *AAA* AGA AAT CGC AAA TGC-3′). The correct mutations were confirmed by complete plasmid DNA sequencing. Gene encoding ADO was originally provided by E. Neil G. Marsh (Department of Biological Chemistry, University of Michigan, USA) in a pET28b-cAD vector which was used for the previous study [30].

pCDF-Ahr

The pCDF-Ahr plasmid was constructed in a pCDFDuet-1 vector (CDF replicon, streptomycin/spectinomycin^R, from Novagen). Gene encoding Ahr (aldehyde reductase from *E. coli* GenBank ID: P27250.2) was PCR-amplified from isolated *E. coli* K-12 genomic DNA, using primers 5′ATTAATCCATGGTCTAGATAATTAATGGATCCA GGAGGAAACATATGTCGATGATAAAAAGCTATGC CGCAAAAG-3′ and 5′-ATTAATCCTAGGAAGCTTC TCGAGTCAAAAATCGGCTTTCAACACCACGCGG-3′, and cloned into using restriction sites NcoI and AvrII [8].

pRSF-PetF

The pRSF-PetF plasmid was constructed in a pRSF-Duet1 vector (RSF replicon, Kanamycin^R, from Novagen) with *fdx* (ferredoxin from *Synechocystis* sp. PCC 6803, GenBank ID: WP_010873424.1) by subcloning synthesized gene from a pUC57 parent vector provided by GenScript into a pRSF-Duet1 vector using NcoI and AvrII restriction sites.

Co-expression and introduction of the pathway in *E. coli*

The *atoB-adhE2* route was introduced into *E. coli* or the knockout cells by co-expressing pACYC-AHCT and pET-AdhE2 vectors. The *atoB-TPC7* route was introduced into *E. coli* or the knockout cells by co-expressing pACYC-AHCT and pET-TPC7 vectors. The *nphT7-adhE2* route was introduced into *E. coli* or the knockout cells by co-expressing pACYC-NHCT and pET-AdhE2 vectors. The *nphT7-TPC7* route was introduced into *E. coli* or the knockout cells by co-expressing pACYC-NHCT and pET-TPC7 vectors in the cells.

E. coli cells containing engineered pathways were further engineered by co-transforming either pCDF-ADO or pCDF-ADO_{A134F} vectors in order to introduce ADO or the ADO_{A134F} variant. The pRSF-PetF vector was co-expressed for Fdx, while pCDF-Ahr was used to introduce Ahr enzyme in the pathway. The presence of all the proteins in each individual plasmid was confirmed by SDS-PAGE and mass spectrometry analysis of the respective SDS-PAGE bands. In the case of proteins with a hexa-histidine tag, Western blotting (using WesternBreeze Chemiluminescent Immunodetection kit from Invitrogen, Carlsbad, CA, USA) was also used to analyse the expression of his-tagged proteins.

SDS-PAGE analysis of protein expression in cells containing plasmids encoding pathway components

Protein expression levels in cells containing plasmids that encode the pathway enzymes were examined by SDS-PAGE. T5 media (20 mL; 12 g tryptone, 24 g yeast extract, 4 mL glycerol, 12.5 g K_2HPO_4, 2.3 g KH_2PO_4, 20 g glucose per litre) was inoculated with 1% (*v*/*v*) transformed *E. coli* cells and incubated at 37°C (180 rpm) until the optical density at 600 nm (OD_{600nm}) reached 0.5. Cultures were then induced with isopropyl β-D-1-thiogalactopyranoside (final concentration of 0.5 mM). Cultures were grown for a further 24 h at 30°C (180 rpm). Samples (200 μL) from the cultures were taken for SDS PAGE analysis. Samples were taken before isopropyl β-D-1-thiogalactopyranoside (IPTG) induction and after 4 or 24 h of IPTG induction and cells harvested by centrifugation. Samples were electrophoresed in 12% RunBlue precast SDS-PAGE gels (Expedeon, Cambridge, UK). Protein bands were visualized by staining with Instant Blue protein stain (Expedeon).

Media, cultivation and detection of propane and butanol

Lysogeny broth (LB) liquid media (10 mL) was inoculated using *E. coli* glycerol stocks (from −80°C) and incubated at 37°C overnight at 180 rpm. Of T5 media, 50 mL (12 g tryptone, 24 g yeast extract, 4 mL glycerol, 12.5 g K_2HPO_4, 2.3 g KH_2PO_4, 20 g glucose per litre) was inoculated with 1% (*v*/*v*) of the inoculum and kept for incubation at 37°C (180 rpm) until the optical density at 600 nm (OD_{600}) reached 0.5. The cultures were then induced with IPTG to a final concentration of 0.5 mM. The cell cultures were further grown for 4 h at 30°C and 180 rpm to prepare the samples for propane detection. In the case of total butanol produced in the butanol pathway, the culture was left at 30°C (180 rpm) for 72 h, while for residual butanol detection, samples were taken from this culture when OD_{600} reached 1.5. Control cultures were made using untransformed *E. coli* strain BL21 DE3 cells.

For propane formation analysis, 50 mL cell culture was centrifuged at 4,000 rpm, and the supernatant was discarded. The cell pellets were resuspended in a 6.25-mL T5 media with 0.5 mM IPTG, and 500 μL of the resuspended culture was transferred into 2-mL crimp sealed GC vial and used for propane analysis. The vials were incubated at 30°C, with shaking at 180 rpm for 3 h. Headspace of 1.0 mL from the cultures grown in the GC vial was manually removed and injected into the GC with a gas tight syringe. Propane detection was carried out using a Varian 3800 GC (Palo Alto, CA, USA) equipped with a DB-WAX column (30 m × 0.32 mm × 0.25 μM film thickness, J&W Scientific, Santa Clara, MA, USA). The injector temperature was 250°C with a split ratio of 10:1. The column temperature was set from 40°C hold for 2 min to 100°C at 20°C/min with helium flow at 1 mL/min and FID temperature at 250°C. Propane peak was identified by comparing it with the analytical propane standard, and quantification was done using a propane calibration curve.

For residual butanol detection, 50 mL liquid culture was spun down at 4,000 rpm for 10 min. Of the supernatant, 500 μL was extracted with 500 μL of ethyl acetate containing 0.2% hexane as internal standard and dried over $MgSO_4$. Of the sample, 1 μL was analysed in GC using a Varian 3800 GC equipped with a HP-5 column (30 m × 0.32 mm × 0.25 μM film thickness, JW Scientific). The injector temperature was 250°C with a split ratio of 20:1. The column temperature was set from 40°C hold for 1 min to 280°C at 20°C/min with helium flow at 1 mL/min and FID temperature at 250°C. Butanol peak was identified by comparing with the analytical butanol standard, and quantification was done using a butanol calibration curve.

Additional file

Additional file 1: Additional information. Figure S1. *SDS-PAGE analysis of overexpressed proteins in E. coli cells transformed with plasmids pACYC-AHCT, pACYC-NHCT and pCDF-Ahr.* The SDS-PAGE gel shows IPTG uninduced and induced *E. coli* cells transformed with pACYC-AHCT, pACYC-NHCT and pCDF-Ahr vectors (sections I, II and III, respectively). Lane 1 is the standard protein marker and in each section, lane 2 shows uninduced cells, lane 3 shows IPTG induced cells after 4 h and lane 4 is the IPTG induced cells after 24 h. AtoB (40.5 kDa); NphT7,(34.6 kDa); Hbd (30.6 kDa); Crt (28.2 kDa); Ter (43.8 kDa) and Ahr (37.8 kDa) are marked. **Figure S2.** *SDS-PAGE analysis of overexpressed proteins in E. coli cells transformed with pET-adhE2 and pET-TPC7 vectors.* The SDS-PAGE gel shows IPTG uninduced and induced *E. coli* cells transformed with pET-AdhE2 and pET-TPC7 vectors (shown in sections I and II, respectively). Lane 1 is the standard protein marker and in each section, lane 2 shows uninduced cells and lane 3 is the IPTG induced cells after 24 h. The band positions of AdhE2 (94.4 kDa); CAR (128.9 kDa); sfp (26.2 kDa) and YciA (16.7 kDa) are shown. **Figure S3.** *SDS-PAGE analysis of overexpressed proteins in E. coli cells transformed with pCDF-ADO and pRSF-PetF vectors.* The SDS-PAGE gel shows IPTG uninduced and induced *E. coli* cells transformed with pCDF-ADO and pRSF-PetF vectors (shown in sections I and II respectively). Lane 1 is the standard protein marker and in each section, lane 2 shows uninduced cells, and lane 3 is the IPTG induced cells after 24 h. The positions of ADO (29.3 kDa) and *Fdx*, ferredoxin (10.4 kDa) are shown. **Figure S4.** *Residual butanol produced in E. coli and knockout strains engineered to harbour the atoB-TPC7 pathway.* Residual butanol produced in the pathway engineered *E. coli* or knockout strains with wild-type ADO or with the ADO_{A134F} variant enzyme are shown. *E. coli and Δahr/ΔyqhD* single or double knockout strains were engineered to contain the *atoB-TPC7* route (indicated by red colour). The effect of wild-type ADO or the ADO_{A134F} variant enzyme with or without *Synechocystis* ferredoxin (Fdx) was also analysed. **Figure S5.** *Residual butanol produced in E. coli and knockout strains engineered to harbour the nphT7-TPC7 pathway.* Residual butanol produced in the pathway engineered *E. coli* or knockout strains with wild-type ADO or with the ADO_{A134F} variant enzyme are shown. *E. coli and Δahr /ΔyqhD* single or double knockout strains were engineered to contain the *nphT7-TPC7* route (indicated by green colour). The effect of wild-type ADO or the ADO_{A134F} variant enzyme with or without *Synechocystis* ferredoxin (Fdx) was also analysed. **Figure S6.** *GC traces and the corresponding propane peak integrated for pathway analysis.* GC traces showing the integrated propane peak for the *AtoB-TPC7* route containing *E. coli BL21* strains coexpressed with ADO and ferredoxin (trace a), *AtoB-TPC7* route containing *E. coli BL21* strains coexpressed with ADO (trace b), *AtoB-TPC7* route engineered in *E. coli BL21* strains coexpressed with ferredoxin in absence of ADO (trace c), *AtoB-TPC7* route engineered in *E. coli BL21* strains (trace d) and control *E. coli BL21* cells without pathway engineering (trace e) are shown. **Figure S7.** *Time course for residual butanol accumulation in* large-scale cultures of the best propane-producing pathways. The *atoB-TPC7* route (indicated by red symbols) and *nphT7-TPC7* (indicated by green symbols) engineered in *ΔyqhD* knockout cells in the presence of the ADO_{A134F} variant and ferredoxin system were analysed at larger scale.

Abbreviations

AdhE2: aldehyde-alcohol dehydrogenase; ADO: aldehyde deformylating oxygenase; ahr: aldehyde reductase; CAR: carboxylic acid reductase; FAS: fatty acid synthesis; Fdx: ferredoxin; GC: gas chromatography; IPTG: isopropyl β-D-1-thiogalactopyranoside; OD_{600}: optical density at 600 nm; Ter: trans-2-enoyl-CoA reductase.

Competing interests

The authors declare that they have no competing interests.

Authors' contributions

NM and BRKM performed the experiments, analysed the results, interpreted the data and produced the initial draft of the manuscript. AP and PK prepared plasmids. KF participated in scale up reaction. NM, AP, PK and DL participated in the design of the study. AP, PK, and MKA helped in drafting the manuscript. NSS and PRJ conceived, coordinated the study and revised the manuscript. All authors approved the final manuscript.

Acknowledgements

The work was supported by a grant from the European Union FP-7 256808 to NSS and PRJ. NSS is an Engineering and Physical Sciences Research Council Established Career Fellow.

Author details

[1]BBSRC/EPSRC Centre for Synthetic Biology of Fine and Speciality Chemicals, Manchester Institute of Biotechnology, Faculty of Life Sciences, 131 Princess Street, The University of Manchester, Manchester M1 7DN. UK. [2]Molecular Plant Biology, Department of Biochemistry, Tykistökatu 6A 6krs, University of Turku, FI 20014 TURUN YLIOPISTO, Turku, Finland. [3]Department of Life Sciences, Imperial College London, Sir Alexander Fleming Building, South Kensington Campus, London SW7 2AZ, UK.

References

1. Hedley M, Jiang W, McLaren R, Singleton DL. Modeling of future-year emissions control scenarios for the lower Fraser Valley: impacts of natural gas and propane vehicle technologies. J Appl Meteorol. 1998;37:1190–204.
2. Shelley C. The story of LPG. London: Poten and Partners: 2003.
3. Laciak M. Properties of artificial gaseous mixtures for their safe use and support the natural gas supply networks. Arch Min Sci. 2012;57:351–62.
4. Thorsteinsson HH, Tester JW. Barriers and enablers to geothermal district heating system development in the United States. Energ Policy. 2010;38:803–13.
5. Keller FJ, Liang HM, Farzad M. Assessment of propane as a refrigerant in residential air-conditioning and heat pump applications. Refrigerants for the 21st Century. 1997. p. 57–65.
6. Velasco I, Rivas C, Martinez-Lopez JF, Blanco ST, Otin S, Artal M. Accurate values of some thermodynamic properties for carbon dioxide, ethane, propane, and some binary mixtures. J Phys Chem B. 2011;115:8216–30.
7. Schirmer A, Rude MA, Li XZ, Popova E, del Cardayre SB. Microbial biosynthesis of alkanes. Science. 2010;329:559–62.
8. Akhtar MK, Turner NJ, Jones PR. Carboxylic acid reductase is a versatile enzyme for the conversion of fatty acids into fuels and chemical commodities. Proc Natl Acad Sci U S A. 2013;110:87–92.
9. Howard TP, Middelhaufe S, Moore K, Edner C, Kolak DM, Taylor GN, et al. Synthesis of customized petroleum-replica fuel molecules by targeted modification of free fatty acid pools in Escherichia coli. Proc Natl Acad Sci U S A. 2013;110:7636–41.
10. Kallio P, Pasztor A, Thiel K, Akhtar MK, Jones PR. An engineered pathway for the biosynthesis of renewable propane. Nat Commun. 2014;5:4731.
11. Lang K, Zierow J, Buehler K, Schmid A. Metabolic engineering of Pseudomonas sp. strain VLB120 as platform biocatalyst for the production of isobutyric acid and other secondary metabolites. Microb Cell Fact. 2014;13:2.
12. Khara B, Menon N, Levy C, Mansell D, Das D, Marsh ENG, et al. Production of propane and other short-chain alkanes by structure-based engineering of ligand specificity in aldehyde-deformylating oxygenase. Chembiochem. 2013;14:1204–8.
13. Coates RC, Podell S, Korobeynikov A, Lapidus A, Pevzner P, Sherman DH, et al. Characterization of cyanobacterial hydrocarbon composition and distribution of biosynthetic pathways. Plos One. 2014;9:e85140.
14. Atsumi S, Higashide W, Liao JC. Direct photosynthetic recycling of carbon dioxide to isobutyraldehyde. Nat Biotechnol. 2009;27:1177–80.
15. Bond-Watts BB, Bellerose RJ, Chang MCY. Enzyme mechanism as a kinetic control element for designing synthetic biofuel pathways. Nat Chem Biol. 2011;7:222–7.
16. Dellomonaco C, Clomburg JM, Miller EN, Gonzalez R. Engineered reversal of the beta-oxidation cycle for the synthesis of fuels and chemicals. Nature. 2011;476:355–U131.
17. Lan EI, Liao JC. Metabolic engineering of cyanobacteria for 1-butanol production from carbon dioxide. Metab Eng. 2011;13:353–63.
18. Lan EI, Liao JC. ATP drives direct photosynthetic production of 1-butanol in cyanobacteria. Proc Natl Acad Sci U S A. 2012;109:6018–23.
19. Lan EI, Ro SY, Liao JC. Oxygen-tolerant coenzyme A-acylating aldehyde dehydrogenase facilitates efficient photosynthetic n-butanol biosynthesis in cyanobacteria. Energ Environ Sci. 2013;6:2672–81.
20. Pasztor A, Kallio P, Malatinszky D, Akhtar MK, Jones PR. A synthetic O-tolerant butanol pathway exploiting native fatty acid biosynthesis in Escherichia coli. Biotechnol Bioeng. 2015;112:120–8.
21. Shen CR, Lan EI, Dekishima Y, Baez A, Cho KM, Liao JC. Driving forces enable high-titer anaerobic 1-butanol synthesis in Escherichia coli. Appl Environ Microb. 2011;77:2905–15.
22. Fontaine L, Meynial-Salles I, Girbal L, Yang XH, Croux C, Soucaille P. Molecular characterization and transcriptional analysis of adhE2, the gene encoding the NADH-dependent aldehyde/alcohol dehydrogenase responsible for butanol production in alcohologenic cultures of Clostridium acetobutylicum ATCC 824. J Bacteriol. 2002;184:821–30.
23. Zhuang ZH, Song F, Zhao H, Li L, Cao J, Eisenstein E, et al. Divergence of function in the hot dog fold enzyme superfamily: the bacterial thioesterase YciA. Biochemistry-Us. 2008;47:2789–96.
24. Rodriguez GM, Atsumi S. Toward aldehyde and alkane production by removing aldehyde reductase activity in Escherichia coli. Metab Eng. 2014;25:227–37.
25. Davis MS, Cronan JE. Inhibition of Escherichia coli acetyl coenzyme A carboxylase by acyl-acyl carrier protein. J Bacteriol. 2001;183:1499–503.
26. James ES, Cronan JE. Expression of two Escherichia coli acetyl-CoA carboxylase subunits is autoregulated. J Biol Chem. 2004;279:2520–7.
27. Yee BC, Delatorre A, Crawford NA, Lara C, Carlson DE, Buchanan BB. The ferredoxin-thioredoxin system of enzyme regulation in a cyanobacterium. Arch Microbiol. 1981;130:14–8.

28. Buckel W, Thauer RK. Energy conservation via electron bifurcating ferredoxin reduction and proton/Na + translocating ferredoxin oxidation. Biochimica Et Biophysica Acta-Bioenergetics. 2013;1827:94–113.
29. Zhang JJ, Lu XF, Li JJ. Conversion of fatty aldehydes into alk (a/e)nes by in vitro reconstituted cyanobacterial aldehyde-deformylating oxygenase with the cognate electron transfer system. Biotechnol Biofuels. 2013;6:86.
30. Eser BE, Das D, Han J, Jones PR, Marsh ENG. Oxygen-independent alkane formation by non-heme iron-dependent cyanobacterial aldehyde decarbonylase: investigation of kinetics and requirement for an external electron donor. Biochemistry-US. 2011;50:10743–50.

Dynamic metabolic profiling of the marine microalga *Chlamydomonas* sp. JSC4 and enhancing its oil production by optimizing light intensity

Shih-Hsin Ho[1], Akihito Nakanishi[1], Xiaoting Ye[1], Jo-Shu Chang[3,4,5], Chun-Yen Chen[5], Tomohisa Hasunuma[1*] and Akihiko Kondo[2,6]

Abstract

Background: Marine microalgae are among the most promising lipid sources for biodiesel production because they can be grown on nonarable land without the use of potable water. Marine microalgae also harvest solar energy efficiently with a high growth rate, converting CO_2 into lipids stored in the cells. Both light intensity and nitrogen availability strongly affect the growth, lipid accumulation, and fatty acid composition of oleaginous microalgae. However, very few studies have systematically examined how to optimize lipid productivity by adjusting irradiance intensity, and the metabolic dynamics that may lead to improved lipid accumulation in microalgae have not been elucidated. Little is known about the mechanism of lipid synthesis regulation in microalgae. Moreover, few studies have assessed the potential of using marine microalgae as oil producers.

Results: In this work, a newly isolated marine microalga, *Chlamydomonas* sp. JSC4, was selected as a potential lipid producer, and the effect of photobioreactor operations on cell growth and lipid production was investigated. The combined effects of light intensity and nitrogen depletion stresses on growth and lipid accumulation were further explored in an effort to markedly improve lipid production and quality. The optimal lipid productivity and content attained were 312 mg L^{-1} d^{-1} and 43.1% per unit dry cell weight, respectively. This lipid productivity is the highest ever reported for marine microalgae. Metabolic intermediates were profiled over time to observe transient changes during lipid accumulation triggered by combined stresses. Finally, metabolite turnover was also assessed using an *in vivo* ^{13}C-labeling technique to directly measure the flow of carbon during lipid biosynthesis under stress associated with light intensity and nitrogen deficiency.

Conclusions: This work demonstrates the synergistic integration of cultivation and dynamic metabolic profiling technologies to develop a simple and effective strategy for enhancing oil production in a marine microalga. The knowledge obtained from this study could be useful in assessing the feasibility of marine microalgae biodiesel production and for understanding the links between dynamic metabolic profiles and lipid biosynthesis during the course of microalgal cultivation.

Keywords: Microalgae, *Chlamydomonas* sp, Lipids, Biodiesel, Metabolite profiling, Carbon flux, Light intensity, Nitrogen depletion

* Correspondence: hasunuma@port.kobe-u.ac.jp
[1]Organization of Advanced Science and Technology, Kobe University, 1-1 Rokkodai, Nada-ku, Kobe 657-8501, Japan
Full list of author information is available at the end of the article

Background

Due to the impact of the fast-growing global demand for fossil fuels on global climate change, considerable attention has been paid to reducing CO_2 emissions by developing new sustainable energy sources as alternatives to fossil fuels [1]. Among the various potential renewable energy sources, biofuels produced from biomass are of the most interest and are expected to play an important role in the near future [2].

Biodiesel is an ideal renewable energy source because it is non-toxic and biodegradable and leads to lower CO_2 emissions [3]. Microalgae-produced biodiesel is one of the most promising candidates, as some oleaginous microalgae can harvest solar energy efficiently due to their high growth rate and convert CO_2 into lipids stored in the cells [4]. Compared with terrestrial plants, microalgae can grow 10 to 50 times faster, resulting in an extremely high CO_2 fixation rate [5,6]. In addition, some oleaginous microalgae, such as *Chlorella, Chlamydomonas, Neochloris,* and *Dunaliella* species, can accumulate lipid up to 50% per unit dry weight of biomass, and the lipid composition is suitable for making biodiesel [7-10]. Furthermore, microalgal growth is not seasonally limited and can take place on nonarable land using municipal wastewater, brackish water, or seawater as a nutrient source [11,12]. The abovementioned characteristics make microalgal lipids promising raw materials for biodiesel synthesis.

Although microalgae seem to have a high potential for biodiesel production, obstacles that may hinder their industrial application remain. For instance, lipid accumulation in microalgae often occurs under environmental stress (for example, nitrogen depletion), which often leads to slower growth that results in lower lipid productivity [7,13]. Moreover, for large-scale microalgal cultivation, a tremendous amount of fresh water is required if the microalgae cannot grow well in brackish water, recycled water, or seawater. This may cause a shortage of available water resources, which are already very limited worldwide [7,14]. Therefore, cultivation strategies that result in optimal lipid content/productivity should be applied, and marine strains should be selected that are capable of rapid growth in saline environments and production of a high content of appropriate lipids.

To enhance the commercial feasibility of microalgae-based biodiesel production, the cultivation of lipid-producing microalgae must be made more efficient, with particular emphasis on optimization of several key factors that significantly affect the yield and properties of biodiesel, such as lipid productivity and composition. Recent studies have noted that suitable light intensity is very effective in promoting growth and lipid accumulation and for improving lipid quality in oleaginous microalgae [15,16]. However, no reports of systematic studies of how to optimize lipid productivity by adjusting irradiance intensity or studies profiling metabolite dynamics associated with improved lipid accumulation in microalgae have been published. In addition, relatively few studies examining the potential use of marine microalgae as oil producers have been published.

In oleaginous microalgae, lipids are synthesized from acetyl-CoA precursors during periods of light exposure through the actions of several enzymes composing the fatty acid synthesis pathway [17]. Although light stress and nitrogen depletion are known to strongly influence growth and lipid accumulation in microalgae [16,18,19], little is known regarding the mechanism of lipid synthesis regulation in microalgae. In addition, a thorough understanding of metabolite flux during lipid synthesis is also lacking.

Metabolic profiling is now a proven and powerful tool for gaining insights into functional biology [20,21]. The comprehensive analysis of a wide range of metabolites in microorganisms using mass spectrometry makes it possible to identify and quantify target metabolites that play key roles in specific biological pathways, thereby revealing the dynamic behavior of biological systems [20,22]. However, this metabolomic technology has yet to be widely applied to microalgae, particularly oleaginous microalgae [7]. To date, microalgal metabolomic studies have primarily focused on the metabolic responses of well-known species (for example, *Chlamydomonas reinhardtii*) to nitrogen/nutrient depletion or variations in CO_2 concentration [23-25]. However, there have been no reports of the metabolic profiling of microalgae cultivated under the combined stress of deficiencies in irradiance and nitrogen availability.

Targeting metabolites using ^{13}C-labeling technologies is another useful strategy for studying biological systems *in vivo* because it provides an overview of metabolite flux through different pathways [26]. However, most studies reported to date have merely obtained 'snapshot' metabolic profiles in microalgae; consequently, little is known about metabolic turnover or pathways in these organisms [22].

Salinity-tolerant microalgae capable of high growth rates and accumulation of high lipid content are rare in algae collection centers around the world. In this study, a newly isolated marine microalga, *Chlamydomonas* sp. JSC4, was used as a candidate lipid producer because the lipid content of this strain can reach over 45% to 50% of the dry cell weight (DCW) under environmental stress, as illustrated in Figure 1. Light supply is one of the most important factors affecting growth rate and lipid accumulation in phototrophic growth of microalgae [16,27]. Therefore, we initially explored the optimal cultivation conditions for microalgae-based biodiesel production by investigating the combined effects of photobioreactor

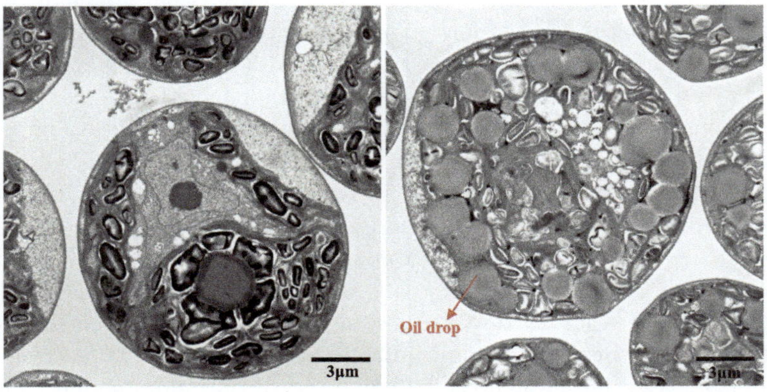

Figure 1 **Morphology of the oil body in *Chlamydomonas* sp. JSC4 cultivated under nitrogen-replete or nitrogen-deficient conditions (light intensity = 250 μmol m^{-2} s^{-1}; CO_2 aeration = 2.0%; CO_2 flow rate = 0.1 vvm).**

(PBR) type, light intensity, and nitrogen depletion. The metabolic profile of *Chlamydomonas* sp. JSC4 under light effect and nitrogen depletion was also monitored to identify the intercorrelations between light intensity, nitrogen depletion, and lipid accumulation. Finally, metabolic turnover in *Chlamydomonas* sp. JSC4 was assessed using an *in vivo* ^{13}C-labeling technique to directly measure the flow of carbon during lipid biosynthesis under conditions of different light intensity and nitrogen deficiency. The aim of this work was to determine the optimal light intensity necessary to maximize the production of lipids of suitable composition for biodiesel production by analyzing the flux of metabolites in *Chlamydomonas* sp. JSC4. The knowledge obtained from this study could be useful in assessing the applicability of this strain as biodiesel feedstock and for enhancing understanding of metabolite dynamics during lipid biosynthesis under stress.

Results and discussion

Effect of PBR type on growth rate, CO_2 fixation, and lipid accumulation in *Chlamydomonas* sp. JSC4

The growth rate and cellular composition of microalgae vary significantly with the illumination conditions during phototrophic growth [16]. In the PBRs used in our previous study [7], light intensity tended to decrease when the biomass concentration increased due to light-shielding effects, causing a significant decrease in growth rate and changes in cellular composition. Therefore, in this study, a new type of PBR was designed (designated a slim-type tubular PBR) to improve the light-receiving area of the PBR from 104.4 to 200.2 cm^2. The target strain, *Chlamydomonas* sp. JSC4, was grown in both conventional and the newly designed PBRs illuminated with a one-side light intensity of 150 μmol m^{-2} s^{-1}, with 2% CO_2 serving as the carbon source.

As shown in Figure 2 and Table 1, the growth rate increased when using the slim-type tubular PBR, resulting in a higher CO_2 fixation rate of 754 mg L^{-1} d^{-1}, which was up to 40% higher than that achieved with a conventional-type tubular PBR. This phenomenon might have been due to the higher surface area to volume (S:V) ratio and shorter light path of the slim-type tubular PBR. The slim-type tubular PBR appears to have a greater light-receiving area and greater light penetration efficiency. Similar results were also observed in other research, demonstrating that a PBR with a high S:V ratio (80 to 100 m^2 m^{-3}) has a shorter light path and eliminates dark areas in the cell broth, thus leading to increased biomass concentration [28]. With respect to lipid production, Figure 2 shows that the type of PBR had no significant effect on the lipid content of *Chlamydomonas* sp. JSC4. Lipid productivity was thus positively proportional to biomass productivity. The maximum lipid productivity achieved with the slim-type tubular PBR was 194 mg L^{-1} d^{-1} (Table 1), which was significantly higher than that obtained with a conventional PBR. The above results indicate that the slim-type tubular PBR is more effective for growing strain JSC4, and thus this type or reactor was used in subsequent experiments.

Light intensity-dependent growth kinetics of *Chlamydomonas* sp. JSC4

For most photoautotrophic microalgae, the growth rate depends heavily on irradiation intensity [15,16]. Considering that *Chlamydomonas* sp. JSC4 was isolated from a coastal area in tropical Taiwan, it was thought that the light intensity of 150 μmol m^{-2} s^{-1} used in the experiments described above (Figure 2) would not be optimal for growth. Therefore, strain JSC4 was cultivated in the slim-type tubular PBR illuminated at different light intensities to determine the optimal light intensity. As shown in Figure 3 and Table 2, the specific growth rate

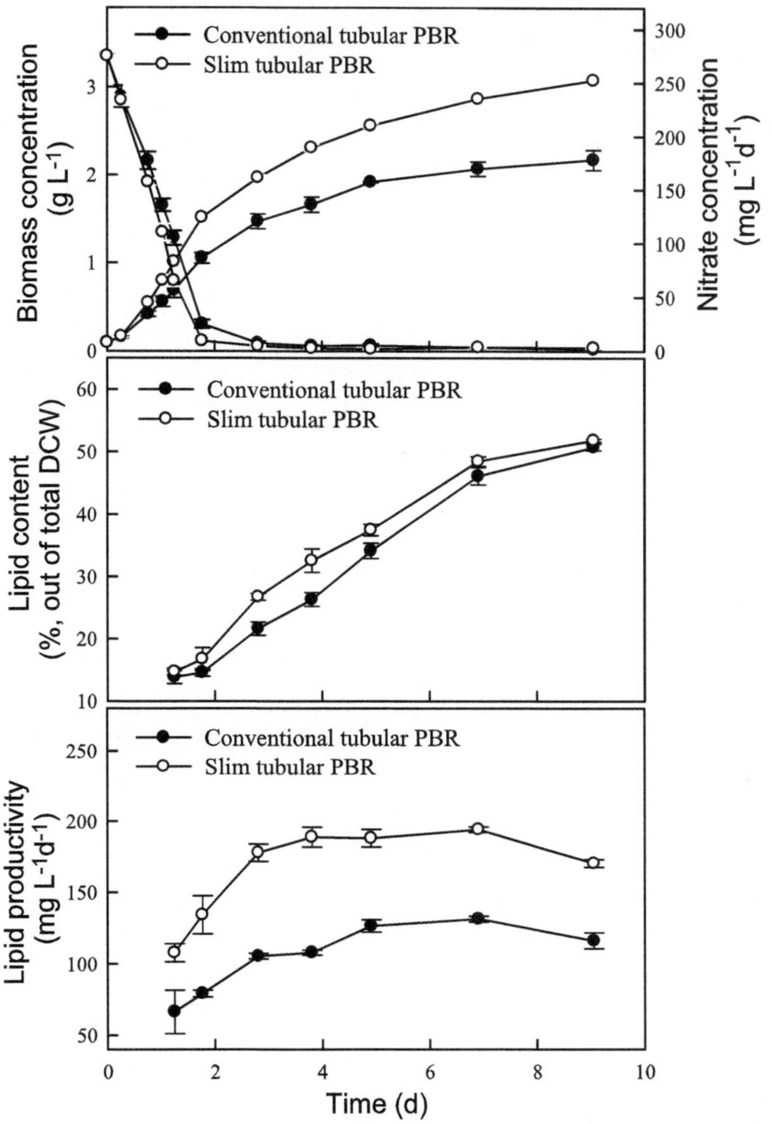

Figure 2 Performance (growth, lipid content, and lipid productivity) of *Chlamydomonas* sp. JSC4 cultivated using different PBR types. Error bars indicate standard deviation of three replicated experiments (light intensity = 150 μmol m^{-2} s^{-1}; CO_2 aeration = 2.0%; CO_2 flow rate = 0.1 vvm).

of strain JSC4 increased rapidly as the light intensity increased from 100 to 300 μmol m^{-2} s^{-1}, reaching a *plateau* at light intensities greater than 300 μmol m^{-2} s^{-1}. The highest specific growth rate was around 3.2 d^{-1} in the light-saturation region (300 to 500 μmol m^{-2} s^{-1}), which was significantly higher than that found in our previous work [7] and other related studies [15,16]. In addition, Table 2 shows that the specific growth rate of strain JSC4 at different light intensities could be fitted via Monod and Haldane models. The results fit the

Table 1 Growth, lipid accumulation, and CO_2-fixation performance of marine *Chlamydomonas* sp. JSC4

PBR type	PBR diameter (cm)	S:V[a] ratio (m^2 m^{-3})	Light-receiving area (cm^2)	Biomass productivity (mg L^{-1} d^{-1})	Lipid content (%)	Lipid productivity (mg L^{-1} d^{-1})	CO_2 fixation rate (mg L^{-1} d^{-1})
Conventional	9.5	41.8	104.4	285 ± 13	46.1 ± 1.4	132 ± 2	537 ± 24
Slim	5.0	80.1	200.2	401 ± 10	48.5 ± 0.8	194 ± 2	754 ± 19

Chlamydomonas sp. JSC4 cultivated for 7 days (5 days of nitrogen depletion) in different photobioreactors (PBRs). Values are the mean ± standard deviation of three replicated experiments (light intensity = 200 μmol m^{-2} s^{-1}; CO_2 aeration = 2%; CO_2 flow rate = 0.1 vvm).
[a]Surface area to volume ratio.

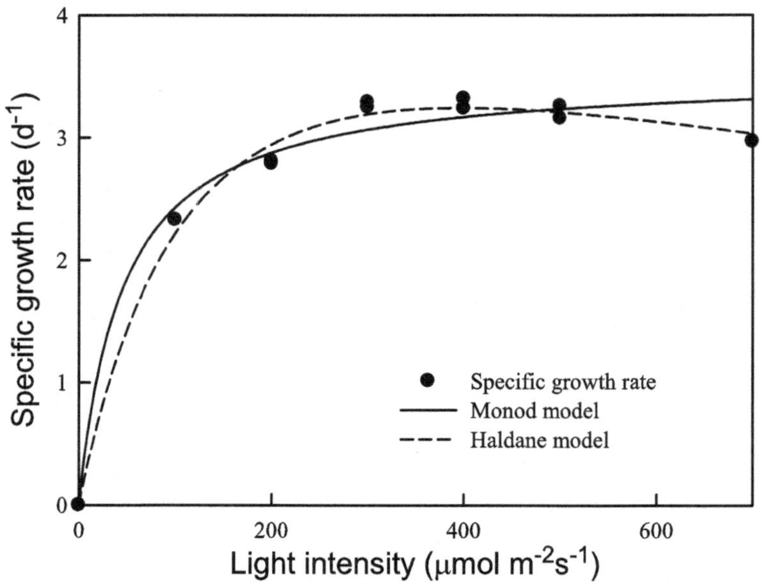

Figure 3 Light intensity-dependent growth kinetics and model simulations of *Chlamydomonas* sp. JSC4 cultivated in a slim-type tubular PBR (CO_2 aeration = 2.0%; CO_2 flow rate = 0.1 vvm).

Haldane model with better agreement (R^2 = 0.991), indicating that slight light inhibition may have occurred. However, due to the high K_I value of 1,097 μmol m^{-2} s^{-1}, strain JSC4 appears to have a high tolerance to the inhibitory effects of high light intensity. These data indicate that strain JSC4 has good potential for outdoor culturing throughout the year in subtropical and tropical areas.

Combined effects of light intensity and nitrogen depletion on growth, CO_2 fixation, and lipid accumulation in *Chlamydomonas* sp. JSC4

To enhance the economic feasibility of microalgae-based biodiesel production, lipid synthesis should be maximized to reach a satisfactory lipid content. In addition, biomass productivity should be enhanced to ensure high lipid productivity. It is known that many microalgae can rapidly accumulate and store large amounts of lipids as an energy source in response to environmental stresses that are unfavorable to growth, such as high light intensity [19], nitrogen depletion [4,18], and high salinity [7]. Both salinity and nitrogen depletion have been shown to be critical factors that affect growth and lipid accumulation in microalgae.

Table 2 Estimated kinetic parameters of Monod- and Haldane-type models

Model	Estimated parameters			
	μ_{max}	K_s	K_I	R^2
Monod	3.52	45.2	N.A.	0.972
Haldane	5.58	143.2	1,097	0.991

Based on the effect of light intensity on the growth of *Chlamydomonas* sp. JSC4 (CO_2 aeration = 2%; CO_2 flow rate = 0.1 vvm).

Our previous study indicated that cultivation of *Chlamydomonas* sp. JSC4 in 2% sea salt for 3 days under nitrogen-depletion conditions can enhance lipid productivity [7]. Furthermore, irradiance also strongly influences growth and lipid accumulation in some microalgae, including *Scenedesmus* [16], *Nannochloropsis* [19], *Pavlova* [29], and *Desmodesmus* [4]. As our previous study showed that strain JSC4 has a high potential to serve as an oil feedstock for biodiesel production [7], optimization of the irradiance conditions for this strain to achieve higher lipid productivity is of great importance. Obtaining a thorough understanding of the synergistic effects of light intensity control and nitrogen depletion is also very important. Another variable that must be evaluated is the cultivation time required to achieve the highest lipid production efficiency.

In this study, strain JSC4 was cultivated under different light intensities for 8 to 9 days, with regular monitoring of biomass and nitrate concentrations, lipid content, lipid productivity, and CO_2 fixation rate. As shown in Figures 4 and 5, the rates of growth and CO_2 fixation increased significantly as the light intensity increased from 150 to 300 μmol m^{-2} s^{-1}. However, a further increase in light intensity to 500 μmol m^{-2} s^{-1} resulted in a slight decrease in both the growth and CO_2 fixation rates, suggesting that excessive illumination inhibits growth and CO_2 fixation in this species, a phenomenon commonly known as the photo-inhibition effect. For instance, Ho *et al.* reported that under high light intensity (540 μmol m^{-2} s^{-1}), the growth and CO_2 fixation rates in *Scenedesmus obliquus* decline markedly [16]. In our experiments, a maximum CO_2 fixation rate of 2,886 mg L^{-1} d^{-1} was obtained at a light intensity of 300 μmol m^{-2} s^{-1} during the optimal time period indicated

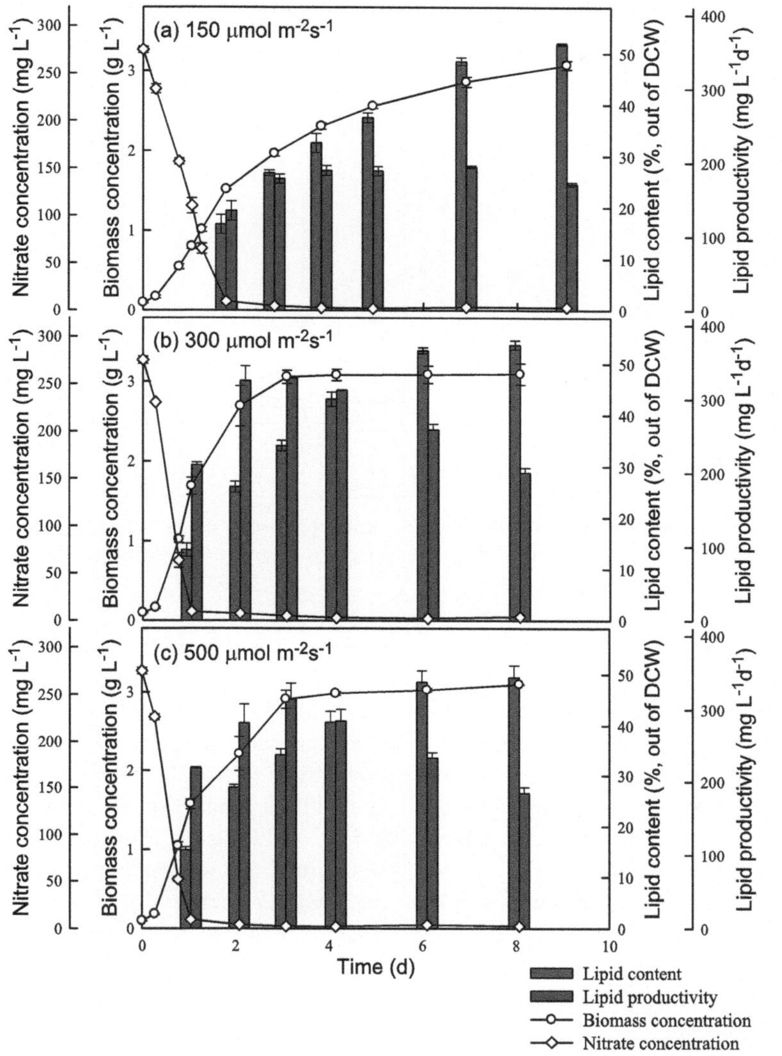

Figure 4 Time-course profiles (a, b, c) of biomass concentration, nitrate concentration, lipid content, and lipid productivity during growth of *Chlamydomonas* sp. JSC4 illuminated with various light intensities. Error bars indicate standard deviation of three replicated experiments (CO_2 aeration = 2.0%; CO_2 flow rate = 0.1 vvm).

in Figure 5. This performance was superior to that obtained both in our previous study [7] and that of most related studies [6]. This excellent CO_2 fixation performance makes strain JSC4 a potential candidate for use in microalgae-based CO_2 mitigation processes for real industrial fuel gases.

With respect to lipid production in strain JSC4, Figure 4 shows that lipid productivity increased slightly as the light intensity increased from 150 to 300 μmol m^{-2} s^{-1}, due primarily to enhanced growth associated with the increase in light intensity. In addition, previous reports showed that high light intensity may lead to an increase in the amount of neutral storage lipids, primarily triacylglycerol (TAG) derived from fatty acids (referred to as lipids in this study) [17,29]. However, in our study, a further increase in light intensity to 500 μmol m^{-2} s^{-1} resulted in a slight drop in lipid content (Figure 4), indicating that excess light intensity inhibits lipid biosynthesis. Moreover, at all light intensities we examined, lipid content increased significantly during stationary phase (during which nitrogen deficiency occurred), suggesting that lipid accumulation in strain JSC4 is not only triggered by the supply of irradiance but also by the stress of nitrogen depletion. Pal *et al.* [19] also demonstrated that exposing microalgae to more than one stress simultaneously (for example, high salinity and nitrogen depletion) may be necessary for further enhancement of lipid accumulation. Therefore, as shown in Figures 4 and 5, appropriately increasing the light intensity to promote growth in combination with limiting the nitrogen supply to trigger lipid accumulation can lead to a simultaneous increase in both the lipid content and productivity in strain JSC4. At a light intensity of 300 μmol m^{-2} s^{-1}, strain

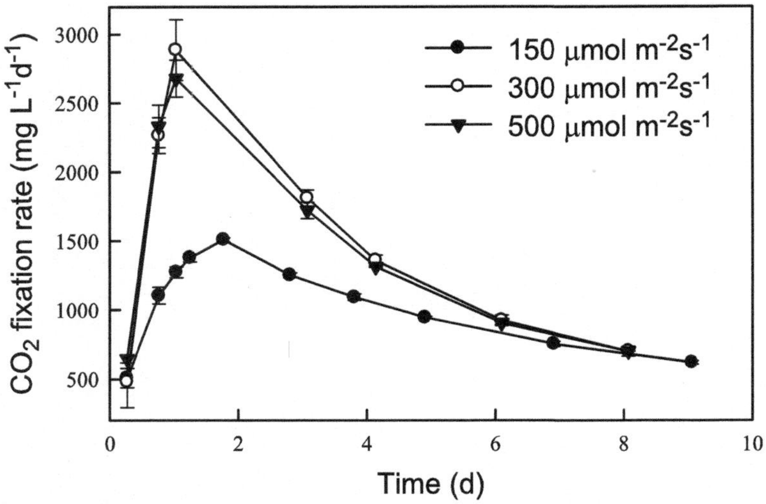

Figure 5 Time-course profiles of CO_2 fixation rate during the growth of *Chlamydomonas* sp. JSC4 illuminated with various light intensities. Error bars indicate standard deviation of three replicated experiments (CO_2 aeration = 2.0%; CO_2 flow rate = 0.1 vvm).

JSC4 displayed a maximum lipid productivity of 328 mg L^{-1} d^{-1} and lipid content of 34.1%.

From an engineering perspective, lipid productivity is absolutely one of the most important performance indicators for evaluating the commercialization potential of lipid-producing microalgae. Moreover, the lipid content is also important when considering the operation costs of downstream lipid extraction and purification processes [16]. Normally, the lipid content of microalgae tends to increase when the nitrogen starvation period is extended, until the maximum lipid content is reached. However, such a prolonged cultivation time leads to a decrease in lipid productivity. Therefore, it was vital to determine the optimal incubation time for strain JSC4 under nitrogen depletion as well as the timing of cell harvesting to ensure the best lipid production performance. At the optimal light intensity determined in this study (300 μmol m^{-2} s^{-1}), the lipid content increased sharply as the nitrogen depletion time increased, until reaching a maximum value of 53.7% after 7 days (Table 3).

Table 3 The effects of duration of nitrogen depletion on lipid content and lipid productivity

Duration of N depletion (d)	Lipid content (% of DCW)	Lipid productivity (mg L^{-1} d^{-1})
0	13.8 ± 1.3	210 ± 4
1	26.0 ± 1.1	325 ± 20
2	34.1 ± 1.1	328 ± 1
3	43.1 ± 1.4	312 ± 2
5	52.6 ± 0.6	259 ± 8
7	53.7 ± 0.8	200 ± 7

Chlamydomonas sp. JSC4 cultured at a light intensity of 300 μmol m^{-2} s^{-1} (CO_2 aeration = 2.0%; CO_2 flow rate = 0.1 vvm.).

However, it is difficult to optimize both lipid content and lipid productivity simultaneously, because during the nitrogen depletion period required for triggering lipid accumulation, growth appeared to slow, leading to lower lipid productivity (Table 3). Therefore, based on these results, optimal lipid production (lipid content = 43.1%; lipid productivity = 312 mg L^{-1} d^{-1}) was obtained at a light intensity of 300 μmol m^{-2} s^{-1} and a 3-day nitrogen depletion period. The resulting lipid productivity of 312 mg L^{-1} d^{-1} was significantly higher than most of the values reported in the literature, as indicated in Table 4.

The effect of light intensity on lipid quality in *Chlamydomonas* sp.

In addition to lipid content and productivity, the fatty acid composition is also important because it provides an indication of a strain's suitability for synthesizing biodiesel [16]. The fatty acid composition of phototrophic microalgae often changes under different irradiance conditions [17]. In this study, the lipid composition of *Chlamydomonas* sp. JSC4 was monitored with respect to light intensity. As shown in Table 5, the major fatty acids present in the lipids produced by strain JSC4 were similar when the cells were cultivated under different light intensities. The predominant fatty acids were palmitic acid (27% to 30%), stearic acid (14% to 15%), oleic acid (22% to 25%), and linoleic acid (16% to 19%). However, the total amount of C16/C18 fatty acids increased slightly, from 90.0% to 92.5%, when the light intensity was increased. Similar results were also reported by Sun *et al.* [18], who noted that the fatty acids with 16- and 18-carbon atoms in *Neochloris oleoabundans* HK-129 increased as the light intensity increased from 50 to 200 μmol m^{-2} s^{-1}. In addition, Table 5 shows that the

Table 4 Comparison of lipid content and lipid productivity

Species	Lipid content (%)	Lipid productivity (mg L^{-1} d^{-1})	Reference
Chlorella sorokiniana CY1	57.7	140.8	[2]
Chlorella vulgaris	17.3	79.1	[31]
Chlorella emersonii	18.6	54.4	[31]
Chlorella protothecoides	43.4	N.A.	[46]
Chlamydomonas sp. JSC4	33.1	169.1	[47]
Chlamydomonas sp. JSC4	59.4	223	[7]
D. tertiolecta UTEX LB999	25.9	N.A.	[48]
D. tertiolecta ATCC30929	67.0	N.A.	[49]
Isochrysis zhangjiangensis	46.0	136.2	[50]
Nannochloropsis sp.	31.3	N.A.	[27]
Nannochloropsis sp.	59.9	74.2	[14]
Nannochloropsis sp. F&M-M24	68.5	110.1	[51]
Nannochloropsis gaditana	73.1	51	[52]
Nannochloropsis sp. F&M-M28	35.7	60.9	[53]
Nannochloropsis sp. F&M-M26	29.6	61.0	[53]
T. pseudonana CCMP1335	20.3	N.A.	[48]
Tetraselmis suecica F&M-M33	39.1	53.1	[51]
Chlamydomonas sp. JSC4	43.1	312	This study

Compared *Chlamydomonas* sp. JSC4 with that of other marine microalgae as reported in the literature.
N.A., not available.

Table 5 Fatty acid composition of *Chlamydomonas* sp. JSC4

	Chlamydomonas sp. JSC4 FA composition		
	150 µmol m^{-2} s^{-1}	300 µmol m^{-2} s^{-1}	500 µmol m^{-2} s^{-1}
Palmitic acid (C16:0)	27.3 ± 0.4	28.5 ± 0.6	29.5 ± 0.5
Palmitoleic acid (C16:1)	2.0 ± 0.2	2.5 ± 0.1	3.1 ± 0.2
Stearic acid (C18:0)	14.0 ± 0.3	14.9 ± 0.5	15.4 ± 0.3
Oleic acid (C18:1)	22.0 ± 0.1	22.9 ± 0.4	24.5 ± 0.9
Linoleic acid (C18:2)	18.7 ± 0.2	18.5 ± 0.3	15.5 ± 0.3
Linolenic acid (C18:3)	5.9 ± 0.1	5.1 ± 0.1	4.9 ± 0.2
Saturated fatty acid	41.9 ± 0.5	43.4 ± 0.9	44.9 ± 0.7
Monounsaturated fatty acid	24.0 ± 0.2	25.4 ± 0.4	27.6 ± 0.9
Polyunsaturated fatty acid	24.6 ± 0.2	23.6 ± 0.4	20.4 ± 0.4
C16 and C18 groups	90.0 ± 1.4	92.3 ± 1.8	92.5 ± 2.1
Lipid content (%)	48.5 ± 0.8	53.1 ± 1.3	48.9 ± 2.8

Chlamydomonas sp. JSC4 cultured for 5 days under nitrogen depleted conditions and different light intensities. Values are the mean ± standard deviation of three replicated experiments (CO_2 aeration = 2.0%; CO_2 flow rate = 0.1 vvm.).

proportions of the main saturated (for example, palmitic and stearic) and monounsaturated (for example, palmitoleic and oleic) fatty acids increased slightly as light intensity increased, indicating that high light intensity alters fatty acid biosynthesis in microalgae, resulting in the production of more of the saturated or monounsaturated fatty acids that primarily make up neutral lipids [17]. These phenomena were also observed in other studies demonstrating that high light intensity triggers the biosynthesis of palmitic, setaric, or oleic acids in microalgae. This lipid composition appears to be adequate for producing biodiesel [18,30].

The effect of light intensity on five important properties (namely, kinematic viscosity, specific gravity, cetane number, iodine number, and average unsaturation) of biodiesel produced from the oil from strain JSC4 is summarized in Table 6. Because an increase in the degree of saturation of the oil used in biodiesel production can increase the cetane number and oxidative ability, whereas a higher degree of unsaturation can improve the fluidity of biodiesel at low temperatures [31,32], the average degree of unsaturation appears to be an essential indicator of biodiesel quality [32]. As shown in Table 6, although the biodiesel obtained from strain JSC4 grown under high light intensity had a slightly lower average degree of unsaturation, the lipid characteristics (that is, kinematic viscosity, specific gravity, cetane number, and iodine number) were not significantly affected by the illumination conditions, as all of these properties showed satisfactory values based on the United States biodiesel standard ASTM D6751 and the European EN 14214 standard. Thus, regardless of the light intensity, the lipids produced by *Chlamydomonas* sp. JSC4 appeared to be of adequate quality for biodiesel synthesis.

Effect of specific light availability on lipid content and lipid production in *Chlamydomonas* sp. JSC4

To further evaluate the feasibility of biodiesel production using *Chlamydomonas* sp. JSC4, the influence of differences in specific light availability on lipid production was investigated. Specific light availability was defined as moles of photons per gram DCW in the PBR over a designated time interval [33]. According to Figure 6, both lipid content and lipid production were positively correlated with the specific light availability. The data clearly show that both lipid content and lipid production increase dramatically initially when the specific light availability is increased until reaching a maximal value. It has been reported that the total amount of energy bound per gram of biomass increases with increasing lipid content because the heating value of the microalgal biomass increases from 18.8 to 27.1 MJ kg^{-1} as the lipid content increases from 10% to 50% [33]. Wilhelm *et al.* [34] also reported that the photon demand increases during lipid

Table 6 Biodiesel properties of *Chlamydomonas* sp. JSC4

Biodiesel property	150 μmol m^{-2} s^{-1}	300 μmol m^{-2} s^{-1}	500 μmol m^{-2} s^{-1}	US standard (ASTM D6751-08)	European standard (EN 14214)
Kinematic viscosity[a]	4.71	4.72	4.74	1.9 to 6.0	3.5 to 5.0
Specific gravity[b]	0.88	0.88	0.88	0.85 to 0.90	0.86 to 0.90
Cetane number[c]	57.60	57.69	57.99	min47	min51
Iodine number[d]	71.54	70.53	67.24	-	max120
Average unsaturation	0.79	0.78	0.73	-	-

The fatty acids produced from dried biomass of *Chlamydomonas* sp. JSC4 cultured for 5 days under conditions of nitrogen depletion and different light intensities. The relationship between average degree of unsaturation of fatty acids and biodiesel properties as reported by Song *et al.* [32] are shown. Values are the mean of three replicated experiments (CO_2 aeration = 2.0%; CO_2 flow rate = 0.1 vvm.).
[a]Calculated from the following equation: Kinematic viscosity (mm^2 s^{-1}, at 40°C) = ¯0.6316 × average unsaturation + 5.2065.
[b]Calculated from the following equation: Specific gravity (kg L^{-1}) = 0.0055 × average unsaturation + 0.8726.
[c]Calculated from the following equation: Cetane number = ¯6.6684 × average unsaturation + 62.876.
[d]Calculated from the following equation: Iodine number (gI_2 g^{-1}) = 74.373 × average unsaturation + 12.71.

biosynthesis in microalgae relative to the demand for primary sugar biosynthesis because of the additional reduction steps. Therefore, an increase in light energy provides more energy for both growth and lipid biosynthesis. However, as shown in Figure 6, increasing the specific light availability to 3 to 4 mol photons g^{-1} results in a slight drop in both lipid content and lipid production, suggesting that the excess illumination inhibits photosynthetic efficiency and lipid synthesis in microalgae [35]. Thus, specific light availability is a simple but

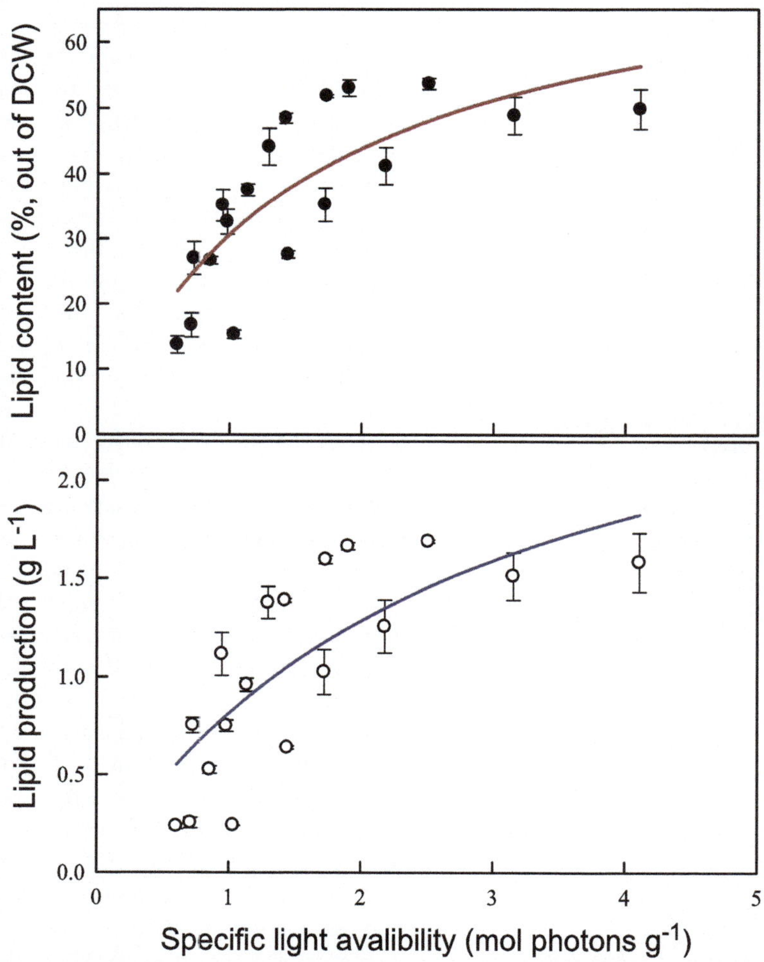

Figure 6 The effect of specific light availability on lipid accumulation over time in *Chlamydomonas* sp. JSC4. Error bars indicate standard deviation of three replicated experiments (CO_2 aeration = 2.0%; CO_2 flow rate = 0.1 vvm).

important indicator of the relationship between light supply and lipid production in microalgae that have the potential for large-scale outdoor culture.

Metabolic profiling of *Chlamydomonas* sp. JSC4 under light stress and nitrogen depletion

Lipid accumulation in microalgae is often affected by environmental stimuli, acting either individually or in combination [17]. Some researchers have postulated that exposing microalgae to combined stresses (such as simultaneous high salinity and nitrogen depletion) may further enhance lipid production efficiency [7,19]. However, little is known about the metabolic details of lipid biosynthesis in response to different light intensities used for cell growth. Metabolic profiling techniques appear to be well-suited for studies aimed at elucidating the mechanisms underlying metabolic changes in lipid biosynthesis in microalgae exposed to various environmental stimuli (such as irradiance and nitrogen deficiency) [23]. To the best of our knowledge, this is the first report demonstrating the metabolic profile of microalgae cultivated under combined light irradiation and nitrogen depletion.

In this study, *Chlamydomonas* sp. JSC4 was cultivated under low and high light intensities (30 and 300 μmol m^{-2} s^{-1}), respectively. Intracellular metabolites were extracted and analyzed using capillary electrophoresis coupled with mass spectrometry (CE/MS), and variations in metabolite amounts over time were monitored as described previously [7]. As shown in Figure 7, cultivation at high light intensity resulted in a higher nitrogen consumption rate and a higher biomass concentration compared with cells cultured under low light intensity, suggesting that 300 μmol m^{-2} s^{-1} is a more suitable light intensity for lipid accumulation in *Chlamydomonas* sp. JSC4. Thus, optimization of the light intensity provides a direct means of obtaining higher cell densities and enhanced lipid content. Notably, the levels of TAG-synthesis-related metabolites (for example, glycerol-3-phosphate [G3P] and acetyl-CoA (AceCoA)) in cells cultured under high light intensity were significantly higher than in cells cultured under low light intensity. This result is in good agreement with previous reports showing that irradiation with high-intensity light provides more energy during cultivation, which may induce microalgae to accumulate more TAG as the neutral storage lipid [17,36]. Moreover, it is known that the photon demand increases during lipid synthesis due to the additional reduction steps [34], and the metabolites of both G3P and AceCoA are extremely important precursors for triacylglycerol formation [17,37]. The positive relationship between G3P, AceCoA, and triacylglycerol synthesis observed in this study in strain JSC4 cultured under high light intensity was therefore expected. Sufficient light supply thus clearly triggers the formation of G3P and AceCoA and then further promotes the synthesis of TAG.

The levels of several key metabolites involved in the Calvin cycle (3-phosphoglyceric acid (3-PG), fructose-6-phosphate (F6P), and glucose-6-phosphate (G6P)) increased significantly along with cell growth, reaching maximal levels on day 2. This suggests that nutrients such as nitrogen either in the medium or inside the cells are sufficient for cell viability. However, after 2 days of nitrogen limitation (day 4), the levels of several metabolites involved in the Calvin cycle (F6P and G6P) decreased, primarily due to transformation of energy compounds into TAG (Figure 7). Moreover, regardless of nitrate status, an increase in the level of the starch precursor glucose-1-phosphate (G1P) was observed in cells irradiated with high-intensity light. This result indicates that strain JSC4 may also store considerable amounts of starch as an energy compound when exposed to certain environmental conditions. Previous reports showed that high light intensity may lead to an increase in the amount of total carbohydrate at microalga *S. obliquus*, which is in good agreement with our study. In addition, several studies have shown that competition between lipid and carbohydrate synthesis occurs in microalgae under stress conditions [38,39]. Therefore, blocking the carbohydrate synthesis pathway may provide an approach for boosting TAG production. Furthermore, in the present study, cultivation under high light intensity led to a dramatic accumulation of intermediates involved in glycolysis, such as 3-PG, phosphoenolpyruvate (PEP), and pyruvate (Pyr). This may have been caused by rate-limiting conversions (that is, conversion of Pyr to AceCoA) that restricted the accumulation of lipids.

CO_2 is assimilated through the Calvin cycle in a reaction catalyzed by ribulose-1,5-biphosphate carboxylase/oxygenase (Rubisco) to form 3-PG. Under low light intensity, the pool size (intracellular concentration) of 3-PG remained low (Figure 7), perhaps because insufficient light energy was available to drive photosynthesis, thus further restricting the growth and ultimately the biosynthesis of lipids.

Analysis of metabolic turnover in *Chlamydomonas* sp. JSC4 under light stress and nitrogen depletion

As shown in Figure 4, cultivation under the dual effects of light irradiation and nitrogen depletion led to a marked increase in lipid content in *Chlamydomonas* sp. JSC4. Metabolic profiling using CE/MS enables the identification of key metabolites of interest and provides a 'snapshot' of the metabolic status of cells at a given point in time. To further understand the dynamics of metabolic flux in JSC4 cells, CE/MS analysis combined with *in vivo* isotope labeling was employed [40]. Recent studies showed that the metabolic kinetics of microalgae can be estimated by measuring changes in the abundance of isotopomers over time [20,22,26].

In the present study, metabolic turnover, defined as the change in the ratio of carbon that is newly incorporated into metabolites to total metabolic carbon, was determined in

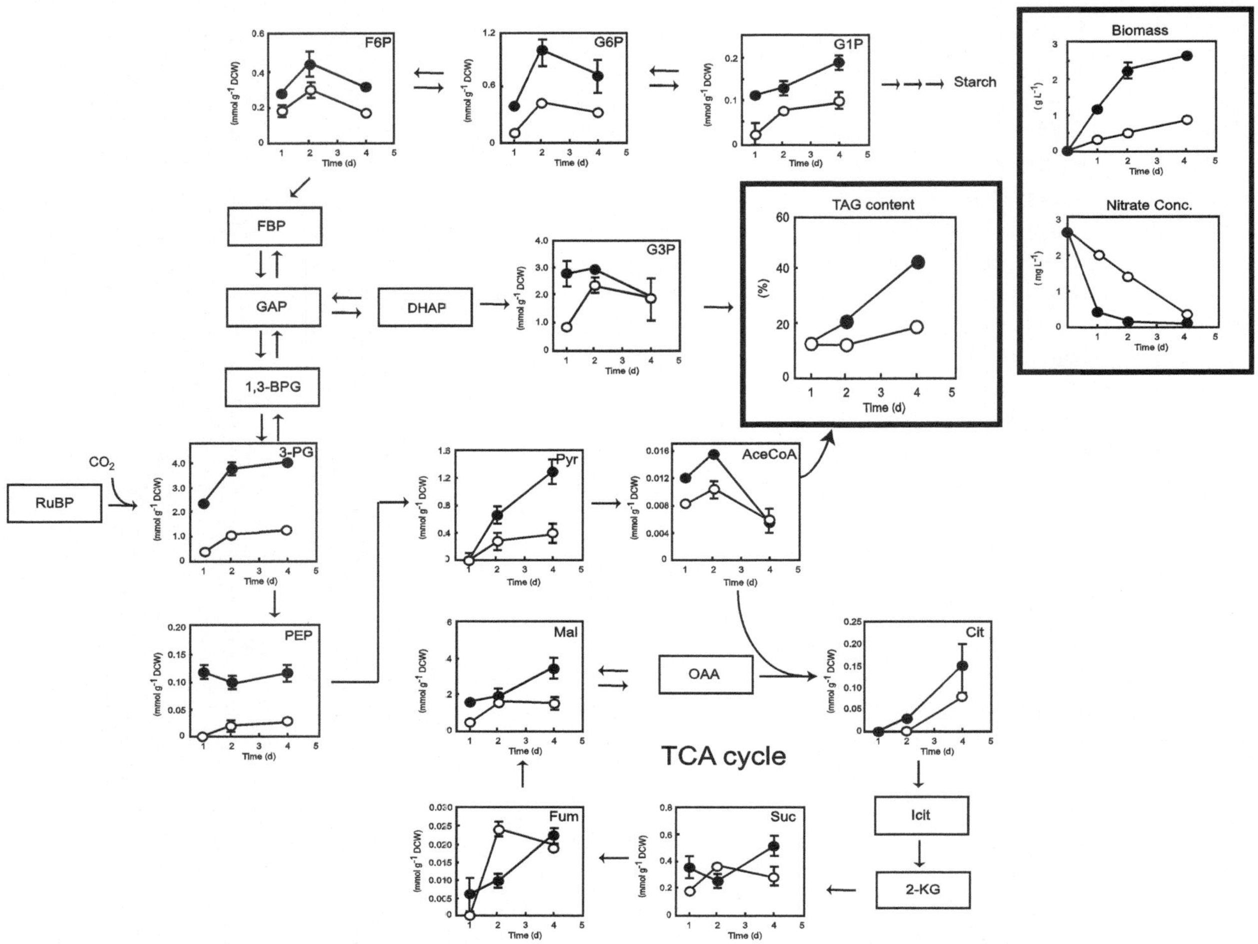

Figure 7 Time-course analysis of metabolite content of *Chlamydomonas* sp. JSC4 cells illuminated with high light intensity (300 µmol m^{-2} s^{-1}; closed circles) or low light intensity (30 µmol m^{-2} s^{-1}; open circles). Error bars indicate standard deviation of three replicated experiments (CO_2 aeration = 2.0%; CO_2 flow rate = 0.1 vvm). Abbreviations: AceCoA, acetyl-CoA; 1,3-BPG, 1,3-bisphosphoglycerate; Cit, citrate; DHAP, dihydroxyacetone phosphate; FBP, fructose-1,6-bisphosphate; F6P, fructose-6-phosphate; Fum, fumarate; GAP, glyceraldehyde-3-phosphate; G1P, glucose-1-phosphate; G3P: glycerol-3-phosphate; G6P, glucose-6-phosphate; 2-KG, 2-ketoglutarate; Mal, malate; OAA, oxaloacetate; PEP, phosphoenolpyruvate; 3-PG, 3-phosphoglycerate; Pyr, pyruvate; RuBP, ribulose; Suc, succinate; TAG, triacylglycerol.

the presence and absence of nitrate under different light intensities using an *in vivo* ^{13}C-labeling assay. The ratio of ^{13}C to total carbon in each metabolite (that is, the ^{13}C fraction (%)) was calculated from mass isotopomer distributions determined by MS. By determining the ratio of the intensity of the monoisotopic ion to that of its isotopic ions, the ratio of stable-isotope-labeled to unlabeled metabolites can be determined. In *in vivo* ^{13}C-labeling, ^{13}C is assimilated through the Calvin cycle enzyme Rubisco in the form of $^{13}CO_2$, which is then used in the production of 3-PG. In the presence of nitrate, the fraction of ^{13}C associated with 3-PG reached a maximum of 80% after 30 min of labeling in cells cultured under high light intensity (300 µmol m^{-2} s^{-1}), but it reached only 65% in cells cultured under low light intensity (30 µmol m^{-2} s^{-1}). Irradiation with high light intensity also resulted in an increase in the ^{13}C-labeling ratio for other sugar phosphates, such as F6P, G6P, G1P, G3P, PEP, and Pyr. These results indicate that sufficient irradiation leads to increases in both biomass concentration and lipid content, as shown in Figures 4 and 6. The ^{13}C fraction did not reach 100% for any metabolites in this study. This result was in agreement with previous reports showing that the ^{13}C fraction rarely reaches 100% for any metabolite in *Arhrospira platensis*, *Nicotiana tabacum*, and *Quercus rubra* [20,41]. This phenomenon may be the result of competition between isotope-labeled carbon and carbon derived from internal storage for assimilation into metabolites ultimately reaching an equilibrium phase.

Under conditions of nitrate depletion, the highest value of the ^{13}C fraction decreased for most metabolites. Notably, by limiting the nitrate supply at a fixed light intensity of 300 µmol m^{-2} s^{-1}, the ^{13}C fractions of malate, citrate, fumarate, and succinate declined from 85.6%, 97.1%, 75.0%, and 54.0% to 45.6%, 42.2%, 20.6%, and 6.2%, respectively, suggesting that availability of ^{13}C for use in the TCA cycle declines when cells are confronted with nitrogen starvation, despite an increase in lipid content (Figure 8). Similar phenomena were also observed in cells cultured under low light intensity (30 µmol m^{-2} s^{-1}). It is not clear whether these responses are common for *Chlamydomonas* sp. and other photosynthetic species. However, several recent reports demonstrated that cultivation under nitrogen-deficient conditions leads to a decrease in turnover of TCA cycle intermediates [20], which is consistent with our findings.

The initial slope of a plot of ^{13}C fraction versus time provides an estimate of the turnover rate of each metabolite, as depicted in Figure 9. The turnover rates of sugar phosphates involved in the Calvin cycle (for example, PEP and 3-PG) were significantly higher than those of organic acids, confirming that in photosynthetic microorganisms, 3-PG is

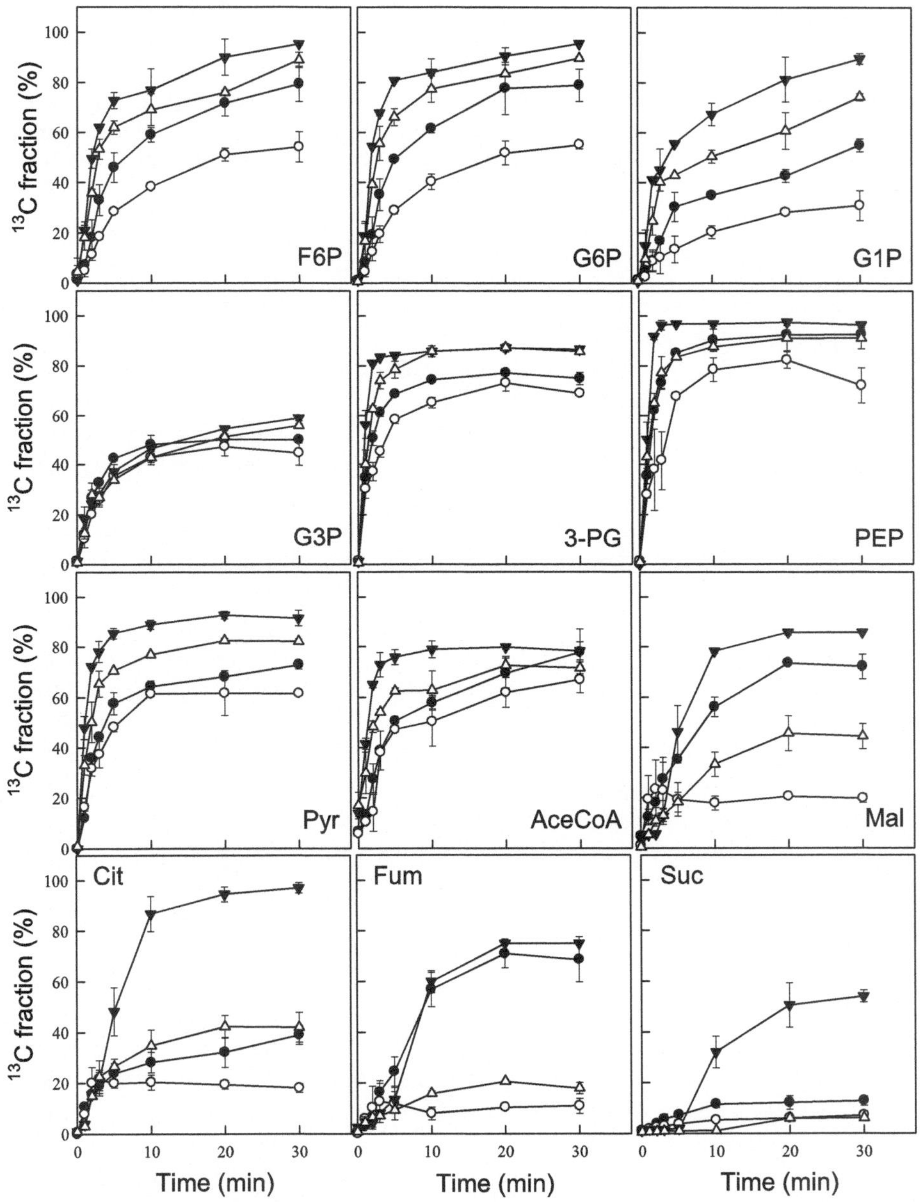

Figure 8 Time-course analysis of the metabolite ^{13}C fraction of *Chlamydomonas* sp. JSC4 cells cultivated under high light intensity of 300 µmol m^{-2} s^{-1}, nitrogen-rich (closed triangles) conditions; high light intensity of 300 µmol m^{-2} s^{-1}, nitrogen-free (open triangles) conditions; low light intensity of 30 µmol m^{-2} s^{-1}, nitrogen-rich (closed circles) conditions; or low light intensity of 30 µmol m^{-2} s^{-1}, nitrogen-free (open circles) conditions. Error bars indicate standard deviation of three replicated experiments. Abbreviations: AceCoA, acetyl-CoA; Cit, citrate; F6P, fructose-6-phosphate; Fum, fumarate; G1P, glucose-1-phosphate; G3P: glycerol-3-phosphate; G6P, glucose-6-phosphate; Mal, malate; PEP, phosphoenolpyruvate; 3-PG, 3-phosphoglycerate; Pyr, pyruvate; Suc, succinate.

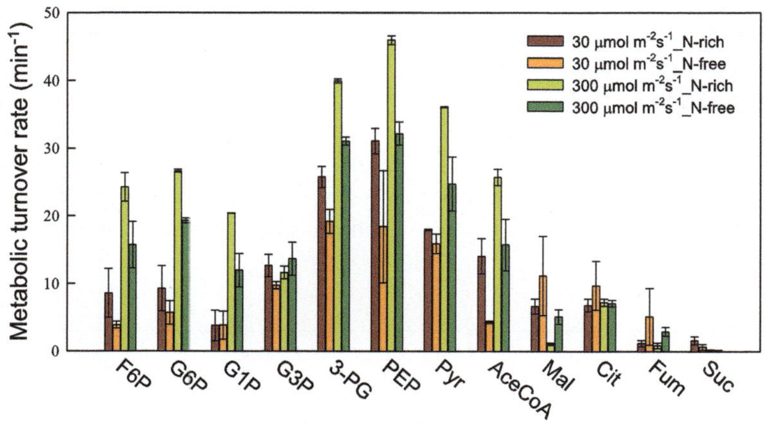

Figure 9 Metabolic turnover rates of key metabolites in *Chlamydomonas* sp. JSC4 cells cultivated under high light intensity of 300 μmol m^{-2} s^{-1}, nitrogen-rich (yellow bars) conditions; high light intensity of 300 μmol m^{-2} s^{-1}, nitrogen-free (green bars) conditions; low light intensity of 30 μmol m^{-2} s^{-1}, nitrogen-rich (red bars) conditions; or low light intensity of 30 μmol m^{-2} s^{-1}, nitrogen-free (orange bars) conditions. Error bars indicate standard deviation of three replicated experiments. Abbreviations: AceCoA, acetyl-CoA; Cit, citrate; F6P, fructose-6-phosphate; Fum, fumarate; G1P, glucose-1-phosphate; G3P: glycerol-3-phosphate; G6P, glucose-6-phosphate; Mal, malate; PEP, phosphoenolpyruvate; 3-PG, 3-phosphoglycerate; Pyr, pyruvate; Suc, succinate.

rapidly generated through the action of the Calvin cycle and then further converted to PEP and other intracellular metabolites (Figure 7). Similar trends were also reported by Hasunuma *et al.* [20], who found that the turnover rates of sugar phosphates are significantly higher than those of organic acids and amino acids in *A. platensis*. In addition, we observed that an increase in light intensity from 30 to 300 μmol m^{-2} s^{-1} led to a sharp rise in the turnover rate of most metabolites (for example, Pyr and AceCoA), which is consistent with the efficient growth of cells and accumulation of lipids demonstrated in Figures 4 and 6. Furthermore, under nitrogen limitation, the turnover rate of most metabolites in strain JSC4 decreased, suggesting that under nitrogen depletion, more energy is required for lipid synthesis, thus restricting the rate of carbon flow. Notably, the pool size of AceCoA (0.01 to 0.03 nmol mg^{-1}), one of the most important TAG-synthesis-related metabolites, was approximately 20 to 50 times smaller than that of Pyr (0.32 to 0.78 nmol mg^{-1}) (as shown in Additional file 1: Table S1), suggesting that lipid accumulation in strain JSC4 could be enhanced further through overexpression of related enzymes. Some efforts have been made to increase the expression of enzymes that are involved in fatty acid biosynthesis pathways, and the results are quite promising [42,43]. As a next step, knockout and overexpression of the genes involved in lipid synthesis could clarify which genes have a significant effect on lipid accumulation and lead to the development of strategies for increasing the lipid content in oleaginous microalgae.

Nitrogen limitation is frequently used to increase the lipid content in *Chlamydomonas* sp. [7]. However, little is known about the mechanism of lipid biosynthesis in these microorganisms. To the best of our knowledge, this is the first report of *in vivo* isotope labeling of *Chlamydomonas* sp. cultivated under different light-intensity conditions coupled with nitrogen limitation. In this study, the combination of metabolic profiling and turnover analysis (collectively referred to as dynamic metabolic profiling) was applied to explore the effects of stress associated with combined light intensity and nitrogen depletion on carbon metabolism in *Chlamydomonas* sp. JSC4. The outcome of this research enhanced current understanding of the relationship between carbon flow and lipid biosynthesis in microalgae cultured under combined light effect and nitrogen depletion. Using the *in vivo* ^{13}C-labeling assay, the dynamic metabolic profile of strain JSC4 was determined, generating a large amount of valuable information that can be applied to enhance the lipid production capacity of this organism through genetic engineering.

Conclusions

This work demonstrated that lipid production in the marine microalga *Chlamydomonas* sp. JSC4 can be enhanced significantly by culturing the organism under conditions of appropriate light intensity (that is, 300 μmol m^{-2} s^{-1}) coupled with nitrogen depletion. The highest lipid productivity achieved in this study was 328 mg L^{-1} d^{-1}, which to the best of our knowledge is the highest value reported in the literature for marine microalgae. The lipids produced under the optimal conditions contained 43.4% saturated, 25.4% monounsaturated, and 23.6% polyunsaturated fatty acids, a composition very suitable for biodiesel production. The data obtained from dynamic metabolic profiling (that is, metabolic profiling and turnover analysis) provided valuable information that increases our understanding of metabolic flow in *Chlamydomonas* sp. JSC4 when cultured in the absence of nitrate and with light of suitable intensity. By integrating cultivation strategies and dynamic metabolic profiling techniques using strain JSC4, this study demonstrated an innovative technology for practically and markedly enhancing microalgal lipid productivity and elucidated the details of carbon metabolism related to lipid biosynthesis.

Methods

Microalgal strain and preculturing medium

The microalga used in this study was *Chlamydomonas* sp. JSC4, which was isolated from a coastal area of southern Taiwan. The 18S rDNA sequence of this strain was deposited in the National Center for Biotechnology Information GenBank database under accession number KF383270. The medium used for preculturing strain JSC4 was Modified Bold 3 N medium (4.4 mM $NaNO_3$, 0.22 mM K_2HPO_4, 0.3 mM $MgSO_4 \cdot 7H_2O$, 0.17 mM $CaCl_2 \cdot 2H_2O$, 0.43 mM KH_2PO_4, and 0.43 mM NaCl; the levels of metals in the medium are described in Berges *et al.* [44]). Cells were routinely cultured at 30°C to 33°C for 3 to 4 days with a continuous supply of 2.0% CO_2 at an aeration rate of 0.05 vvm. The culture was illuminated continuously at a light intensity of approximately 250 μmol m^{-2} s^{-1}. The light intensity was measured using a Li-250 Light Meter with a Li-190SA pyranometer sensor (Li-COR Inc., Lincoln, NE, USA).

PBR operation

Both conventional-type (9.5 cm in diameter; height to diameter (H:D) ratio of 0.8) and slim-type (5.0 cm in diameter; H:D ratio of 10) PBRs equipped with external light sources were used in this study. The S:V ratio of the conventional and slim-type PBRs was 41.8 and 80.1, respectively. *Chlamydomonas* sp. JSC4 was precultured in Modified Bold 3 N medium and inoculated into the PBRs at a concentration of 90 to 100 mg L^{-1}. For batch cultures, microalgae were cultivated at 32°C to 35°C and pH 6.5 to 7.5 with an agitation rate of 400 rpm under various light intensities in the range of approximately 30 to 500 μmol m^{-2} s^{-1}. As the sole carbon source, 2.0% CO_2 was continuously fed into the culture at a rate of 0.05 vvm. Liquid samples were collected from the sealed glass vessel at designated time intervals to determine the culture density, pH, and residual nitrate concentration. The amount of CO_2 reduced was determined by measuring the difference between the CO_2 concentrations in the influent and effluent streams of the PBR using a GM70 CO_2 detector (Vaisala, Tokyo, Japan).

Determination of culture density

The density of microalgae in the PBR was determined regularly by measuring the optical density (OD) at a wavelength of 682 nm (denoted as OD_{682}) using a UVmini-1240 UV/Vis spectrophotometer (Shimadzu, Kyoto, Japan) after proper dilution with deionized water to give a range of measurement between 0.1 and 0.9. The DCW was determined by filtering 50-mL aliquots of culture through a cellulose acetate membrane filter (0.45-μm pore size, 47 mm diameter). Each loaded filter was then freeze-dried until the weight was invariant. The dry weight of a blank filter was subtracted from that of the loaded filter to obtain the DCW of microalgae. The OD_{682} values were also converted to biomass concentration via appropriate calibration between OD_{682} and DCW, and the conversion factor was determined as 1.0 OD_{682} = 0.7 ~ 0.9 g DCW L^{-1}.

Measurement of residual nitrate content

Nitrate concentration was determined according to our previously reported method [45]. In general, each sample collected from the PBR was filtered with a 0.22-μm pore-size filter and then diluted 20-fold with deionized water. The residual nitrate content of the diluted samples was determined according to the OD_{220} with appropriate calibrations. The conversion factor was determined as 1.0 OD_{220} = ~389.3 mg nitrate L^{-1}.

Determination of growth parameters and CO_2 fixation performance

Change in biomass concentration (mg L^{-1}) over time was used to calculate the specific growth rate (d^{-1}) based on a plot of DCW (on a logarithmic scale) versus time. Biomass productivity (mg L^{-1} d^{-1}) was calculated according to the following equation:

$$P = \frac{\Delta X}{\Delta t},$$

where ΔX represents the variation in biomass concentration (mg L^{-1}) over cultivation time (d).

The CO_2 fixation rate (P_{CO2}; mg L^{-1} d^{-1}) was calculated according to the following equation:

$$P_{CO_2}(\mathrm{mg/L/d}) = 1.88 \times P_{\mathrm{biomass}},$$

where $P_{biomass}$ represents biomass productivity (mg L^{-1} d^{-1}), as described above. The typical molecular formula for microalgal biomass ($CO_{0.48}H_{1.83}N_{0.11}P_{0.01}$) [10] was used in this study.

To evaluate the effect of irradiance along with biomass concentration on the lipid accumulation process, the specific light availability ($I_{PAR,\ spec}$) as reported by Munkel *et al.* [33] was calculated according to the following equation:

$$I_{\mathrm{PAR,\,spec}} = \frac{\int I_{\mathrm{PAR}}\,dt}{\varnothing \mathrm{DWV}},$$

where I_{PAR} represents irradiance on the PBR surface integrated over a defined time interval. ∅ DWrepresents the mean biomass concentration during the given time interval, V represents the total working volume, and t represents the cultivation time.

Determination of the lipid content and fatty acid profile

After an appropriate amount of time for consuming nitrogen, strain JSC4 cells were harvested from the culture medium by centrifugation (7,000 rpm for 2 min). The cells were washed twice with deionized water, lyophilized, and weighed. The lipid composition was determined as fatty acid methyl esters (FAMEs) following direct transesterification of lipids according to the method described in Ho *et al.* [7]. FAMEs were analyzed by gas chromatography/mass spectrometry (GC/MS) on a GCMS-QP2010 Plus instrument (Shimadzu). Samples were injected onto a DB-23 capillary column (60 m, 0.25 mm internal diameter, 0.15-μm film thickness; Agilent Technologies, Palo Alto, CA, USA). Helium was used as the carrier gas at a flow rate of 2.3 mL min^{-1}. The injector, ion source, and interface source temperatures were set at 230°C, 230°C, and 250°C, respectively. The oven temperature was initially set at 50°C for 1 min, increased from 50°C to 175°C at a rate of 25°C/min, increased from 175°C to 230°C at a rate of 4°C/min, and held at 230°C for 5 min. Purified FAMEs were identified based on retention time and the pattern of fragmentation by electron impact analysis. Supelco 37 Component FAME Mix (Sigma-Aldrich Co., St. Louis, MO, USA) was utilized as a quantitative standard, and pentadecanoic acid (Sigma-Aldrich Co.) was used as an internal standard.

Sampling procedures for metabolic profiling

Cell sampling was performed according to our previously reported method [20], with minor modifications. Strain JSC4 cells (equivalent to 5 to 10 mg dry weight) were removed from cultivation vessels and added to a fourfold volume of prechilled (−30°C) quenching solution consisting of Modified Bold 3 N medium containing 32.5% methanol, then filtered using 1-μm pore-size Omnipore filter disks (Millipore, Billerica, MA, USA). After washing with 20 mM ammonium bicarbonate prechilled to 4°C, cells retained on the filters were immediately placed in 1 mL of prechilled (−30°C) methanol containing 12.4 μM piperazine-1,4-bis(2-ethanesulfonic acid) as an internal standard for mass analysis. Intracellular metabolites were extracted using cold 10:3:1 (*v*/*v*/*v*) methanol:chloroform:water combined with 30 min of bead-beater cell-breaking, as described previously [23]. Cells were suspended by vortexing and then 1 mL of the cell suspension was mixed with 100 μL of prechilled (4°C) water and 300 μL of chloroform. The cell suspension was shaken at 1,200 rpm in a model MBR-022UP incubator (TAITEC, Saitama, Japan) for 30 min at 4°C in the dark before centrifugation at 14,000 *g* for 5 min at 4°C. Next, 980 μL of the cell extract obtained as the supernatant was transferred to a clean tube. After adding 440 μL of water, the aqueous and organic layers were phase separated by centrifugation at 14,000 *g* for 5 min at 4°C. After filtration through a Millipore 5-kDa cut-off filter for the removal of solubilized proteins, the aqueous-layer extracts were evaporated under vacuum using a FreeZone 2.5 Plus freeze dry system (Labconco, Kansas City, MO, USA). Dried extracts were stored at −80°C until analysis by CE/MS.

CE/MS metabolite analysis

Metabolites were determined according to the CE/MS method described by Hasunama *et al.* [20]. Dried metabolites were dissolved in 20 μL of Milli-Q water before analysis. The CE/MS experiments were performed using an Agilent G7100 CE system, an Agilent G6224AA LC/MSD time-of-flight (TOF) mass spectrometer, and an Agilent 1200 series isocratic HPLC pump equipped with a 1:100 splitter for delivery of the sheath liquid.

Agilent ChemStation software for CE and MassHunter software for the Agilent TOF-MS system were used for system control and data acquisition, respectively. The analytical conditions for anionic metabolite analyses were as described previously [20]. CE separations were performed using a fused silica capillary (1 m × 50 μm i.d.) filled with 50 mM ammonium acetate (pH 9) for anionic metabolite analyses. The CE polarity was such that the electrolyte vial (inlet) was at the anode, and the electrospray ionization (ESI) probe (outlet) was at the cathode. Samples were injected into the CE system at a pressure of 50 mbar for 30 s. The voltage applied to the CE capillary was set at 30 kV, with a ramp time of 0.3 min. For anionic metabolite analyses, the electrolyte solution was passed through the capillary using an air pump and was delivered at a pressure of 10 mbar between 0.4 and 30 min and 100 mbar between 30.1 and 49.5 min. The flow rate of the sheath liquid was set at 8 μL min^{-1}. The ESI-MS analyses were conducted in either the positive or negative ion mode using a capillary voltage of −3.5 or 3.5 kV, respectively. The TOF-MS fragmenter, skimmer, and Oct RFV were set to 100, 65, and 750 V, respectively. The flow of heated drying nitrogen gas (300°C) was maintained at 10 L min^{-1}. Mass data were acquired at a rate of 1 spectrum s^{-1} over the mass to charge ratio (*m*/*z*) range of 70 to 1,000.

^{13}C-labeling

To analyze metabolic turnover in *Chlamydomonas* sp. JSC4, *in vivo* ^{13}C-labeling was performed using sodium ^{13}C-bicarbonate ($NaH^{13}CO_3$) as a carbon source. Strain JSC4 was precultivated in Modified Bold 3 N medium or Modified Bold 1 N medium at a light intensity of 30 μmol m^{-2} s^{-1} or 300 μmol m^{-2} s^{-1}, respectively, for approximately 2 to 3 days and then resuspended in Modified Bold

3 N or Modified Bold 0 N medium containing 25 mM $NaH^{13}CO_3$. After labeling for 1 to 30 min, 5 to 10 mg of strain JSC4 cells were collected by filtration and processed as described in the previous section. Extracted intracellular metabolites were then analyzed using CE/MS. Mass spectral peaks of biological origin were identified manually by searching for mass shifts between ^{12}C- and ^{13}C-mass spectra. The ^{13}C fraction of each metabolite and the metabolic turnover rate were determined as described previously [40].

Additional file

Additional file 1: Table S1. Comparison of key metabolite pools. Comparison of key metabolite pools in *Chlamydomonas* sp. JSC4 under cells cultivated under low light intensity of 30 μmol m^{-2} s^{-1}, nitrogen-rich conditions; high light intensity of 300 μmol m^{-2} s^{-1}, nitrogen-rich conditions; or high light intensity of 300 μmol m^{-2} s^{-1}, nitrogen-free conditions. Values are the mean ± standard deviation of three replicated experiments.

Abbreviations

3-PG: 3-phosphoglycerate; AceCoA: Acetyl-CoA; CE/MS: Capillary electrophoresis coupled with mass spectrometry; DCW: Dry cell weight; ESI: Electrospray ionization; F6P: Fructose-6-phosphate; FAMEs: Fatty acid methyl esters; G1P: Glucose-1-phosphate; G3P: Glycerol-3-phosphate; G6P: Glucose-6-phosphate; GC/MS: Gas chromatography/mass spectrometry; OD: Optical density; PBR: Photobioreactor; PEP: Phosphoenolpyruvate; Pyr: Pyruvate; TAG: Triacylglycerol; TOF: Time-of-flight.

Competing interests

The authors declare that they have no competing interests.

Authors' contributions

SHH performed and analyzed all experiments and drafted the manuscript. AN assisted in analysis of the metabolic experimental data. XY participated in metabolite analysis and contributed to the draft manuscript. JSC assisted in drafting the manuscript and revising it critically before submission. CYC assisted in analysis of growth kinetics and model simulations. TH and AK co-coordinated the study and helped in drafting the manuscript. All authors read and approved the final manuscript.

Acknowledgements

The authors would like to thank Mrs. Mika Kukuda for technical support. This work was supported through project P07015 from the New Energy and Industrial Technology Development Organization (NEDO).

Author details

[1]Organization of Advanced Science and Technology, Kobe University, 1-1 Rokkodai, Nada-ku, Kobe 657-8501, Japan. [2]Department of Chemical Science and Engineering, Graduate School of Engineering, Kobe University, 1-1 Rokkodai, Nada-ku, Kobe 657-8501, Japan. [3]Department of Chemical Engineering, National Cheng Kung University, Tainan 701, Taiwan. [4]Research Center for Energy Technology and Strategy, National Cheng Kung University, Tainan 701, Taiwan. [5]Center for Bioscience and Biotechnology, National Cheng Kung University, Tainan 701, Taiwan. [6]Biomass Engineering Program, RIKEN, 1-7-22 Suehiro-cho, Tsurumi-ku, Yokohama, Kanagawa 230-0045, Japan.

References

1. Ho S-H, Li P-J, Liu C-C, Chang J-S. Bioprocess development on microalgae-based CO_2 fixation and bioethanol production using *Scenedesmus obliquus* CNW-N. Bioresour Technol. 2013;145:142–9.
2. Chen C-Y, Chang J-S, Chang H-Y, Chen T-Y, Wu J-H, Lee W-L. Enhancing microalgal oil/lipid production from *Chlorella sorokiniana* CY1 using deep-sea water supplemented cultivation medium. Biochem Eng J. 2013;77:74–81.
3. Gouveia L, Oliveira AC. Microalgae as a raw material for biofuels production. J Ind Microbiol Biotechnol. 2009;36:269–74.
4. Ho S-H, Chang J-S, Lai Y-Y, Chen C-NN. Achieving high lipid productivity of a thermotolerant microalga *Desmodesmus* sp. F2 by optimizing environmental factors and nutrient conditions. Bioresour Technol. 2014;156:108–16.
5. Wang B, Li YQ, Wu N, Lan CQ. CO_2 bio-mitigation using microalgae. Appl Microbiol Biotechnol. 2008;79:707–18.
6. Ho S-H, Chen C-Y, Lee D-J, Chang J-S. Perspectives on microalgal CO_2-emission mitigation systems — a review. Biotechnol Adv. 2011;29:189–98.
7. Ho S-H, Nakanishi A, Ye X, Chang J-S, Hara K, Hasunuma T, et al. Optimizing biodiesel production in marine *Chlamydomonas* sp. JSC4 through metabolic profiling and an innovative salinity-gradient strategy. Biotechnol Biofuels. 2014;7:97.
8. Chen C-Y, Yeh K-L, Aisyah R, Lee D-J, Chang J-S. Cultivation, photobioreactor design and harvesting of microalgae for biodiesel production: a critical review. Bioresour Technol. 2011;102:71–81.
9. Gouveia L, Marques AE, da Silva TL, Reis A. *Neochloris oleabundans* UTEX# 1185: a suitable renewable lipid source for biofuel production. J Ind Microbiol Biotechnol. 2009;36:821–6.
10. Chisti Y. Biodiesel from microalgae. Biotechnol Adv. 2007;25:294–306.
11. Cho S, Luong TT, Lee D, Oh Y-K, Lee T. Reuse of effluent water from a municipal wastewater treatment plant in microalgae cultivation for biofuel production. Bioresour Technol. 2011;102:8639–45.
12. Siaut M, Cuiné S, Cagnon C, Fessler B, Nguyen M, Carrier P, et al. Oil accumulation in the model green alga *Chlamydomonas reinhardtii*: characterization, variability between common laboratory strains and relationship with starch reserves. BMC Biotechnol. 2011;11:7.
13. Ho S-H, Chen W-M, Chang J-S. *Scenedesmus obliquus* CNW-N as a potential candidate for CO_2 mitigation and biodiesel production. Bioresour Technol. 2010;101:8725–30.
14. Jiang L, Luo S, Fan X, Yang Z, Guo R. Biomass and lipid production of marine microalgae using municipal wastewater and high concentration of CO_2. Appl Energ. 2011;88:3336–41.
15. Liu J, Yuan C, Hu G, Li F. Effects of light intensity on the growth and lipid accumulation of microalga *Scenedesmus* sp. 11–1 under nitrogen limitation. Appl Biochem Biotechnol. 2012;166:2127–37.
16. Ho S-H, Chen C-Y, Chang J-S. Effect of light intensity and nitrogen starvation on CO_2 fixation and lipid/carbohydrate production of an indigenous microalga *Scenedesmus obliquus* CNW-N. Bioresour Technol. 2012;113:244–52.
17. Hu Q, Sommerfeld M, Jarvis E, Ghirardi M, Posewitz M, Seibert M, et al. Microalgal triacylglycerols as feedstocks for biofuel production: perspectives and advances. Plant J. 2008;54:621–39.
18. Sun X, Cao Y, Xu H, Liu Y, Sun J, Qiao D, et al. Effect of nitrogen-starvation, light intensity and iron on triacylglyceride/carbohydrate production and fatty acid profile of *Neochloris oleoabundans* HK-129 by a two-stage process. Bioresour Technol. 2014;155:204–12.
19. Pal D, Khozin-Goldberg I, Cohen Z, Boussiba S. The effect of light, salinity, and nitrogen availability on lipid production by *Nannochloropsis* sp. Appl Microbiol Biotechnol. 2011;90:1429–41.
20. Hasunuma T, Kikuyama F, Matsuda M, Aikawa S, Izumi Y, Kondo A. Dynamic metabolic profiling of cyanobacterial glycogen biosynthesis under conditions of nitrate depletion. J Exp Bot. 2013;64:2943–54.
21. Baran R, Reindl W, Northen TR. Mass spectrometry based metabolomics and enzymatic assays for functional genomics. Curr Opin Microbiol. 2009;12:547–52.
22. Ito T, Sugimoto M, Toya Y, Ano Y, Kurano N, Soga T, et al. Time-resolved metabolomics of a novel trebouxiophycean alga using $^{13}CO_2$ feeding. J Biosci Bioeng. 2013;116:408–15.
23. Bölling C, Fiehn O. Metabolite profiling of *Chlamydomonas reinhardtii* under nutrient deprivation. Plant Physiol. 2005;139:1995–2005.
24. Renberg L, Johansson AI, Shutova T, Stenlund H, Aksmann A, Raven JA, et al. A metabolomic approach to study major metabolite changes during acclimation to limiting CO_2 in *Chlamydomonas reinhardtii*. Plant Physiol. 2010;154:187–96.
25. Ito T, Tanaka M, Shinkawa H, Nakada T, Ano Y, Kurano N, et al. Metabolic and morphological changes of an oil accumulating trebouxiophycean alga in nitrogen-deficient conditions. Metabolomics. 2013;9:178–87.
26. Shinohara M, Sakuragi H, Morisaka H, Miyake H, Tamaru Y, Fukusaki E, et al. Fixation of CO_2 in Clostridium cellulovorans analyzed by ^{13}C-isotopomer-based target metabolomics. AMB Express. 2013;3:1–6.

27. Wahidin S, Idris A, Shaleh SRM. The influence of light intensity and photoperiod on the growth and lipid content of microalgae *Nannochloropsis* sp. Bioresour Technol. 2013;129:7–11.
28. Posten C. Design principles of photo-bioreactors for cultivation of microalgae. Eng Life Sci. 2009;9:165–77.
29. Guedes AC, Meireles LA, Amaro HM, Malcata FX. Changes in lipid class and fatty acid composition of cultures of *Pavlova lutheri*, in response to light intensity. J Am Oil Chem Soc. 2010;87:791–801.
30. Guihéneuf F, Mimouni V, Ulmann L, Tremblin G. Combined effects of irradiance level and carbon source on fatty acid and lipid class composition in the microalga *Pavlova lutheri* commonly used in mariculture. J Exp Mar Biol Ecol. 2009;369:136–43.
31. Talebi AF, Mohtashami SK, Tabatabaei M, Tohidfar M, Bagheri A, Zeinalabedini M, et al. Fatty acids profiling: a selective criterion for screening microalgae strains for biodiesel production. Algal Res. 2013;2:258–67.
32. Song M, Pei H, Hu W, Ma G. Evaluation of the potential of 10 microalgal strains for biodiesel production. Bioresour Technol. 2013;141:245–51.
33. Münkel R, Schmid-Staiger U, Werner A, Hirth T. Optimization of outdoor cultivation in flat panel airlift reactors for lipid production by Chlorella vulgaris. Biotechnol Bioeng. 2013;110:2882–93.
34. Wilhelm C, Jakob T. From photons to biomass and biofuels: evaluation of different strategies for the improvement of algal biotechnology based on comparative energy balances. Appl Microbiol Biotechnol. 2011;92:909–19.
35. Richmond A, Cheng-Wu Z, Zarmi Y. Efficient use of strong light for high photosynthetic productivity: interrelationships between the optical path, the optimal population density and cell-growth inhibition. Biomol Eng. 2003;20:229–36.
36. Khotimchenko SV, Yakovleva IM. Lipid composition of the red alga *Tichocarpus crinitus* exposed to different levels of photon irradiance. Phytochemistry. 2005;66:73–9.
37. Husic HD, Tolbert N. Effect of osmotic stress on carbon metabolism in *Chlamydomonas reinhardtii* accumulation of glycerol as an osmoregulatory solute. Plant Physiol. 1986;82:594–6.
38. Li Y, Han D, Hu G, Sommerfeld M, Hu Q. Inhibition of starch synthesis results in overproduction of lipids in *Chlamydomonas reinhardtii*. Biotechnol Bioeng. 2010;107:258–68.
39. Ramazanov A, Ramazanov Z. Isolation and characterization of a starchless mutant of *Chlorella pyrenoidosa* STL-PI with a high growth rate, and high protein and polyunsaturated fatty acid content. Phycol Res. 2006;54:255–9.
40. Hasunuma T, Harada K, Miyazawa S-I, Kondo A, Fukusaki E, Miyake C. Metabolic turnover analysis by a combination of in vivo ^{13}C-labelling from $^{13}CO_2$ and metabolic profiling with CE-MS/MS reveals rate-limiting steps of the C3 photosynthetic pathway in *Nicotiana tabacum* leaves. J Exp Bot. 2010;61:1041–51.
41. Delwiche C, Sharkey T. Rapid appearance of ^{13}C in biogenic isoprene when $^{13}CO_2$ is fed to intact leaves. Plant Cell Environ. 1993;16:587–91.
42. Radakovits R, Jinkerson RE, Darzins A, Posewitz MC. Genetic engineering of algae for enhanced biofuel production. Eukaryot Cell. 2010;9:486–501.
43. Dunahay TG, Jarvis EE, Roessler PG. Genetic transformation of the diatoms *Cyclotella cryptica* and *Navicula saprophila*. J Phycol. 1995;31:1004–12.
44. Berges JA, Franklin DJ, Harrison PJ. Evoluation of an artificial sea water medium: improvements in enriched seawater, articial water over the last two decades. J Phycol. 2001;37:1138–45.
45. Ho S-H, Kondo A, Hasunuma T, Chang J-S. Engineering strategies for improving the CO_2 fixation and carbohydrate productivity of *Scenedesmus obliquus* CNW-N used for bioethanol fermentation. Bioresour Technol. 2013;143:163–71.
46. Campenni L, Nobre BP, Santos CA, Oliveira A, Aires-Barros M, Palavra A, et al. Carotenoid and lipid production by the autotrophic microalga *Chlorella protothecoides* under nutritional, salinity, and luminosity stress conditions. Appl Microbiol Biotechnol. 2013;97:1383–93.
47. Nakanishi A, Aikawa S, Ho S-H, Chen C-Y, Chang J-S, Hasunuma T, et al. Development of lipid productivities under different CO_2 conditions of marine microalgae *Chlamydomonas* sp. JSC4. Bioresour Technol. 2014;152:247–52.
48. Jiang Y, Yoshida T, Quigg A. Photosynthetic performance, lipid production and biomass composition in response to nitrogen limitation in marine microalgae. Plant Physiol Biochem. 2012;54:70–7.
49. Takagi M, Takagi M, Karseno Yoshida T. Effect of salt concentration on intracellular accumulation of lipids and triacylglyceride in marine microalgae *Dunaliella* cells. J Biosci Bioeng. 2006;101:223–6.
50. Feng D, Chen Z, Xue S, Zhang W. Increased lipid production of the marine oleaginous microalgae *Isochrysis zhangjiangensis* (Chrysophyta) by nitrogen supplement. Bioresour Technol. 2011;102:6710–6.
51. Bondioli P, Della Bella L, Rivolta G, Chini Zittelli G, Bassi N, Rodolfi L, et al. Oil production by the marine microalgae *Nannochloropsis* sp. F&M-M24 and *Tetraselmis suecica* F&M-M33. Bioresour Technol. 2012;114:567–72.
52. San Pedro A, González-López CV, Acién FG, Molina-Grima E. Marine microalgae selection and culture conditions optimization for biodiesel production. Bioresour Technol. 2013;134:353–61.
53. Rodolfi L, Chini Zittelli G, Bassi N, Padovani G, Biondi N, Bonini G, et al. Microalgae for oil: strain selection, induction of lipid synthesis and outdoor mass cultivation in a low-cost photobioreactor. Biotechnol Bioeng. 2009;102:100–12.

7

Systematic evaluation of the degraded products evolved from the hydrothermal pretreatment of sweet sorghum stems

Shaolong Sun, Jialong Wen, Shaoni Sun and Run-Cang Sun*

Abstract

Background: Conversion of plant cell walls to bioethanol and bio-based chemicals requires pretreatment as a necessary step to reduce recalcitrance of cell walls to enzymatic and microbial deconstruction. In this study, the sweet sorghum stems were subjected to various hydrothermal pretreatment processes (110°C to 230°C, 0.5 to 2.0 h), and the focus of this work is to systematically evaluate the degraded products of polysaccharides and lignins in the liquor phase obtained during the pretreatment process.

Results: The maximum yield of xylooligosaccharides (52.25%) with a relatively low level of xylose and other degraded products was achieved at a relatively high pretreatment temperature (170°C) for a short reaction time (0.5 h). Higher temperature (>170°C) and/or longer reaction time (>0.5 h at 170°C) resulted in a decreasing yield of xylooligosaccharides, but increased the concentration of arabinose and galactose. The xylooligosaccharides obtained are composed of xylopyranosyl residues, together with lower amounts of 4-*O*-Me-*α*-D-GlcpA units. Meanwhile, the concentrations of the degraded products (especially furfural) increased as a function of pretreatment temperature and time. Molecular weights of the water-soluble polysaccharides and lignins indicated that the degradation of the polysaccharides and lignins occurred during the conditions of harsh hydrothermal pretreatment. In addition, the water-soluble polysaccharides (rich in xylan) and water-soluble lignins (rich in *β*-*O*-4 linkages) were obtained at 170°C for 1.0 h.

Conclusions: The present study demonstrated that the hydrothermal pretreatment condition had a remarkable impact on the compositions and the chemical structures of the degraded products. An extensive understanding of the degraded products from polysaccharides and lignins during the hydrothermal pretreatment will be beneficial to value-added applications of multiple chemicals in the biorefinery for bioethanol industry.

Keywords: Sweet sorghum stem, Hydrothermal pretreatment, Degraded products, Heteronuclear single quantum coherence (HSQC)

Background

Lignocelluloses have huge growth potential due to their contribution to the bioethanol production and to decrease CO_2 emission and global warming in recent years [1,2]. Sweet sorghum is considered as one of the potential renewable sources of energy for economic development and environmental sustainability, owing to its wide adaptability and high concentration of soluble sugars [3]. This plant could be effectively used as a source of fodder (stem), food (grains), and also as feedstock (cellulose, hemicelluloses, and lignins) for the production of industrial chemicals [3,4]. However, in the production of bioethanol, the direct enzymatic hydrolysis of cellulose is greatly hampered by the dense and complex architectural structure of biomass [5]. Since pretreatment accelerates enzymes accessing the rigid structure of biomass for maximum yield of fermentable sugar, pretreatment is considered as a crucial link prior to biomass enzymatic hydrolysis [5,6]. By now, a variety of physical, chemical, mechanical, and biological pretreatment techniques have been developed, including the developments of steam explosion, organosolv pretreatment, and hydrothermal

* Correspondence: rcsun3@bjfu.edu.cn
No. 35 Tsing Hua East Road, Haidian District, Beijing Key Laboratory of Lignocellulosic Chemistry, Beijing Forestry University, Beijing 100083, China

pretreatment (HTP) [7-9]. Among them, HTP is a handy and eco-friendly technology for the pretreatment of lignocelluloses as compared to others, since the medium only contains water, avoiding corrosion of equipment [10]. In addition, HTP can result in structural changes of lignins and cellulose as well as solubilization of hemicelluloses, which in turn contribute to the improved enzyme accessibility in pretreated biomass [10]. Furthermore, HTP can also produce some important by-products, such as xylo-oligosaccharides (XOS), and some other chemicals (for example, furfural, formic acid, and acetic acid) [11]. The fundamental understanding of the degraded products can increase high-value utilization of biomass in a biorefinery process [12]. Therefore, it is necessary to characterize the degraded products in the liquors recovered during the HTP, which will enhance biorefinery viability in the future.

In this study, the sweet sorghum stems (SSS) were subjected to different HTP conditions (110°C to 230°C, 0.5 to 2.0 h). For the substrates obtained after the HTP process, the enzymatic hydrolysis efficiency and structural properties have been thoroughly investigated in a recent publication [13]. This work mainly focuses on chemical characteristics of the degraded products in the liquor phase produced during the HTP process. Among them, the XOS, monosaccharides, and degree of polymerization of XOS were analyzed by high-performance anion exchange chromatography (HPAEC). Structural features of the XOS were identified by two-dimensional heteronuclear single quantum coherence (2D-HSQC). The chemical features (composition, molecular weights, and structural features) of the water-soluble polysaccharides and lignins (WSPs and WSLs) obtained from the liquor phase during the HTP process were systematically evaluated by HPAEC, gel permeation chromatography (GPC), and 2D-HSQC NMR, which will contribute to the potential applications in bioethanol industry under the biorefinery scenario. The schematic diagram of the experimental procedure is illustrated in Figure 1.

Results and discussion

The liquor pH

The pH of the initial liquor before heating was 6.80, and it decreased to 3.13 with the increasing HTP temperature and time (Table 1). This is expected that the hydronium ions released by the water during the HTP cause depolymerization of hemicelluloses by selective hydrolysis

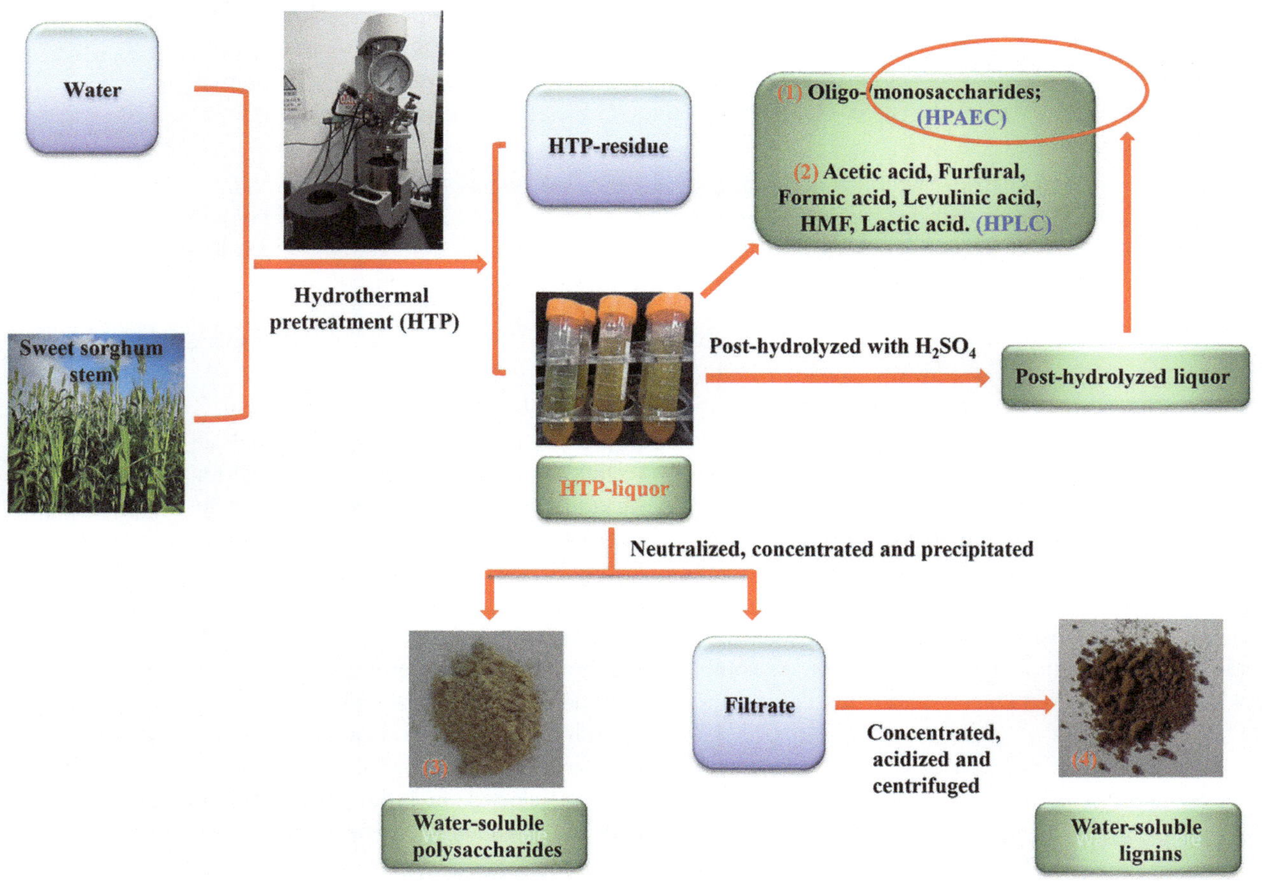

Figure 1 Schematic illustration of the experimental procedure.

Table 1 The liquor pH and the degraded products of polysaccharides from sweet sorghum stem during the hydrothermal pretreatment process

Temperature (°C)-time (h)	Degraded products (g/L)						pH
	Formic acid	Acetic acid	Levulinic acid	Lactic acid	HMF	Furfural	
110-1.0	ND	ND	ND	ND	ND	ND	5.18
130-1.0	ND	ND	ND	ND	ND	ND	4.50
150-1.0	1.10	0.38	0.52	ND	0.06	ND	3.78
170-0.5	0.85	0.48	0.74	0.28	0.12	0.18	3.65
170-1.0	1.96	0.63	1.39	0.37	0.21	0.85	3.48
170-2.0	2.26	1.01	2.26	0.73	0.32	2.11	3.30
190-0.5	2.56	1.26	2.43	1.65	0.38	3.32	3.22
210-0.5	3.16	1.64	2.95	2.20	1.23	3.50	3.18
230-0.5	3.76	4.03	3.65	2.41	2.27	4.53	3.13

ND, not detectable. Data represented are the averages of the results obtained from the duplicated experiments.

of glycosidic linkages, liberating *O*-acetyl group and other acid moieties to form various acids (such as acetic and uronic acids). The release of these acids is thought to catalyze the hydrolysis of hemicelluloses and oligosaccharides, resulting in more hemicelluloses that were degraded and solubilized in the liquor as the pretreatment temperature and time increased [14]. This is also the main reason for decreasing solid yields from 92.4% to 51.5% as the pretreatment temperature and time increased. Similarly, the strong positive correlations between the pH of the liquor and the solid yield have been also observed in a previous study [15].

Xylo-oligosaccharides and monosaccharides

Hemicelluloses (xylan-type) in lignocelluloses are hydrolyzed to xylose or xylo-oligomers during the HTP [6,16,17]. In general, the degree of xylan hydrolysis increases as the HTP severity increases [14]. In this study, the hydrolysis products in the liquor were mainly XOS and monosaccharides. Figure 2 shows the yields of XOS and monosaccharides as the pretreatment temperature and time increased during the HTP. As can be seen (Figure 2A), the yields of XOS (based on the initial xylan in raw material) steadily increased from 3.18% to 47.20% with an increase in the pretreatment temperature from 110°C to 170°C for a constant time (1.0 h). The reason for this is that more xylan was degraded and solubilized in the liquor at the elevated temperatures. Specially, the maximum yield of XOS (52.25%) was achieved in a relatively short period (0.5 h) at 170°C. When the pretreatment time was further prolonged to 1.0 and 2.0 h at 170°C, the yield of XOS decreased constantly with the prolongation of reaction time. The fact suggested that the prolongation of reaction time was adverse to the production of XOS at 170°C for 0.5 h. As expected, when the pretreatment temperature was higher than 170°C at a constant short time (0.5 h), the yields of XOS constantly decreased from 52.25% to 0.37% (in the experiment performed at 230°C). This was attributed to the fact that almost all the xylan was degraded into the xylose and other small molecules, resulting in only a small amount of XOS detected in the liquor phase under this condition [18]. Taken together, the optimal yield of XOS was achieved in the case of the material pretreated at 170°C for 0.5 h. Hence, this condition was considered as a promising pathway to obtain high-yield of XOS from SSS by the HTP process.

Figure 2A also illustrates that the yield of xylose (based on the initial xylan in raw material) was obtained at various pretreatment conditions. When the pretreatment time was 1.0 h, a significant enhancement of xylose concentration was observed as the pretreatment temperature increased from 110°C to 170°C, and a higher pretreatment temperature and longer pretreating period resulted in a rapid raise of xylose. The maximum yield of xylose was achieved (21.47% of the initial xylan) at 170°C for 2.0 h. The concentration variations of arabinose and galactose at various pretreatment conditions are shown in Figure 2B. The change of galactose concentration was quite similar to that of xylose, whose maximum concentration was also obtained (0.44 g/L) at 170°C for 2.0 h. The highest arabinose concentration was observed (2.36 g/L) at 170°C for 0.5 h. Under the optimal pretreatment condition for XOS (170°C for 0.5 h), the concentrations of arabinose and galactose were 2.36 and 0.21 g/L, respectively. Moreover, low concentrations of glucose were detected in the liquors pretreated at harsh conditions (that is, 0.59 g/L at 210°C for 0.5 h and 0.48 g/L at 230°C for 0.5 h, respectively), indicating that the part of non-crystalline cellulose may be hydrolyzed under the conditions given.

Degree of polymerization and HSQC analysis of xylo-oligosaccharides

A systematic study was proposed for the production, purification, and applications of XOS as food additives and nutraceuticals. Considered as food ingredients, XOS have favorable modulation on the intestinal function [19]. In the present study, the degree of polymerization (DP) of XOS obtained during the HTP from SSS was determined. Figure 3 presents the concentration and DP distribution of XOS. The concentrations of xylobiose, xylotriose, xylotetraose, xylopentaose, and xylohexaose were determined by HPAEC. The total concentration of XOS was determined by an increasing concentration of monosaccharides after post-hydrolysis.

The DP of XOS illustrated that it was influenced by the pretreatment temperature and time. As the pretreatment temperature increased (110°C to 170°C, 1.0 h), more low-DP XOS (DP 2 to 6) were obtained. However, when the pretreatment temperature further increased from 170°C

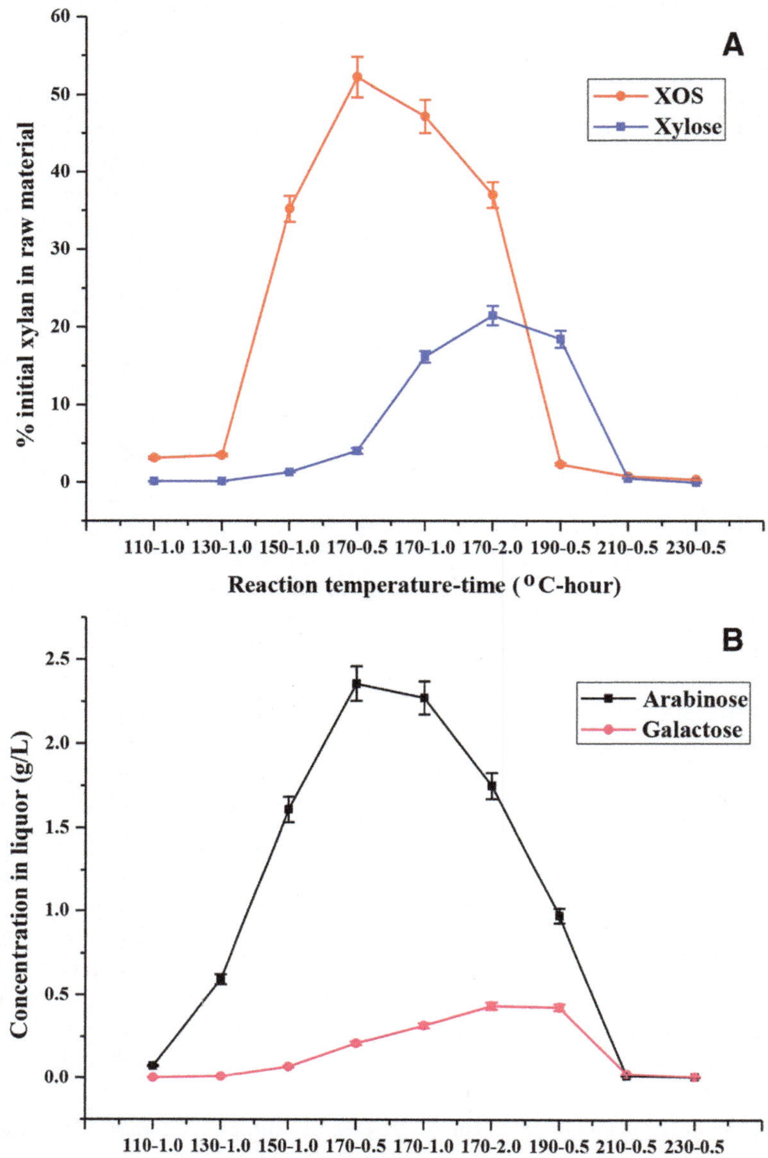

Figure 2 Experimental values for (A) XOS and xylose, (B) arabinose, and galactose obtained during the hydrothermal pretreatment at 110°C to 230°C. The yields of XOS and xylose are expressed as the percentage of the xylan in raw materials. The concentrations of arabinose and galactose are expressed as the concentration in the liquor phase. The error bars are standard deviations from experiments performed in duplicate. XOS, xylo-oligosaccharides.

to 230°C, the yields of low-DP XOS (DP 2 to 6) rapidly declined with the pretreatment severity. The maximum yield of XOS was achieved with 14.05% xylobiose (2.12 g/L), 9.61% xylotriose (1.45 g/L), 4.71% xylotetraose (0.71 g/L), 16.10% xylopentaose (2.43 g/L), 5.04% xylohexaose (0.76 g/L), and 50.49% high-DP XOS (DP > 6) (7.63 g/L) at 170°C for 0.5 h. However, when the reaction time was prolonged to 1.0 h at 170°C, the yield of low-DP XOS (DP 2 to 6) increased but the high-DP XOS (DP > 6) decreased. After further prolonging the reaction time to 2.0 h, the low and high-DP XOS simultaneously reduced. These phenomena were probably due to the XOS that was further degraded into other small molecules (such as furfural) under the harsh conditions, as revealed by the subsequent degraded products analysis in the liquors produced during the HTP from SSS.

A 2D-HSQC spectrum (Figure 4) was used to identify the various structural features of XOS obtained under the pretreated at 170°C for 0.5 h. The signals of the XOS were interpreted as follows: (a) reducing end xylopyranosyl (Xylp) residues of α and β conformations (δ_C/δ_H 92.0/5.13 and 96.5/4.52); (b) internal Xylp residues $(1 \rightarrow 4)$-β-D-Xylp,

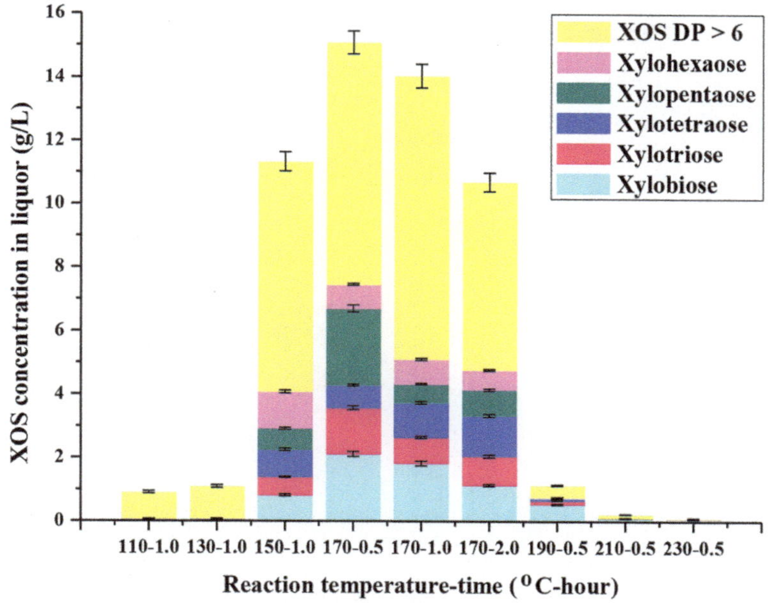

Figure 3 The concentration of xylo-oligosaccharides with different degrees of polymerization. The error bars are standard deviations from experiments performed in duplicate.

that is, δ_C/δ_H 101.6/4.40 (C_1-H_1), 72.6/3.23 (C_2-H_2), 73.6/3.49 (C_3-H_3), 76.3/3.72 (C_4-H_4), 62.9/4.04 (C_{5eq}-H_{5eq}), and 62.9/3.32 (C_{5ax}-H_{5ax}); (c) terminal (non-reducing end) Xylp residues (*β*-D-Xylp-(1-, that is, δ_C/δ_H 75.6/3.37 and 69.2/3.56); and (d) a small amount of Xylp units attached with 4-*O*-Me-*α*-D-GlcpA units (δ_C/δ_H 72.6/3.60 and 59.9/3.40). Many groups exhibited unequal intensities, indicating that the XOS contained a mixture of several molecules. The XOS showed a low degree of branching according to the weak signals of 4-*O*-Me-*α*-D-GlcpA residues. Note that the structural features of XOS are dependent not only on the origin of the xylan-rich hemicelluloses but also on the process [20]. For example, XOS prepared from xylan-rich hemicelluloses, isolated using aqueous potassium hydroxide from delignified sugarcane bagasse, by hydrolysis with crude xylanase secreted by *Pichia stipitis* contained Araf and 4-*O*-Me-*α*-D-GlcpA residues [21].

Degraded products of polysaccharides

The main degraded products of polysaccharides from lignocelluloses during the HTP were acetic acid, furfural, hydroxymethylfurfural (HMF), and formic acid, which were described in detail in a recent publication [18]. As can be seen, hemicelluloses are partially acetylated and the acetyl ester bonds are hydrolyzed to produce acetic acid under the HTP process. Furfural is a pentose dehydration product and can be further degraded through hydrolytic fission of the aldehyde group to formic acid under the acidic hydrothermal condition [22]. HMF is a dehydration product of hexose sugars (glucose, galactose, and mannose) and can be further transformed into levulinic and formic acids under the acidic condition [23]. These degraded products are undesirable, since they act as inhibitors of the fermentation process [24].

In this study, acetic acid, furfural, HMF, formic acid, levulinic acid, and lactic acid were found to be the main degraded products detected in the liquor fractions. The concentrations of the degraded products of polysaccharides in the liquor phase were listed in Table 1. The concentrations of acetic acid (4.03 g/L) and furfural (4.53 g/L) were significantly higher than other degraded products (<3.76 g/L) at the severest pretreatment condition. It was interesting to note that the release of all degraded products was closely related to the pretreatment temperature and time, that is, the concentrations of all degraded products increased with increasing temperature at a constant time or increasing time at a constant temperature. As expected, the maximum concentrations of all degraded products were observed in the case of the material pretreated at 230°C for 0.5 h, corresponding to the severest experimental condition in this work. It appears that the HTP condition under the higher HTP temperature at a short reaction time (that is, 230°C for 0.5 h) favors the dehydration of xylose to furfural as compared to that obtained at other experiment conditions. The concentration of HMF remained low throughout among the whole range of pretreatment severities due to limited cellulose solubilization and consequent glucose formation. The concentration of levulinic acid was much lower than that of formic acid, suggesting that most of formic acid was produced from furfural and

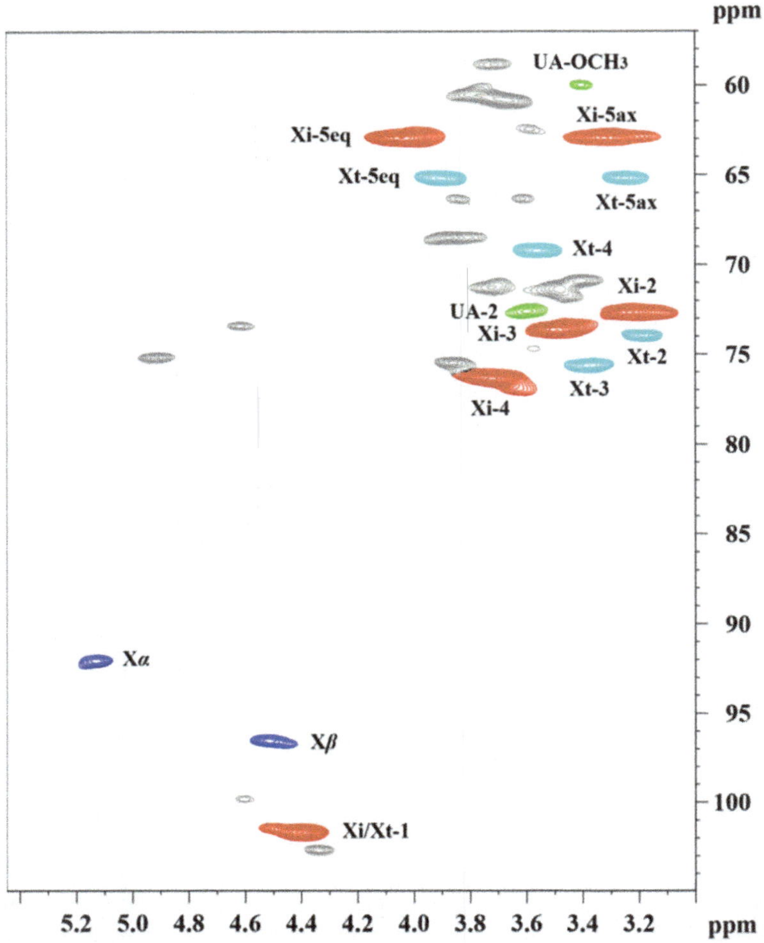

Figure 4 HSQC analysis of xylo-oligosaccharides obtained during the hydrothermal pretreatment at 170°C for 0.5 h. HSQC, heteronuclear single quantum coherence; Xα and Xβ, xylopyranosyl reducing ends; Xi and Xt, internal and nonreducing ends; UA, 4-*O*-methyglucuronic acid.

not from HMF degradation, since the latter would lead to the production of equimolar amounts of levulinic and formic acids [18].

Water-soluble polysaccharides

The composition of the WSPs obtained from the HTP-liquor phase is presented in Table 2. These WSPs contained most xylose and glucose together with small amount of arabinose, galactose, rhamnose, mannose, and uronic acids (glucuronic and galacturonic acids). Under the HTP temperature lower than 210°C, the content of xylose was apparently higher than glucose in the WSPs. In contrast, as the HTP temperature over 190°C, the content of glucose predominates over xylose in the WSPs. The high content of glucose in the two WSPs ($WSP_{210(0.5)}$ and $WSP_{230(0.5)}$) was probably related to the hydrolysis of amorphous cellulose under the harsh conditions. The amorphous cellulose may be hydrolyzed to the cellulosic fragment, resulting in part of cellulose that was co-precipitated in the WSPs. Table 3 exhibits weight average (M_w) and number average (M_n) molecular weights along with polydispersity indexes (PDIs, M_w/M_n) of the WSPs. The M_w and M_n of the WSPs gradually reduced as the pretreatment temperature and time increased, implying that the C-O bonds of the WSP was gradually cleaved, especially at the harsh conditions (≥170°C). In addition, these WSPs with lower PDIs (1.10 to 2.01) suggested that the HTP process yielded more homogeneous polysaccharide polymers.

Structural features of the WSPs ($WSP_{130(1.0)}$ and $WSP_{170(1.0)}$) were determined by 2D-HSQC NMR technique (Figure 5). Comparison of HSQC spectra of the WSPs obtained from the liquor phase can help to better understand the structural variation of the WSPs during the HTP process. For $WSP_{130(1.0)}$, cross signals of (1→4)-linked *β*-D-Xyl*p* units at δ_C/δ_H 102.0/4.31 (C_1-H_1), 73.2/3.15 (C_2-H_2), 74.5/3.38 (C_3-H_3), 76.0/3.63 (C_4-H_4), 62.9/3.94 (C_{5eq}-H_{5eq}), and 63.3/3.25 (C_{5ax}-H_{5ax}) were clearly observed. The signals of *α*-L-arabinofuranosyl units were found at δ_C/δ_H 109.4/5.15,

Table 2 Relative content of sugars and uronic acids of the water-soluble polysaccharides isolated from the hydrothermally pretreated liquor phase under the various processing conditions

Temperature (°C)-time (h)	Neutral sugars and uronic acids						
	Rhamnose	Arabinose	Galactose	Glucose	Xylose	Mannose	Uronic acid
110-1.0	1.76	21.34	10.13	27.48	33.69	0.94	4.66
130-1.0	0.95	21.02	10.24	26.63	37.86	0.29	3.01
150-1.0	ND	6.27	6.07	21.04	65.56	ND	1.06
170-0.5	ND	3.42	4.17	14.59	77.28	ND	0.54
170-1.0	ND	6.72	5.49	21.02	66.77	ND	ND
170-2.0	ND	6.50	4.17	20.23	69.10	ND	ND
190-0.5	ND	5.76	3.02	18.64	72.58	ND	ND
210-0.5	ND	ND	2.63	81.16	16.21	ND	ND
230-0.5	ND	ND	ND	82.68	17.32	ND	ND

ND, not detectable. Data represented are the averages of the results obtained from the duplicated experiments.

80.1/3.93, 78.3/3.67, 86.3/4.09, and 61.2/3.65, which are characteristic of C_1-H_1, C_2-H_2, C_3-H_3, C_4-H_4, and C_5-H_5, respectively. Some weak signals at δ_C/δ_H 97.4/5.16, 71.5/3.42, 74.0/3.73, 82.4/3.08, 72.1/4.19, and 59.6/3.33 are assigned to C_1-H_1, C_2-H_2, C_3-H_3, C_4-H_4, C_5-H_5, and -OCH_3 of 4-*O*-Me-*α*-D-Glc*p*A units, respectively [25,26]. As the HTP temperature was further increased to 170°C and kept for 1.0 h, only the cross signals of *β*-D-Xyl*p* units were observed, implying that the xylan-rich polymers ($WSP_{170(1.0)}$) were collected. This is because that the side chains of the WSP macromolecules were significantly cleaved under this condition. These results suggested that the degradation degree of xylans increased with the HTP temperature increase, as revealed by the aforementioned molecular weight analysis.

Water-soluble lignins

Lignin is considered the most persistent component of the plant cell wall biopolymers [27]. It is found primarily in the secondary cell wall and plays a major role in pathogen resistance, water regulation, and conferring strength for the integrity of the cell wall structure. The effects of lignin on biomass enzymatic digestibility have received extensive attention [10,28]. It is generally recognized that the existence of lignins in lignocelluloses restricts enzymatic saccharification by physically impeding the accessibility of cellulase to cellulose [10]. HTP causes fragmentation of lignins, which results in the redistribution of lignins and its structural changes depending on the HTP severity [29]. It has been reported that a decrease in molecular weight of native lignins would facilitate its dissolution and/or migration to the surface in the reaction media during the HTP [10]. The spherical droplets found in the pretreated hardwood are thought to be mainly composed of lignins based on previous reports [30,31]. In addition, pseudo-lignin derived from the dehydrated carbohydrates during severe HTP was also suggested to be responsible for the formation of droplets [32,33]. In this study, scanning electron microscopy (SEM) images demonstrated that the harsh HTP process (210°C for 0.5 h) prompted migration or formation of lignins or pseudo-lignin (Figure 6A). Figure 6B shows a partial enlargement, revealing in more spherical droplets with different sizes. To sum up, SEM images suggested that the lignin or pseudo-lignin during the harsh HTP has inhomogeneous morphological features. To prove the fragmentation of lignins during the HTP, the molecular weights of the WSLs ($WSL_{110(1.0)}$, $WSL_{130(1.0)}$, $WSL_{150(1.0)}$, and $WSL_{170(1.0)}$) were determined by GPC analysis (Table 4). It can be seen that an elevated HTP temperature caused the tremendous decrease of molecular weight in the WSLs, especially at higher temperature condition. Several researchers have investigated the evaluation of structural changes of lignins during the HTP process, and the results suggest that there

Table 3 Weight-average (M_w) and number-average (M_n) molecular weights and polydispersits (M_w/M_n) of the water-soluble polysaccharides isolated from the hydrothermally pretreated liquor phase under the various processing conditions

	Water-soluble polysaccharides (temperature (°C)-time (h))								
	110-1.0	130-1.0	150-1.0	170-0.5	170-1.0	170-2.0	190-0.5	210-0.5	230-0.5
M_w	31,220	25,920	16,830	14,320	7,110	6,710	7,080	6,190	4,780
M_n	15,550	14,160	8,520	7,670	6,240	5,930	6,150	5,620	4,020
M_w/M_n	2.01	1.83	1.98	1.88	1.14	1.13	1.15	1.10	1.19

Data represented are the averages of the results obtained from the duplicated experiments.

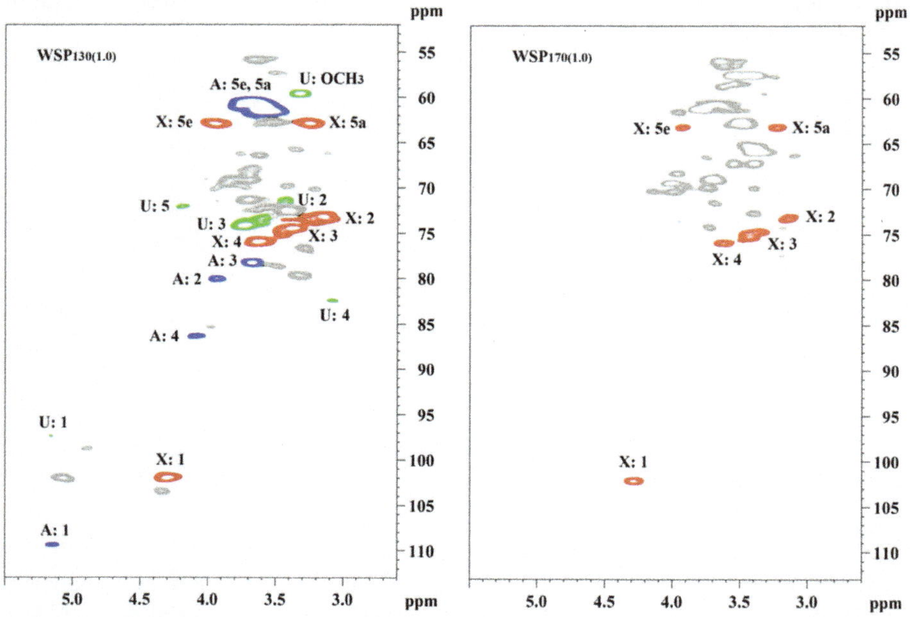

Figure 5 HSQC spectra of water-soluble polysaccharides ($WSP_{130(1.0)}$ and $WSP_{170(1.0)}$). HSQC, heteronuclear single quantum coherence; WSP, water-soluble polysaccharide; X, xylopyranosyl units; A, arabinofuranosyl units; U, 4-*O*-methyglucuronic acid.

is no significant decrease in the molecular weight of lignins. This was attributed to the partial protection of lignins by polysaccharides in the compact biomass matrix, which may facilitate the simultaneous re-condensation process occurring along with depolymerization [34]. Moreover, the polydispersity indices of the WSLs were slightly decreased (1.48 to 1.13) with an increasing pretreatment temperature from 110°C to 170°C for 1.0 h, suggesting that more homogeneous lignin fragments were obtained under the harsh condition (170°C for 1.0 h).

To further reveal structural differences of the WSLs ($WSL_{130(1.0)}$ and $WSL_{170(1.0)}$) obtained from the liquor phase, the more detailed structural information was investigated by 2D-HSQC NMR spectra (Figure 7) [35-37]. The inter-unit linkages in the WSLs, β-ether (A, β-*O*-4) and phenylcoumaran (C, β-5) were identified due to the presence of cross-peaks at δ_C/δ_H 71.5/4.82 (A_α), 60.2/3.60 (A_γ), and 62.5/3.67 (C_γ), respectively. Moreover, the cross-peaks of methoxy groups in the WSLs ($-OCH_3$, δ_C/δ_H 55.5/3.72) were clearly observed. The aromatic lignin units syringyl (S), guaiacyl (G) and *p*-hydroxyphenyl (H) units showed prominent correlations at δ_C/δ_H 103.6/6.63 ($S_{2,6}$), 110.7/6.93 (G_2), 115.4/6.78 (G_5), 118.8/6.73 (G_6), 128.0/7.19 ($H_{2,6}$), respectively. Minor amounts of oxidized S units (S') were detected due to the presence of a correlation at δ_C/δ_H 106.8/7.20 ($S'_{2,6}$). In addition, the WSLs obtained appeared to contain small amounts of *p*-hydroxycinnamates (*p*-coumaric and ferulic acids). The *p*-coumaric acid (PCA) was characterized by some relatively intense correlations at δ_C/δ_H 129.8/7.49 ($PCA_{2,6}$), 115.0/6.27 (PCA_8), and 144.0/7.47 (PCA_7), while ferulic acid (FA) was found at δ_C/δ_H 110.9/7.31 (FA_2), 115.0/6.27 (FA_8), and 144.0/7.47 (FA_7). As can be seen, when the pretreatment temperature reached 170°C, some expected changes appeared in the HSQC spectrum of $WSL_{170(1.0)}$ as compared to that of $WSL_{130(1.0)}$. For example, the correlated signals (δ_C/δ_H 71.5/4.82, 60.2/3.60, 83.6/4.29, and 85.2/4.08) of β-*O*-4 were dramatically increased in $WSL_{170(1.0)}$. Meanwhile, the signal strength of polysaccharides reduced in $WSL_{170(1.0)}$. This was mainly attributed to the degradation of polysaccharides, especially under the higher temperatures. The facts showed that more pure lignins were achieved under this condition. Besides, no signal of S' was detected in $WSL_{170(1.0)}$, implying that the oxidized S units occurred at a relatively lower temperature for a shorter time of pretreatment. The correlated signals of PCA were increased while the signals of FA were decreased as the pretreatment temperature increased from 110°C to 170°C for 1.0 h.

Process mass balance

In this study, the maximum yield of XOS was achieved at 170°C for 0.5 h with a relatively low level of xylose and other by-products. Prolonged HTP time and higher temperature reduced the yield of XOS and enhanced the concentrations of monosaccharides and by-products. Therefore, a process mass balance of the HTP process was developed for the two HTP conditions (0.5 and 2.0 h at 170°C) (Figure 8). Process yield was normalized to a common basis of 100 kg of dried SSS as the starting material [38]. It should be noted that, after the HTP, the liquor and solid were separated by filtration. The yields of the HTP

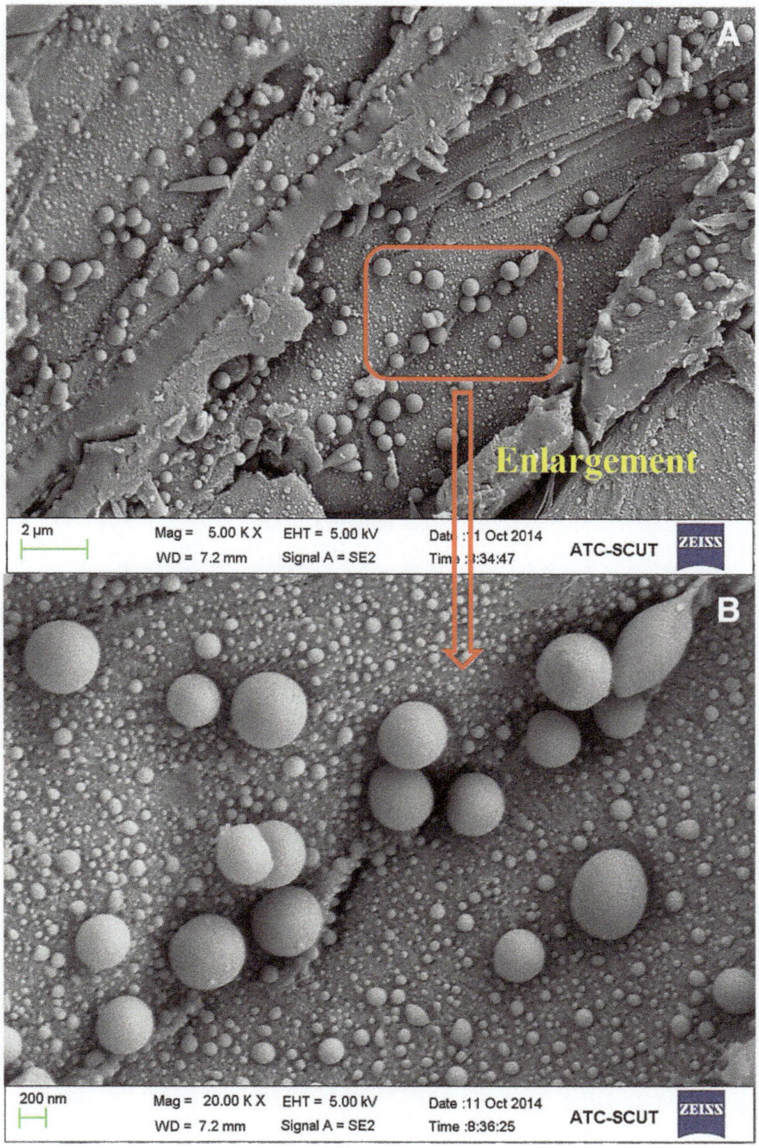

Figure 6 SEM images of the hydrothermally pretreated substrate at 210°C for 0.5 h. (A) Magnification × 5,000. **(B)** Magnification × 20,000. SEM, scanning electron microscopy.

residues decreased from 68.60 to 61.90 kg with an increasing HTP time from 0.5 to 2.0 h at 170°C [39]. It was found that 15.10 and 10.70 kg XOS could be obtained when the HTP processes were performed at 170°C for 0.5 and 2.0 h, respectively (Figure 8). In addition, other more degraded products appeared in the liquor fractions for the prolonged HTP period, which was attributed to the significant degradation of hemicelluloses under the harsh HTP conditions.

Conclusions

The present study demonstrated that the hydrothermally pretreated conditions had significant influences on the compositions and their chemical structures of the degraded products. It was found that the maximum yield of XOS of 52.25% was achieved at 170°C for 0.5 h with a relatively low level of xylose and other degraded products. Higher temperature and/or longer reaction time reduced the yield of XOS and enhanced concentrations of monosaccharides (arabinose and galactose). In addition, as the pretreatment temperature and time increased, the concentrations of the degraded products (such as acetic acid and furfural) were dramatically increased. Furthermore, the 2D-HSQC spectrum of the XOS suggested that the XOS obtained mainly contained xylopyranosyl residues together with small amounts of 4-*O*-Me-*α*-D-GlcpA units. Molecular weights of the water-soluble polysaccharides and lignins implied that the degradation of the polysaccharides and lignins occurred at the harsh

Table 4 Weight-average (M_w) and number-average (M_n) molecular weights and polydispersity indices (M_w/M_n) of the water-soluble lignins isolated from the hydrothermally pretreated liquor phase under the various processing conditions

	Water-soluble lignins (temperature (°C)-time (h))			
	110-1.0	130-1.0	150-1.0	170-1.0
M_w	1,610	1,590	1,050	950
M_n	1,090	1,220	900	840
M_w/M_n	1.48	1.30	1.17	1.13

Data represented are the averages of the results obtained from the duplicated experiments.

conditions. Interestingly, 2D-HSQC spectra suggested that the water-soluble polysaccharides (rich in xylan) and water-soluble lignins (rich in *β-O-*4 linkages) were obtained at 170°C for 1.0 h. In short, systematic evaluation of the degraded products during the hydrothermal pretreatment of SSS will be beneficial to value-added applications of multiple chemicals in a biorefinery process for bioethanol industry.

Methods

Raw materials

SSS was obtained from the experimental farm of the North-Western University of Agricultural and Forest Sciences and Technology (Yangling, Shaanxi, China). As determined by the National Renewable Energy Laboratory (NREL) using a two-step acid hydrolysis procedure, the extractive-free SSS was found to contain 41.1% glucan, 25.9% xylan, 2.3% arabinan, 0.7% galactan, and 21.4% lignin (19.5% Klason lignin and 1.9% acid-soluble lignin) [40]. More details of the materials used in this study were given in a recent literature [26]. XOS (DP 2 to 6) standards were obtained from Megazyme (Megazyme International Ireland Ltd., Wicklow, Ireland). All chemicals were analytical grade and purchased from Sigma-Aldrich (Beijing, China).

Hydrothermal pretreatment

The HTP process was carried out in a 1,000-mL stainless steel autoclave (Parr Instrument Company, Moline, IL, USA) with a magnetic stirrer at a solid to liquor ratio of 1:10 (g/mL) by a PID controller (model 4848). The HTP experiments were divided into nine batches, in each batch, 15.0 g the extractive-free material was mixed with 150 mL deionized water. Then, the mixture was heated at 110°C, 130°C, or 150°C for 1.0 h, 170°C for different periods (0.5, 1.0, or 2.0 h), or 190°C, 210°C, or 230°C for 0.5 h. After the reaction, the reactor was cooled by cold water.

On the one hand, a liquor sample of 3 mL was post-hydrolyzed with 4% H_2SO_4 at 121°C for 1 h to determine the total concentration of XOS and their degree of substitution by acetyl group. The increased concentrations of monosaccharides and acetic acid after post-hydrolysis was regarded as the concentration of XOS and the amount of acetyl group attached on XOS, respectively [41,42]. According to this procedure, saccharides of DP 2 or higher were considered as XOS. This is different from the common definition of XOS which refers to a DP of 2 to 10. All the liquor samples were filtered through 0.22-μm filter before analysis. The concentration of XOS was expressed as monosaccharides equivalents.

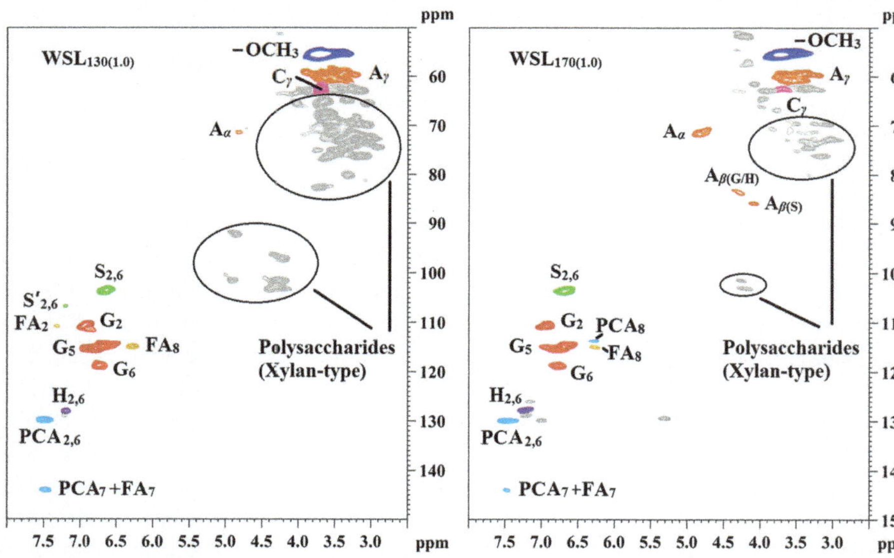

Figure 7 HSQC spectra of water-soluble lignins ($WSL_{130(1.0)}$ and $WSL_{170(1.0)}$). HSQC, heteronuclear single quantum coherence; WSL, water-soluble lignin; A, *β*-aryl-ether units (*β*-*O*-4); C, phenylcoumaran substructures (*β*-5); PCA, free *p*-coumaric acid; FA, ferulate; H, *p*-hydroxyphenyl units; G, guaiacyl units; S, syringyl units; S′, oxidized syringyl units bearing a carbonyl at $C_α$.

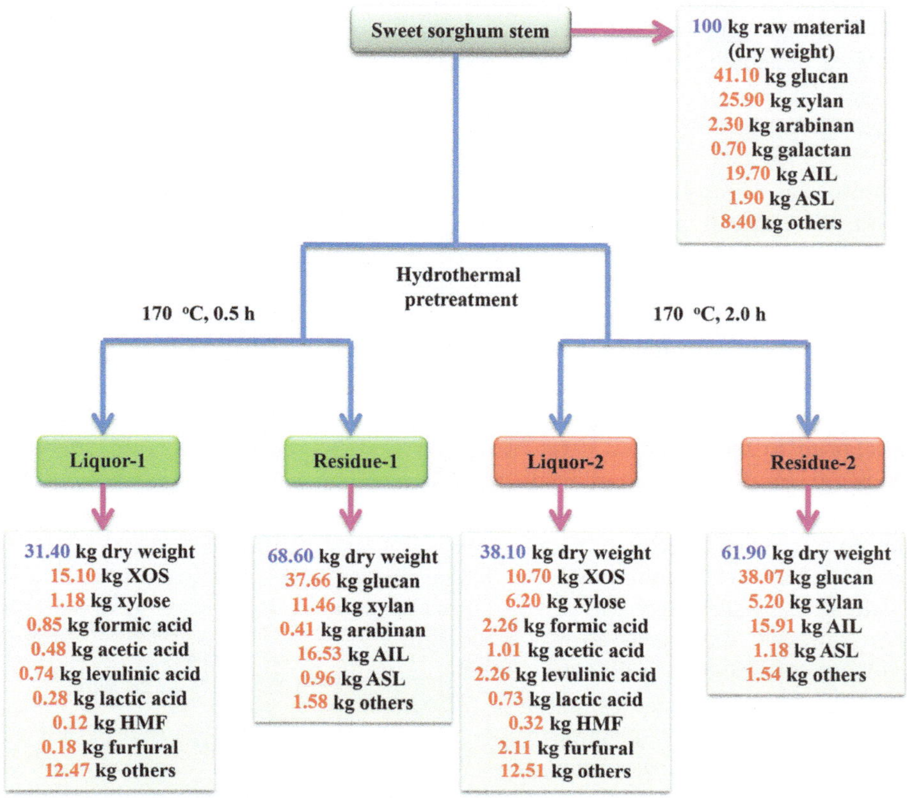

Figure 8 Mass balance during the pretreatment at 170°C for 0.5 and 2.0 h.

On the other hand, the liquor (100 mL) was neutralized to pH 5.5 to 6.0 with 6 M NaOH and were further concentrated to 40 mL under vacuum. Subsequently, each concentrated solution was poured into 95% ethanol (120 mL) with vigorous stirring, and the WSP was obtained by filtering, washing with 70% aqueous ethanol. The supernatants were concentrated to 30 mL and poured in 150 mL acidic water (pH = 2.0, adjusted by HCl) to precipitate the WSL. The detailed procedures are previously described in a publication [43]. The WSP and WSL obtained were freeze-dried using a lyophilizer (Thermo Modulyo Freeze Dryer; Thermo Scientific, Waltham, MA, USA) at −50°C under vacuum for 48 h.

Analysis procedures

HPAEC was used to quantify the monosaccharides and oligosaccharides of the liquor phase. An ICS-3000 HPAEC system with an AS50 autosampler, comprising a Carbopac PA-20 column (4 × 250 mm, Dionex, Sunnyvale, CA, USA) for monosaccharides and a Carbopac PA-100 column (4 × 250 mm, Dionex) for oligosaccharides were used. The quantitative analysis of the degraded products (acetic acid, furfural, formic acid, levulinic acid, HMF, lactic acid) was determined on a high-performance liquid chromatography (HPLC, Agilent 1200 series, Agilent Technologies Inc., Santa Clara, CA, USA) using the method described in a previous paper with minor modification [44]. The XOS powder obtained in the case of the material pretreated at 170°C for 0.5 h was characterized by HSQC NMR. HSQC experiment was conducted with 20 mg sample dissolved in 0.5 ml D_2O. The number of collected complex points was 1,024 for the ^{1}H-dimension with a relaxation of 1.5 s. The number of scans was 128, and 256 time increments were recorded in ^{13}C-dimension. The $1J_{C-H}$ used was 146 Hz. Prior to Fourier transformation, the data matrixes were zero filled up to 1,024 points in the ^{13}C-dimension. Substrate (hydrothermally substrate pretreated at 210°C for 0.5 h) was analyzed by SEM. SEM images were performed with a ZEISS EVO 18 (Carl Zeiss, Inc., Oberkochen, Germany). The accelerating voltage of the instrument was maintained at 5.00 and 20.00 kV.

The chemical compositions of the WSPs were analyzed by HPAEC. Each of the fractions (5 mg) was hydrolyzed with 10% H_2SO_4 at 105°C for 2.5 h. Then, the hydrolysate was filtrated, diluted 50-fold, and injected into a HPAEC system (Dionex ICS 3000) with an amperometric detector, an AS50 autosample, a CarbopacTM PA-20 column (4 × 250 mm, Dionex), and a guard PA-20 column (3 × 30 mm, Dionex). Neutral sugars and uronic acids were separated in a 5-mM NaOH isocratic (carbonate free and purged with nitrogen) for 20 min, followed by a 0 to 75 mM NaAc gradient in a 5-mM NaOH for 15 min.

Then, the columns were washed with 200 mM NaOH for 10 min to remove the carbonate and followed by a 5-min elution with 5 mM NaOH to re-equilibrate the column before the next injection [45]. The M_w and M_n molecular weights of all the WSPs were determined by GPC, using a PL aquagel-OH 50 column (300 × 7.7 mm, Polymer Laboratories Ltd., Church Stretton, Shropshire, UK). The data were calibrated with PL pullulan polysaccharide standards. The detector used was a differential refractive index detector (RID). The eluent was 0.02 M NaCl in 0.005 M sodium phosphate buffer (pH 7.5), and the flow rate was 0.5 ml/min. Soluble-state NMR spectra were conducted on a Bruker AVIII 400 MHz spectrometer (Bruker AXS, Inc., Madison, WI, USA). HSQC NMR technique has also been applied to the structural determination of the WSPs based on the aforementioned conditions (XOS). The M_w and M_n of the WSLs were determined by GPC with an ultraviolet detector (UV) at 240 nm. The column used was a PL gel 10 mm mixed-B 7.5 mm i.d. column, which was calibrated with PL polystyrene standards. For 2D-HSQC spectra, the Bruker pulse program "hsqcetgpsi" was used and the parameters used is listed as below: the number of collected complex points was 1 K for the ^1H dimension with d_1 (2 s), number of scanning is 64, and 256 time increments were always recorded.

Abbreviations
2D-HSQC: two-dimensional heteronuclear single quantum coherence; DP: degree of polymerization; FA: ferulic acid; G: guaiacyl; GPC: gel permeation chromatography; H: *p*-hydroxyphenyl; HMF: hydroxymethylfurfural; HPAEC: high-performance anion exchange chromatography; HPLC: high-performance liquid chromatography; HTP: hydrothermal pretreatment; M_n: number-average; M_w: weight-average; NREL: National Renewable Energy Laboratory's; PCA: free *p*-coumaric acid; PDI: polydispersity indexes; RID: refractive index detector; S: syringyl; SEM: scanning electron microscopy; SSS: sweet sorghum stem; WSL: water-soluble lignins; WSP: water-soluble polysaccharide; XOS: xylo-oligosaccharides.

Competing interests
The authors declare that they have no competing interests.

Authors' contributions
SLS completed performed the major experiments, analyzed the data, and prepared the manuscript. JLW and SNS helped with the overall pretreatment experiments of the pretreatment and the instrumental analysis. RCS designed the project, supervised the experiments, and finalized the manuscript. All authors read and approved the final manuscript.

Acknowledgements
The authors are extremely grateful for the financial support from the National Science Foundation of China (31430092) and the State Forestry Administration (20124803).

References
1. Fargione J, Hill J, Tilman D, Polasky S, Hawthorne P. Land clearing and the biofuel carbon debt. Science. 2012;319:1235–8.
2. Himmel ME, Ding SY, Johnson DK, Adney WS, Nimlos MR, Brady JW, et al. Biomass recalcitrance: engineering plants and enzymes for biofuels production. Science. 2007;315:804–7.
3. Prakasham RS, Nagaiah D, Vinutha KS, Uma A, Chiranjeevi T, Umakanth AV, et al. Sorghum biomass: a novel renewable carbon source for industrial bioproducts. Biofuels. 2012;5:159–74.
4. Massacci A, Battistelli A, Loreto F. Effect of drought stress on photosynthetic characteristics, growth and sugar accumulation of field-grown sweet sorghum. Funct Plant Biol. 1996;23:331–40.
5. Alvira P, Tomás-Pejó E, Ballesteros M, Negro M. Pretreatment technologies for an efficient bioethanol production process based on enzymatic hydrolysis: a review. Bioresour Technol. 2010;10:4851–61.
6. Sun SN, Cao XF, Sun SL, Xu F, Song XL, Jones GL. Improving the enzymatic hydrolysis of thermo-mechanical fiber from *Eucalyptus urophylla* by a combination of hydrothermal pretreatment and alkali fractionation. Biotechnol Biofuels. 2014;7:116.
7. Moreno AD, Tomás-Pejó E, Ibarra D, Ballesteros M, Olsson L. Fed-batch SSCF using steam-exploded wheat straw at high dry matter consistencies and a xylose-fermenting *Saccharomyces cerevisiae* strain: effect of laccase supplementation. Biotechnol Biofuels. 2013;6:160.
8. Park N, Kim HY, Koo BW, Yeo H, Choi IG. Organosolv pretreatment with various catalysts for enhancing enzymatic hydrolysis of pitch pine (*Pinus rigida*). Bioresour Technol. 2010;101:7046–53.
9. Zhang T, Wyman CE, Jakob K, Yang B. Rapid selection and identification of *Miscanthus* genotypes with enhanced glucan and xylan yields from hydrothermal pretreatment followed by enzymatic hydrolysis. Biotechnol Biofuels. 2012;5:56.
10. Pu YQ, Hu F, Huang F, Davison BH, Ragauskas AJ. Assessing the molecular structure basis for biomass recalcitrance during dilute acid and hydrothermal pretreatments. Biotechnol Biofuels. 2013;6:15.
11. Romaní A, Garrote G, Alonso JL, Parajó JC. Bioethanol production from hydrothermally pretreated *Eucalyptus globulus* wood. Bioresour Technol. 2010;101:8706–12.
12. Liu S. Woody biomass: niche position as a source of sustainable renewable chemicals and energy and kinetics of hot-water extraction/hydrolysis. Biotechnol Adv. 2010;28:563–82.
13. Sun SL, Sun SN, Wen JL, Zhang XM, Peng F, Sun RC. Assessment of integrated process based on hydrothermal and alkaline treatments for enzymatic saccharification of sweet sorghum stems. Bioresour Technol. 2014;175:473–9.
14. Garrote G, Kabel MA, Schols HA, Falqué E, Domínguez H, Parajó JC. Effects of Eucalyptus globulus wood autohydrolysis conditions on the reaction products. J Agric Food Chem. 2007;55:9006–13.
15. Kabel MA, Bos G, Zeevalking J, Voragen AG, Schols HA. Effect of pretreatment severity on xylan solubility and enzymatic breakdown of the remaining cellulose from wheat straw. Bioresour Technol. 2007;98:2034–42.
16. Marzialetti T, Valenzuela Olarte MB, Sievers C, Hoskins TJ, Agrawal PK, Jones CW. Dilute acid hydrolysis of Loblolly pine: a comprehensive approach. Ind Eng Chem Res. 2008;47:7131–40.
17. Xiao X, Bian J, Peng XP, Xu H, Xiao B, Sun RC. Autohydrolysis of bamboo (*Dendrocalamus giganteus* Munro) culm for the production of xylo-oligosaccharides. Bioresour Technol. 2013;138:63–70.
18. Nitsos CK, Matis KA, Triantafyllidis KS. Optimization of hydrothermal pretreatment of lignocellulosic biomass in the bioethanol production process. ChemSusChem. 2013;6:110–22.
19. Moure A, Gullón P, Domínguez H, Parajó JC. Advances in the manufacture, purification and applications of xylo-oligosaccharides as food additives and nutraceuticals. Process Biochem. 2006;41:1913–23.
20. Akpinar O, Erdogan K, Bakir U, Yilmaz L. Comparison of acid and enzymatic hydrolysis of tobacco stalk xylan for preparation of xylooligosaccharides. LWT-Food Sci Technol. 2010;43:119–25.
21. Bian J, Peng F, Peng XP, Peng P, Xu F, Sun RC. Structural features and antioxidant activity of xylooligosaccharides enzymatically produced from sugarcane bagasse. Bioresour Technol. 2013;127:236–41.
22. Dunlop A. Furfural formation and behavior. Ind Eng Chem. 1948;40:204–9.
23. Horvat J, Klaić B, Metelko B, Šunjić V. Mechanism of levulinic acid formation. Tetrahedron Lett. 1985;26:2111–4.
24. Palmqvist E, Hahn-Hägerdal B. Fermentation of lignocellulosic hydrolysates. II: inhibitors and mechanisms of inhibition. Bioresour Technol. 2000;74:25–33.
25. Peng P, Peng F, Bian J, Xu F, Sun R. Studies on the starch and hemicelluloses fractionated by graded ethanol precipitation from bamboo *Phyllostachys bambusoides* f. shouzhu Yi. J Agric Food Chem. 2011;59:2680–8.

26. Sun SL, Wen JL, Ma MG, Song XL, Sun RC. Integrated biorefinery based on hydrothermal and alkaline treatments: investigation of sorghum hemicelluloses. Carbohyd Polym. 2014;111:663–9.
27. Siqueira G, Milagres AMF, Carvalho W, Koch G, Ferraz A. Topochemical distribution of lignin and hydroxycinnamic acids in sugar-cane cell walls and its correlation with the enzymatic hydrolysis of polysaccharides. Biotechnol Biofuels. 2011;4:7.
28. Rezende CA, de Lima MA, Maziero P, DeAzevedo ER, Garcia W, Polikarpov I. Chemical and morphological characterization of sugarcane bagasse submitted to a delignification process for enhanced enzymatic digestibility. Biotechnol Biofuels. 2011;4:54.
29. El Hage R, Chrusciel L, Desharnais L, Brosse N. Effect of autohydrolysis of *Miscanthus x giganteus* on lignin structure and organosolv delignification. Bioresour Technol. 2010;101:9321–9.
30. Selig MJ, Viamajala S, Decker SR, Tucker MP, Himmel ME, Vinzant TB. Deposition of lignin droplets produced during dilute acid pretreatment of maize stems retards enzymatic hydrolysis of cellulose. Biotechnol Progr. 2007;23:1333–9.
31. Donohoe BS, Decker SR, Tucker MP, Himmel ME, Vinzant TB. Visualizing lignin coalescence and migration through maize cell walls following thermochemical pretreatment. Biotechnol Bioeng. 2008;101:913–25.
32. Sannigrahi P, Ragauskas AJ, Miller SJ. Effects of two-stage dilute acid pretreatment on the structure and composition of lignin and cellulose in loblolly pine. BioEnerg Res. 2008;1:205–14.
33. Kumar R, Hu F, Sannigrahi P, Jung S, Ragauskas AJ, Wyman CE. Carbohydrate derived-pseudo-lignin can retard cellulose biological conversion. Biotechnol Bioeng. 2013;110:737–53.
34. Samuel R, Cao S, Das BK, Hu F, Pu Y, Ragauskas AJ. Investigation of the fate of poplar lignin during autohydrolysis pretreatment to understand the biomass recalcitrance. RSC Adv. 2013;3:5305–9.
35. del Río JC, Rencoret J, Marques G, Li J, Gellerstedt G, Jiménez-Barbero J, et al. Structural characterization of the lignin from jute (Corchorus capsularis) fibers. J Agric Food Chem. 2009;57:10271–81.
36. Wen JL, Xue BL, Xu F, Sun RC, Pinkert A. Unmasking the structural features and property of lignin from bamboo. Ind Crop Prod. 2013;42:332–43.
37. Ralph J, Lundquist K, Brunow G, Lu F, Kim H, Schatz PF, et al. Lignins: natural polymers from oxidative coupling of 4-hydroxyphenyl-propanoids. Phytochem Rev. 2004;3:29–60.
38. Percival Zhang YH, Berson E, Sarkanen S, Dale BE. Sessions 3 and 8: pretreatment and biomass recalcitrance: fundamentals and progress. Appl Biochem Biotech. 2009;153:80–3.
39. Sun SL, Wen JL, Ma MG, Sun RC. Structural elucidation of sorghum lignins from integrated biorefinery process based on hydrothermal and alkaline treatments. J Agric Food Chem. 2014;62:8120–8.
40. Sluiter A, Hames B, Ruiz R, Scarlata C, Sluiter J, Templeton D, et al. Determination of structural carbohydrates and lignin in biomass. Golden CO: National Renewable Energy Laboratory; 2008. Technical Report NREL/TP-510-42618.
41. Gullón B, Yáñez R, Alonso J, Parajó J. Production of oligosaccharides and sugars from rye straw: a kinetic approach. Bioresour Technol. 2010;101:6676–84.
42. Nabarlatz D, Farriol X, Montané D. Kinetic modeling of the autohydrolysis of lignocellulosic biomass for the production of hemicellulose-derived oligosaccharides. Ind Eng Chem Res. 2004;43:4124–31.
43. Sun RC, Lawther JM, Banks WB. Effects of extraction time and different alkalis on the composition of alkali-soluble wheat straw lignins. J Agric Food Chem. 1996;44:3965–70.
44. Yang HY, Wang K, Xu F, Sun RC, Lu YB. H_2SO_4-catalyzed hydrothermal pretreatment of triploid poplar to enhance enzymatic hydrolysis. Ind Eng Chem Res. 2012;51:11598–604.
45. Bian J, Peng F, Peng P, Xu F, Sun RC. Isolation and fractionation of hemicelluloses by graded ethanol precipitation from *Caragana korshinskii*. Carbohyd Res. 2010;345:802–9.

8

Assembly of lipase and P450 fatty acid decarboxylase to constitute a novel biosynthetic pathway for production of 1-alkenes from renewable triacylglycerols and oils

Jinyong Yan[1*], Yi Liu[1,2], Cong Wang[1], Bingnan Han[3] and Shengying Li[1*]

Abstract

Background: Biogenic hydrocarbons (biohydrocarbons) are broadly accepted to be the ideal 'drop-in' biofuel alternative to petroleum-based fuels due to their highly similar chemical composition and physical characteristics. The biological production of aliphatic hydrocarbons is largely dependent on engineering of the complicated enzymatic network surrounding fatty acid biosynthesis.

Result: In this work, we developed a novel system for bioproduction of terminal fatty alkenes (1-alkenes) from renewable and low-cost triacylglycerols (TAGs) based on the lipase hydrolysis coupled to the P450 catalyzed decarboxylation. This artificial biosynthetic pathway was constituted using both cell-free systems including purified enzymes or cell-free extracts, and cell-based systems including mixed resting cells or growing cells. The issues of high cost of fatty acid feedstock and complicated biosynthesis network were addressed by replacement of the *de novo* biosynthesized fatty acids with the fed cheap TAGs. This recombinant tandem enzymatic pathway consisting of the *Thermomyces lanuginosus* lipase (Tll) and the P450 fatty acid decarboxylase $OleT_{JE}$ resulted in the production of 1-alkenes from purified TAGs or natural oils with 6.7 to 46.0% yields.

Conclusion: Since this novel hydrocarbon-producing pathway only requires two catalytically efficient enzymatic steps, it may hold great potential for industrial application by fulfilling the large-scale and cost-effective conversion of renewable TAGs into biohydrocarbons. This work highlights the power of designing and implementing an artificial pathway for production of advanced biofuels.

Keywords: Lipase, P450 fatty acid decarboxylase, Alkenes, Cell-free catalysis, Whole cell catalysis

Introduction

Shortage of petroleum-based fuels and increasing environmental concerns have led to great efforts to develop sustainable and clean biofuels from renewable feedstocks. Bioethanol manufactured by fermentation of sugars and biodiesel produced *via* transesterification of vegetable oils, animal fat, or waste oils are regarded as the two major first-generation biofuels, which are dominating the current global market of biofuels [1]. However, some undesirable properties such as high miscibility with water and low energy density (for bioethanol), incompatibility with current engine systems, and problems associated with storage and distribution have limited their further popularization [2]. Therefore, advanced biofuels with better fuel properties have been receiving ever-increasing attention, among which biohydrocarbons, especially the medium- to long-chain fatty alkanes/alkenes, are considered the ideal alternatives to petroleum-based transportation fuels due to their highly similar chemical composition and physical characteristics. Thus, biohydrocarbons hold the potential to be developed into 'drop-in' biofuels compatible with the existing distribution infrastructures used for gasoline,

* Correspondence: yjiny@126.com; lishengying@qibebt.ac.cn
[1]Key Laboratory of Biofuels, and Shandong Provincial Key Laboratory of Energy Genetics, Qingdao Institute of Bioenergy and Bioprocess Technology, Chinese Academy of Sciences, No. 189 Songling Road, 266101 Qingdao, Shandong, China
Full list of author information is available at the end of the article

diesel, and jet fuels in terms of their fatty acyl chain length [3].

At present, chemical hydrotreatment of acylglycerides, fatty acids, or fatty acyl esters, and *de novo* microbial biosynthesis are the two major strategies for production of biohydrocarbons. The former strategy requires expensive metal catalysts (for example, Pd and Pt), high temperature (250 to 450°C), and high pressure (20 to 70 bar), hence being energy intensive and environmentally unfriendly [4]. In comparison, the latter bioproduction strategy highlighted by a limited number of metabolically engineered systems is undoubtedly 'greener' [5-10]. However, all these biosynthetic systems suffer greatly from low yield of hydrocarbons, preventing them from industrialization.

For instance, the first engineered alkane biosynthetic pathway consisting of the acyl-acyl carrier protein (ACP) reductase Orf1594 from *Synechococcus elongates* PCC7942 and the aldehyde decarbonylase from *Nostoc punctiforme* PCC73102 was constructed in *Escherichia coli* and gave rise to a total alkane titer of approximately 300 mg/l [5]. Later, two similar hybrid biosynthetic routes by coupling a carboxylic acid reductase or a fatty acid reductase complex to the same fatty aldehyde decarbonylase were also reconstituted, however, leading to lower alkane yields [6,7]. Recently, based on the *in vivo* activity of the long-chain fatty acid P450 decarboxylase $OleT_{JE}$, our laboratory engineered a series of *E. coli* strains capable of *de novo* biosynthesis of 1-alkenes from glucose with the highest total alkene titer of 96.7 mg/l [10].

All these engineered pathways were constructed through diverting the fatty acid biosynthetic pathway toward hydrocarbon synthesis via different mechanisms of deoxygenation (Figure 1A). Since fatty acid biosynthesis plays a central role in energy metabolism of all living organisms and involves multiple interplaying enzymatic steps [11-17], its regulation is highly complicated and hence difficult to be manipulated. Metabolic engineering efforts aimed at overproduction of fatty acids or their derivatives (for example, fatty acyl-ACPs or fatty acyl CoAs) for high production of hydrocarbons *via* genetically reprogramming the fatty acid biosynthetic system turn out to be intrinsically difficult and technically challenging.

To overcome the limit of *de novo* biosynthesis of hydrocarbons, in this study, we propose a novel strategy for long-chain fatty alkene biosynthesis through coupling the glyceride hydrolytic activity of the *Thermomyces lanuginosus* lipase (Tll) to the fatty acid decarboxylation activity of the P450 enzyme $OleT_{JE}$ (Figure 1B). Tll has been well characterized as a biocatalyst with remarkable capacity of hydrolyzing triacylglycerols (TAGs) to free fatty acids (FFAs) involved in oil modifications [18]. $OleT_{JE}$ was recently discovered as a novel P450 enzyme with the unique ability to decarboxylate long-chain FFAs, forming terminal alkenes (1-alkenes) [19]. It is evident that FFAs should be able to couple these two naturally unrelated reactions. Thus, we selected these two enzymes to construct an artificial two-step biosynthetic pathway for biological production of 1-alkenes.

In the pilot biosynthetic system based on purified enzymes, the exogenous feedstock including pure TAGs and natural oils were first efficiently hydrolyzed by Tll, and the released free fatty acids (FFAs) were decarboxylated by $OleT_{JE}$ resulting in 1-alkenes (Graphical abstract in Additional file 1). Upon the proof-of-concept by using pure enzymes, the cell-based systems including the mixed resting cells that express Tll and $OleT_{JE}$, respectively, and the resting cells that co-express Tll and $OleT_{JE}$ were also evaluated. Moreover, the *in situ* catalytic system that is capable of mediating the biotransformation of TAGs → FFAs → 1-alkenes during cell growth was further customized from the perspective of cost-effective production of alkenes at industrial scale. This innovative tandem biotransformation process for producing olefin hydrocarbons from TAGs broadens the application of lipase in biofuel synthesis, which has long been only focused on biodiesel production [20-22]. The simple two-step (hydrolysis followed by decarboxylation) enzymatic pathway appears to be efficient and more manageable compared to the complex fatty acid biosynthetic network. Since TAGs are widely present in plant oils, animal fats, and oleaginous microbial cells [23], the feedstock for future large-scale application of this new biohydrocarbon-producing strategy should be abundant and cost-effective.

Results and discussion

Conversion of TAGs → FFAs → 1-alkenes catalyzed by purified Tll and $OleT_{JE}$

The two enzymes $OleT_{JE}$ and Tll were functionally expressed in *E. coli* and purified to homogeneity (Figure S1 in Additional file 2). In the lipase-catalyzed hydrolysis of pure TAGs (0.5 mM) that only have C12, C14, and C16 fatty acyl chains (that is, trilaurin, trimyristin, and tripalmitin), Tll released 0.80 mM lauric acid (C12), 0.78 mM myristic acid (C14), and 1.15 mM palmitic acid (C16), corresponding to 53.3 ± 1.3, 52.0 ± 1.1, and 76.7 ± 1.5% yield (Figure 2A), respectively.

To test whether the fatty acid decarboxylation activity of $OleT_{JE}$ can be coupled to the lipase-catalyzed hydrolysis in one pot, 2 μM $OleT_{JE}$ and 1 mM H_2O_2 as cofactor were added into the TAG hydrolytic reactions. As expected, 0.35 mM 1-undecene (C11), 0.49 mM 1-tridecene (C13), and 0.40 mM 1-pentadecene (C15) were produced (Figure 2B), corresponding to 23.3 ± 0.9, 32.7 ± 1.4, and 26.7 ± 0.9% of the theoretically maximal conversion from TAG to corresponding 1-alkene. When only taking the second decarboxylation step into account, as

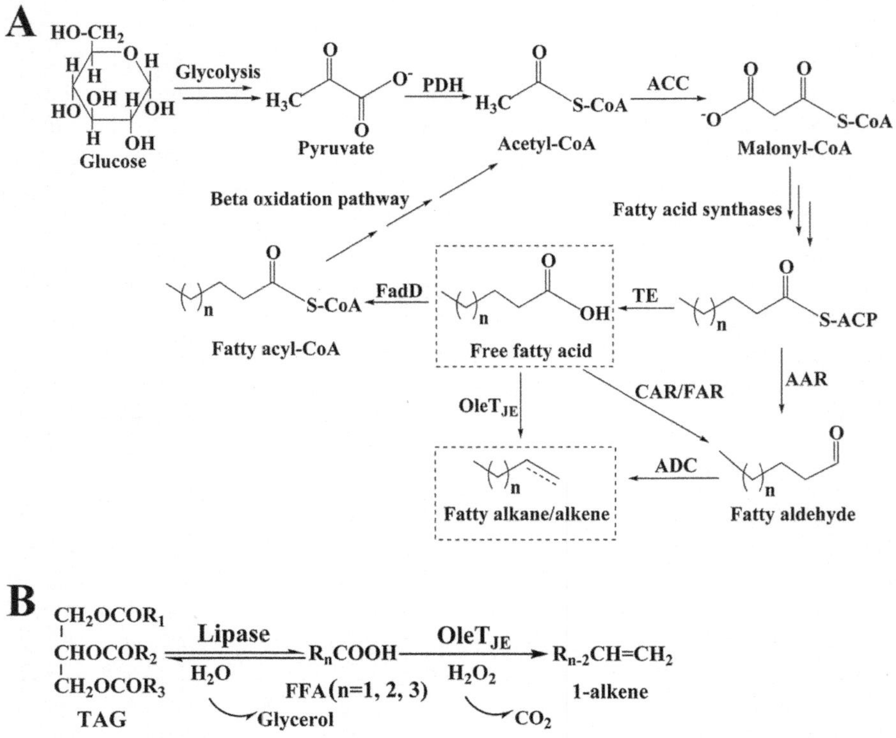

Figure 1 Multi-step and artificial two-step pathways. (A) Native and engineered multi-step pathways for hydrocarbon biosynthesis based on fatty acid metabolic network. **(B)** The artificial two-step 1-alkene biosynthetic pathway developed in this work consisting of the lipase Tll and the P450 fatty acid decarboxylase $OleT_{JE}$. AAR, acyl-ACP reductase; ACC, acetyl-CoA carboxylase; ACP, acyl carrier protein; ADC, aldehyde decarbonylase; CAR, carboxylic acid reductase; FAR, fatty acid reductase; PDH, pyruvate dehydrogenase; TAG, triacylglycerol; TE, thioesterase.

shown in Figure 2B, the conversions of the C12, C14, and C16 FFA to corresponding 1-alkene were 39.8 ± 1.5, 58.3 ± 1.3, and 33.3 ± 1.1%, respectively. These results evidently demonstrate the compatibility between the reactions mediated by Tll and $OleT_{JE}$. Among the tested TAG hydrolysis reactions, Tll displayed the highest activity toward tripalmitin (C16 TAG), which indicates the fatty acyl chain length preference of Tll being C16 over C14 and C12. By contrast, in the tandem hydrolysis-decarboxylation process, the alkene product with the greatest yield turned out to be 1-tridecene (Figure 2B), decarboxylated from the C14 FFA-myristic acid. In the one-pot catalytic relay, the consumption of fatty acids by $OleT_{JE}$ (forming alkenes) was expected to pull the equilibrium of TAGs ↔ fatty acids toward fatty acids (Figure 1B). Consistently, the total FFA yield of the tested dual enzymatic systems calculated by summing up the reacted FFA (equals to the produced alkene in mole) and the remaining FFA (Figure 2B) was slightly improved compared to that of the single lipase hydrolytic system (Figure 2A).

In a mixed biocatalytic system of dual enzymes, the ratio between the two enzymes is often a key factor for the overall conversion ratio. Thus, we elected to use trimyristin (the TAG with three C14 fatty acyl chains), which gave the highest 1-alkene yield at a 1:1 enzyme ratio (Figure 2B), as substrate to optimize the proportion of lipase to decarboxylase. As shown in Figure 2C, the C13 alkene yield was improved to 0.69 mM (corresponding to 46.0 ± 1.8% conversion of TAG to alkene) at a molar ratio of 1:2 (Tll:$OleT_{JE}$). It was reported that the extensive metabolic engineering efforts aimed at overproduction of FFAs in *E. coli* have so far achieved less than 30% of the maximum theoretical yield *via de novo* biosynthesis from the starting carbon source glucose [16]. The reported yields of FFA downstream products such as hydrocarbons were even lower [5-7,10]. Thus, the 46.0% overall conversion from TAG to 1-alkene achieved by the cell-free system of tandem lipase-$OleT_{JE}$ in this work appears significant. The *in vitro* manipulation *via* enzyme ratio optimization held another significant advantage over the less accurate *in vivo* metabolic engineering, which likely requires a delicate coordination of the complicated regulation networks of transcription, translation, and metabolite fluxes.

Subsequently, three natural oils other than pure TAGs including olive oil, soybean oil, and microalgae oil were investigated as substrates for the novel Tll-$OleT_{JE}$ catalytic relay system. The Tll-catalyzed hydrolysis gave rise to 1.21, 1.12, and 1.37 mM total FFAs from olive oil,

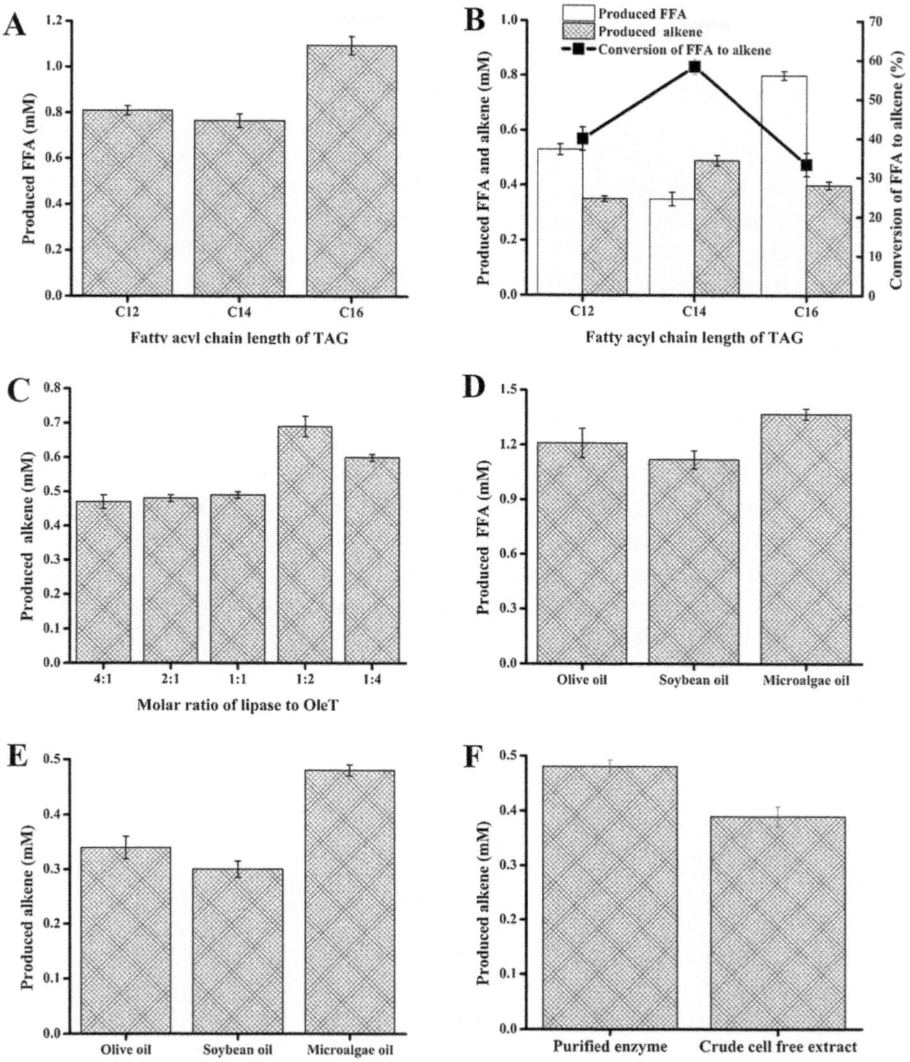

Figure 2 Production of FFAs and 1-alkenes and optimization of the enzyme ration between Tll and $OleT_{JE}$. (A) Production of FFAs from hydrolysis of different TAGs by purified Tll. **(B)** Production of FFAs and 1-alkenes from TAGs catalyzed by the coupled Tll-$OleT_{JE}$. **(C)** Optimization of the enzyme ratio between Tll and $OleT_{JE}$ for improvement of the overall alkene yield from the TAG with C14 fatty acyl chain length. **(D)** Production of FFAs from hydrolysis of different natural oils by purified Tll. **(E)** Production of 1-alkenes from natural oils catalyzed by the coupled Tll-$OleT_{JE}$. **(F)** Production of 1-alkenes from microalgae oil mediated by purified enzymes and cell-free extracts. FFA, free fatty acid; TAG, triacylglycerol.

soybean oil, and microalgae oil (Figure 2D), corresponding to 80.7 ± 2.3, 74.7 ± 2.1, and 91.3% ± 3.3 conversion of TAGs to FFAs, respectively. The profiles of produced FFAs are described in Table S1 in Additional file 3. In the coupled hydrolysis-decarboxylation reactions, 0.34, 0.30, and 0.48 mM total alkenes including 1-tridecene (for microalgae oil only), 1-pentadecene, and 1-heptadecene, corresponding to 21.3 ± 0.7, 20.0 ± 0.9, and 32.0 ± 1.0 overall conversion from TAGs to alkenes (Figure 2E; Table S1 in Additional file 3), were produced from olive oil, soybean oil, and microalgae oil, respectively. Interestingly, no higher alkenes were observed in this coupled reaction system, although FFAs of C20:5 and C22:6 were released from microalgae oil. This indicates the relatively lower substrate flexibility of $OleT_{JE}$ than that of Tll, which is consistent with the reported fatty acid chain length selectivity of saturated C12 to C20 [10,19]. Comparatively, microalgae oil appeared to be a more suitable feedstock for 1-alkene production *via* tandem biocatalysis. Since microalgae oils could be an excellent feedstock for this biosynthetic pathway to produce advanced hydrocarbon biofuels, all following experiments used this feedstock.

To reduce the biocatalyst cost by omitting the enzyme purification step, the crude *E. coli* cell-free extracts of Tll and $OleT_{JE}$ were applied for the production of 1-alkene using microalgae oil as feedstock. As a result, 0.39 mM

1-alkenes were generated, comparable to the 0.48 mM yield when using purified enzymes (Figure 2F).

Production of 1-alkenes mediated by mixed resting *E. coli* recombinant cells

Biotransformation mediated by whole cell catalysts represents a promising strategy for industrial manufacture because it skips the costly enzyme purification and immobilization steps. To reduce the cost on biocatalysts, the two *E. coli* whole cell catalysts of Tll and $\mathrm{OleT_{JE}}$ were mixed and evaluated. Importantly, in the mixed cell system, the enzyme ratio can be easily controlled by adjusting cell dosage. However, multiple events of substrate/product transfer across cell membrane must be considered.

For the TAG hydrolytic reactions mediated by overexpressed cytoplasmic Tll in *E. coli*/pRSFDuet-*tll* whole cells, freeze-dried cells hydrolyzed 70% microalgae oil to form 1.05 mM FFAs (Figure 3A), which are more catalytically efficient than non-treated cells (0.9 mM, 60 ± 2.3%). This indicates freeze-drying is a simple but effective way to improve substrate accessibility and hence catalytic activity.

To further improve mass transfer of TAG substrates, Tll was periplasmically overexpressed in *E. coli*/pET-22(b)-*tll* in order to hydrolyze TAGs more efficiently. As expected, a higher yield of 80% (1.2 mM FFAs produced) was achieved (Figure 3A). For periplasmic overexpression of Tll in *E. coli*, fusion of the Tll lipase to the pelB leader sequence directs the lipase to *E. coli* periplasm, where the leader peptide is removed by a signal peptidase [24]. Considering that the two types of cells used for biotransformations contained the same total activities (normalized in terms of cell-free extract), we reason that the enhancement of FFA production catalyzed by periplasmic Tll overexpression cells was likely attributed to improved substrate accessibility by localization of enzymes in periplasm. The freeze-dried *E. coli*/pET-22(b)-*tll* barely increased the degree of hydrolysis (Figure 3A).

In the case of Tll-$\mathrm{OleT_{JE}}$ coupled reaction in the mixed cell system, *E. coli*/pET-22(b)-*tll* (for periplasmic expression of Tll) plus *E. coli*/pACYCDuet-*oleT*$_{JE}$ (for cytoplasmic expression of $\mathrm{OleT_{JE}}$) gave higher alkene yield (0.21 mM, 14 ± 0.4%) than that (0.10 mM, 7 ± 0.3%) of *E. coli*/pRSFDuet-*tll* (for cytoplasmic expression of Tll) plus *E. coli*/pACYCDuet-*oleT*$_{JE}$ (for cytoplasmic expression of $\mathrm{OleT_{JE}}$) (Figure 3B). Also shown in Figure 3B, the mixed freeze-dried cells displayed higher conversion than the non-treated counterparts for the group with cytoplasmically overexpressed Tll (*E. coli*/pRSFDuet-*tll* + *E.*

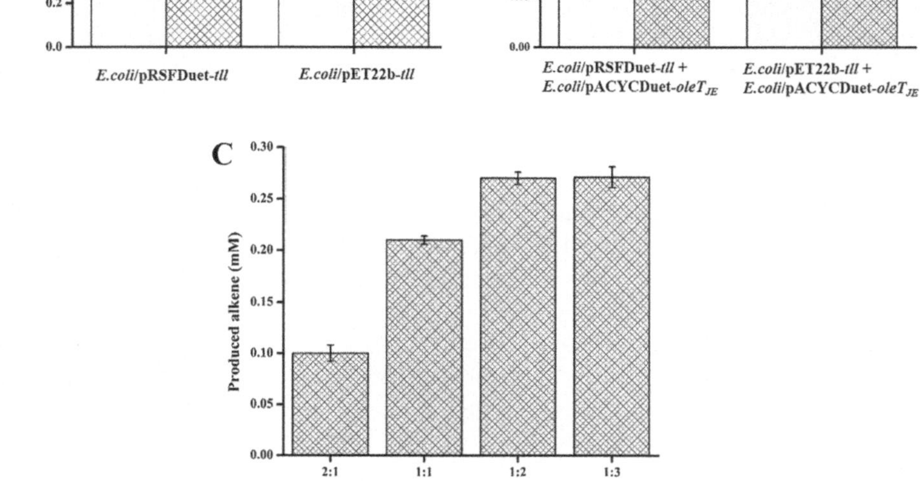

Figure 3 Production of FFAs and 1-alkenes from microalgae oil and optimization of enzyme ratio. (A) Production of FFAs from microalgae oil by resting *E. coli* cells with cytoplasmic and periplasmic overexpression of Tll. **(B)** Production of 1-alkenes from microalgae oil catalyzed by mixed resting *E. coli* cells. **(C)** Optimization of the enzyme ratio between Tll and $\mathrm{OleT_{JE}}$ in the form of freeze-dried whole cell catalysts for improvement of the overall alkene yield from microalgae oil.

coli/pACYCDuet-*oleT*$_{JE}$), but there was only a slight improvement for the group with periplasmic overexpression of Tll (*E. coli*/pET-22(b)-*tll* + *E. coli*/pACYCDuet-*oleT*$_{JE}$).

Further optimization of the enzyme ratio in the form of freeze-dried cells (*E. coli*/pET-22(b)-*tll* + *E. coli*/pACYCDuet-*oleT*$_{JE}$) enhanced the yield to 18 ± 0.8% (0.27 mM) at a 1:2 molar ratio (Tll:OleT$_{JE}$) (Figure 3C). Although the ratio between the two enzymes can be adjusted in the mixed cell system, the catalytic efficiency of the whole system apparently would suffer from complicated transportation of intermediate FFAs from one type of cells to another. This is likely the major reason for the low conversions from TAGs to 1-alkenes. Therefore, engineering of an *E. coli* strain with co-expression of the two enzymes seems necessary.

Biosynthesis of 1-alkenes with a whole cell catalyst co-expressing Tll and OleT$_{JE}$

The approach of co-expressing Tll (for oil hydrolysis) and OleT$_{JE}$ (for alkene formation) in the same *E. coli* host could potentially overcome complex cross-membrane transport of intermediate FFAs between different types of cells, thereby gaining more efficient substrate channeling. Thus, we constructed two types of co-expression strains including *E. coli*/pRSFDuet-*tll* + pACYCDuet-*oleT*$_{JE}$ and *E. coli*/pET-22(b)-*tll* + pACYCDuet-*oleT*$_{JE}$ and evaluated their alkene-producing abilities. The non-treated *E. coli* cells with cytoplasmic co-expression of Tll and OleT$_{JE}$ (*E. coli*/pRSFDuet-*tll* + pACYCDuet-*oleT*$_{JE}$) converted 11 ± 0.7% microalgae oil into 1-alkenes (0.16 mM), whereas the freeze-dried cells gave a higher yield of 17 ± 1.1% (0.26 mM) (Figure 4A), both of which are greater than that of corresponding mixed cell systems. This improvement can be explained by better channeling of intermediates between adjacent enzyme active sites because of co-localization of two enzymes inside a single cell.

A similar profile was observed for the periplasmic co-overexpression cell system (*E. coli*/pET-22(b)-*tll* + pACYCDuet-*oleT*$_{JE}$). Specifically, the strain of *E. coli*/pET-22(b)-*tll* + pACYCDuet-*oleT*$_{JE}$ demonstrated improved hydrocarbon yield (0.35 mM, 24 ± 1.3%) compared to the corresponding cytoplasmic overexpression cell system (*E. coli*/pRSFDuet-*tll* + pACYCDuet-*oleT*$_{JE}$). Again, this result indicates the importance of the contact between TAGs and Tll for the whole alkene-producing pathway. The freeze-dried periplasmic overexpression cells (*E. coli*/pET-22(b)-*tll* + pACYCDuet-*oleT*$_{JE}$) did not enhance the conversion (0.36 mM, 24 ± 1.4%) (Figure 4A), indicating that freeze-drying treatment cannot further improve the cell permeabilization that has already been optimized by periplasmic expression of Tll. We infer that periplasmic expression of Tll probably leads to better permeability than the corresponding cytoplasmic expression; thus, the freeze-drying treatment showed a positive permeabilization effect on the cytoplasmic co-expression cell type, but not on the periplasmic co-expression type.

Measurement of the activity of the two enzymes contained in co-expression cells showed molar ratios of 1:1.7 and 1:1.9 (Tll:OleT$_{JE}$), respectively, close to the optimal ratio of 1:2. More significantly, the periplasmic co-expression cells retained more than 70 ± 3.2% alkene-producing activity after recycling in three successive batches, displaying great potential for industrial application.

Biosynthesis of 1-alkenes by co-expressed Tll and OleT$_{JE}$ *in situ* during cell growth

Integration of enzyme generation and enzymatic transformation into a single process would be significantly more energy efficient and of industrial simplification by making better use of fermentation energy and skipping enzyme purification. We previously developed an integrated bioprocess for *in situ* biodiesel production that occurs simultaneously with lipase generation in a *Pichia*

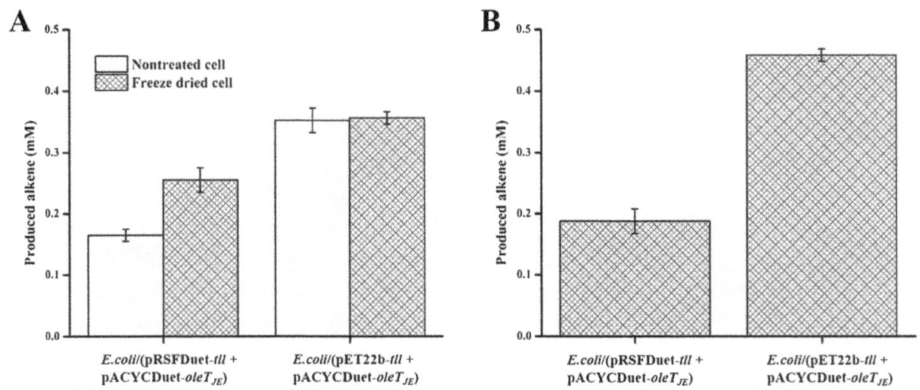

Figure 4 Production of 1-alkenes from microalgae oil. (A) Production of 1-alkenes from microalgae oil catalyzed by *E. coli* resting cells with co-expression of Tll and OleTJE. **(B)** Production of 1-alkenes from microalgae oil catalyzed by growing *in situ E. coli* cell catalysts.

pastoris yeast system [25]. Inspired by this strategy, in the present study, in parallel with combining hydrolysis of TAGs and decarboxylation of FFAs into a single *E. coli* host, we tested the strategy of developing *E. coli in situ* catalytic system for biosynthesis of 1-alkenes from TAGs. This *in situ* system integrated the process of enzyme production with the enzyme-catalyzed tandem biotransformations into a single process in one pot, which could significantly save energy consumption and simplify operational procedures.

Based on the tandem hydrolysis-decarboxylation *in situ* catalyzed by Tll and $OleT_{JE}$ during cell growth, the two types of Tll-$OleT_{JE}$ co-expression *E. coli* cells (*E. coli*/pRSFDuet-*tll* + pACYCDuet-*oleT*$_{JE}$ and *E. coli*/pET-22(b)-*tll* + pACYCDuet-*oleT*$_{JE}$) gave 0.19 mM (12 ± 0.6%) and 0.46 mM (31 ± 1.1%) alkene yield from the fed microalgae oils, respectively (Figure 4B). Interestingly, periplasmic overexpression cells (*E. coli*/pET-22(b)-*tll* + pACYCDuet-*oleT*$_{JE}$) even secreted some lipases into the extracellular space (that is, culture medium) during the *in situ* process, which may significantly facilitate the hydrolysis-decarboxylation coupled reactions, as reflected by enhanced alkene production. Thus, the *in situ* catalytic system based on periplasmic overexpression could take advantage of both cell-free enzymes and whole cell catalysts. The extracellular and intracellular enzymes of growing *E. coli* cells were likely to be simultaneously utilized to produce alkenes *in situ* and in one pot. Moreover, we further evaluated this *in situ* system by recovering the whole cells and applied them for a new batch reaction as resting whole cell catalysts, which displayed 61.0 ± 2.2 and 73.0 ± 2.9% relative yield for cytoplasmic overexpression cells (*E. coli*/pRSFDuet-*tll* + pACYCDuet-*oleT*$_{JE}$) and periplasmic overexpression cells (*E. coli*/pET-22(b)-*tll* + pACYCDuet-*oleT*$_{JE}$), respectively.

Notably, higher concentrations of the substrate oil (1 to 5 mM) or H_2O_2 (1 to 15 mM) were also tested; however, the alkene yields did not increase (data not shown). It is possible that the low activity or low tolerance of $OleT_{JE}$ toward the oil-FFA mixture may limit higher alkene yield. Thus, based on the crystal structure of $OleT_{JE}$ recently reported by Belcher et al. [26], the protein engineering of this enzyme to improve its activity or tolerance in the context of two-step reactions could be helpful for developing a more efficient converter microorganism.

Compared to the *in vivo* construction and regulation of metabolic pathways, the *in vitro* strategy of cell-free enzyme-based and whole cell-based systems represents an important opportunity for bioproduction of hydrocarbons. These *in vitro* systems could be readily controlled through preparation of enzyme or whole cell cocktails by simple mixing procedures or accurate modification of reaction conditions [27,28]. These multiple cell-free enzyme systems (purified enzymes, cell-free extracts) and cell-based systems (mixed resting cells, whole cells with co-expressed enzymes, growing *in-situ* cell catalysts) provide more flexible choices for specific purpose and situation (Figure 5).

Conclusion

A novel biosynthetic pathway for production of biohydrocarbon 1-alkenes from renewable oils was assembled, which consists of the lipase Tll to release FFAs from TAGs and the P450 fatty acid decarboxylase $OleT_{JE}$ responsible for decarboxylation of FFAs to generate 1-alkenes. The biotransformation of TAGs → FFAs → 1-alkenes was achieved using a variety of biocatalytic systems, including purified enzymes, cell-free extracts, resting whole cells, and growing *in situ* cell catalysts. These one-pot tandem biocatalyzes only required two sequential enzymatic steps, but allowed the efficient production of 1-alkenes from TAGs with 6.7 to 46.0% conversion ratios from 0.5 mM renewable feedstock including pure TAGs and diverse natural oils. Especially, the biosynthesis of 1-alkenes based on resting whole cells and growing *in situ* cell catalysts exhibits potential for industrial application.

Finally, it is worth noting that this novel artificial biosynthetic route could be adapted for other microorganisms, especially for those genetically tractable oleaginous microorganisms if considering the abundant endogenous TAGs as potential cellular substrates for the Tll-$OleT_{JE}$ coupled system.

Materials and methods

Plasmids, strains, and reagents

The plasmids pACYCDuet-1, pRSFDuet-1, and pET-22 (b) were obtained from Novagen (Darmstadt, Germany). Recombinant plasmid pET-28(b)-*oleT*$_{JE}$ for expression of the P450 fatty acid decarboxylase $OleT_{JE}$ was constructed in our laboratory previously [10]. *E. coli* DH5α and BL21 (DE3) strains were preserved in our laboratory. Various authentic standards of TAGs (C12, trilaurin, C14, trimyristin, and C16, tripalmitin), fatty acids (lauric acid, myristic acid, palmitic acid, stearic acid, and heptadecanoic acid), and fatty 1-alkenes (1-undecene, 1-tridecene, 1-pentadecene, and 1-heptadecene) were purchased from TCI (Shanghai, China). Kanamycin, chloramphenicol, thiamine, and isopropyl β-D-1-thiogalactopyranoside (IPTG) were products of Solarbio Science & Technology Co., Ltd (Beijing, China). All used restricted enzymes were supplied by Thermo Scientific (Shanghai, China). PrimeSTAR GXL DNA polymerase and dNTPs were from Takara Bio Inc. (Otsu, Japan). DNA manipulation kits were bought from Omega Bio-Tek (Norcross, GA, USA) and Promega (Madison, WI, USA). Ni-NTA resin was from Qiagen (Venlo, Netherlands). PD-10 desalting columns were products of GE Healthcare (Pewaukee, WI, USA). Ultra centrifugal

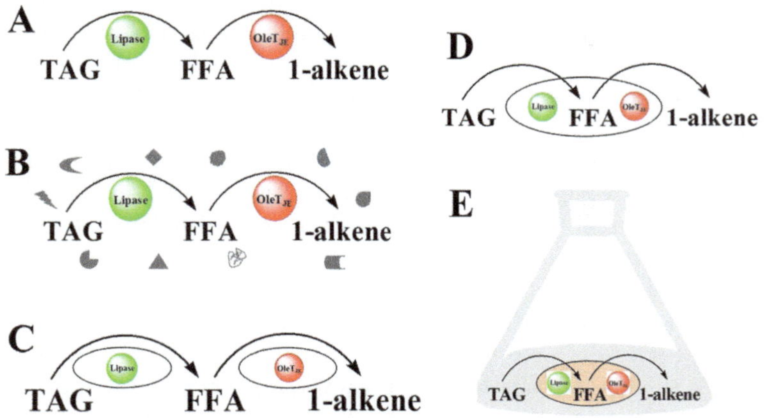

Figure 5 Catalytic systems for the conversion of triacylglycerol to 1-alkene. Various lipase-OleT$_{JE}$ coupling catalytic systems for the conversion of triacylglycerol to 1-alkene including **(A)** purified enzymes, **(B)** cell-free extract, **(C)** mixed resting cells, **(D)** co-expression cells, and **(E)** *in situ* growing cells. FFA, free fatty acid; TAG, triacylglycerol.

filters were purchased from Millipore (Billerica, MA, USA). Bradford Protein Assay kit was bought from Beyotime Institute of Biotechnology (Jiangsu, China). Other routine reagents were commercially available products of analytical grade. The *tll* gene was synthesized by GenScript (Piscataway, NJ, USA). Oligo primers were synthesized by Sangon Biotech (Shanghai, China). Soybean oil and olive oil were purchased from local market. Microalgae oil was kindly donated by Professor Tianzhong Liu at the Qingdao Institute of Bioenergy and Bioprocess Technology, Chinese Academy of Sciences.

TB medium was composed of 1.2% tryptone, 2.4% yeast extract, 0.5% glycerol, 0.23% KH_2PO_4, and 1.25% K_2HPO_4. One millimolar of thiamine was supplemented for the expression of OleT$_{JE}$. For maintaining corresponding plasmids, 25 μg/ml chloramphenicol or 50 μg/ml kanamycin was added.

The lysis buffer (pH 8.0) was composed of NaH_2PO_4 50 mM, NaCl 300 mM, glycerol 10%, amd imidazole 10 mM. The washing buffer (pH 8.0) comprised NaH_2PO_4 50 mM, NaCl 300 mM, glycerol 10%, and imidazole 20 mM. Elution buffer (pH 8.0) was composed of NaH_2PO_4 50 mM, NaCl 300 mM, glycerol 10%, and imidazole 250 mM. Desalting buffer (pH 7.4) was a mixture of NaH_2PO_4, EDTA 1 mM, and glycerol 10%.

Cloning and expression of enzymes

The *oleT$_{JE}$* gene was amplified using the previously constructed plasmid pET-28(b)-*oleT$_{JE}$* as template and the primer pair as follows: *Bam*HI-*oleT$_{JE}$*, CGC*GGATCC*G ATGGCAACACTTAAGAGGGATAAGGGCTTA (the *Bam*HI restriction site is italicized); and *Hin*dIII-*oleT$_{JE}$*, CAATG*AAGCTT*TTATGTTCTGTCTAC AACTTCGC (the italicized nucleotides denote the *Hin*dIII cutting site). For *tll* gene cloning, the synthetic *tll* gene (Genbank accession number AF054513.1) was used as template for PCR amplification with primers including *Bam*HI-*tll*, AGCCA*GGATCC*GAGTCCTATTCGTCGAGAGGTCT CG and *Hin*dIII-*tll*, GCCGC*AA GCTT*CTAAAGACAT GTCCCAATTAACCC. These amplified *oleT$_{JE}$* and *tll* fragments were double digested with *Bam*HI/*Hin*dIII and ligated into pACYCDuet-1 and pRSFDuet-1 to create the recombinant plasmids pACYCDuet-*oleT$_{JE}$* and pRSFDuet-*tll*, respectively. The recombinant plasmids were first transformed into competent cells of *E. coli* DH5α for DNA sequence confirmation, and then transformed into *E. coli* BL21 (DE3) for cytoplasmic overexpression of OleT$_{JE}$ and Tll. For construction of engineered *E. coli* strains with periplasmic overexpression of lipase, the plasmid pET-22(b) carrying an N-terminal *pelB* signal sequence was used. The *tll* fragment from *Bam*HI/*Hin*dIII double digested pRSFDuet-*tll* was inserted into corresponding sites of pET-22(b), resulting in the recombinant plasmid pET-22(b)-*tll*. The engineered *E. coli* BL21 (DE3) strains harboring corresponding recombinant plasmids were similarly constructed as described above.

A single colony of the *E. coli* BL21 (DE3) strain carrying pACYCDuet-*oleT$_{JE}$*, pRSFDuet-*tll*, or pET-22(b) based counterparts were used to inoculate the seed culture in LB, and this culture was grown overnight at 37°C, 200 rpm. A 1% volume of pre-culture was inoculated into 1 l of TB medium with appropriate antibiotics (plus 0.5 mM 5-aminolevulinic acid and 1 mM thiamine for OleT$_{JE}$ expression). When OD_{600} reached approximately 0.8, IPTG was added to the final concentration of 0.2 mM for induction of enzyme expression at 18°C for 20 h.

Purification of enzymes

The harvested post-induced cells by centrifugation were resuspended with 40 ml lysis buffer and disrupted using ultra-sonication. Cell slurry was separated into cell-free lysate and cell debris at 10,000 × *g* for 30 min at 4°C.

The resulted cell-free lysate was gently mixed with 1 ml 50% Ni-NTA slurry at 4°C for 1 h. The mixture was loaded onto a column and washed with 100 to 200 ml washing buffer until no protein eluted in the flow-through. The 6 × His tagged proteins bound to Ni-NTA resin were eluted with 5 ml elution buffer. The eluted protein fraction was concentrated using Millipore ultra-filter with 30-kDa molecular weight cutoff (2,000 × *g*, 30 min, 4°C) and loaded onto a pre-equilibrated PD-10 column for buffer exchange using desalting buffer. All protein fractions were monitored using Bradford Protein Assay kit. The aliquots of collected fractions were flash frozen by liquid nitrogen and stored at −80°C for later use.

Determination of enzyme concentration

Diluted purified $OleT_{JE}$ solution was reduced by adding sodium dithionite prior to the initial scan at 400 to 600 nm to record the baseline spectrum. Then, the P450 enzyme solution was subjected to CO bubbling for 40 s before the second scan to get the reduced CO-bound P450 spectrum, based on which the functional $OleT_{JE}$ was quantified from the absorbance difference between A_{450} and A_{490} using the molar extinction coefficient of 91 mM^{-1} cm^{-1} [29]. The lipase activity was measured as described elsewhere [25].

Whole cell catalyst preparation

The induced culture broth was centrifuged to harvest fresh wet cells, termed as non-treated cells. These fresh wet cells were subjected to freezing at −80°C for 12 h, and then freeze-drying for 24 h with lyophilizer to obtain freeze-dried cells.

Enzymatic assays and whole cell biotransformation

For cell-free enzyme biotransformation, the standard 100-μl hydrolysis reaction assay containing 0.5 mM substrate, 2 μM purified Tll in sodium phosphate buffer (100 mM, pH 7.4) was carried out in a 1.5-ml Eppendorf tube at 30°C for 6 h. The tested substrates included pure TAGs (fatty acyl chain length of C12, C14, and C16) and oils (microalgae oil, soybean oil, and olive oil). For the tandem reactions of hydrolysis followed by decarboxylation, 2 μM $OleT_{JE}$ and 1 mM H_2O_2 were supplemented into the hydrolysis reaction while maintaining the total volume of 100 μl. During optimization of the molar ratio between these two enzymes, the amount of each enzyme was adjusted appropriately.

For resting whole cell biotransformation, various amounts of resting recombinant cells (approximate 50 mg) with the same activities (in terms of cell-free extract) were added into 500 μl sodium phosphate buffer containing 0.5 mM microalgae oil. In the case of lipase-$OleT_{JE}$ coupled catalysis reaction, 1 mM H_2O_2 was supplied. During optimization of the enzyme ratio in the whole cell catalysis, the cell amounts were accordingly adjusted. In the whole cell-mediated recycling biotransformations, *E. coli* cells were recovered and washed after each batch reaction, and then applied in a new batch reaction. These resting whole cell-mediated biotransformations were performed at 30°C for 24 h.

For biotransformation mediated by growing cells *in situ*, 12-h post-induction cultures were supplemented with 0.5 mM microalgae oil and 1 mM H_2O_2 for additional 36-h reaction in shaking flask at 28°C. All experiments were repeated three times.

Determination of FFA and 1-alkene

The reactions were quenched and extracted with 100 μl ethyl acetate containing 0.5 mM heptadecenoic acid as internal standard for GC-MS analysis. The GC oven equipped with the HP-INNOWAX (Agilent Technologies, Inc., Santa Clara, CA, USA; 30 m × 0.25 mm × 0.25 μm) column was heated using the program as follows: isothermal at 40°C for 4 min, 40 to 250°C at the rate of 10°C/min, and 250°C for 15 min. The mass fragment range of 50 to 500 m/z was monitored under electron ionization conditions (1741 eV). FFAs and alkenes were quantified using corresponding standard compounds and the internal standard as references.

Additional files

Additional file 1: Graphical abstract. An artificial two-step biosynthetic pathway for biological production of 1-alkenes using three kinds of natural oils.

Additional file 2: Figure S1. Plasmid construction and purification of Tll (A) and $OleT_{JE}$ (B).

Additional file 3: Table S1. Profiles of produced FFAs via lipase Tll mediated hydrolysis and produced 1-alkenes via lipase-$OleT_{JE}$ tandem biotransformation. All experiments were performed three times.

Abbreviations

TAGs: triacylglycerols; Tll: *Thermomyces lanuginosus* lipase; FFAs: free fatty acids; IPTG: isopropyl β-D-1-thiogalactopyranoside.

Competing interests

The authors declare that they have no competing interests.

Authors' contributions

JY and SL conceived of the study, designed the experiments, analyzed the data, and wrote the manuscript. JY, YL, CW, and BH performed the experiments. All authors read and approved the final manuscript.

Acknowledgements

This work was supported by the Qingdao Institute of Bioenergy and Bioprocess Technology Director Innovation Foundation for Young Scientists (Grant Number: Y37203410U), the Applied Basic Research Programs of Science and Technology of Qingdao (Grant Number: 14-2-4-10-jch), the National Natural Science Foundation of China (Grant Number: NSFC 31270855), and the National High Technology Research and Development Program of China (Grant Number: 2014AA093510). We are also grateful for the funding from 'Recruitment Program of Global Experts, 2012.' We thank Professor Tianzhong Liu at the Qingdao Institute of Bioenergy and Bioprocess Technology, Chinese Academy of Sciences, for providing us with the microalgae oils.

Author details

[1]Key Laboratory of Biofuels, and Shandong Provincial Key Laboratory of Energy Genetics, Qingdao Institute of Bioenergy and Bioprocess Technology, Chinese Academy of Sciences, No. 189 Songling Road, 266101 Qingdao, Shandong, China. [2]University of Chinese Academy of Sciences, No. 19A Yuquan Road, 100049 Beijing, China. [3]Key Laboratory for Marine Drugs, Department of Pharmacy, Renji Hospital, Shanghai Jiao Tong University School of Medicine, No. 145 Shandongzhong Road, 200127 Shanghai, China.

References

1. Keasling JD, Chou H. Metabolic engineering delivers next-generation biofuels. Nat Biotechnol. 2008;26:298–9.
2. Atsumi S, Liao JC. Metabolic engineering for advanced biofuels production from *Escherichia coli*. Curr Opin Biotechnol. 2008;19:414–9.
3. Lee SK, Chou H, Ham TS, Lee TS, Keasling JD. Metabolic engineering of microorganisms for biofuels production: from bugs to synthetic biology to fuels. Curr Opin Biotechnol. 2008;19:556–63.
4. Serrano-Ruiz JC, Ramos-Fernández EV, Sepúlveda-Escribano A. From biodiesel and bioethanol to liquid hydrocarbon fuels: new hydrotreating and advanced microbial technologies. Energy Environ Sci. 2012;5:5638–52.
5. Schirmer A, Rude MA, Li X, Popova E, del Cardayre SB. Microbial biosynthesis of alkanes. Science. 2010;329:559–62.
6. Akhtar MK, Turner NJ, Jones PR. Carboxylic acid reductase is a versatile enzyme for the conversion of fatty acids into fuels and chemical commodities. Proc Natl Acad Sci U S A. 2013;110:87–92.
7. Thomas PH, Sabine M, Karen M, Christoph E, Dagmara MK, George NT, et al. Synthesis of customized petroleum-replica fuel molecules by targeted modification of free fatty acid pools in *Escherichia coli*. Proc Natl Acad Sci U S A. 2013;110:7636–41.
8. Beller HR, Goh EB, Keasling JD. Genes involved in long-chain alkene biosynthesis in *Micrococcus luteus*. Appl Environ Microbiol. 2010;76:1212–23.
9. Sukovich DJ, Seffernick JL, Richman JE, Gralnick JA, Wackett LP. Widespread head-to-head hydrocarbon biosynthesis in bacteria and role of OleA. Appl Environ Microbiol. 2010;76:3850–62.
10. Liu Y, Wang C, Yan JY, Zhang W, Guan WN, Lu XF, et al. Hydrogen peroxide-independent production of α-alkenes by $OleT_{JE}$ P450 fatty acid decarboxylase. Biotechnol Biofuels. 2014;7:28–39.
11. Kung Y, Runguphan W, Keasling JD. From fields to fuels: recent advances in the microbial production of biofuels. ACS Synth Biol. 2012;1:498–513.
12. Peralta-Yahya PP, Zhang FZ, del Cardayre SB, Keasling JD. Microbial engineering for the production of advanced biofuels. Nature. 2012;488:320–8.
13. Steen EJ, Kang Y, Bokinsky G, Hu Z, Schirmer A, McClure A, et al. Microbial production of fatty-acid-derived fuels and chemicals from plant biomass. Nature. 2010;463:559–62.
14. Liu TG, Vora H, Khosla C. Quantitative analysis and engineering of fatty acid biosynthesis in *E. coli*. Metab Eng. 2010;12:378–86.
15. Lu XF, Vora H, Khosla C. Overproduction of free fatty acids in *E. coli*: implications for biodiesel production. Metab Eng. 2008;10:333–9.
16. Lennen RM, Pfleger BF. Engineering *Escherichia coli* to synthesize free fatty acids. Trends Biotechnol. 2012;30:659–67.
17. Lennen RM, Max AK, Kritika K, Robert AZ, Kristin EB, Mary SL, et al. Membrane stresses induced by overproduction of free fatty acids in *Escherichia coli*. Appl Environ Microbiol. 2011;77:8114–28.
18. Fernandez-Lafuente R. Lipase from *Thermomyces lanuginosus*: uses and prospects as an industrial biocatalyst. J Mol Catal B Enzym. 2010;62:197–212.
19. Rude MA, Baron TS, Brubaker S, Alibhai M, del Cardayre SB, Schirmer A. Terminal olefin (1-alkene) biosynthesis by a novel P450 fatty acid decarboxylase from *Jeotgalicoccus* species. Appl Environ Microbiol. 2011;77:1718–27.
20. Yan JY, Yan YJ, Liu SX, Hu J, Wang GL. Preparation of cross-linked lipase-coated micro-crystals for biodiesel production from waste cooking oil. Bioresour Technol. 2011;102:4755–8.
21. Yan JY, Li AT, Xu Y, Ngo Thao PN, Phua SC, Li Z. Efficient production of biodiesel from waste grease: one-pot esterification and transesterification with tandem lipases. Bioresour Technol. 2012;123:332–7.
22. Yan JY, Zheng XL, Li SY. A novel and robust recombinant *Pichia pastoris* yeast whole cell biocatalyst with intracellular overexpression of a *Thermomyces lanuginosus* lipase: preparation, characterization and application in biodiesel production. Bioresour Technol. 2014;151:43–8.
23. Luisa G, Ana CO. Microalgae as a raw material for biofuels production. J Ind Microbiol Biotechnol. 2009;36:269–74.
24. Brundiek HB, Evitt AS, Kourist R, Bornscheuer UT. Creation of a lipase highly selective for *trans* fatty acids by protein engineering. Angew Chem Int Ed. 2012;51:412–4.
25. Yan JY, Zheng XL, Du L, Li SY. Integrated lipase production and *in situ* biodiesel synthesis in a recombinant *Pichia pastoris* yeast: an efficient dual biocatalytic system composed of cell free enzymes and whole cell catalysts. Biotechnol Biofuels. 2014;7:55–62.
26. Belcher J, McLean KJ, Matthews S, Woodward LS, Fisher K, Rigby SE, et al. Structure and biochemical properties of the alkene producing cytochrome P450 $OleT_{JE}$ (CYP152L1) from the *Jeotgalicoccus* sp. 8456 bacterium. J Biol Chem. 2014;289:6535–50.
27. You C, Zhang YHP. Self-assembly of synthetic metabolons through synthetic protein scaffolds: one-step purification, co-immobilization, and substrate channeling. ACS Synth Biol. 2013;2:102–10.
28. Zhang YHP, Sun JB, Zhong JJ. Biofuel production by *in vitro* synthetic enzymatic pathway biotransformation. Curr Opin Biotechnol. 2010;21:663–9.
29. Johnston WA, Huang WL, De Voss JJ, Hayes MA, Gillam EMJ. Quantitative whole-cell cytochrome P450 measurement suitable for high-throughput application. J Biomol Screen. 2008;13:135–41.

9

From wheat straw to bioethanol: integrative analysis of a separate hydrolysis and co-fermentation process with implemented enzyme production

Vera Novy[1], Karin Longus[1] and Bernd Nidetzky[1,2*]

Abstract

Background: Lignocellulosic ethanol has a high potential as renewable energy source. In recent years, much research effort has been spent to optimize parameters involved in the production process. Despite that, there is still a lack of comprehensive studies on process integration. Single parameters and process configurations are, however, heavily interrelated and can affect the overall process efficiency in a multitude of ways. Here, we present an integrative approach for bioethanol production from wheat straw at a representative laboratory scale using a separate hydrolysis and co-fermentation (SHCF) process. The process does not rely on commercial (hemi-) cellulases but includes enzyme production through *Hypocrea jecorina* (formerly *Trichoderma reesei*) on the pre-treated feedstock as key unit operation. Hydrolysis reactions are run with high solid loadings of 15% dry mass pre-treated wheat straw (DM WS), and hydrolyzates are utilized without detoxification for mixed glucose-xylose fermentation with the genetically and evolutionary engineered *Saccharomyces cerevisiae* strain IBB10B05.

Results: Process configurations of unit operations in the benchtop SHCF were varied and evaluated with respect to the overall process ethanol yield ($Y_{Ethanol\text{-}Process}$). The highest $Y_{Ethanol\text{-}Process}$ of 71.2 g ethanol per kg raw material was reached when fungal fermentations were run as batch, and the hydrolysis reaction was done with an enzyme loading of 30 filter paper units (FPU)/$g_{DM\ WS}$. 1.7 ± 0.1 FPU/mL were produced, glucose and xylose were released with a conversion efficiency of 67% and 95%, respectively, and strain IBB10B05 showed an ethanol yield of 0.4 g/$g_{Glc\ +\ Xyl}$ in 15% hydrolyzate fermentations. Based on the detailed process analysis, it was further possible to identify the enzyme yield, the glucose conversion efficiency, and the mass losses between the unit operations as key process parameters, exhibiting a major influence on $Y_{Ethanol\text{-}Process}$.

Conclusions: $Y_{Ethanol\text{-}Process}$ is a measure for the efficiency of the lignocellulose-to-bioethanol process. Based on mass balance analysis, the correlations between single process parameters and $Y_{Ethanol\text{-}Process}$ were elucidated. The optimized laboratory scale SHCF process showed efficiencies similar to pilot scale plants. The herein presented process analysis can serve as effective and simple tool to identify key process parameters, bottlenecks, and future optimization targets.

* Correspondence: bernd.nidetzky@tugraz.at
[1]Institute of Biotechnology and Biochemical Engineering, Graz University of Technology, NAWI Graz, Petersgasse 12/I, 8010 Graz, Austria
[2]Austrian Centre of Industrial Biotechnology, Petersgasse 14, A-8010 Graz, Austria

Background

Utilization of biomass as renewable and sustainable energy source has called attention from politics and R&D facilities around the world [1-5]. Second-generation bioethanol produced from lignocellulosic waste streams constitutes the most feasible technical option. Reasons are, among many, the geographically evenly distributed and inexpensive feedstock as well as the neat or blended application in the transportation sector without the requirement of major technical adaptation [1-3,6].

The lignocellulose-to-bioethanol process consists of five unit operations; a) pre-treatment of the feedstock, b) production of the (hemi-) cellulolytic enzymes, c) enzymatic hydrolysis of the pre-treated feedstock, d) fermentation of the hydrolyzate to bioethanol, and e) down-stream processing [3,6-10]. In the last two decades, all five unit operations have been subjected to intensive research activities, which resulted in key technologies with improved yields and efficiencies. Thus, different biological, chemical, and physical pre-treatment methods have been developed and are applied alone or in combination to increase the enzyme accessibility and to facilitate cellulose depolymerization [7,9,11]. Further process intensification was achieved by combination of two or more process steps into one single unit operation [5,9], most importantly the simultaneous saccharification and (co-) fermentation (SS(C)F) process which has been applied in bioethanol production at laboratory [12,13] and pilot plant [1,14] scale. Genetic and evolutionary engineering enabled *Saccharomyces cerevisiae* to convert both glucose and xylose, the major hemicellulose-derived sugar, and it enhanced the organism's robustness towards inhibitory compounds (e.g., furans, acids, and phenolic compounds) which are by-products formed during pre-treatment of the feedstock [6,8,10,15-21].

Unit operations within a process are often strongly interlinked so that variation of process parameters in one unit operation can impact the overall process efficiency through indirect effects on other unit operations. Optimization studies, therefore, must implement a complete mass balance-based process analysis and not only focus on single unit operations isolated from the respective others in the process. Despite the extensive research efforts made in R&D facilities in the recent past, there is currently still a lack in comprehensive analyses done at the level of the whole process [1,3]. Besides techno-economic analyses published since the mid-1980s [1,2,4,5,22], data are mainly available from a few pilot scale plants such as the SEKAB plant in Sweden [1,23,24]. The scale of pilot plants, however, already excludes high-throughput analysis of different process configurations. In contrast, data acquired at laboratory scale might not be characteristic and lack transferability for larger scale productions.

In this study, we present an integrative analysis of a lignocellulose-to-bioethanol process on representative laboratory scale (90 mL to 4 L). The process was run as separate hydrolysis and co-fermentation (SHCF), which ensures optimal conditions for both hydrolysis and mixed glucose-xylose fermentation [19,25]. Further, SHCF, as compared to simultaneous saccharification and co-fermentation (SSCF), reduces the complexity of the process, which is important for larger scale applications. Consequently, lignocellulose-to-bioethanol processes which are heading towards industrial scale (e.g., POET-DSM 'Project Liberty,' Clariant AG 'SunLiquid') are run as SHCF. An overview of the three-step process is given in Figure 1. Firstly, a fraction of the pre-treated wheat straw was used for cultivation of a *Trichoderma reesei* strain SVG17 for the production of (hemi-) cellulolytic enzymes. 'On-site' enzyme production is economically advantageous and efficient [26] and it reduces the cost and supply dependency on commercial enzyme suppliers. Despite that, research and pilot scale studies almost exclusively rely on commercially available enzymes [1,2,14,23,26]. The enzyme solution is then utilized for the saccharification of the pre-treated wheat straw (Figure 1). Since fungal cultivation was conducted on the same feedstock, the enzyme mixture is highly optimized for this substrate. The resulting hydrolyzate contains glucose and xylose, and both sugars are converted efficiently to ethanol by the application of the genetically and evolutionary engineered *S. cerevisiae* strain IBB10B05 [19,27].

As shown in Figure 1, the single unit operations were integrated to one SHCF process and mass balance analysis was performed. The resulting overall process ethanol yield ($Y_{\text{Ethanol-Process}}$), the amount of ethanol produced from 1 kg of dry mass feedstock, thereby quantifies the process efficiency, and it is the key parameter to assess and compare different processes [1,2,14,23]. Consequently, mass balance analysis and $Y_{\text{Ethanol-Process}}$ were used in this study to investigate different process configurations, and it generated an in-depth knowledge of the SHCF process.

The aim of this study was to establish a tool for process analysis already on laboratory scale. Integration of unit operations and balance-based analysis of process configurations identified bottlenecks and potential optimization targets within the SHCF process. An integrative process analysis at an early stage can render process development more efficient and therefore might contribute to the success of second-generation bioethanol production.

Results and discussion

The feedstock

The substrate presented in this study was Austrian wheat straw pre-treated with steam explosion. Wheat straw has a high potential as sustainable biomass source in Europe based on its abundance and low cost [28].

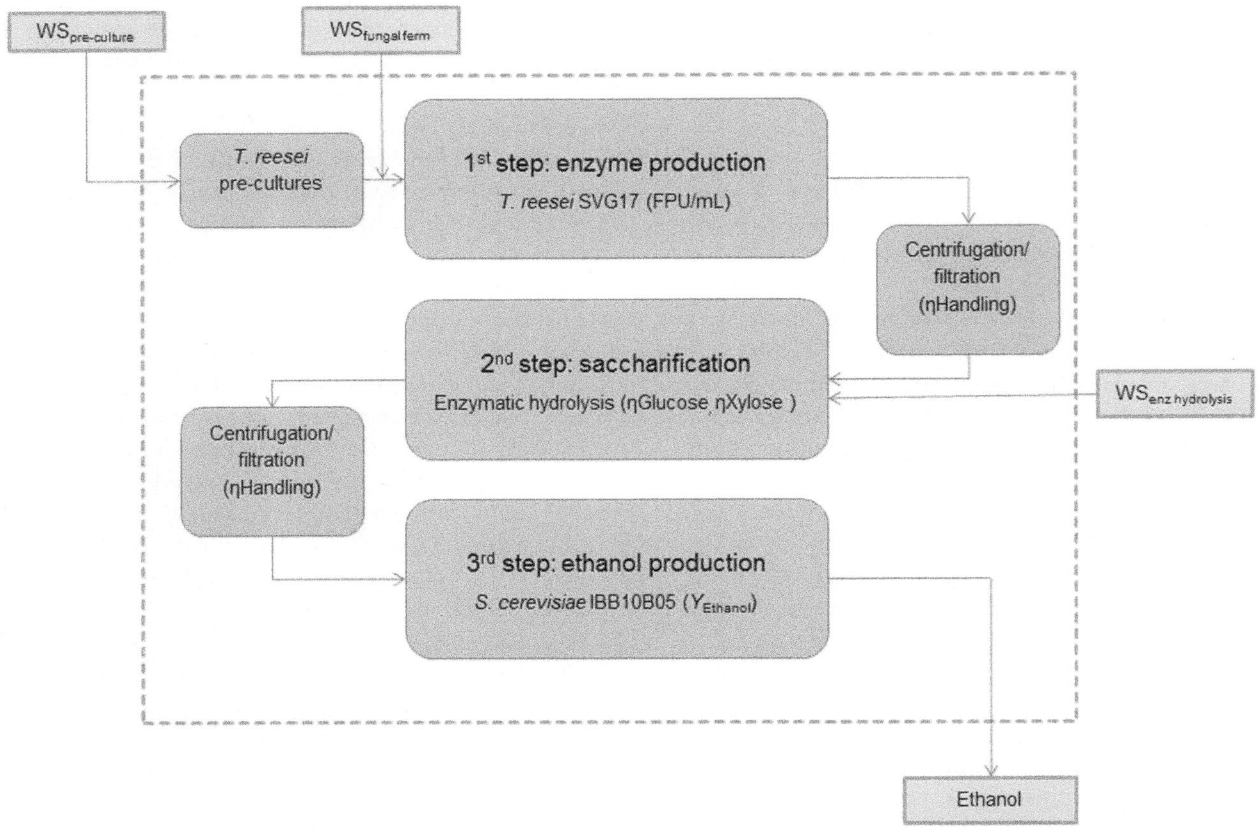

Figure 1 An overview of the laboratory scale SHCF process.

Steam explosion, in combination with chemicals or alone, has been described as an efficient and cost-effective method for pre-treatment of wheat straw [9]. The raw material in this study was treated with a simple method based on steam explosion only. The pre-treated wheat straw had a dry mass (DM) content of approximately 90% and thereof the water insoluble fraction was 69%. The compositional analysis is depicted in Table 1. The acid hydrolyzate contained majorly glucose and xylose. Other hemicellulose-derived sugars (e.g., L-arabinose, galactose) were only present in amounts below the detection limit of the high performance liquid chromatography (HPLC) system and are not mentioned in Table 1. Although processed under the same conditions, the pre-treated wheat straw composition showed batch-to-batch variability. Thus, the glucose and xylose content varied by 5% and 13%, respectively (Table 1). Variation in the pre-treated feedstock composition, especially in xylose content, has been observed before [25]. The amorphous nature of the hemicellulose and different levels of degradation during the pre-treatment as well as seasonal variations are likely explanations. Since both batches of wheat straw were utilized throughout this study, analysis of unit operation and mass balanced process analysis were based on averaged values (Table 1). The raw material before pre-treatment had a xylose content of 24.9 ± 0.4% and a glucose content of 36.1 ± 0.1% dry matter (data not shown). Mass losses caused by pre-treatment were 10% on average.

Table 1 Compositional analysis of the pre-treated wheat straw

Components in dry matter	Percentage [%]		
	Batch 1	Batch 2	Mean
Carbohydrates			
Glucose	43.7	48.7	46.2 ± 2.5
Xylose	17.2	13.1	15.2 ± 2.0
Non-carbohydrates			
Acid-soluble lignin	1.3	2.0	1.7 ± 0.3
Acid-insoluble lignin	27.7	30.0	28.9 ± 1.2
Ashes	4.5	4.4	4.5 ± 0.0
Others	5.6	4.3	4.9 ± 0.6

Experimental analysis of unit operations

The three unit operations of the SHCF process, enzyme production, hydrolysis, and fermentation, were analyzed under varying process conditions. As shown in Figure 1, the process streams between the unit operations were treated by centrifugation, filtration, and concentration. The resulting losses were included into the process analysis with the efficiency factor of conditioning steps (ηHandling), and it was determined to be 75% on average.

Production of (hemi-) cellulases by T. reesei SVG17

T. reesei is the majorly applied organism for (hemi-) cellulase production, and it has been studied and continuously improved since the 1960s [29-33]. The herein presented *T. reesei* SVG17 is a mutant of the QM9414 strain, and previous studies have described it as useful enzyme producer at both laboratory and pilot scale [31]. Fungal cultivations were run as batch fermentation with a substrate loading of 30 g DM pre-treated wheat straw per L ($g_{DM\ WS}$/L). In 7 days of fermentation, a total volumetric cellulolytic activity of 1.7 ± 0.1 FPU/mL was reached. The beta-glucosidase activity was determined to be 0.6 ± 0.1 U/mL. To increase the enzyme yields, fermentations were also run as fed-batch. Per 2 L of fermentation, 30 $g_{DM\ WS}$ was added three times after 66, 94, and 138 h of fermentation. The time course of the fed-batch fermentation is depicted in Figure 2A. In 210 h of fermentation, the volumetric cellulase activity reached 2.7 ± 0.02 FPU/mL. Similar to the batch fermentation, the beta-glucosidase activity was approximately half of the FPU/mL value and it was determined to be 1.5 ± 0.02 U/mL. Addition of feedstock was described to prolong the phase in which enzyme production is most active by freshly inducing both biomass growth and enzyme expression [30,32]. Consequently, the cellulase activity is increasing after each addition of WS, and only towards the end (180 to 210 h), the FPU/mL-time curve is stagnating. The time courses of beta-glucosidase activity and protein concentration show a similar pattern with a less pronounced effect of the feed. In the fed-batch fermentation, it was possible to improve the volumetric cellulase activity 1.6-fold as compared to the batch fermentation. This increase, however, required a 2.5-fold higher substrate loading. The impact of both process configurations on $Y_{Ethanol\text{-}Process}$ was evaluated with mass balance analysis and will be discussed hereinafter.

Despite the difficult substrate conditions, *T. reesei* SVG17 was able to grow on the pre-treated wheat straw and showed efficient and reproducible production of (hemi-) cellulolytic enzymes. A drawback of strain SVG17 is the relatively low beta-glucosidase activity in the enzyme solution which did not exceed 50% of the overall cellulase activity (measured in FPU). A beta-glucosidase activity to FPU ratio of 1 was found to be the lower limit for efficient biomass conversion [34,35]. Low beta-glucosidase activity is an often observed problem in the production of cellulolytic enzymes by *T. reesei*. To overcome this problem, genetically engineered strains overexpressing heterologous beta-glucosidase genes have been described in the literature [34-37]. Although the availability of a robust *T. reesei* strain producing a beta-glucosidase-boosted enzyme solution could be important in view of an optimized process output, the aim of this study was not primarily optimization itself, but rather provision of a *basis for optimization* through integrative mass balance analysis of a representative SHCF process. Therefore, the use of *T. reesei* strain SVG17 was fully in line with the concept of the study, and the limitation in beta-glucosidase noted was not considered to restrict the relevance of the current investigation. Moreover,

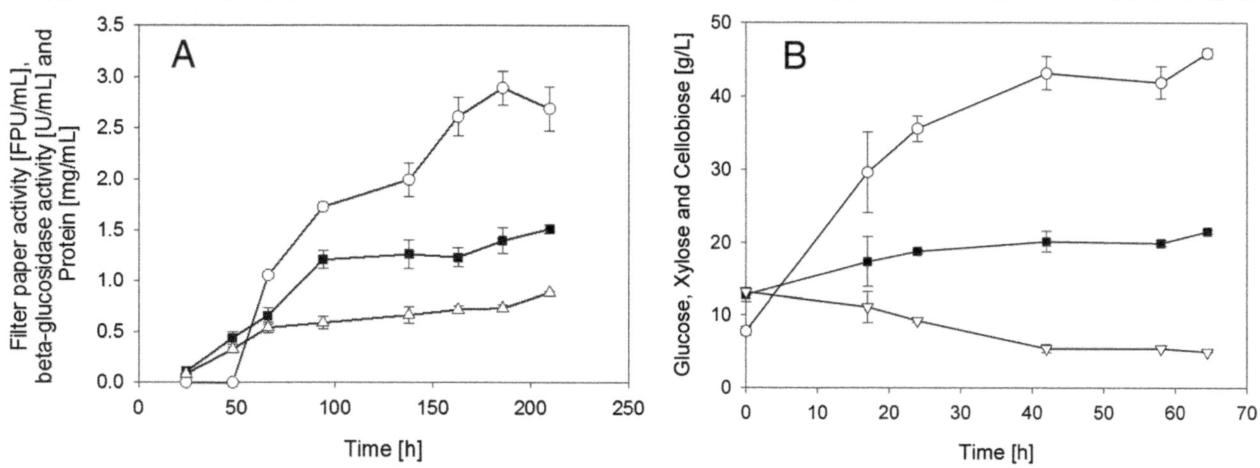

Figure 2 Time courses of the *T. reesei* SVG17 fed-batch fermentation (A) and the enzymatic hydrolysis (B). Thirty grams DM WS was added after 66, 94, and 138 h to the fungal fermentation. Hydrolysis reaction was run with 30 FPU/$g_{DM\ WS}$. Both time courses represent mean values from two experiments. Symbols: **(A)** Total cellulase activity (empty circles), beta-glucosidase activity (filled squares), and protein concentration (empty triangles). **(B)** Glucose (empty circles), xylose (filled squares), and cellobiose (empty triangles).

the effects of enhanced beta-glucosidase activity on hydrolysis yield and process performance are described later in this manuscript.

Enzymatic hydrolysis

To render lignocellulose-to-bioethanol processes economically feasible, a high ethanol titer is crucial [17]. Consequently, hydrolysis reactions must aim for high solid loadings to increase the sugar content. This, in turn, does increase the content of compounds, which are potentially toxic for the fermentation organism. In a previous study, we have shown that the solid loading in the hydrolysis reaction can be increased from 5% to 15% DM WS without introducing severe inhibition effects on *S. cerevisiae* strain IBB10B05 [19]. Enzyme loadings were varied. Firstly, reactions were run with 25 FPU/$g_{DM\ WS}$ [19,27] and the final hydrolyzate contained 40.6 ± 5.7 g/L glucose and 18.0 ± 2.3 g/L xylose. Based on the compositional analysis of the feedstock (Table 1), this equals a conversion efficiency of 60% for glucose (ηGlucose) and 81% for xylose (ηXylose). To improve the conversion efficiencies, the enzyme loading was increased to 30 FPU/$g_{DM\ WS}$, and ηGlucose of 67% and ηXylose of 95% were reached. The 15% hydrolyzate contained 46.1 ± 0.2 g/L glucose and 21.5 ± 0.2 g/L xylose. The time course of the 30 FPU/$g_{DM\ WS}$ hydrolysis reaction is depicted in Figure 2B. In addition to glucose and xylose, a considerable amount of cellobiose (5.0 g/L, Figure 2) was released into the hydrolyzate. Cellobiose is known to have an inhibitory impact on cellulases (e.g., [38,39]). Accumulation of cellobiose during hydrolysis is caused by limitation in beta-glucosidase activity in the enzyme mixture used. To evaluate the impact of enhanced beta-glucosidase activity, hydrolysis reactions (30 FPU/$g_{DM\ WS}$) were additionally performed with supplemented Novozyme188. The FPU to beta-glucosidase activity was 1, which was chosen according to the literature [34,35]. The resulting hydrolyzate had a glucose and xylose concentration of 51.8 ± 1.1 g/L and 23.0 ± 1.0 g/L, respectively. The cellobiose concentration was below 1.5 g/L. Addition of beta-glucosidase increased the ηGlucose from 67% to 78%. The already high ηXylose was further increased and reached full conversion.

Fermentation to ethanol with S. cerevisiae strain IBB10B05

S. cerevisiae IBB10B05 proved to be a sturdy and efficient fermentation strain for mixed glucose-xylose fermentation in spent sulfite liquor, wheat straw hydrolyzates, and a combination thereof [19,27]. Strain IBB10B05 performs excellently under most basic process and substrate conditions. Thus, fermentations were run in simple batch cultures without process monitoring and control (e.g., pH adjustment). The 15% hydrolyzate was applied without pre-treatment (e.g., detoxification) and yeast extract was added as a sole substrate supplement. An in-depth physiological characterization of strain IBB10B05 in fermentation of 15% hydrolyzates has been published recently [19]. The fermentation time course is depicted in Additional file 1. Both glucose (~37 g/L) and xylose (~19 g/L) were depleted within 50 h of fermentation, and in total, 22 g/L ethanol was produced within this time frame (Additional file 1, [19]). This represents an ethanol yield of ~0.4 g/$g_{Glc\ +\ Xyl}$.

Integration of unit operations and process analysis

The single unit operations, enzyme production, hydrolysis, and fermentation, were integrated to one SHCF process as depicted in Figure 1. To evaluate and compare the different process configurations in context of the complete SHCF process, mass balance analysis was performed. The different process configurations are described based on critical output parameters (FPU/mL, ηGlucose, ηXylose, ηHandling, and ethanol yield ($Y_{Ethanol}$)), and the corresponding mass balance analyses are summarized in Additional file 2. A comparison of process configurations based on $Y_{Ethanol\text{-}Process}$ is depicted in Figure 3. Note that to account for losses caused by pre-treatment and to facilitate comparison of the laboratory scale SHCF process with data from the literature, mass balance analyses were based on the raw material. It was assumed that the required input of raw material was 10% higher as the input calculated for the pre-treated wheat straw (Additional file 2). Throughout this study, $Y_{Ethanol\text{-}Process}$ is given in g ethanol produced per kg raw material ($g_{Ethanol}$/$kg_{DM\ RM}$).

Enzyme production, the first unit operation of the SHCF process, was run as batch or as fed-batch. The latter approach resulted in a higher volumetric activity but also required a higher substrate loading ('Production of (hemi-) cellulases by *T. reesei* SVG17'). When enzymes were produced in batch fermentations (Config.1), the resulting $Y_{Ethanol\text{-}Process}$ was 71.2 $g_{Ethanol}$/$kg_{DM\ RM}$. In fed-batch fermentations (Config.2), $Y_{Ethanol\text{-}Process}$ was 58.0 $g_{Ethanol}$/$kg_{DM\ RM}$, which is a 19% decrease as compared to Config.1. This clearly emphasizes the need for evaluating process parameters in context of the complete process. Despite the extensive research on cellulase production in *T. reesei* (e.g., [29,32,40,41]), the influence of substrate or process conditions on the success of the overall bioethanol production process is scarcely considered. Enzymatic hydrolysis, the second unit operation, was analyzed with two different enzyme loadings, 25 FPU/$g_{DM\ WS}$ (Config.3) and 30 FPU/$g_{DM\ WS}$ (Config.1). Config.3 resulted in an overall process ethanol yield of 67.5 $g_{Ethanol}$/$kg_{DM\ RM}$ (Additional file 2, Figure 3). This is 5% less as compared to the reaction with 30 FPU/$g_{DM\ WS}$ ($Y_{Ethanol\text{-}Process}$ 71.2 $g_{Ethanol}$/$kg_{DM\ RM}$). The last unit operation, the fermentation to ethanol, was accomplished with the

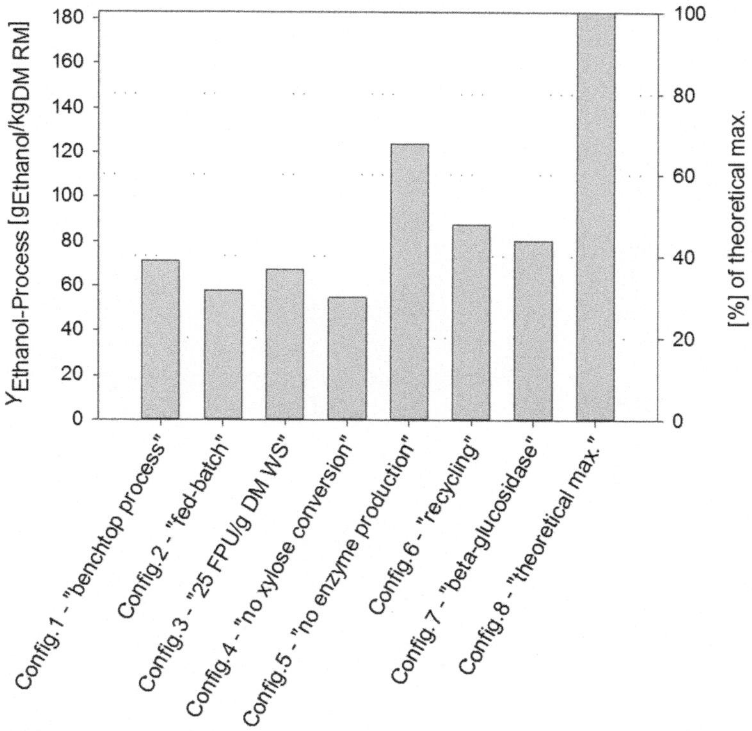

Figure 3 The influence of different process configurations on $Y_{Ethanol\text{-}Process}$. Detailed description of process parameters are summarized in Additional file 2.

xylose-fermenting *S. cerevisiae* strain IBB10B05 (Figure 1). To compare the efficiencies of a SHCF and a separate hydrolysis and fermentation (SHF) process, mass balance analysis was additionally performed with glucose conversion only (Config.4). With an ethanol on glucose yield of 0.45 g/g_{Glc} [19,27], the $Y_{Ethanol\text{-}Process}$ was 54.6 $g_{Ethanol}/kg_{DM\ RM}$ (Additional file 2, Figure 3). This is 23% less as compared to Config.1. Although, genetically engineered xylose-fermenting *S. cerevisiae* strains have been described extensively in the literature (e.g., [10,15]), pilot plants often still operate with non-GM yeasts and rely on glucose conversion only [1,23]. This integrative process analysis, however, clearly highlights the importance of efficient conversion of all sugars in the hydrolyzates and strain IBB10B05 proved to be an excellent candidate.

Integration of unit operations and analyses of the different process configurations (Config.1 to 3) showed that Config.1 has the highest $Y_{Ethanol\text{-}Process}$ (Figure 3) and the mass balance analysis is depicted in Table 2. Thus, enzyme production in batch fermentation (working volume 4 L) and processing of the enzyme solution (ηHandling 75%) resulted in a total of 5,100 FPU. With an enzyme loading of 30 $FPU/g_{DM\ WS}$, it was possible to hydrolyze 170 g DM WS. After treatment of the 15% hydrolyzate (ηHandling 75%), 39.5 g glucose and 18.4 g xylose were available for fermentation, which was converted to 23.1 g of ethanol. A total process ethanol yield of 71.2 $g_{Ethanol}/kg_{DM\ RM}$ was reached (Table 2).

To assess the efficiency, the benchtop SHCF process was compared to currently available data from pilot (SEKAB, IBUS, BCyL) and commercial (Clariant, 'SunLiquid') scale plants. A summary of process configurations, plant capacities, and $Y_{Ethanol\text{-}Process}$ is depicted in Table 3. The pilot scale plants summarized in Table 3 operate without 'on-site' production of (hemi-) cellulolytic enzymes. To still render a comparison of the process efficiencies possible, mass balance analysis of the laboratory scale SHCF process was performed excluding enzyme production (Config.5, Figure 3 and Additional file 2). Without the loss of feedstock required for the fungal cultivation, the $Y_{Ethanol\text{-}Process}$ was 123.7 $g_{Ethanol}/kg_{DM\ RM}$, which is 1.7-fold higher as compared to Config.1. This is already within the range of $Y_{Ethanol\text{-}Process}$ reported for the pilot scale plants (118 to 157.8 3 $g_{Ethanol}/kg_{DM\ RM}$; Table 3), clearly highlighting the usefulness of the herein presented laboratory scale SHCF process as a model for establishing an integrative process analysis. However, direct comparison of Config.1 and Config. 5, solely based on $Y_{Ethanol\text{-}Process}$, is not sufficient. 'On-site' enzyme production has been described to entail several advantages, such as being cost-effective and

Table 2 Mass balance analysis of the benchtop SHCF process

	Input	Output
1st step: enzyme production (*T. reesei* SVG17)		
Pre-cultures ($WS_{pre\text{-}culture}$)	5.6 g DM WS	
Batch cultivation ($WS_{fungal\ ferm}$)	120 g DM WS	
Total cellulolytic activity		1.7 FPU/mL
ηHandling 75%		5,100 FPU total
2nd step: saccharification (enzymatic hydrolysis)		
Substrate loading ($WS_{enz\ hydrolysis}$)	170 g DM WS	
Glucose (ηGlucose 67%, ηHandling 75%)		39.5 g
Xylose (ηXylose 95%, ηHandling 75%)		18.4 g
3rd step: ethanol production (*S. cerevisiae* IBB10B05)		
Ethanol ($Y_{Ethanol}$ 0.4 g/$g_{Glc+Xyl}$)		23.1 g
Total		
Substrate loading - pre-treated	295.6 g DM WS	
Substrate loading - raw material	325.2 g DM RM	
Ethanol		23.1 g
$Y_{Ethanol\text{-}Process}$	71.2 $g_{Ethanol}/kg_{DM\ RM}$	

An overview of the process is depicted in Figure 1. Boundary conditions: batch fermentations with 30 $g_{DM\ WS}$/L, 15% DM WS, and 30 FPU/$g_{DM\ WS}$.

efficient [26,42]. Thus, for further evaluation of the feasibility of the process configurations, a detailed economic analysis, as published for other process [2,14,22], must be conducted.

The by far most efficient process is the 'SunLiquid' process, which is operated as SHCF with implemented 'on-site' enzyme production. The reported $Y_{Ethanol\text{-}Process}$ is 222.2 $g_{Ethanol}/kg_{DM\ RM}$ (Table 3) [42]. This is 18% higher as compared to the maximum theoretical yield of the benchtop SHCF process which was determined to be 183.1 $g_{Ethanol}/kg_{DM\ RM}$ (Config.8, Figure 3 and Additional file 2). There are several factors that could explain the exceptionally high $Y_{Ethanol\text{-}Process}$ described for the 'SunLiquid' process. The first factor is the pre-treatment method applied. Ideally, pre-treatment of lignocellulosic biomass enriches the structural carbohydrate cellulose and hemicellulose by removal of the lignin and enhances the accessibility of the partially crystalline cellulose. It therefore has an impact on the efficiency of the fungal fermentation and the enzymatic hydrolysis and is directly affecting $Y_{Ethanol\text{-}Process}$ [7,11]. In this study, the steam explosion was performed as batch and running the process continuously could reduce mass losses caused by pre-treatment. However, evaluation of varying pre-treatment methods was beyond the scope of this study and is not considered in more detail hereinafter. The second factor influencing $Y_{Ethanol\text{-}Process}$ is the potential application of enzyme or solid recycling to boost the hydrolysis efficiency [43,44]. This process option was also analyzed in context of this study. With an estimated increase in glucose released per enzyme loading of 35% (e.g., [43,44]), the $Y_{Ethanol\text{-}Process}$ of the benchtop SHCF process would improve to 87.8 $g_{Ethanol}/kg_{DM\ RM}$ (Config.6, Figure 3 and Additional file 2). The third factor is the organism employed for enzyme production. The importance of a *T. reesei* strain secreting a balanced enzyme mixture has been described in the literature [34-37]. In this study, the application of an enzyme solution with a beta-glucosidase to FPU ratio of 1 was investigated. The resulting $Y_{Ethanol\text{-}Process}$ was 80.3 $g_{Ethanol}/kg_{DM\ RM}$ (Config.7, Figure 3 and Additional file 2). In comparison to Config.1, the altered process configurations resulted in

Table 3 $Y_{Ethanol\text{-}Process}$ for commercial, pilot, and laboratory scale processes

Substrate	Capacity [1,000 $t_{DM\ RM}$/year]	Process	$Y_{Ethanol\text{-}Process}$ [$kg_{Ethanol}/t_{DM\ RM}$]	Reference
Cereal straw	150	SHCF + E[a]	222.2	Clariant 'SunLiquid' [42]
Wheat straw	8.8	SSF	123	'IBUS' [23]
Forestry residues	0.7	SSF	118 to 157.8	'SEKAB' [1]
Wheat straw	25.6	SHF	154	'BCyL' [1]
Wheat straw	-	SHCF/+E[a]	123.7/71.2	This study

[a]'On-site' enzyme production.

an increase in $Y_{Ethanol\text{-}Process}$ of 19% (Config.6) and 11% (Config.7). However, the process yields did not exceed 50% of the reported $Y_{Ethanol\text{-}Process}$ of the 'SunLiquid' process. Therefore, we took the process analysis one step further and analyzed the benchtop SHCF process towards potential bottlenecks.

Based on the detailed mass balance analysis of the integrated benchtop SHCF process and the resulting understanding of the process streams, it was possible to identify three key parameters which exhibit a significant influence on $Y_{Ethanol\text{-}Process}$, the enzyme yield, ηGlucose, and ηHandling (Figure 1). The correlations between $Y_{Ethanol\text{-}Process}$ and these parameters are depicted in Figure 4. Firstly, $Y_{Ethanol\text{-}Process}$ was plotted over the enzyme yield (FPU/mL) (Figure 4A). To increase the $Y_{Ethanol\text{-}Process}$ of Config.1 to 100 $g_{Ethanol}/kg_{DM\ RM}$, for an instance, a 3.1-fold increase in total cellulolytic activity is required (from 1.7 to 5.3 FPU/mL, Figure 4A). The importance of an efficient enzyme production is further highlighted, by comparing the maximum theoretic $Y_{Ethanol\text{-}Process}$ observed in this study with the 'SunLiquid' process. To enhance the $Y_{Ethanol\text{-}Process}$ of Config.8 to 222 $g_{Ethanol}/kg_{DM\ RM}$ [42], the enzyme yield must be increased twofold, from 1.7 FPU/mL to 3.4 FPU/mL. In addition to the above discussed factors, a higher enzyme yield during fungal fermentation might therefore explain for the high $Y_{Ethanol\text{-}Process}$ of the 'SunLiquid' process.

As shown in Figure 4B, the overall process efficiency is further influenced by the two parameters ηGlucose and ηHandling. Thus, $Y_{Ethanol\text{-}Process}$ can vary between 22.4 (ηGlucose 25% and ηHandling 50%) and 127.2 $g_{Ethanol}/kg_{DM\ RM}$ (ηGlucose 95% and ηHandling 95%) under else same boundary conditions. Possible improvement of ηGlucose has already been described within this study. Data summarized in Table 3 suggest that $Y_{Ethanol\text{-}Process}$ increases with the scale of the plant ('Capacity'). A correlation between plant capacity and overall yield is further supported by the literature [1-5,14,22-24], and it is suggested that processes are becoming more efficient with increasing scale [1-3,5]. Explicit information on ηHandling for pilot and commercial scale plants are not given. The fact that $Y_{Ethanol\text{-}Process}$ of Config.5 ("no enzyme production") is within the range of to the pilot scale plants (Table 3), however, might indicate that the ηHandling are similar. To further increase $Y_{Ethanol\text{-}Process}$, improvement of ηHandling will be a target of future optimization studies of the benchtop SHCF process.

Conclusions

In this study, an integrative process analysis of a benchtop SHCF is presented. Based on mass balance analysis, the influence of varying process configurations on $Y_{Ethanol\text{-}Process}$ was analyzed. Thereby, a fundamental understanding of the complete process was established. This allowed for identification of the process parameters, which have the highest impact on $Y_{Ethanol\text{-}Process}$, the enzyme yield, ηGlucose, and ηHandling. It was further shown that, under comparable process conditions (Conifg.5 - no enzyme production), $Y_{Ethanol\text{-}Process}$ of the benchtop SHCF process is equal to pilot scale plants. We therefore believe that the benchtop-scale analysis described herein presents an important and useful tool to identify bottlenecks and optimization targets within the process with reasonable effort and expenditure.

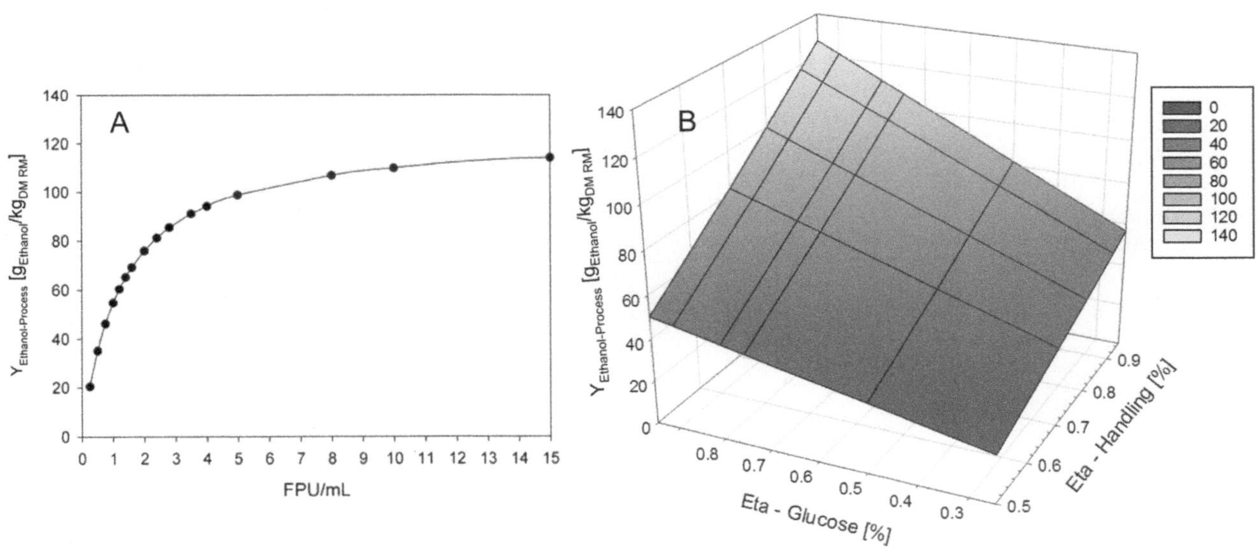

Figure 4 The influence of key parameters on $Y_{Ethanol\text{-}Process}$. **(A)** $Y_{Ethanol\text{-}Process}$ in dependence of the total cellulolytic activities produced by fungal fermentation (boundary conditions: batch with 30 $g_{DM\ WS}/L$, 30 FPU/$g_{DM\ WS}$, ηGlucose 67%, ηXylose 95%, ηHandling 75%, and $Y_{Ethanol}$ 0.4 $g/g_{Glc+Xyl}$). **(B)** Influence of ηGlucose and ηHandling on $Y_{Ethanol\text{-}Process}$ (boundary conditions: batch with 30 $g_{DM\ WS}/L$, 1.7 FPU/mL, 30 FPU/$g_{DM\ WS}$, ηXylose 95%, and $Y_{Ethanol}$ 0.4 $g/g_{Glc+Xyl}$).

Findings of this study stress the importance to establish a mass balance-based understanding of the lignocellulose-to-bioethanol process and to optimize unit operations or process parameters with respect to $Y_{Ethanol\text{-}Process}$.

Methods

Chemicals and media used

Unless mentioned differently, all chemicals were from Carl Roth + Co KG (Karlsruhe, Germany). Mineral (M-) media for cultivation of *T. reesei* SVG17 contained 5 g/L yeast extract, 5 g/L KH_2PO_4, 3.75 g/L $(NH_4)_2SO_4$, 0.3 g/L $MgSO_4 \times 7H_2O$, 0.3 g/L $CaCl_2 \times 2H_2O$, and 1 mL/L trace elements (5 g/L $FeSO_4 \times 7H_2O$, 1.6 g/L $MnSO_4 \times H_2O$, 1.4 g/L $ZnSO_4 \times 7H_2O$, 0.2 g/L $CoCl_2 \times 6H_2O$, 15 g/L EDTA disodium chloride salt). Additionally, 0.2 mL/L Tween 80 (Sigma-Aldrich, St. Louis, MO, USA) and 0.5 g/L rapeseed oil were supplied. Potato-dextrose-agar (PDA) was prepared as described by the manufacturer. Yeast extract peptone glucose (YPD) medium for cultivation of *S. cerevisiae* contained 10 g/L yeast extract, 20 g/L peptone from casein, and 20 g/L glucose. In 15% hydrolyzate conversion experiments, 10 g/L of yeast extract was added as a sole medium supplement.

Preparation of the lignocellulosic feedstock

Pre-treated wheat straw from Austria was kindly provided by the University of Applied Sciences in Upper Austria (FH Wels). Throughout this study, two batches of feedstock were received and they were both pre-treated as described in the following. The wheat straw was air-dried to a water content of approximately 8% (*w/w*), and the fibers were chaffed to reduce the length of the fibers to below 8 cm. Further, the wheat straw was treated by steam explosion at 200°C, 15 bar for 10 min, with a water to wheat straw ratio of 1. After cooling, the wheat straw was dried to a dry mass content of 90% and stored at 4°C in plastic bags. Per batch, DM and water insoluble (WIS) content was analyzed in triplets. For the former, a moisture analyzer (MA 50, Sartorius AG, Göttingen, Germany) operated at 105°C was used. For determination of the WIS content, 2 g of wheat straw was washed with 50 mL of 50°C warm water, dried at 105°C for 24 h, and weighted. The content of structural carbohydrates, lignin, and ash in the wheat straw was analyzed following the protocol of the National Renewable Energy Laboratory (NREL) [45].

Strains

For the production of the (hemi-) cellulolytic enzymes, the *T. reesei* strain SVG17 was applied. The strain background has been described in a previously published study [31]. For conversion of the 15% hydrolyzate to ethanol, the *S. cerevisiae* strain IBB10B05 was utilized. It is a descendant of the BP10001 strain, which was further improved by evolutionary engineering. Detailed descriptions of the construction of the BP10001 strain and the evolution strategy have been published elsewhere [18,20].

Bioreactor cultivations for the production of the (hemi-) cellulolytic enzymes

T. reesei SVG17 was maintained on PDA plates at 4°C. The fungus was revitalized by transfer of a piece of overgrown agar onto fresh PDA plates. Incubation was for 3 days at 30°C. Starter cultures were prepared in 300-mL wide mouth shake flasks without baffles filled with 200 mL of mineral medium (M-medium) containing 14 $g_{DM\ WS}$/L. Fermentation was started by transfer of a piece of overgrown PDA agar to the fermentation media. Incubation was for 5 days at 30°C, 200 rpm, and pH 4.5 in a Certomat BS-1 orbital incubator shaker (Sartorius AG, Göttingen, Germany). Subsequently pre-cultures were pooled to ensure homogeneous biomass composition.

Main cultivations were run in batch or fed-batch mode in benchtop bioreactors (Labfors III, Infors AG, Bottmingen, Switzerland) with 2 L as working volume. The bioreactors are equipped with two six-plated impellers. The reactor diameter to impeller diameter ratio was 3 and the reactor height to reactor diameter ratio was 1.5. Cultivations were accomplished in duplicates (two parallel bioreactors). M-media was supplemented with 30 $g_{DM\ WS}$/L, and the fermentation was started by transfer of 200 mL of the pooled pre-cultures into the bioreactor. Fermentation conditions were as follows: 30°C, pH 4.5 (adjusted online with 1 M KOH and 3%, by volume NH_3), and a dissolved oxygen concentration of 20%. The latter was adjusted continuously with an agitation (200 to 800 rpm) and aeration (0.16 to 1.5 L/min pressurized air) cascade. Incubation was for 7 days. In fed-batch mode, 30 g DM WS and media supplements for 1 L of fermentation media was added per bioreactor after 66, 94, and 138 h of fermentation. Fed-batch experiments were prolonged to 9 days of fermentation time. During the fermentation, samples were frequently removed and centrifuged (15,700*g*, 4°C for 10 min, Eppendorf 5415 R, Eppendorf, Hamburg, Germany) and the supernatant stored at 4°C for analysis. Enzyme activity was evaluated in beta-glucosidase and total cellulolytic activity (given in filter paper units, FPU/mL). FPU activity was measured following International Union of Pure and Applied Chemistry (IUPAC) recommendations [46]. Beta-glucosidase activity was determined as described previously [47], with the following alterations. Two hundred fifty microliters of diluted enzyme solution was mixed with 250 μL of 2.0 mM para-nitrophenyl-β-D-glucopyranoside solution (in 50 mM sodium acetate buffer, pH 5.0) and incubated at 50°C for 10 min. The reaction was stopped with 1 M Na_2CO_3. Protein concentration was determined with the Bradford method [48]. Proteins were precipitated and quantified utilizing the Roti-Quant kit (Roth) and following the manufacturer's instructions.

The enzyme solution was harvested and prepared for the subsequent enzymatic hydrolysis as described in the following. The supernatant of the fungal cultivation was collected by centrifugation (4,420*g*, 4°C; 20 min; Sorvall RC-5B; Thermo Fisher Scientific Inc., Waltham, MA, USA), concentrated by evaporation (45°C, 40 mbar, Laborota 4000, Heidolph Instruments GmbH & Co. KG, Schwabach, Germany) to one tenth of its original volume, filtrated sterile (Whatman Klari-Flex System; GE Healthcare, Little Chalfont, UK), and stored at 4°C. The loss of enzyme solution was included into the mass balance analysis with the efficiency factor ηHandling (Figure 1) and was determined by measuring volume and mass before and after the processing steps.

Enzymatic hydrolysis of the wheat straw

The substrate loading for the hydrolysis reaction was 15%, by weight dry mass wheat straw. Two different enzyme loadings were applied, 25 and 30 FPU/$g_{DM\ WS}$. Additionally, one hydrolysis reaction (30 FPU/$g_{DM\ WS}$) was performed with supplemented beta-glucosidase (Novozyme188; Sigma-Aldrich) to a total activity of 30 U/$g_{DM\ WS}$. Reactions were performed in 10 mM sodium acetate buffer (pH 4.8) in 500-mL shake flaks with ground-in glass stoppers. Total mass of the reaction was 260 g. The wheat straw suspension was autoclaved and the enzyme added aseptically. Incubation was 50°C, 200 rpm for 72 h (Certomat BS-1). During the hydrolysis, samples were taken. Immediate sample work-up included the boiling of the reaction mixture (100°C, 15 min. 300 rpm, Thermomixer 'comfort' Eppendorf) and centrifugation and storage of the supernatant at −20°C for HPLC analysis.

The hydrolyzate was prepared for the subsequent conversion to bioethanol as described in the following. Firstly, it was heated to 100°C for 15 min in a water bath. Remaining solids were removed by centrifugation (4,420*g*, 4°C, 10 min, Sorvall RC-5B) and discarded. The pH of the supernatant was adjusted to 6.5 with 1 M NaOH, and it was filtrated sterile ('Klari-Flex') and stored at 4°C. Mass and volume loss was determined before and after the processing of the hydrolyzate and included into the mass balance analysis as ηHandling.

Shaken-bottle cultivation for bioethanol fermentation

Fermentation of 15% hydrolyzate to ethanol utilizing the *S. cerevisiae* strain IBB10B05 was accomplished as described previously [19]. In brief, seed and starter cultures were prepared aerobically in shaken flask cultures in YPD media at 30°C. Main cultivations were performed anaerobically at 30°C in glass bottles (working volume 90 mL) tightly sealed with rubber septa. Eighty percent of the total volume was wheat straw hydrolyzate, and the remainder volume was composed of 10% yeast extract and 10% inoculum. The starting optical density at 600 nm (OD_{600}) was 5. Incubation was at 30°C and 180 rpm (Certomat BS-1) for 7 days. During the fermentation, samples (1.5 mL) were frequently removed from the fermentation media and centrifuged (10 min, 15,700*g*, 4°C, Eppendorf 5415R) and the supernatant stored at −20°C for HPLC analysis.

HPLC analysis of sugars and metabolites

Sugars (glucose, xylose, arabinose, galactose, and cellobiose) as well as extracellular metabolites (ethanol, glycerol, xylitol, and acetate) were analyzed by HPLC (Merck-Hitachi LaChrome system equipped with an L-7250 autosampler, a Merck L-7490 RI detector, and a Merck L-7400 UV detector). The system was equipped with an Aminex HPX-87H column and an Aminex Cation H guard column (both Bio-Rad Laboratories, Hercules, CA, USA). The operation temperature was 65°C for both columns and the flow rate of the mobile phase (5 mM sulfuric acid) was 0.6 mL/min. The hemicellulose-derived sugars (galactose and arabinose) were only present in minor amounts (<0.5 g/L) and are therefore not further mentioned.

Additional files

Additional file 1: Mixed glucose-xylose fermentation in 15% hydrolyzate utilizing *S. cerevisiae* strain IBB10B05. Time course was derived from a previous publication [19]. Symbols: glucose (crosses), xylose (full diamonds), glycerol (empty triangles), xylitol (empty squares), ethanol (empty circles).

Additional file 2: Summary of in- and output parameters and mass balance analyses of the different process configurations.

Abbreviations

SHF: separate hydrolysis and fermentation; SSF: simultaneous saccharification and fermentation; SHCF: separate hydrolysis and co-fermentation; SSCF: simultaneous saccharification and co-fermentation; DM: dry mass; WS: pre-treated wheat straw; RM: raw material; FPU: filter paper unit; Glc: glucose; Xyl: xylose; HPLC: high performance liquid chromatography; IUPAC: International Union of Pure and Applied Chemistry; M-medium: mineral medium; NREL: National Renewable Energy Laboratory; OD_{600}: optical density at 600 nm; WIS: water-insoluble; $Y_{Ethanol}$: ethanol yield; $Y_{Ethanol\text{-}Process}$: overall process ethanol yield; YPD: yeast extract peptone glucose; ηHandling: efficiency factor of conditioning steps; ηGlucose/Xylose: conversion efficiency of enzymatic hydrolysis.

Competing interests

The authors declare that they have no competing interests.

Authors' contributions

VN and BN designed the research. VN planned the experiments. VN and KL performed the experiments and analyzed the data. The manuscript was written from contributions of all authors. VN and BN wrote the paper. The manuscript has been read and approved by all authors.

Acknowledgements
We would like to thank Alexander Jäger and Heike Kahr from the University of Applied Sciences Upper Austria for providing us with the pre-treated wheat straw.

References

1. Gnansounou E. Production and use of lignocellulosic bioethanol in Europe: current situation and perspectives. Bioresour Technol. 2010;101:4842–50.
2. Gnansounou E, Dauriat A. Techno-economic analysis of lignocellulosic ethanol: a review. Bioresour Technol. 2010;101:4980–91.
3. Chen H, Qiu W. Key technologies for bioethanol production from lignocellulose. Biotechnol Adv. 2010;28:556–62.
4. Sukumaran RK, Surender VJ, Sindhu R, Binod P, Janu KU, Sajna KV, et al. Lignocellulosic ethanol in India: prospects, challenges and feedstock availability. Bioresour Technol. 2010;101:4826–33.
5. Cardona CA, Sánchez ÓJ. Fuel ethanol production: process design trends and integration opportunities. Bioresour Technol. 2007;98:2415–57.
6. Hahn-Hägerdal B, Galbe M, Gorwa-Grauslund MF, Lidén G, Zacchi G. Bio-ethanol – the fuel of tomorrow from the residues of today. Trends Biotechnol. 2006;24:549–56.
7. Govumoni SP, Koti S, Kothagouni SY, Venkateshwar S, Linga VR. Evaluation of pretreatment methods for enzymatic saccharification of wheat straw for bioethanol production. Carbohydr Polym. 2013;91:646–50.
8. Sims REH, Mabee W, Saddler JN, Taylor M. An overview of second generation biofuel technologies. Bioresour Technol. 2010;101:1570–80.
9. Talebnia F, Karakashev D, Angelidaki I. Production of bioethanol from wheat straw: an overview on pretreatment, hydrolysis and fermentation. Bioresour Technol. 2010;101:4744–53.
10. van Maris A, Abbott D, Bellissimi E, van den Brink J, Kuyper M, Luttik M, et al. Alcoholic fermentation of carbon sources in biomass hydrolysates by *Saccharomyces cerevisiae*: current status. Antonie Van Leeuwenhoek. 2006;90:391–418.
11. Alvira P, Tomás-Pejó E, Ballesteros M, Negro MJ. Pretreatment technologies for an efficient bioethanol production process based on enzymatic hydrolysis: a review. Bioresour Technol. 2010;101:4851–61.
12. Erdei B, Barta Z, Sipos B, Reczey K, Galbe M, Zacchi G. Ethanol production from mixtures of wheat straw and wheat meal. Biotechnol Biofuels. 2010;3:16.
13. Olofsson K, Palmqvist B, Liden G. Improving simultaneous saccharification and co-fermentation of pretreated wheat straw using both enzyme and substrate feeding. Biotechnol Biofuels. 2010;3:17.
14. Sassner P, Galbe M, Zacchi G. Techno-economic evaluation of bioethanol production from three different lignocellulosic materials. Biomass Bioenergy. 2008;32:422–30.
15. Hahn-Hägerdal B, Karhumaa K, Fonseca C, Spencer-Martins I, Gorwa-Grauslund M. Towards industrial pentose-fermenting yeast strains. Appl Microbiol Biotechnol. 2007;74:937–53.
16. Matsushika A, Inoue H, Kodaki T, Sawayama S. Ethanol production from xylose in engineered *Saccharomyces cerevisiae* strains: current state and perspectives. Appl Microbiol Biotechnol. 2009;84:37–53.
17. Viikari L, Vehmaanperä J, Koivula A. Lignocellulosic ethanol: from science to industry. Biomass Bioenergy. 2012;46:13–24.
18. Klimacek M, Kirl E, Krahulec S, Longus K, Novy V, Nidetzky B. Stepwise metabolic adaption from pure metabolization to balanced anaerobic growth on xylose explored for recombinant *Saccharomyces cerevisiae*. Microb Cell Fact. 2014;13:37.
19. Novy V, Krahulec S, Wegleiter M, Muller G, Longus K, Klimacek M, et al. Process intensification through microbial strain evolution: mixed glucose-xylose fermentation in wheat straw hydrolyzates by three generations of recombinant *Saccharomyces cerevisiae*. Biotechnol Biofuels. 2014;7:49.
20. Petschacher B, Nidetzky B. Altering the coenzyme preference of xylose reductase to favor utilization of NADH enhances ethanol yield from xylose in a metabolically engineered strain of *Saccharomyces cerevisiae*. Microb Cell Fact. 2008;7:9.
21. Chundawat SPS, Beckham GT, Himmel ME, Dale BE. Deconstruction of lignocellulosic biomass to fuels and chemicals. Annu Rev Chem Biomol Eng. 2011;2:121–45.
22. Macrelli S, Mogensen J, Zacchi G. Techno-economic evaluation of 2nd generation bioethanol production from sugar cane bagasse and leaves integrated with the sugar-based ethanol process. Biotechnol Biofuels. 2012;5:22.
23. Larsen J, Østergaard Petersen M, Thirup L, Wen Li H, Krogh IF. The IBUS process – lignocellulosic bioethanol close to a commercial reality. Chem Eng Technol. 2008;31:765–72.
24. Maas RH, Bakker RR, Boersma AR, Bisschops I, Pels JR, de Jong E, et al. Pilot-scale conversion of lime-treated wheat straw into bioethanol: quality assessment of bioethanol and valorization of side streams by anaerobic digestion and combustion. Biotechnol Biofuels. 2008;1:14.
25. Erdei B, Franko B, Galbe M, Zacchi G. Separate hydrolysis and co-fermentation for improved xylose utilization in integrated ethanol production from wheat meal and wheat straw. Biotechnol Biofuels. 2012;5:12.
26. Sorensen A, Teller PJ, Lubeck PS, Ahring BK. Onsite enzyme production during bioethanol production from biomass: screening for suitable fungal strains. Appl Biochem Biotechnol. 2011;164:1058–70.
27. Novy V, Krahulec S, Longus K, Klimacek M, Nidetzky B. Co-fermentation of hexose and pentose sugars in a spent sulfite liquor matrix with genetically modified *Saccharomyces cerevisiae*. Bioresour Technol. 2013;130:439–48.
28. Kim S, Dale B. Global potential bioethanol production from wasted crops and crop residues. Biomass Bioenergy. 2004;4:361–75.
29. Peterson R, Nevalainen H. *Trichoderma reesei* RUT-C30–thirty years of strain improvement. Microbiology. 2012;158:58–68.
30. Schuster A, Schmoll M. Biology and biotechnology of *Trichoderma*. Appl Microbiol Biotechnol. 2010;87:787–99.
31. Esterbauer H, Steiner W, Labudova I, Hermann A, Hayn M. Production of *Trichoderma* cellulase in laboratory and pilot scale. Bioresour Technol. 1991;36:51–65.
32. Persson I, Tjerneld F, Hahn-Hägerdal B. Fungal cellulolytic enzyme production: a review. Process Biochem. 1991;26:65–74.
33. Mandels M, Reese ET. Induction of cellulase in *Trichoderma viride* as influenced by carbon sources and metals. J Bacteriol. 1957;73:269–78.
34. Chen M, Qin Y, Liu Z, Liu K, Wang F, Qu Y. Isolation and characterization of a β-glucosidase from *Penicillium decumbens* and improving hydrolysis of corncob residue by using it as cellulase supplementation. Enzyme Microb Technol. 2010;46:444–9.
35. Ma L, Zhang J, Zou G, Wang C, Zhou Z. Improvement of cellulase activity in *Trichoderma reesei* by heterologous expression of a beta-glucosidase gene from *Penicillium decumbens*. Enzyme Microb Technol. 2011;49:366–71.
36. Murray P, Aro N, Collins C, Grassick A, Penttila M, Saloheimo M, et al. Expression in *Trichoderma reesei* and characterisation of a thermostable family 3 beta-glucosidase from the moderately thermophilic fungus *Talaromyces emersonii*. Protein Expr Purif. 2004;38:248–57.
37. Nakazawa H, Kawai T, Ida N, Shida Y, Kobayashi Y, Okada H, et al. Construction of a recombinant *Trichoderma reesei* strain expressing *Aspergillus aculeatus* beta-glucosidase 1 for efficient biomass conversion. Biotechnol Bioeng. 2012;109:92–9.
38. Gruno M, Väljamäe P, Pettersson G, Johansson G. Inhibition of the *Trichoderma reesei* cellulases by cellobiose is strongly dependent on the nature of the substrate. Biotechnol Bioeng. 2004;86:503–11.
39. Murphy L, Bohlin C, Baumann MJ, Olsen SN, Sørensen TH, Anderson L, et al. Product inhibition of five *Hypocrea jecorina* cellulases. Enzyme Microb Technol. 2013;52:163–9.
40. Ahamed A, Vermette P. Effect of culture medium composition on *Trichoderma reesei*'s morphology and cellulase production. Bioresour Technol. 2009;100:5979–87.
41. Domingues FC, Queiroz JA, Carbral JMS, Fonseca LP. The influence of culture conditions on mycelial structure and cellulase production by *Trichoderma reesei* Rut C-30. Enzyme Microb Technol. 2000;26:394–401.
42. Clariant AG. Sunliquid® - Cellulosic ethanol from agricultural residues. [http://www.clariant.com/en/Solutions/Products/2014/10/16/16/16/sunliquid] Accessed 22 Jan 2015.
43. Weiss N, Borjesson J, Pedersen LS, Meyer AS. Enzymatic lignocellulose hydrolysis: improved cellulase productivity by insoluble solids recycling. Biotechnol Biofuels. 2013;6:5.
44. Yang J, Zhang X, Yong Q, Yu S. Three-stage hydrolysis to enhance enzymatic saccharification of steam-exploded corn stover. Bioresour Technol. 2010;101:4930–5.

45. Sluiter A, Hames B, Ruiz R, Scarlata C, Sluiter J, Templeton D, et al. Determination of structural carbohydrates and lignin in biomass. National Renewable Energy Laboratory: Golden, CO, USA; 2011.
46. Ghose T. Measurement of cellulase activities. Pure Appl Chem. 1987;2:257–68.
47. Takashima S, Nakamura A, Hidaka M, Masaki H, Uozumi T. Molecular cloning and expression of the novel fungal beta-glucosidase genes from *Humicola grisea* and *Trichoderma reesei*. J Biochem. 1999;125:728–36.
48. Bradford MM. A rapid and sensitive method for the quantitation of microgram quantities of protein utilizing the principle of protein-dye binding. Anal Biochem. 1976;72:248–54.

10

Lipid recovery from a vegetable oil emulsion using microbial enrichment cultures

Jelmer Tamis[1*], Dimitry Y Sorokin[1,2], Yang Jiang[1], Mark C M van Loosdrecht[1] and Robbert Kleerebezem[1]

Abstract

Background: Many waste streams have a relatively high vegetable oil content, which is a potential resource that should be recovered. Microbial storage compound production for the recovery of lipids from lipid-water emulsions with open (unsterilized) microbial cultures was investigated in a sequencing batch reactor using a diluted vegetable oil emulsion as model substrate.

Results: After feeding, triacylglycerides (TAG) were accumulated intracellular by the microbial enrichment culture and subsequently used for growth in the remainder of the sequencing batch cycle. Roughly 50% of the added TAG could be recovered as intracellular lipids in this culture. The maximum lipid storage capacity of the enrichment culture was 54% on volatile suspended solids (VSS) mass basis in a separate fed-batch accumulation experiment. The microbial community was dominated by a lipolytic fungus, *Trichosporon gracile*, that was responsible for intracellular lipid accumulation but also a significant fraction of lipolytic and long chain fatty-acid-utilizing bacteria was present.

Conclusion: Herewith, we demonstrate an effective strategy for enrichment of a microbial community that can accumulate significant amounts of lipids from wastewaters without the need for sterilization of substrates or equipment. Further optimization of this process will make recovery of lipids from wastewater possible.

Keywords: Vegetable oil, Storage compounds, Microbial enrichment culture, Feast-famine, Resource recovery

Introduction

The use of agro-industrial organic residues for the production of valuable commodities is a logical step towards a bio-based economy. Traditionally, these wastes are used for biogas or compost production. Recently, several alternatives for the production of more valuable compounds have been proposed, for example, the production of biopolymers [1,2], volatile fatty acids [3], or medium chain length fatty acids [4]. One of the critical issues related to resource recovery from wastewater is the efficient up-concentration and purification of the valuable compounds that are present in diluted form in the wastewater. In general, the production of storage compounds with microbial cultures from diluted waste streams improves the efficiency of resource recovery since the product is concentrated inside the biomass and can be readily separated from the water using standard sludge settling/separation methods.

A relatively well studied example of a process based on storage compound production by microbial enrichment cultures is the production of polyhydroxyalkanoate (PHA) and contents up to 90% on VSS mass basis have been obtained by application of selective pressure in the form of feast-famine conditions [5,6]. This process is currently further evaluated in industrial environments [7-9]. Interestingly, it has been found that different carbon sources in a feast-famine process can result in formation of different storage compounds: microbial cultures enriched on glucose or starch, produced polyglucose as storage compound [10,11], a microbial culture enriched on glycerol produced a mixture of polyglucose and PHA [12], microbial communities enriched on different types of volatile fatty acids (VFA) produced different varieties of PHA [13,14], and a microbial culture enriched on methanol did not produce any storage compounds [15]. While the above examples illustrate the many substrate

* Correspondence: j.tamis@tudelft.nl
[1]Department of Biotechnology, Delft University of Technology, Julianalaan 67, 2628 BC Delft, The Netherlands
Full list of author information is available at the end of the article

types that have been investigated, the fate of lipids as substrate in a feast-famine process remains unclear.

Lipids are an important constituent of many types of wastewater, especially triacylglycerides (TAG), present in (among others) effluents from vegetable oil crop processing industries. For example, in the wastewater from the palm oil industry, high concentrations of emulsified oil are found that are hard to recover using physical-chemical methods [16]. Palm oil is a fast growing market with a current production of around 50 Mton/year [17] accompanied by the production of 3 m^3 wastewater per ton oil palm oil produced [18] resulting in an estimated palm oil mill effluent flow of 150 million m^3/year.

It was proposed that lipids found in wastewater may be recovered through the use of lipid-accumulating microorganism for the production of biodiesel [19]. The economic feasibility of lipid extraction for diesel production from sludge will depend on the lipid content accumulated in the sludge but a cost of around 1 US $/l was reported from sludge containing 10% lipids [20]. In this approach, the recovery of lipids from wastewater requires a microorganism with an as high TAG storage capacity as possible. However, not all microorganisms store TAG from wastewater in an appropriate way for biodiesel production.

A variety of metabolic pathways related to the microbial conversion of TAG have been described in literature (Figure 1). Due to the relatively large molecule size (M_w around 800 to 1,000 g/mol) and strongly hydrophobic nature of TAG, the uptake mechanism involves an extracellular hydrolysis step yielding long chain fatty acids (LCFA) and glycerol, which are subsequently transported into the cell [21]. Once the substrate is taken up, different routes are possible for the production of storage compounds: for example, PHA production by pure cultures of prokaryotic strains was reported [22-25] as well as the intracellular storage of TAG by pure cultures, for example, by oleaginous yeasts [26,27] or bacteria [28,29].

One approach for the recovery of lipids from wastewater is the cultivation of pure cultures of strains with very high lipid storage capacity. This approach has been evaluated in laboratory experiments with synthetic substrates and sterilized molasses [27]. However, since sterilization of diluted wastewater streams for biodiesel production is likely economically unfeasible, the use of unsterilized wastewater was evaluated [30,31]. In these cases, a precultivated pure culture of oleaginous yeast was used to inoculate a batch experiment with acidic wastewater of a distillery as substrate. It was observed that a biomass with a relatively high lipid content (respectively 44% and 14%) could be grown by this method.

A different strategy was proposed by Santamauro *et al.* [32] in which an oleaginous yeast could be selectively enriched in a low pH environment. In general, oleaginous yeasts thrive relatively well in low pH environments. Nevertheless, many non-oleaginous microorganisms are known to be able to grow under low pH conditions [33]. The selective pressure imposed by the low pH (or temperature) does not intrinsically provide a competitive

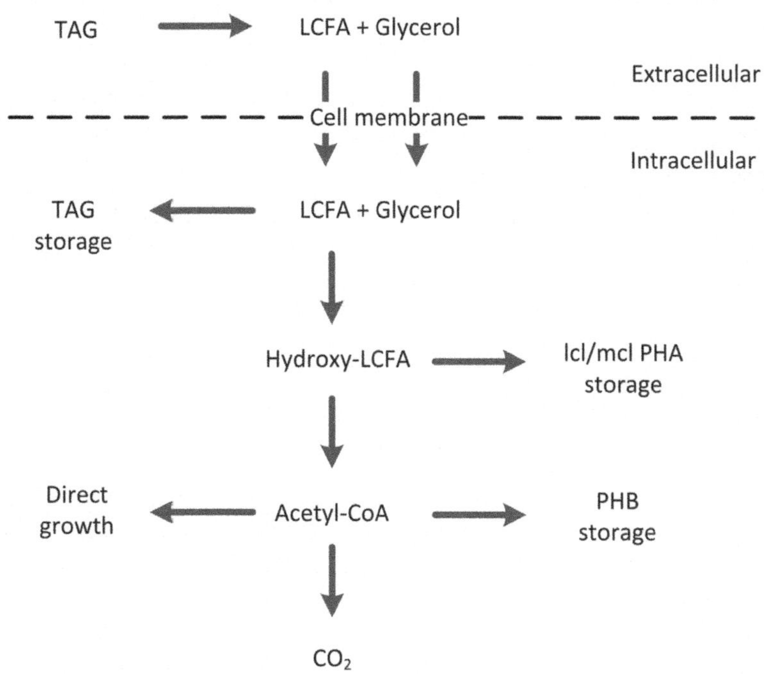

Figure 1 Simplified metabolic pathways of storage compound formation and growth on TAG based on literature reports [21-29,38]. LCFA, long chain fatty acid; PHB, polyhydroxybutyrate; TAG, triacylglycerides.

advantage for lipid-accumulating microorganisms, which makes the enrichment culture unstable and likely not optimal for lipid accumulation [34].

Since lipid accumulation has been reported as a microbial survival strategy to balance carbon and energy requirements in periods of absence of external substrate [35], in this study, we propose the application of a feast-famine strategy as a novel approach for the enrichment of lipid-accumulating organisms. This strategy is analogous to the feast-famine process for PHA production [6] and is based on ecological principles that provide an intrinsic competitive advantage to lipid-accumulating species. This process potentially enables the recovery of lipids from a broader spectrum of wastewaters, can be operated as sequenced batch or continuously, and alleviates the need for single batch operations with pre-cultivated pure culture inocula. An open (non-sterilized) reactor system was inoculated with wastewater sludge and pulse-fed with a model substrate (soybean oil) to evaluate the potential of the feast-famine principle for the recovery of lipids from diluted lipid-containing emulsions.

Results

Reactor operation

After inoculation of the sequencing batch reactor (SBR) with activated sludge (WWTP Dokhaven, The Netherlands) a typical feast-famine pattern became apparent in less than 5 days (10 cycles). The length of the feast phase (that is, time it took to take up the lipid substrate added with the influent) was initially more than 3 h and became more stable after 25 days (50 cycles) with a feast phase length of around 1 to 2 h (Figure 2). Variations in the length of the feast phase are due to the formation of biofilm in the reactor, which had to be periodically removed.

After the reactor had reached a stable operational performance, the carbon mass balance over the system was evaluated by analysis of the influent and effluent in the period of day 68 until day 76 of operation (16 cycles). The average effluent total suspended solids (TSS) concentration was 1.0 ± 0.16 g/l (standard deviation over the dataset of 16 cycles) with an ash fraction of 0.14 ± 0.03, resulting in a volatile suspended solids (VSS) concentration of 0.86 ± 0.15 g/l. Additionally, the amount of biofilm that was formed on the reactor wall was removed and measured, indicating an average biofilm growth of 0.25 gVSS per cycle. TOC measurements indicated that the carbon content of the organic solids in the effluent was 0.59 (±0.03) gC/gVSS resulting in a total organic solids production equivalent to 59 ± 10 Cmmol per cycle. Off-gas measurements showed a CO_2 production of 17 ± 3 mmol per cycle. Herewith, most of the carbon present in the influent substrate (83 Cmmol/cycle) could be accounted for by carbon in the produced organic solids and CO_2 (76 ± 10 Cmmol/cycle). The 10% deviation in the balance is partly due to the complications arising from the reactor wall growth.

To get further insight into the conversions during the cycle, three experiments were performed. For all three cycle experiments, the biomass concentrations and length of feast phase were within the range determined during the steady operational period. In Figure 3, profiles from a typical cycle are shown. For practical purposes and in order to establish the substrate amount accurately, the substrate in these experiments was dosed manually in one short pulse, and the amount of substrate was evaluated by the weight of the dosing syringe before and after addition of the oil.

After dosing the vegetable oil, the reactor liquid became opaque, indicating that the mixture of microorganisms

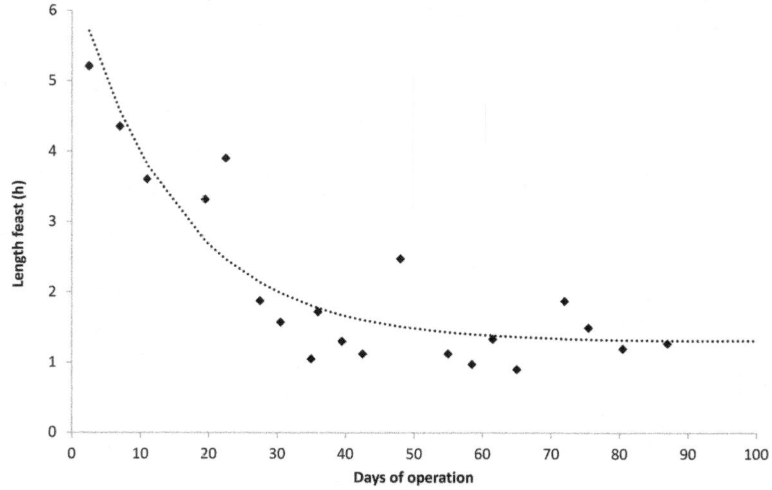

Figure 2 Monitoring of the start-up of the sequencing batch reactor by the evolution of the length of the period needed for TAG substrate uptake (feast phase) by the enrichment culture in the reactor. h, hours.

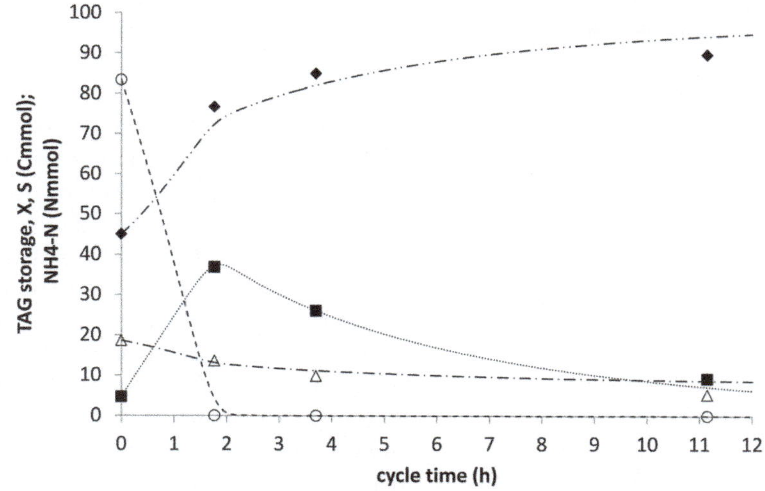

Figure 3 Amounts of TAG substrate (○, ——--), lipid storage (■, ·····), active biomass (♦, –·· – ··–), and ammonium (Δ, –· – ·–) during a cycle in the TAG-fed sequencing batch reactor. Symbols represent measurement and lines represent model-based calculations. h, hours; TAG, triacylglycerides.

and vegetable oils formed an emulsion; this opaque feature disappeared again at the end of the feast phase (after 2 h) when the TAG was depleted from the medium. Gas chromatography (GC) analysis showed that the lipid content of the biomass increased from 6% ± 0.6% on VSS mass basis, at the start of the cycle, to 25% ± 1.5% on VSS mass basis at the end of the feast phase (average ± standard deviation over three cycle experiments), indicating significant storage of lipids (this was also confirmed by microscope images: Figure 4, right). The composition of the accumulated microbial lipids was almost identical to the substrate oil (Table 1), indicating direct utilization of the externally hydrolyzed LCFA and glycerol without their *de novo* synthesis or conversion. GC analysis showed only peaks that corresponded with the LCFA standard, and no significant peaks corresponding with the PHA standards, suggesting that TAG was the only relevant storage compound present in the culture at any time during the cycle. The lipid content of the cells decreased during the famine phase indicating growth on the stored lipids. Nitrogen was present throughout the cycle, and it was observed that nitrogen was taken up both during the feast and the famine phase, indicating significant microbial growth during the whole SBR cycle. The increase in VSS concentration in the feast phase was higher than that would be expected from TAG storage only, supporting the indication of growth of microorganisms in both the feast phase and the famine phase. A model was calibrated to the experimental data to identify characteristic process parameters (Additional file 1: Appendix C). It was estimated that about half of the substrate was used for growth while the other half was accumulated as intracellular lipids during the feast phase. An overall biomass-specific uptake rate of 1.0 Cmol/(Cmol h), a yield of lipid storage on substrate of 0.9 Cmol/Cmol and biomass on lipids of 0.6 to 0.7 Cmol/Cmol were estimated from the cycle measurements. A summary of the experimental data obtained during steady operation and cycle measurements

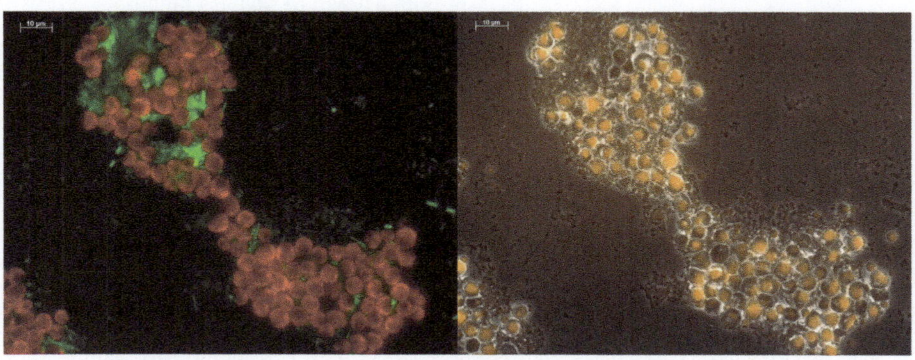

Figure 4 Microscope images of a sample from the reactor at the end of the feast phase. Left: FISH staining with the EUK516 probe to stain eukaryotes (red) and the EUB338 probe to stain prokaryotes (green). Right: Nile blue A staining to identify lipids (orange).

Table 1 Comparison of the LCFA constituents from literature, used substrate and intracellular storage compounds

	Soybean oil [47]	Soybean oil (this study)	Intracellular storage (this study)
	Percentage of LCFA mass		
Oleic acid	23%	23%	25% ± 5%
Linoleic	53%	55%	50% ± 6%
α-Linolenic	8%	6%	7% ± 2%
Palmitic	11%	12%	11% ± 3%
Stearic	4%	4%	4% ± 2%

For the composition of the intracellular storage, the average and the standard deviation over a dataset of six individual samples are provided. LCFA, long chain fatty acid.

in terms of carbon flows is provided in the (Additional file 1: Appendix B).

The maximum lipid storage capacity of the culture was evaluated using fed-batch experiments in which soybean oil was available in excess. The maximum lipid content appeared to be dependent on the availability of nitrogen. In the absence of a nitrogen source, only 24% lipid storage on VSS basis was observed while in another experiment with ammonium present in the reactor liquid (C:N molar ratio was 4/1 Cmol/Nmol), a lipid accumulation of 54% on VSS basis was observed. While the initial lipid accumulation was relatively fast, the lipid content of the culture stabilized after 4 to 5 h of operation, indicating the maximum storage capacity of the culture under these conditions was reached (Figure 5).

Lipase activity assays

The lipase activity was measured using a standard test with 4-nitrophenyl-palmitate [36] during the feast and the famine phases. An activity of 0.80 ± 0.28 mmol/(l min) was measured during the feast phase (average ± standard deviation over a dataset of six measurements). During the famine phase, an activity of 0.73 ± 0.27 mmol/(l min) was measured (average ± standard deviation over a dataset of three measurements). This indicated that there was no significant difference in lipase activity between the feast and the famine phases. For all cases, we evaluated the activity of both the reactor broth (with biomass) and the reactor supernatant. The average activity measured in the reactor broth was 0.99 ± 0.18 mmol/(l min) and in the supernatant 0.57 ± 0.14 mmol/(l min). This indicated that a significant part of the enzymatic activity was due to extracellular enzymes but also that part of the lipase activity was cell-bound. The measured lipase activity was relatively high compared to the lipid uptake rate observed during the feast phase. Using an uptake of 83 Cmmol (approximately 4 mmol ester bonds) in 2 h and a reactor volume of 2 l, the minimal required lipase activity would be 0.01 to 0.02 mmol/(l min). This indicated that extracellular lipase activity was not limiting the conversion rates.

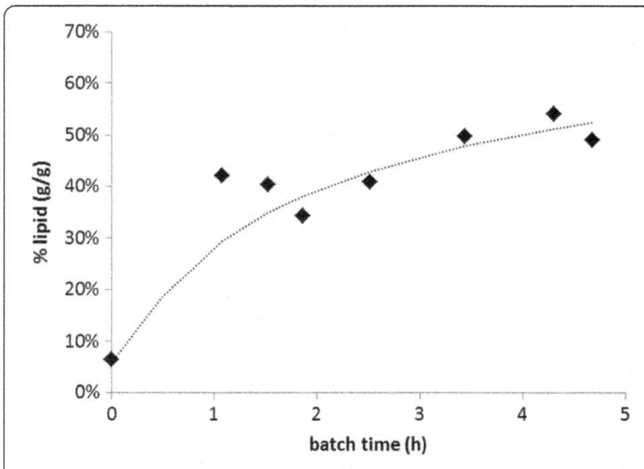

Figure 5 The evolution of lipid content of the enriched microbial culture over time in a fed-batch experiment with nutrients present. h, hours.

Microbial community structure

The molecular and microscopic analyses of the reactor biomass showed a presence of a dimorphic fungus and several bacterial morphotypes. The direct microscopy results were confirmed by fluorescence *in situ* hybridization (FISH) analysis which indicated the presence of both eukaryotes and prokaryotes (Figure 4, left). Furthermore, lipid-specific Nile blue A staining indicated that the fungus was involved in lipid storage while the majority of the prokaryotes in the reactor did not contain significant amounts of lipid-like intracellular storage compounds (Figure 4, right).

Denaturing gradient gel electrophoresis (DGGE) was used to characterize the communities. Eukaryotic primer-based PCR products indicated the presence of *Trichosporon gracile* (100% similarity) from a genus known to include several oleaginous species [27,37]. The bacterial-specific DGGE analysis of the reactor biomass indicated the presence of a few proteobacterial phylotypes closely related to the *Acetinobacter* and *Gordonia* genera (Figure 6), which are both known to possess metabolic pathways for degradation of hydrophobic substrates [38,39].

Cultivation efforts have been made to try to resolve the role of the microbial species present in the reactor biomass in lipid degradation and utilization of the products. Incubations, with and without antibiotics, allowed the isolation of the eukaryote *Trichosporon* and four prokaryote microorganisms. The eukaryotic dimorphic fungus *T. gracile* was isolated on lipid plates with antibiotics; it formed mycelium-containing colonies surrounded by zones of lipid hydrolysis (Figure 7). This organism was most versatile in its metabolism, being able to degrade the TAG, utilize the products of hydrolysis (LCFA and glycerol), and reform the TAG inside the cells. The bacterial

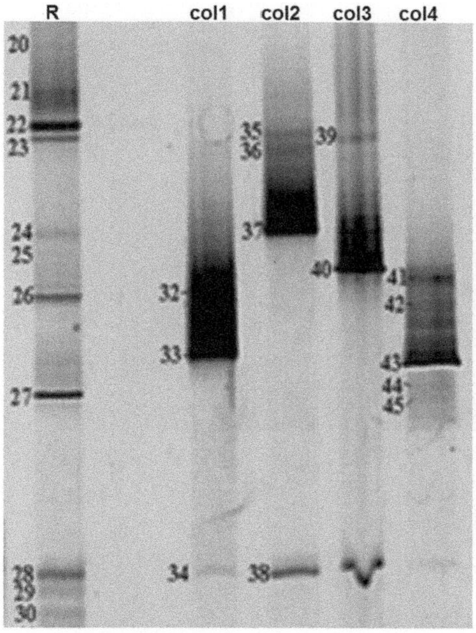

Identification results

Band	Sample	Closest match	% similarity
B20	Reactor	*Acinetobacter seohaensis*	98-99
B21			
B22			
B23			
B24		*Pseudomonas alcaligenes.*	98
B27		*Gordonia sp*	99
B28		*Acinetobacter sp.*	99
B30		*Gordonia sp.*	97
B32	Col 1	*Diaphorobacter sp.*	97
B33			
B34			
B35	Col 2	*Pseudomonas alcaligenes*	99
B36			
B37			
B38			
B39	Col 3	*Shewanella putrefaciens*	99
B40			
B41	Col 4	*Hydrogenophaga intermedia*	99
B43			
B44			

Figure 6 Bacterial-specific DGGE analysis of the reactor biomass and isolates. Col 1, LCFA isolate, long rods; Col 2, lipolytic, short motile rods; Col 3, LCFA isolate, small rods, Col 4: glycerol enrichments; DGGE, denaturing gradient gel electrophoresis; LCFA, long chain fatty acid; R, reactor biomass.

enrichments with three different substrates - lipid, LCFA, and glycerol, yielded four isolates: (i) a small motile rod-shaped bacterium able to grow with lipid and LCFA with strong lipolytic activity (col2, identified as a *Pseudomonas alcaligenes* from the betaproteobacteria). Isolates (ii) and (iii) are two different types of LCFA utilizers. The first LCFA degrader (col1) is a long rod with the ability to store lipids inside the cells. This microorganism was identified as a representative of the genus *Diaphorobacter* from the betaproteobacterial lineage. Although the genus *Diaphorobacter* is quite common for activated sludge, our isolate is very different in cell morphology from the three species described of this genus. The ability to utilize LCFA has never been tested for this genus. The second LCFA degrader is a small rod with no apparent storage polymers (col3) identified as a representative of the genus *Shewanella* in the gammaproteobacteria. (iv) Glycerol-specific enrichment resulted in a domination of *Hydrogenophaga* sp. (betaproteobacterium) represented by small motile rods. Summarizing, the cultivation results showed two functional parts - a eukaryotic dimorphic fungus, doing a complete job of lipid degradation and lipid accumulation, and a bacterial block, basically doing equivalent work but divided between different species (Figure 7).

Despite that two of the bacterial species (*Diaphorobacter* and *Pseudomonas*) were clearly recognizable by microscopic inspection of the reactor biomass, only the latter was detected directly by DGGE analysis of the reactor biomass. One of the possible explanations is a DNA extraction bias in the presence of lipids.

Discussion

A microbial culture capable of accumulating lipids from a dilute emulsion was enriched in a sequencing batch reactor with a feast-famine regime. The development of this enrichment from an activated sludge inoculum occurred relatively fast (25 days). A lipid-storing fungus and a variety of proteobacterial species co-existed in the enrichment culture. Microscopic observation of the reactor biomass indicated that the bacterial phenotypes (except a small population of large cells identified as a *Diaphorobacter* sp.) did not store significant amounts of lipids but instead used lipids or its immediate hydrolysis products directly as growth substrates.

Analysis of the microbial community by DGGE indicated the presence of the eukaryotic fungus *T. gracile*, which could be readily isolated by microbial plating techniques with antibiotics and appeared to be able to accumulate significant amount of lipids intracellularly. The bacterial population analysis was more complicated since *Gordonia* and *Acinetobacter* species that were detected by DGGE could not be isolated using microbial plating techniques. *Gordonia* and *Acinetobacter* species are important species in the lipid-degrading populations that cause foaming in wastewater treatment plants [40], and especially *Gordonia* species are known to accumulate large amounts of intracellular lipids [41]. Furthermore, while *Diaphorobacter* was observed in the biomass (based on its morphology) and could be readily isolated by microbial plating; it could not be detected by DGGE analysis. This shows that microbial analysis tools such as microbial

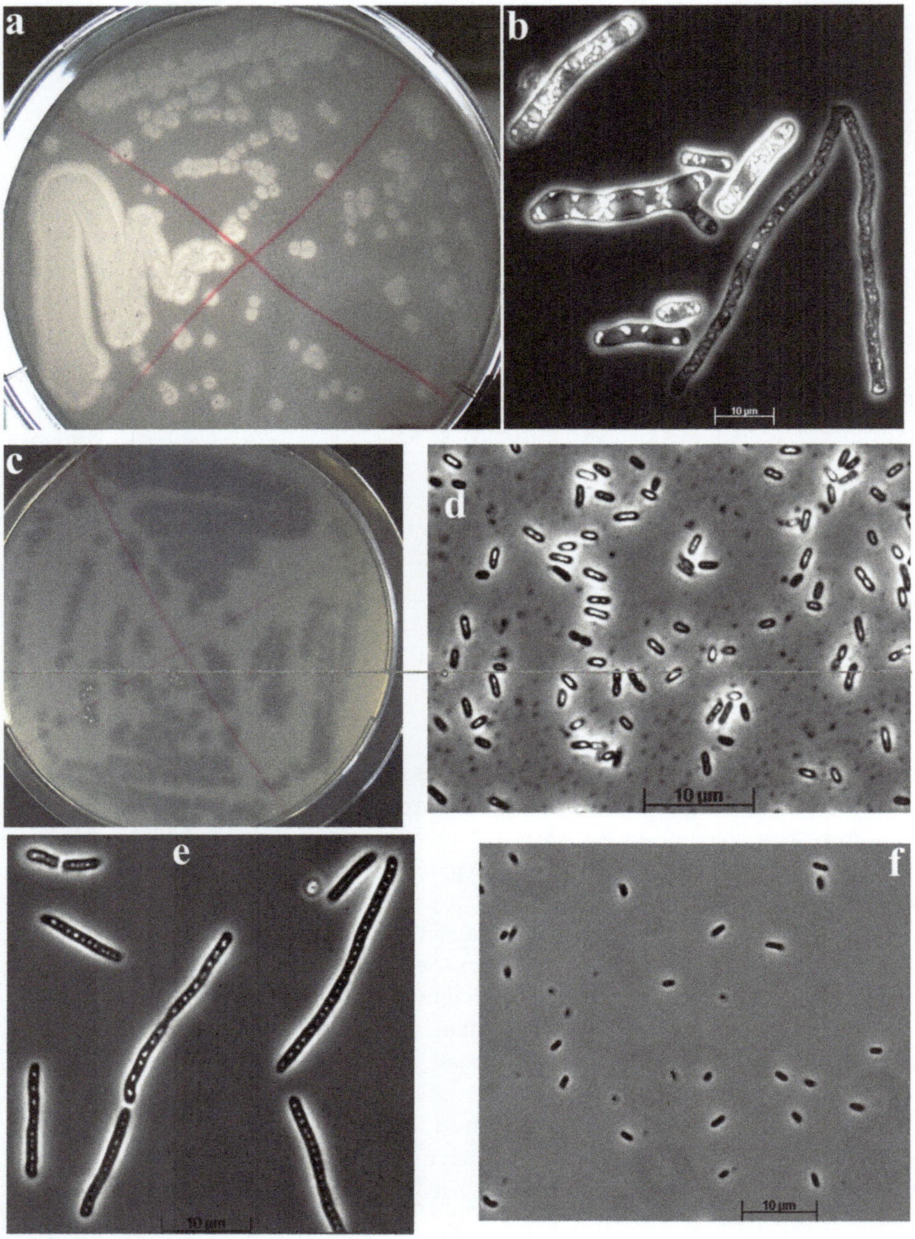

Figure 7 Different functionalities from isolates obtained by plating of the reactor culture. Upper panel: *Trichosporon gracile* (**(a)** lypolytic colonies, **(b)** lipid-storing hyphal cells); middle panel: lipolytic *Pseudomonas alcaligenes* (col2) (**(c)** lipolytic colonies, **(d)** cells in liquid culture with lipid); bottom panel: LCFA-utilizing bacterial isolates (**(e)** *Diaphorobacter* sp. (col1) and **(f)** *Hydrogenophaga* sp. (col3)).

plating and DGGE can easily introduce bias in the population analysis of a reactor. In conclusion, we could identify and culture the dominant microorganism in the culture, a yeast, but the major bacteria could not be exclusively identified by DGGE or microbial plating.

The presence of the bacterial side population in the reactor may be partly explained by the role of bacteria that by themselves were not able to hydrolyse and use TAG as substrate, but could thrive in the reactor by using the LCFA or glycerol that was liberated through the extracellular lipase activity facilitated by, for example, *Trichosporon.* Furthermore, the coexistence of storing and non-storing microbes indicates that lipid uptake and storage by *T. gracile* was not fast enough to outcompete direct growth by bacterial lipolytic and LCFA utilizers. The specific uptake rate of the microbial culture was with 1 Cmol/(Cmol h) relatively low compared to similar feast-famine systems on other substrates, for example, an uptake rate of around 4 Cmol/(Cmol h) was reported for a feast-famine culture cultivated using acetate as carbon source [42].

The measured hydrolytic activity was more than an order of magnitude higher than the observed uptake rate suggesting that hydrolysis was not the rate limiting step. It should be noted that the values obtained from the enzyme assay only serve as rough indication since the lipase activity may vary when TAG is the substrate instead of 4-nitrophenol-palmitate [36,43]. Factors that potentially limit the lipase activity include transport limitations due to formation of an emulsion during the feast phase [44].

In this study, we show the use of a feast-famine based enrichment strategy for the enrichment of a mixed microbial culture with enhanced lipid accumulation capacity. Remarkably, the maximum accumulation of intracellular lipids of 54% on VSS basis was only achieved when excess nutrients were supplied. This is in contrast with PHA-producing enrichment cultures, which were reported to achieve higher storage contents during nitrogen-deficient fed-batch experiments [45]. Possibly, the major lipid-storing species could not function under nitrogen-limited conditions. Further research may include investigation of the behavior of pure cultures of *T. gracile* to evaluate this hypothesis.

The lipid content of the microbial enrichment culture obtained in this study was around 50% on VSS mass basis and was mainly related to the presence of the oligeanous yeast *T. gracile*. The *Trichosporon* genus is known to be able to accumulate high amount of lipids, for example, TAG contents higher than 60% were reported in pure culture experiments [27]. The high TAG content of these cultures suggest that further improvement may be achieved by development of strategies that provide a competitive advantage for *Trichosporon* over the bacterial side population.

Nevertheless, the achieved lipid content of around 50% is already significantly more than the 10% lipid content in sludge that was used as reference content to evaluate the feasibility biodiesel production [20] or and even more than the oil content of soybeans [46] or olives, implicating the potential feasibility of lipid recovery for the production of biodiesel. Herewith, we show that ecological principles can effectively be employed for the enrichment of lipid-accumulating species, enabling potential application on a broad variety of waste streams. The use of enrichment cultures decreases capital and operating costs because (semi-continuous) open cultivation techniques can be used without the need for sterilization or re-inoculation.

Conclusion

A microbial enrichment culture of lipid-storing organisms in co-existence with microorganisms that used TAG for direct formation of biomass was observed in a soybean-oil-fed reactor system. The former was represented by a lipolytic dimorphic fungus, while the latter consisted of a consortium of proteobacteria capable of utilizing either the lipid itself or its hydrolysis products, mainly as a substrate for direct growth. The maximum lipid accumulation capacity of the culture was 54% on VSS mass basis. Results were obtained from an enrichment reactor that was not optimized for maximum storage capacity, suggesting that future optimization may lead to microbial enrichment cultures with even higher lipid storage capacity.

Methods

Enrichment reactor setup and operational parameters

A SBR with a liquid volume of 2 l was operated aerobically, similar to Johnson *et al.* An operational cycle comprised of a feed phase of 30 min, during which TAG and nutrients were dosed, followed by a reaction phase of 11 h. At the end of the cycle, half of the volume of the reactor was discharged, to be replaced with new feed in the next cycle, resulting in a total cycle length of 12 h, a hydraulic retention time (HRT) and solid retention time (SRT) of 24 h. The reactor was completely mixed (1,000 rpm) throughout the cycle, and biofilms were structurally removed (several times per week) to ensure that the SRT equaled the hydraulic retention time. The temperature was controlled at 30°C ± 1°C using a thermostat, and the pH was maintained at 7.0 ± 0.1 by pH control using 1-M NaOH and 1-M HCl. The gas flow into the reactor was 1.5 l/min consisting of 0.3 l/min air and 1.2 l/min recycled off-gas. The off-gas was cooled to 5°C by a condenser to minimize water evaporation.

Medium composition

Soybean oil (Markant merk, Beesd, Gelderland, The Netherlands) was chosen as representative model substrate for vegetable oil waste. The substrate soybean oil was analyzed by GC as described in the 'Sampling and measurement' section, and the results can be found in the 'Results' section (Table 1); the composition was similar to that of soybean oil found in literature [47]. In each cycle, 1.3 g (equivalent to 83 Cmmol) soybean oil was dosed into the reactor using a precision syringe pump. Additionally, 200 ml/cycle nutrient solution containing 3.6 g/l NH_4Cl, 3.4 g/l KH_2PO_4, 1.4 g/l $MgSO_4 \cdot 7H_2O$, 0.54 g/l KCl, 1.0 g/l EDTA Titriplex III (Sigma-Aldrich, St. Louis, MO, USA), 0.033 g/l $ZnSO_4.7H_2O$, 0.024 g/l $CoCl_2.6H_2O$, 0.024 g/l $CuSO_4.5H_2O$, 0.075 g/l $FeSO_4.7H_2O$, 0.017 g/l $(NH_4)_6Mo_7O_{24}.4H_2O$, 0.110 g/l $CaCl_2.H_2O$ according to Vishniac and Santer [48], and 0.050 g/l allylthiourea (to prevent nitrification) was added together with 800 ml/cycle of water.

Sampling and measurement

The temperature, pH, and off-gas composition in terms of CO_2 and O_2 concentrations were monitored online using a standard sensor (Mettler Toledo, Columbus, OH, USA) and an off-gas analyzer (NGA 2000, Rosemount Inc., Shakopee, MN, USA). In addition, samples were taken regularly at the end of the cycle to measure the TSS and the VSS concentrations according to standard methods [49] and the ammonium concentration using spectrophotometric method (LCK-348, Hach-Lange, Düsseldorf, Germany). For evaluation of the carbon balance, the organic carbon content of the organic solids was measured using a colorimetric method (LCK 381, Hach-Lange, Düsseldorf, Germany).

In order to characterize the dynamics during a cycle, samples were taken to measure the intracellular lipid concentrations, TSS and VSS production, ammonium uptake, CO_2 production, and O_2 consumption. Intracellular storage compounds were analyzed as described by Johnson *et al.* [6] except that a modified gas chromatography method was used: analysis of extracted PHA and lipids derivatives was performed using direct injection into a polyethylene glycol column (HP-INNOWax, 60 m × 250 × 0.15 μm) installed on a gas chromatograph (model 6890 N, Agilent Technologies, Santa Clara, CA, USA) equipped with a flame ionization detector. The column temperature program comprised an initial oven temperate of 100°C, operated for 2 min, followed by an increase to 170°C at a rate of 20°C/min, and a further increase to 240°C at a rate of 5°C/min resulting in a total runtime of 30 min per sample. The temperatures of the injector and detector were 230°C and 250°C, respectively. Calibration standards included PHB, PHV, hydroxyhexanoate, and hydroxypalmitate for identification of PHA, and oleic-, linoleic-, a-linolenic-, palmitic-, and stearic-acid for identification of LCFAs. For each standard, three different concentrations were used to make a calibration curve. For each GC spectrum, it was checked whether there were unknown peaks indicating missing compounds.

The lipase activity in the culture was evaluated using a colorimetric method based on the hydrolysis of an artificial lipid, 4-nitrophenyl-palmitate [36] both in untreated samples and in samples in which biomass had been removed by centrifugation for 10 min at 2,500 rpm. The hydrolysis product, 4-nitrophenyl was measured spectrometrically at 410 nm.

Microbial community structure

The general microbial community composition was investigated using FISH analysis as described by Johnson *et al.* [6], using the EUB338 and EUK516 probe mixtures for identification of bacterial and eukaryote species, respectively [50]. Intracellular storage compounds were visualized using Nile blue A staining. The microbial diversity was further analyzed by DGGE using eukaryotic and prokaryotic primers [51]. Additionally, traditional cultivation techniques were used to separate the functional groups of microorganisms from the reactor and to isolate pure cultures. For this, serial dilutions in liquid medium containing three different substrates were used: soybean oil, the free LCFA fraction of soybean oil, and glycerol. To separate the eukaryotic community from bacteria, a mixture of kanamycine and streptomycine (100 mg/l each) was applied in a dilution series. The final positive dilutions on each substrate were plated on solid media with corresponding substrates, except that LCFA and soybean oil were first emulsified by sonication with gum arabic to allow detection of the LCFA-utilizing and lipolytic colonies [52].

Fed-batch accumulation experiments

The storage capacity of the microbial enrichment culture was assessed using a fed-batch accumulation experiment. Apart from the feeding regime, all other operational parameters were kept identical to the enrichment reactor: pH was maintained at 7.0 ± 0.1, the temperature was maintained at 30°C ± 1°C, and the reactor liquid was stirred with 1,000 rpm. The biomass from the enrichment reactor was used as inoculum, and the oxygen profile was used to monitor the activity in the reactor; when the oxygen consumption decreased, a new pulse of soybean oil substrate was dosed. The biomass was characterized in terms of lipid content over time.

Data analysis and interpretation

It was not possible to take a representative sample for measurement of the substrate concentration from the oil-water emulsion that formed quickly after substrate dosage. Instead the soybean oil uptake rate in the medium was estimated by assuming that all the oil had been taken up after the feast phase. This assumption was verified using microscopy to make sure that all oil had been taken up and was not absorbed to the biomass. The length of the feast phase was estimated using the off-gas data, with the feast phase characterized by relatively high oxygen consumption and CO_2 production rates and the onset of the famine phase by a relatively sharp decline in both these rates (Additional file 1: Appendix A). The active biomass concentration was calculated by subtracting the amount of storage compounds from the amount of VSS in the system. Active biomass concentrations were verified by the evaluation of the ammonium uptake and nitrogen content of the biomass. Conversion yields and rates were obtained using a modified version of a model used for modeling PHA-producing feast-famine cultures [53]. The only modifications in the model were replacement of PHA storage by storage of TAG and inclusion of a fraction of the substrate that was used for direct growth. A description of the model is provided in the (Additional file 1: Appendix C).

Additional file

Additional file 1: Appendix A: Off-gas profile. Appendix B: Carbon flow sheet. Appendix C: Model.

Abbreviations

GC: gas chromatography; HRT: hydraulic retention time; LCFA: long chain fatty acid; PHA: polyhydroxyalkanoate; SBR: sequenced batch reactor; SRT: solid retention time; TAG: triacylglyceride; TSS: total suspended solids; VFA: volatile fatty acid; VSS: volatile suspended solids.

Competing interests

The authors declare that they have no competing interests.

Authors' contributions

JT operated the reactors and performed most sampling and measurement and data analysis, and wrote the lion share of the manuscript. DS performed the main part of the characterization of the microbial composition. YJ performed the FISH analysis. RK and ML conceived of the study, participated in its design and coordination, and drafted the manuscript. All authors read and approved the final manuscript.

Acknowledgements

This research was funded by the Dutch Foundation for Technical Sciences (STW) as part of the Waste to Resource program (W2R project nr. 11605).

Author details

[1]Department of Biotechnology, Delft University of Technology, Julianalaan 67, 2628 BC Delft, The Netherlands. [2]Winogradsky Institute of Microbiology, RAS, Leninskii avenue, 14, Leninskii avenue, 32a, Moscow 119991, Russia.

References

1. Kleerebezem R, Van Loosdrecht MCM. Mixed culture biotechnology for bioenergy production. Curr Opin Biotechnol. 2007;18:207–12.
2. Reis MAM, Serafim LS, Lemos PC, Ramos AM, Aguiar FR, Van Loosdrecht MCM. Production of polyhydroxyalkanoates by mixed microbial cultures. Bioprocess Biosyst Eng. 2003;25:377–85.
3. Agler MT, Wrenn BA, Zinder SH, Angenent LT. Waste to bioproduct conversion with undefined mixed cultures: the carboxylate platform. Trends Biotechnol. 2011;29:70–8.
4. Spirito CM, Richter H, Rabaey K, Stams AJM, Angenent LT. Chain elongation in anaerobic reactor microbiomes to recover resources from waste. Curr Opin Biotechnol. 2014;27:115–22.
5. Jiang Y, Marang L, Kleerebezem R, Muyzer G, Van Loosdrecht MCM. Polyhydroxybutyrate production from lactate using a mixed microbial culture. Biotechnol Bioeng. 2011;108:2022–35.
6. Johnson K, Jiang Y, Kleerebezem R, Muyzer G, Van Loosdrecht MCM. Enrichment of a mixed bacterial culture with a high polyhydroxyalkanoate storage capacity. Biomacromolecules. 2009;10:670–6.
7. Anterrieu S, Quadri L, Geurkink B, Dinkla I, Bengtsson S, Arcos-Hernandez M, et al. Integration of biopolymer production with process water treatment at a sugar factory. N Biotechnol. 2014;31:308–23.
8. Jiang Y, Marang L, Tamis J, Van Loosdrecht M, Dijkman H, Kleerebezem R. Waste to resource: converting paper mill wastewater to bioplastic. Water Res. 2012;46:5517–30.
9. Liu H-Y, Hall PV, Darby JL, Coats ER, Green PG, Thompson DE, et al. Production of polyhydroxyalkanoate during treatment of tomato cannery wastewater. Water Environ Res. 2008;80:367–72.
10. Carta F, Beun JJ, Van Loosdrecht MCM, Heijnen JJ. Simultaneous storage and degradation of PHB and glycogen in activated sludge cultures. Water Res. 2001;35:2693–701.
11. Karahan Ö, Van Loosdrecht MCM, Orhon D. Modeling the utilization of starch by activated sludge for simultaneous substrate storage and microbial growth. Biotechnol Bioeng. 2006;94:43–53.
12. Moralejo-Gárate H, Mar'atusalihat E, Kleerebezem R, Loosdrecht MM. Microbial community engineering for biopolymer production from glycerol. Appl Microbiol Biotechnol. 2011;92:631–9.
13. Albuquerque MGE, Martino V, Pollet E, Avérous L, Reis MAM. Mixed culture polyhydroxyalkanoate (PHA) production from volatile fatty acid (VFA)-rich streams: effect of substrate composition and feeding regime on PHA productivity, composition and properties. J Biotechnol. 2011;151:66–76.
14. Jiang Y, Hebly M, Kleerebezem R, Muyzer G, Van Loosdrecht M. Metabolic modeling of mixed substrate uptake for polyhydroxyalkanoate (PHA) production. Water Res. 2011;45:1309–21.
15. Marang L, Jiang Y, van Loosdrecht M, Kleerebezem R. Impact of non-storing biomass on PHA production: an enrichment culture on acetate and methanol. International journal of biological macromolecules. 2014;71:74–80.
16. Igwe J, Onyegbado C. A review of palm oil mill effluent (POME) water treatment. Global J Environ Res. 2007;1:54–62.
17. FAO: FAOSTAT. FAOSTAT. Food & Agriculture Organization of the United Nations; 2014.
18. Borja R, Banks C. Anaerobic digestion of palm oil mill effluent using an up-flow anaerobic sludge blanket reactor. Biomass Bioenergy. 1994;6:381–9.
19. Muller EE, Sheik AR, Wilmes P. Lipid-based biofuel production from wastewater. Curr Opin Biotechnol. 2014;30:9–16.
20. Mondala A, Liang K, Toghiani H, Hernandez R, French T. Biodiesel production by *in situ* transesterification of municipal primary and secondary sludges. Bioresour Technol. 2009;100:1203–10.
21. Wakelin NG, Forster CF. An investigation into microbial removal of fats, oils and greases. Bioresour Technol. 1997;59:37–43.
22. Eggink G, van der Wal H, Huijberts GN, De Waard P. Oleic acid as a substrate for poly-3-hydroxyalkanoate formation in *Alcaligenes eutrophus* and *Pseudomonas putida*. Ind Crop Prod. 1992;1:157–63.
23. Lageveen RG, Huisman GW, Preusting H, Ketelaar P, Eggink G, Witholt B. Formation of polyesters by Pseudomonas oleovorans: effect of substrates on formation and composition of poly-(R)-3-hydroxyalkanoates and poly-(R)-3-hydroxyalkenoates. Appl Environ Microbiol. 1988;54:2924–32.
24. Lee SH, Oh DH, Ahn WS, Lee Y, Choi J, Lee SY. Production of poly (3-hydroxybutyrate-co-3-hydroxyhexanoate) by high-cell-density cultivation of Aeromonas hydrophila. Biotechnol Bioeng. 2000;67:240–4.
25. Ng K-S, Ooi W-Y, Goh L-K, Shenbagarathai R, Sudesh K. Evaluation of jatropha oil to produce poly (3-hydroxybutyrate) by *Cupriavidus necator* H16. Polym Degrad Stab. 2010;95:1365–9.
26. Liang M-H, Jiang J-G. Advancing oleaginous microorganisms to produce lipid via metabolic engineering technology. Prog Lipid Res. 2013;52:395–408.
27. Zhu L, Zong M, Wu H. Efficient lipid production with *Trichosporon fermentans* and its use for biodiesel preparation. Bioresour Technol. 2008;99:7881–5.
28. Alvarez HA, Steinbüchel AS. Triacylglycerols in prokaryotic microorganisms. Appl Microbiol Biotechnol. 2002;60:367–76.
29. Slijkhuis HVG, Johan W, Kylstra, Dirk J. Microthrix parvicella, a filamentous bacterium from activated sludge: growth on Tween 80 as carbon and energy source. J Gen Microbiol. 1984;130:2035–42.
30. Ling J, Nip S, Shim H. Enhancement of lipid productivity of Rhodosporidium toruloides in distillery wastewater by increasing cell density. Bioresource technology. 2013;146:301–9.
31. W-f P, Huang C, Chen X-f, Xiong L, Chen Y, Ma L-l. Microbial conversion of wastewater from butanol fermentation to microbial oil by oleaginous yeast Trichosporon dermatis. Renewable Energy. 2013;55:31–4.
32. Santamauro F, Whiffin FM, Scott RJ, Chuck CJ. Low-cost lipid production by an oleaginous yeast cultured in non-sterile conditions using model waste resources. Biotechnology for biofuels. 2014;7:34.
33. Magan N. Fungi in extreme environments. In: Kubicek CP, Esser K, Druzhinina IS, editors. Environmental and microbial relationships. New York: Springer; 2007. p. 85.
34. Mooij PR, Stouten GR, Van Loosdrecht MCM, Kleerebezem R. Ecology-based selective environments as solution to contamination in microalgal cultivation. Curr Opin Biotechnol. 2015;33:46–51.
35. Andreasen K, Nielsen PH. Growth of *Microthrix parvicella* in nutrient removal activated sludge plants: studies of *in situ* physiology. Water Res. 2000;34:1559–69.
36. Vorderwulbecke T, Kieslich K, Erdmann H. Comparison of lipases by different assays. Enzyme Microb Technol. 1992;14:631–9.
37. Holdsworth JE, Ratledge C. Lipid turnover in oleaginous yeasts. J Gen Microbiol. 1988;134:339–46.
38. Arenskötter M, Bröker D, Steinbüchel A. Biology of the metabolically diverse genus Gordonia. Appl Environ Microbiol. 2004;70:3195–204.
39. Juni E. Genetics and physiology of Acinetobacter. Annu Rev Microbiol. 1978;32:349–71.

40. Chipasa K, Mędrzycka K. Behavior of lipids in biological wastewater treatment processes. J Ind Microbiol Biotechnol. 2006;33:635–45.
41. Gouda MK, Omar SH, Aouad LM. Single cell oil production by Gordonia sp. DG using agro-industrial wastes. World J Microbiol Biotechnol. 2008;24:1703–11.
42. Jiang Y, Marang L, Kleerebezem R, Muyzer G, Van Loosdrecht MCM. Effect of temperature and cycle length on microbial competition in PHB-producing sequencing batch reactor. lInternational Society for Microbial Ecology. 2011;5:896–907.
43. Sigurgísladóttir S, Konráðsdóttir M, Jónsson Á, Kristjánsson J, Matthiasson E. Lipase activity of thermophilic bacteria from icelandic hot springs. Biotechnol Lett. 1993;15:361–6.
44. Cirne DG, Paloumet X, Björnsson L, Alves MM, Mattiasson B. Anaerobic digestion of lipid-rich waste - effects of lipid concentration. Renew Energy. 2007;32:965–75.
45. Johnson K, Kleerebezem R, Van Loosdrecht M. Influence of ammonium on the accumulation of polyhydroxybutyrate (PHB) in aerobic open mixed cultures. J Biotechnol. 2010;147:73–9.
46. FAO. Technology of production of edible flours and protein products from soybeans. FAO agricultural services bulletin no 97. 1992.
47. Babayan VK. Medium chain triglycerides and structured lipids. Lipids. 1987;22:417–20.
48. Vishniac W, Santer M. Thiobacilli. Bacteriol Rev. 1957;21:195–213.
49. Clesceri LS, Greenberg AE, Eaton AD. Standard methods for the examination of water and wastewater. 20th ed. USA: American Public Health Association; 1999.
50. Alm EW, Oerther DB, Larsen N, Stahl DA, Raskin L. The oligonucleotide probe database. Appl Environ Microbiol. 1996;62:3557.
51. Muyzer G. DGGE/TGGE a method for identifying genes from natural ecosystems. Curr Opin Microbiol. 1999;2:317–22.
52. Sorokin DY, Jones BE. Improved method for direct screening of true lipase-producing microorganisms with particular emphasis on alkaline conditions. Microbiology. 2009;78:125–30.
53. Tamis J, Marang L, Jiang Y, Van Loosdrecht MC, Kleerebezem R. Modeling PHA-producing microbial enrichment cultures - towards a generalized model with predictive power. N Biotechnol. 2014;31:324–34.

A novel arabinose-inducible genetic operation system developed for *Clostridium cellulolyticum*

Jie Zhang[1,3,4], Ya-Jun Liu[1,3*], Gu-Zhen Cui[1,3] and Qiu Cui[1,2,3*]

Abstract

Background: *Clostridium cellulolyticum* and other cellulolytic *Clostridium* strains are natural producers of lignocellulosic biofuels and chemicals via the consolidated bioprocessing (CBP) route, and systems metabolic engineering is indispensable to meet the cost-efficient demands of industry. Several genetic tools have been developed for *Clostridium* strains, and an efficient and stringent inducible genetic operation system is still required for the precise regulation of the target gene function.

Results: Here, we provide a stringent arabinose-inducible genetic operation (ARAi) system for *C. cellulolyticum*, including an effective gene expression platform with an oxygen-independent fluorescent reporter, a sensitive MazF-based counterselection genetic marker, and a precise gene knock-out method based on an inducible ClosTron system. A novel arabinose-inducible promoter derived from *Clostridium acetobutylicum* is employed in the ARAi system to control the expression of the target gene, and the gene expression can be up-regulated over 800-fold with highly induced stringency. The inducible ClosTron method of the ARAi system decreases the off-target frequency from 100% to 0, which shows the precise gene targeting in *C. cellulolyticum*. The inducible effect of the ARAi system is specific to a universal carbon source L-arabinose, implying that the system could be used widely for clostridial strains with various natural substrates.

Conclusions: The inducible genetic operation system ARAi developed in this study, containing both controllable gene expression and disruption tools, has the highest inducing activity and stringency in *Clostridium* by far. Thus, the ARAi system will greatly support the efficient metabolic engineering of *C. cellulolyticum* and other mesophilic *Clostridium* strains for lignocellulose bioconversion.

Keywords: Consolidated bioprocessing, Counterselection marker, Inducible gene expression, Metabolic engineering, Targetron

Background

Lignocellulosic biomass is the most abundant and renewable raw material on Earth, and its sustainability and effective cost make it an attractive feedstock of carbon source [1-4]. Because of the complex composition and recalcitrant structure, the application of lignocellulose is difficult [5,6]. Consolidated bioprocessing (CBP) is considered an optimal strategy for lignocellulose conversion because it integrates enzyme production, cellulose hydrolysis, and fermentation in one step to reduce cost and simplify processing [6-8]. *Clostridium cellulolyticum* and many other cellulolytic and solventogenic *Clostridium* species are promising CBP candidates because of their capability to degrade lignocellulose, but systems metabolic engineering of clostridial strains is still necessary to satisfy industrialization [9,10].

Although complex cell wall and anaerobic growth conditions of *Clostridium* make genetic manipulation difficult, a few genetic manipulation methods have been developed recently, including gene disruption methods via either homologous recombination or intron retrohoming mechanism [11-14] and heterologous gene expression methods using replicative or integrative plasmids [15,16]. Inducible gene expression tools are required in the metabolic engineering for *Clostridium* strains because the precise regulation of the target gene function is crucial for either the native pathway engineering or heterologous gene

* Correspondence: liuyj@qibebt.ac.cn; cuiqiu@qibebt.ac.cn
[1]Shandong Provincial Key Laboratory of Energy Genetics, Qingdao Institute of Bioenergy and Bioprocess Technology, Chinese Academy of Sciences, 189, Songling Road, Qingdao 266101, People's Republic of China
Full list of author information is available at the end of the article

introduction in chassis strains [17,18]. Moreover, several clostridial genetic tools in hand can be improved with an effective inducible gene expression system. For example, toxic gene-derived counterselection markers can be developed in *Clostridium* strains by using an inducible promoter for seamless and successive genome editing [19,20]. ClosTron is a gene targeting method derived from a mesophilic mobile group II intron *Ll.ltrB* [21-23]. It has been extensively used in the gene disruption of *Clostridium* strains [24-28], but its high off-targeting activity affects precise genetic engineering [29]. According to the ribozyme-based DNA integration mechanism of targetron [30], the DNA targeting specificity can be improved by precise management of the expression of intron RNA and intron-encoded protein (IEP) [22,31] using a proper inducible gene expression system.

An ideal inducible gene expression system should have high inducing efficiency, stringency, and specificity to an exogenous inducer innocuous to the host cell, with which the expression level of the target gene as well as the function of the related pathway can be easily regulated theoretically [32]. Various inducible gene expression systems have been constructed, such as the well-known isopropyl-β-d-thiogalactoside (IPTG)/lactose-inducible *lac* system, tetracycline-inducible *tet* system, and arabinose-inducible *ara* system [33-35]. Both *lac* system and *tet* system use a repressor binding to the operator *lacO*/*tetO* in the absence of specific inducers, and the addition of the inducer can activate downstream gene expression [34]. Although relatively high background expression has been observed [36], a typical *lac* system LacI-*lacO* is used to develop the inducible T7 gene expression system, which is widely used in *Escherichia coli* strains for heterologous protein expression [37]. When applied in *Clostridium* strains, the inducing efficiencies of lactose-inducible promoters are relatively low [18,38]. In addition to the lactose-inducible promoter, *tet* system has also been used in the controlled gene expression of non-cellulolytic *Clostridium acetobutylicum*, termed Pcm-2tetO1 [20,39]. Pcm-2tetO1 can up-regulate gene expression by 313-fold with anhydrotetracycline as an inducer, which is the highest inducing efficiency among the reported inducible promoters applied in the genus *Clostridium* [18,20,38-42]. The optimal working condition of Pcm-2tetO1 requires a high inducer dosage, but elevated concentration of anhydrotetracycline shows significant inhibitive effects on cell growth [20]. AraC-P_{BAD} derived from *E. coli* is the most well-known *ara* inducible system, which uses AraC as either an activator or a repressor to control the expression of downstream genes [33], and has been used in several Gram-negative and Gram-positive bacteria [43-46], but not in *Clostridium* strains so far. Thus, novel inducible gene expression systems with high inducing capability as well as proper inducers are required for the metabolic engineering of *Clostridium* strains.

No inducible promoter has been developed for the cellulolytic species *C. cellulolyticum* by far [39]. It is reported that the metabolism of L-arabinose is regulated by *C. acetobutylicum* at the transcriptional level using an AraR-mediated regulation system [47,48]. In the present study, we develop an arabinose-inducible promoter based on the AraR-regulon of *C. acetobutylicum*. Using this inducible promoter, a novel arabinose-inducible genetic operation system, termed ARAi, is constructed for *C. cellulolyticum*. Our results confirm that the ARAi system can be used to regulate gene expression in *C. cellulolyticum* with high stringency and activity. Furthermore, the ARAi system provides an inducible ClosTron method for precise gene knock-out in *C. cellulolyticum*, which can facilitate the systems metabolic engineering of *C. cellulolyticum* and other *Clostridium* strains for commodity and industry.

Results

Construction of the arabinose-inducible expression system

Sensitive regulation of L-arabinose metabolism with AraR as the key repressor has been observed and characterized in *Bacillus subtilis* [49] as well as *C. acetobutylicum* [47,48]. The AraR-regulon members of *C. acetobutylicum* include 11 genes involved in arabinose utilization, transport, and conversion via the phosphate pentose pathway [47], in which the expression of gene *ptk*, encoding a phosphoketolase, showed the most sensitivity to the inducer L-arabinose [47]. Hence, to obtain enhanced induction activity, we selected the promoter of *ptk* and the *araR* regulator expression cassette of *C. acetobutylicum* to construct the ARAi system (Additional file 1). Then, 370-bp from the upstream region of *ptk* was used as the promoter P_{ptk} to make sure that the binding site of AraR was covered [47]. The self-regulation of AraR was observed [47], so the predicted native promoter and terminator regions of *araR* were employed in the *araR* expression cassette (Additional file 1).

Controlled expression of an oxygen-independent green fluorescent protein PpFbFPm in *C. cellulolyticum* using the ARAi system

Although the *araR*-binding sequences in *C. acetobutylicum* and *C. cellulolyticum* were distinct [47], whether the ARAi system could be interfered by the indigenous arabinose-inducible system of *C. cellulolyticum* was not certain. Thus, the availability of ARAi system in *C. cellulolyticum* was tested using an oxygen-independent green fluorescent protein PpFbFPm as the reporter [14]. The plasmid pARA-PpFbFPm including both P_{ptk}-PpFbFPm and AraR expression cassettes was constructed for ARAi-regulated expression of PpFbFPm. The control plasmid pPTK-PpFbFPm was also constructed for constitutive expression of PpFbFPm without the expression of AraR repressor (Table 1, Additional file 1).

C. cellulolyticum strains H10::pARA-PpFbFPm and H10::pPTK-PpFbFPm, carrying pPTK-PpFbFPm or pARA-PpFbFPm, respectively, were cultivated with or without addition of L-arabinose, and the intracellular expression of PpFbFPm was investigated by green fluorescence imaging. As shown in Figure 1A, intense green fluorescence was observed in cells of H10::pPTK-PpFbFPm with or without L-arabinose, which confirmed that the *ptk* promoter originating from *C. acetobutylicum* was not recognized and regulated by native AraR in *C. cellulolyticum*. In contrast, H10::pARA-PpFbFPm showed no green fluorescence emission without L-arabinose (Figure 1B), providing a consistent result with previous reports that AraR functioned as a repressor rather than an activator [47]. With addition of L-arabinose, cells of H10::pARA-PpFbFPm showed bright green fluorescence similar to that of H10::pPTK-PpFbFPm, which was indicative of strongly induced expression of PpFbFPm (Figure 1B). These results suggested that the ARAi system was available for controlled expression of target protein in *C. cellulolyticum*. The expression of green fluorescence protein was observed in *E. coli* host cells containing pARA-PpFbFPm whether the inducer was added or not. This indicated the expression leakage in *E. coli*, and suggested that the ARAi system should not be used in *E. coli* strains.

The efficiency and stringency of ARAi system in *C. cellulolyticum*

The crude enzyme of H10::pARA-GusA was used to determine the efficiency of the ARAi system in *C. cellulolyticum* by β-glucuronidase (GusA) assay using both chromogenic substrate 5-bromo-4-chloro-3-indolyl-β-D-glucuronic acid (X-gluc) and fluorogenic substrate 4-methylumbelliferyl-β-D-glucuronic acid (MUG) [20]. H10::pARA-PpFbFPm was used as a negative control (Table 1).

With X-gluc as a substrate, the enzymatic reaction mixture of H10::pARA-PpFbFPm showed no change, as well as that of H10::pARA-GusA without induction of L-arabinose, indicating no expression of GusA. However, the crude enzyme of H10::pARA-GusA turned the

Table 1 Bacterial strains and plasmids used in this study

Bacterial strains and plasmids	Relevant characteristic	Sources
Strains		
E. coli		
DH5α	*f80dlacZΔM15, Δ(lacZYA-argF)U169, deoR, recA1, endA1, hsdR17(r_k–, m_k+), phoA, supE44, l–, thi-1, gyrA96, relA1*	Transgen
BL21(DE3)	*ompT gal dcm lon hsdS$_B$(r_B– m_B–) l (DE3 [lacl lacUV5-T7 gene 1 ind1 sam7 nin5])*	Transgen
C. cellulolyticum		
H10	ATCC35319, wild type stain	ATCC
H10Δ*pyrF*	Derived from H10, *Ccel_0614* disrupted via homologous recombination	[11]
H10::MspI297s	Derived from H10, *mspI*$^-$, *Ccel_2866*:: MspI297s	[14]
H10Δ*pyrF*::CipC117a	Derived from H10Δ*pyrF*, *Ccel_0728*::CipC117a	[11]
H10Δ*pyrF*Δ*mspI*	Derived from H10Δ*pyrF*, *Ccel_2866*::MspI297s	This work
H10Δ*pyrF*Δ*cipC*	Derived from H10Δ*pyrF*, *Ccel_0728* ::CipC117a	This work
Plasmids		
pMTC6	Mls^R, Amp^R, *E. coli-C. cellulolyticum* shuttle vector, containing *PpFbFPm*, *thl* promoter, *thl* terminator	[14]
pPTK-PpFbFPm	Derived from pMTC6, containing *ptk* promoter instead of *thl* promoter	
pARA-PpFbFPm	Derived from pMTC6, containing ARAi-*PpFbFPm* cassette	This work
pGusA2-2tetO1	pIMP1 containing Pcm-2tetO1 promoter and *gusA* gene	[20]
pARA-GusA	Derived from pARA-PpFbFPm, containing ARAi-*gusA* cassette	This work
pARA-MazE	Derived from pARA-PpFbFPm, containing T7-*mazE* cassette	This work
pARA-MazE/F	Derived from pARA-MazE, containing ARAi-*mazF* cassette	This work
pSY6-*mspI*	Mls^R, Amp^R, *E. coli-C. cellulolyticum* shuttle vector, *ptb* promoter, containing *mspI* targetron	[14]
pARA-*mspI*	Derived from pSY6-*mspI*, containing ARAi system instead of *ptb* promoter	This work
pARA-PyrF-*mspI*	Derived from pARA-*mspI*, containing P_{fd}-PyrF expression cassette	This work
pGZ-*cipC*	Mls^R, Amp^R, *E. coli-C. cellulolyticum* shuttle vector, *ptb* promoter, containing *cipC* targetron	[11]
pARA-*cipC*	Derived from pGZ-*cipC*, containing ARAi system instead of *ptb* promoter	This work
pARA-PyrF-*cipC*	Derived from pARA-*cipC*, containing P_{fd}-PyrF expression cassette	This work

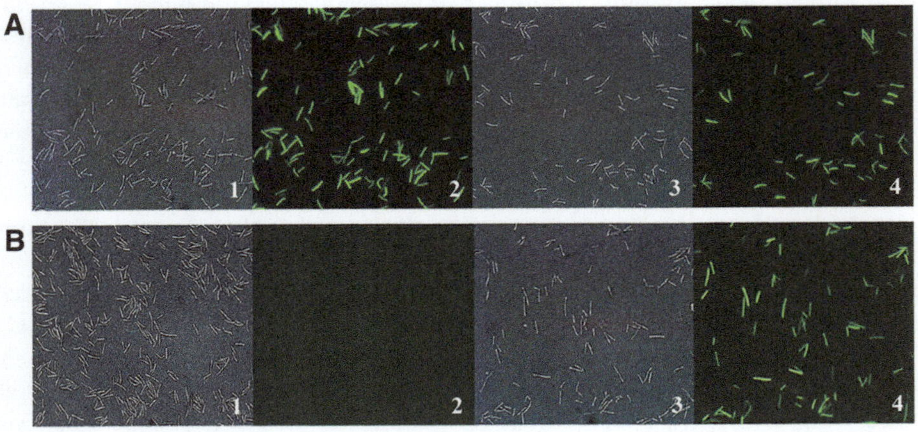

Figure 1 Confirmation of PpFbFPm expression in *C. cellulolyticum* H10 by *in vivo* fluorescent imaging. *C. cellulolyticum* cells containing pPTK-PpFbFPm **(A)** or pARA-PpFbFPm **(B)** grown at mid-log phase were induced with 10 g/L L-arabinose at 34°C for 2 h or without induction and were used to detect the intracellular fluorescence. 1, 3 corresponds to cell morphology; 2, 4 corresponds to fluorescent imaging; 1, 2 shows cells without induction; 3, 4 shows cells with induction.

colorless solution into dark blue when L-arabinose was added (Figure 2A). These results suggested strict and efficient induction activity of ARAi system in *C. cellulolyticum*. When tested with MUG as the substrate, the GusA activity of H10::pARA-GusA was 106.1 ± 8.7 U/mg and was stimulated 100-fold after induction for 0.5 h, indicating that the ARAi system was efficient in *C. cellulolyticum*. H10::pARA-PpFbFPm without L-arabinose induction also showed 26.4 ± 2.8 U/mg activity, which might be a slight expression leakage or the fluorescent background of the reaction system.

To further test the activity of the ARAi system response to induction duration, GusA activity in H10::pARA-GusA was examined during a time course (0 to 6 h) with

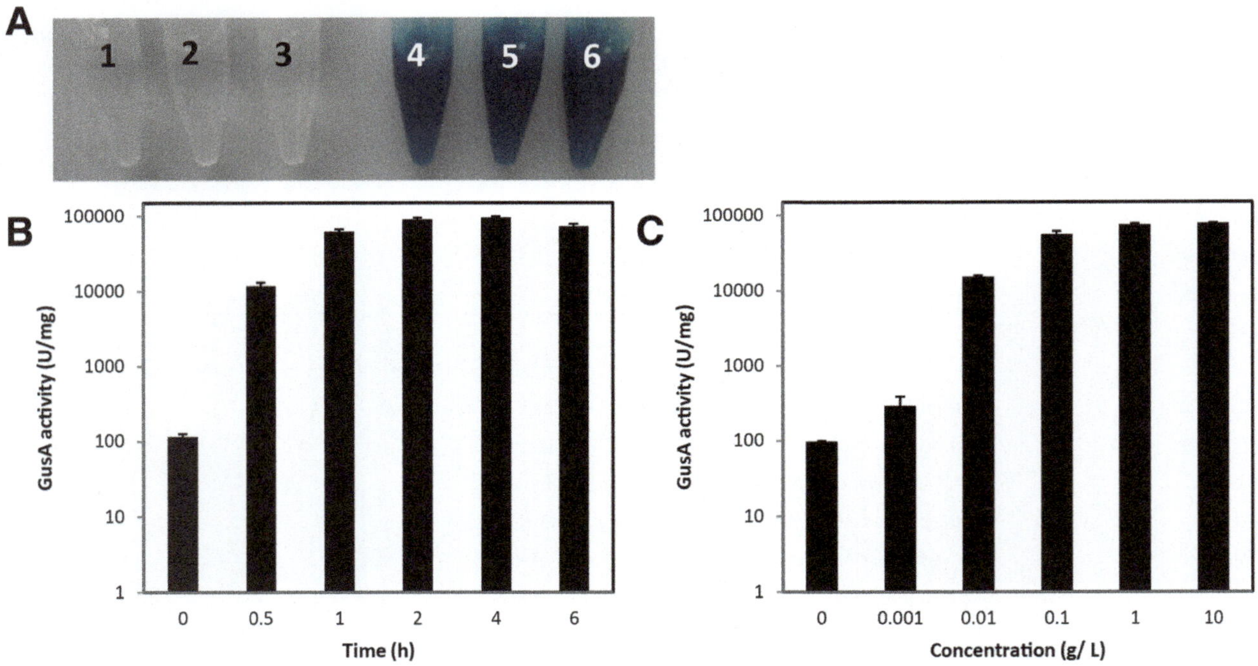

Figure 2 Investigation of the efficiency and stringency of the ARAi system in *C. cellulolyticum* H10. (A) Determination of inducible expression of GusA in H10::pARA-GusA with X-gluc as a substrate. The blue colors of the reaction solutions indicate the induced expression of GusA. The numbered tubes contained: 1, negative control without crude enzyme sample; 2, H10::pARA-PpFbFPm; 3, H10::pARA-GusA without L-arabinose; 4 to 6, H10::pARA-GusA with 0.1, 1 and 10 g/L L-arabinose, respectively. **(B)** Determination of GusA activity during a time course (0 to 6 h) using MUG as a substrate. Then 1 g/L L-arabinose was used as the inducer. **(C)** Determination of the influence of inducer dosage (0 to 10 g/L) on GusA activity using MUG as a substrate. The induction time was 2 h.

addition of 1 g/L L-arabinose. GusA activity increased dramatically and reached a maximum level of approximately 9.0×10^4 U/mg after a 2-h induction (Figure 2B). Then, the influence of the inducer dosage on GusA activity was measured by supplementing 0 to 10 g/L L-arabinose as an inducer. The induction time was 2 h. As shown in Figure 2C, GusA activity was significantly stimulated by over 800-fold with increased dosage of the inducer, whereas further addition of L-arabinose from 0.1 g/L up to 10 g/L barely enhanced GusA activity. Thus, the optimum induction condition of ARAi system was 0.1 g/L L-arabinose with 2-h incubation, indicating high inducibility and efficiency of the ARAi system.

The inducer specificity and inhibition effect on the ARAi system in *C. cellulolyticum*

In addition to L-arabinose, several sugars including D-arabinose, D-glucose, D-xylose, D-fructose, D-galactose, and D-mannose were used to investigate the inducer specificity of the ARAi system, and the induced GusA activity in H10::pARA-GusA cells was monitored to determine the inducing activity. All selected sugars could be accumulated by *C. cellulolyticum* H10 except D-arabinose. L-arabinose showed over 1000-fold higher inducing activity than the other sugars, which indicated that L-arabinose was the specific inducer of the ARAi system (Figure 3). Surprisingly, the addition of D-xylose, D-glucose, D-galactose, or D-mannose significantly inhibited the induction activity of L-arabinose by approximately 40- to 500-fold, whereas D-fructose showed slight influence (Figure 3).

L-arabinose can be utilized by *C. cellulolyticum* as a carbon source, thus the inducing activity of ARAi may be influenced if the induction period is prolonged. To test the stability of the ARAi system in *C. cellulolyticum*, L-arabinose induction of H10::pARA-GusA was performed for 3 days and both GusA activity and residual L-arabinose and cellobiose in broth were monitored. The results showed that 1 g/L L-arabinose could be completely utilized by H10::pARA-GusA in 12 h, at which time point the GusA activity decreased by 36% (Additional file 2). This indicated that the consumption of L-arabinose affected the inducing activity of the ARAi system. GusA activity was also detected and slightly decreased after 12 h of incubation, which might be explained by the stability of the GusA protein in *C. cellulolyticum* cells or the persistent binding of intracellular L-arabinose with AraR.

We further investigated whether the L-arabinose utilization could be affected by the potential carbon catabolize repression (CCR) effect in *C. cellulolyticum* H10 by the addition of cellobiose, D-glucose, or D-xylose with L-arabinose simultaneously. Although no typical CCR diauxie growth was observed, we detected slower utilization of L-arabinose than D-glucose or D-xylose by *C. cellulolyticum* H10 in batch fermentation (Figure 4). L-arabinose (5 g/L) was completely used in 60 h as a single carbon source, whereas the addition of D-glucose or D-xylose decreased the assimilation rate of L-arabinose, and no more utilization was detected after 80 h of cultivation. In contrast, supplementation of cellobiose showed a slight effect on the utilization rate of L-arabinose (Figure 4). The results indicated competitive transport of L-arabinose with other monoses, which might cause the inhibition effect of ARAi system in *C. cellulolyticum*.

Establishment of a counterselection genetic marker MazF in *C. cellulolyticum* using the ARAi system

Counterselection markers are important genome editing tools for systems metabolic engineering, and one of the prerequisites of developing a toxin-based counterselection

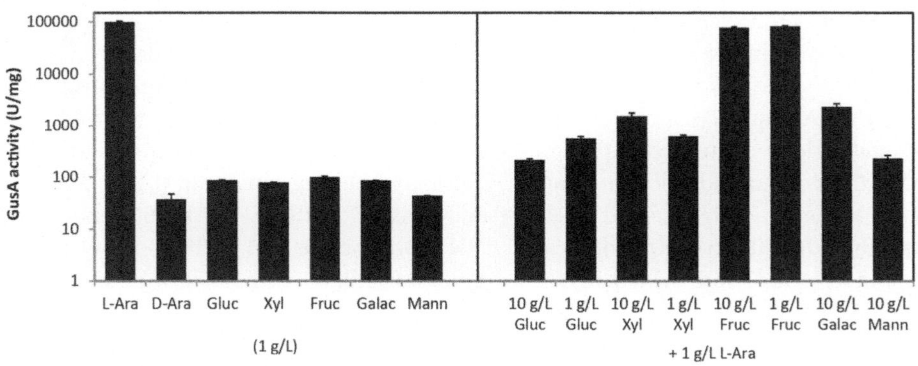

Figure 3 The inducer specificity and inhibition effect of the ARAi system. H10::pARA-GusA was first cultivated with cellobiose to mid-log phase, and 1 g/L of various sugars were used to test the inducer specificity of the ARAi system, including L-arabinose (L-Ara), D-arabinose (D-Ara), D-glucose (Gluc), D-xylose (Xyl), D-fructose (Fruc), D-galactose (Galac), and D-mannose (Mann). The inhibition effect of the ARAi system was detected with a mixture of 1 g/L L-arabinose with 1 or 10 g/L of other sugars as inducers. A blank control without inducer was prepared, and the values of all samples were standardized by subtracting the value of the blank control. The induction time was 2 h. Three independent experiments were performed to calculate the average values and standard errors.

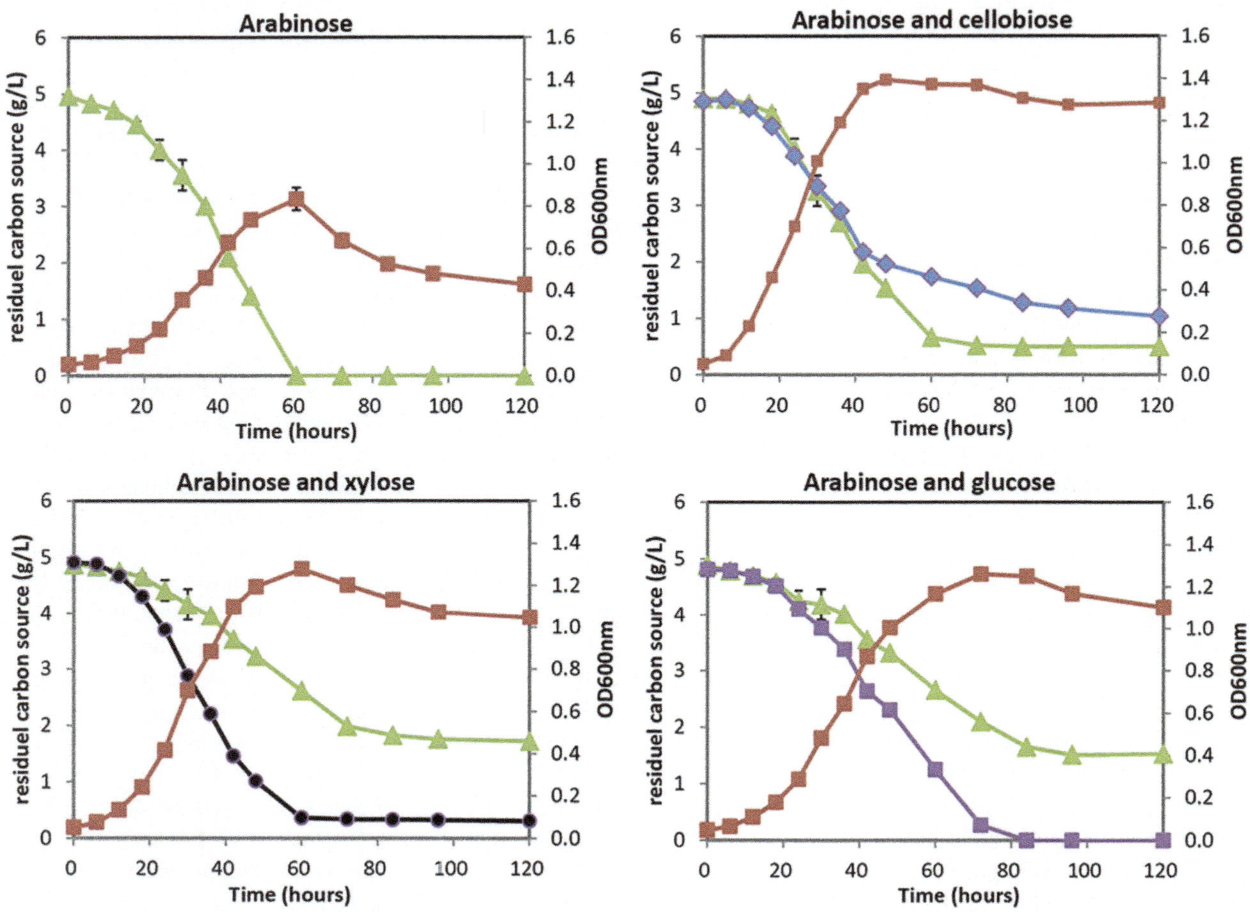

Figure 4 The growth analysis of *C. cellulolyticum* H10 using various sugars as carbon sources. 5 g/L L-arabinose or sugar mixtures (L-arabinose + cellobiose, L-arabinose + D-xylose, or L-arabinose + D-glucose) were used as carbon sources. Cell growth was determined by monitoring the optical density at 600 nm (OD_{600nm}, red square). The residual carbon sources in broth were measured by HPLC. Green triangle, L-arabinose; blue diamond, cellobiose; black cycle, xylose; purple square, glucose. Three independent cultivations were used to calculate average values and standard errors.

marker, such as MazF, is a stringent inducible regulation system to tightly control the toxic gene expression [19]. Using the developed inducible promoter, a counterselection marker MazF was established in the ARAi system for *C. cellulolyticum*.

MazF cleaves mRNA at ACA sequences and is toxic to host cell growth [50], and strictly controlled expression of MazF is essential for the construction of the genetic marker [19,51]. We failed to construct the plasmid pARA-MazF by ligating *mazF* gene directly to the downstream of the ARAi promoter in *E. coli* DH5α, which may be explained by the expression leakage of the ARAi system in *E. coli*. In the MazEF toxin-antitoxin system, MazE acts as the antitoxin and interacts with MazF. Thus, we planned to firstly introduce the expression of *mazE* under the control of a T7 promoter/terminator to protect the host cell from the released MazF [52]. In principle, MazE would only be expressed in strains possessing the T7 RNA polymerase system, such as the *E. coli* BL21(DE3) strain, with the presence of the inducer IPTG, and would not interfere with the function of MazF as a counterselection maker in *C. cellulolyticum*. The plasmid pARA-MazE was successfully constructed by cloning a T7-*mazE* cassette into pARA-PpFbFPm. However, no BL21(DE3) transformants were obtained with or without addition of IPTG when *mazF* was ligated into pARA-MazE. Surprisingly, the target plasmid pARA-MazE/F was successfully created in the *E. coli* DH5α strain containing no T7 RNA polymerase and without IPTG induction (Table 1, Additional file 1), with no clear explanation.

pARA-MazE/F was then transformed into *C. cellulolyticum* H10. The obtained strain H10::pARA-MazE/F was plated onto erythromycin-containing solid GS-2 medium with or without addition of L-arabinose to test the availability of the counterselection marker MazF in *C. cellulolyticum*. H10::pARA-PpFbFPm was also tested as a control. In principle, the induced expression of MazF would be lethal to the host cell unless the expression vector was cured. We observed that the growth of control strain H10::pARA-PpFbFPm showed no difference with the presence of either erythromycin (Ery^+, Ara^-), L-arabinose (Ery^-, Ara^+), or both (Ery^+, Ara^+) (Figure 5B). H10::pARA-MazE/F grew regularly on the erythromycin supplemented agar plate without L-arabinose (Ery^+, Ara^-), but did not grow when both L-arabinose and erythromycin were added (Ery^+, Ara^+) (Figure 5A), suggesting that the expression of toxic MazF was strictly induced by L-arabinose.

In addition, H10::pARA-MazE/F cells grew normally on the agar plate containing only L-arabinose but no erythromycin (Ery^-, Ara^+) (Figure 5A). The colonies were randomly selected and inoculated into fresh liquid GS-2 medium with (Ery^+) or without (Ery^-) erythromycin, and no growth was observed in Ery^+ medium. This indicated that without the selection stress of antibiotics, the plasmid pARA-MazE/F was cured due to the induced expression of MazF. Hence, the establishment of the counterselection system pARA-MazE/F not only verified the stringency of ARAi system in *C. cellulolyticum* but also provided an efficient inducible counterselection marker for further seamless genetic manipulation.

Development of an inducible ClosTron using ARAi system

Because of its high efficiency, ClosTron has been extensively used in the gene disruption of mesophilic *Clostridium* strains [24-26,28], but a deficiency of low target specificity still remains in the targetron methods. We hypothesized that off-target integrations of targetron occurred due to sequence similarity to the desired target site and continuous expression of targetron elements. Therefore, the off-target frequency might decrease if the expression of targetron is controlled using an inducible gene expression system instead of the strong *ptb* (phophotransbutyrylase) promoter [25].

To verify this hypothesis, we modified two reported targetron plasmids pSY6-*mspI* [14] and pGZ-PyrF-*cipC* [11] using the developed arabinose-inducible promoter, and obtained pARA-PyrF-*mspI* and pARA-PyrF-*cipC* targeting *mspI* and *cipC* genes, respectively (Table 1, Additional file 1). The new targetron plasmids were transformed into the chassis strain H10Δ*pyrF*, and the desired mutants H10Δ*pyrF*Δ*mspI* and H10Δ*pyrF*Δ*cipC* were obtained following a two-step procedure as described [11], except that, in the first step, transformants were inoculated and cultivated in antibiotic-free medium and L-arabinose induction was performed for controlled expression of intron RNA and IEP. The effect of L-arabinose induction was investigated by PCR before 5-fluoroorotic acid (FOA) screening. The PCR results of transformants carrying pARA-PyrF-*mspI* or pARA-PyrF-*cipC* showed both bands with or without a full-length (0.9 kb) intron sequence (Figure 6), indicating that the colonies were a mixture of host strain H10Δ*pyrF* and mutant strain H10Δ*pyrF*Δ*cipC* or H10Δ*pyrF*Δ*mspI* before plasmid curing (Figure 6). The bands indicating intron insertion at the desired sites became more apparent with longer induction time, which suggested that targetron expression was regulated by

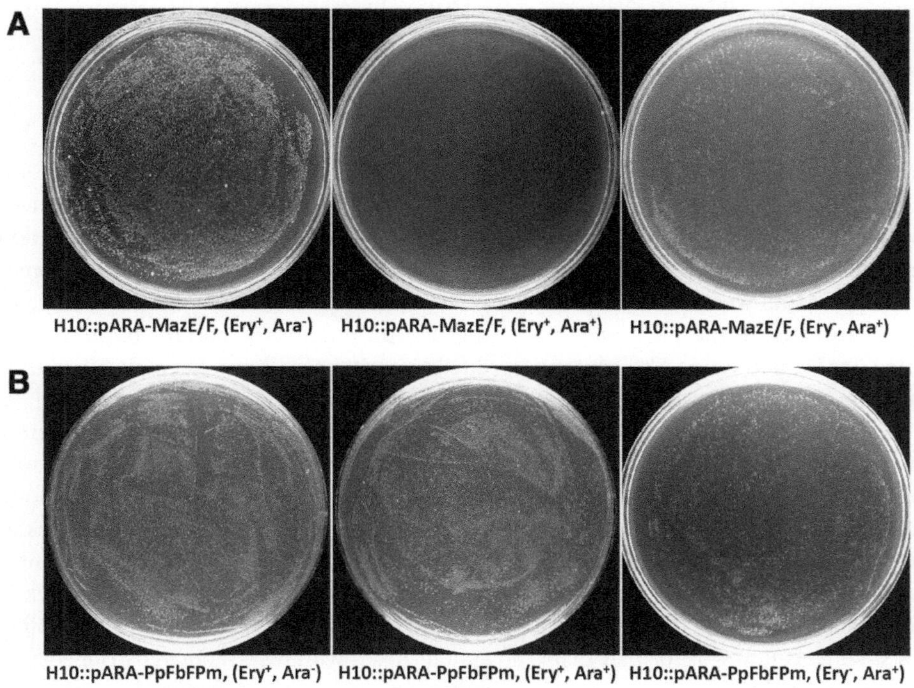

Figure 5 Availability of the counterselection marker MazF in *C. cellulolyticum* H10. Cells of H10::pARA-MazE/F **(A)** and H10::pARA-PpFbFPm **(B)** were plated onto solid GS-2 medium with (superscript +) or without (superscript −) addition of erythromycin (Ery) or L-arabinose (Ara).

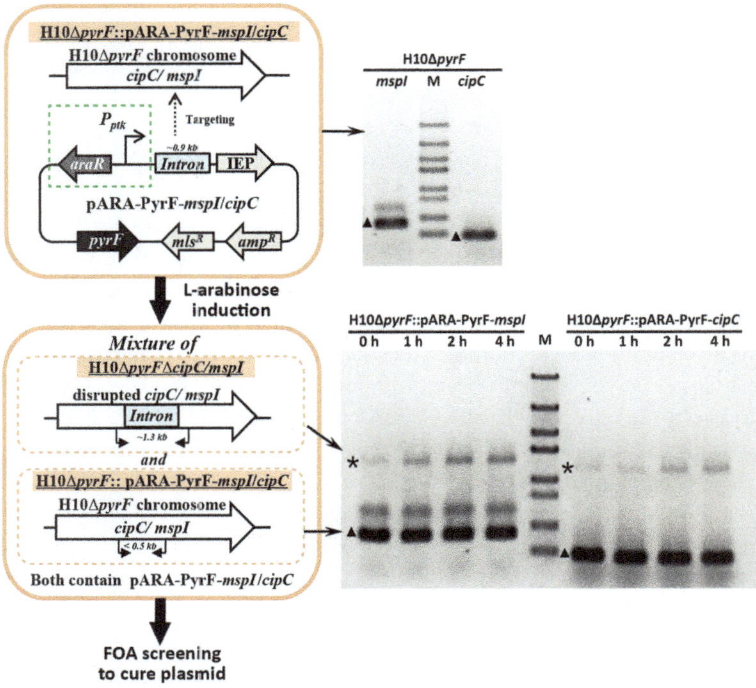

Figure 6 Schematic representation and PCR confirmation of H10Δ*pyrF* transformants carrying pARA-PyrF-*mspI* or pARA-PyrF-*cipC* before FOA screening. Green dashed boxes indicated that the developed arabinose-inducible expression system containing a P_{ptk} promoter and an AraR expression cassette (Additional file 1). The transformants were inoculated and cultivated in antibiotic-free medium until mid-log phase, and 1 g/L L-arabinose was added for induction for 0, 1, 2, and 4 h. Primer sets Ccel2866-F/R and Ccel0728-F/R were used for PCR flanking the intron target sites for gene *mspI* and *cipC*, respectively. The asterisks indicate the PCR products of mutants containing the 0.9-kb intron sequence, and the bands of 300 to 400 bp marked by triangles indicate the PCR products of host strain H10Δ*pyrF*. M, DNA marker (from top to bottom, 5000, 3000, 2000, 1500, 1000, 800, 500, and 300 bp).

the ARAi system, and a novel inducible ClosTron was developed.

The FOA screening was performed to cure targetron plasmids for transformants of pARA-PyrF-*mspI* or pARA-PyrF-*cipC* after 0, 2, or 4 h of L-arabinose induction (Figure 6). Forty-eight or 96 colonies of each transformant were randomly chosen for colony PCR screening, and those colonies with a single band indicating the intron insertion were identified as mutants (Additional file 3). Two to six verified mutants were analyzed using southern hybridization to investigate the off-target frequency of the inducible ClosTron system. The mutants containing single intron insertion were determined as target mutants with no off-target integration, and the number of target mutants was counted to calculate the target specificity by dividing the number of all tested mutants. Both *C. cellulolyticum* mutants H10::MspI297s and H10Δ*pyrF*::CipC117a (Table 1) constructed following the reported ClosTron method [11,14] contained three off-targeted intron insertions in the genomic DNA (Additional file 4). H10Δ*pyrF*Δ*mspI* and H10Δ*pyrF*Δ*cipC* were constructed using the inducible ClosTron. No off-target integration was observed in H10Δ*pyrF*Δ*mspI*, and only one of H10Δ*pyrF*Δ*cipC* mutants showed an additional intron insertion after the arabinose induction for 4 h (Additional file 4). This result indicated that high target specificity could be obtained using the inducible ClosTron.

Discussion

Controllable regulation of gene expression is essential for either native pathway engineering or heterologous function introduction. Recognizing the importance of inducible gene expression system in metabolic engineering toward the overproduction of certain bioproducts, various inducible promoters have been developed using chemicals, light, or radiation as inducers [32], including the widely used inducible T7 expression system based on the *lac* promoter [33]. This inducible system requires the simultaneous regulation of the expression of T7 RNA polymerase [37], and its expression leakage and cost-inefficient inducer IPTG make it not suitable for difficult microorganisms such as *Clostridium* strains [36]. In this study, we provide a novel arabinose-inducible system ARAi with high inducing stringency and efficiency, and a specific inducible effect to L-arabinose.

L-arabinose is a universal carbon source, and has been used as an efficient inducer in the AraC-P_{BAD} system derived from *E. coli* [33]. The promoter of AraC-P_{BAD} system could be repressed by supplementation of glucose because of the CCR effect [33]. For the arabinose-inducible system ARAi developed in this study, the addition of L-arabinose was able to 'turn on' the gene expression. The addition of some other monoses can 'turn down' the inducing activity in *C. cellulolyticum* (Figure 4), which may benefit the large range of expression regulation. Thus, the ARAi system can be used to flexibly regulate the function of endogenous or heterologous gene for metabolic engineering. In addition, the gene expression can be up-regulated over 800-fold using the ARAi system, which has the highest inducing activity in *Clostridium* so far according to our knowledge. The low inducer dosage will contribute to cost reduction for industrialization. Whereas the influence of other sugars in the medium on the inductive efficiency and the accumulation of L-arabinose by the host cell should also be considered in further applications of the ARAi system.

Counterselection markers are required for efficient chromosomal gene deletion and integration [19], especially in those strains whose genetic manipulations rely on replicative plasmid owing to the low transformation efficiency, such as *C. cellulolyticum*. Two types of counterselection markers are mainly used in *Clostridium* strains. The first type relies on the creation of auxotrophic chassis, such as *pyrF*-disrupted auxotrophic mutant for uracil, and the chassis strains are usually constructed by homologous recombination [11,53]. The second type requires the strict control of toxin expression (for example, *codA* and *mazF*) and the absence of relative antitoxins (for example, *upp* and *mazE*, respectively) [19,20,54]. Benefiting from the high efficiency and stringency of the ARAi system, we successfully constructed an MazF-based counterselection marker. This inducible genetic marker provided the same function of promoting the curing of targetron plasmid (Figure 5) as the reported *pyrF*-based screening system [11], and could be used to achieve seamless genome editing via homologous recombination for systems metabolic engineering of *C. cellulolyticum* and other mesophilic *Clostridium* strains [19].

ClosTron is a popular and convenient tool for gene inactivation, and it has been modified previously for high targeting efficiency. For example, Jia et al. provided a scarless gene deletion approach by combining homologous recombination with ClosTron [55]. Cui et al. disrupted the MspI encoding gene in *C. cellulolyticum* to construct a cell chassis that requires no methylation of heterologous DNA [14], and further developed a *pyrF*-based assistant system for ClosTron to promote plasmid curing and enable successive gene targeting [11]. However, the low target specificity of ClosTron is still a challenge for researchers, and tedious and ineffective analyses with multiple steps are generally required and reinforced to confirm single intron insertion [29]. Highly active off-targeting has also been observed in a newly developed thermotargetron method for thermophiles [12,56]. We consider the off-targeting of targetron to be mainly caused by the short recognition sequence of intron RNA and the persistent expression of intron RNA and IEP, thus the precisely controlled expression of targetron elements may solve the off-target integration problem in *C. cellulolyticum*. Although a modest decrease in target specificity was observed, the application of the ARAi system dramatically enhanced the target specificity of the resulting inducible ClosTron method, and thus solved the problem of off-target integration.

Although *C. cellulolyticum* and *C. acetobutylicum* have a close phylogenetic relationship, and the ARAi system developed in this work as well as several other genetic elements used in *C. cellulolyticum* are derived from *C. acetobutylicum* [11,14,20,25,28], apparent difference in physiological and biochemical properties are present in these two species. For instance, *C. acetobutylicum* cannot degrade cellulose, whereas *C. cellulolyticum* is a typical cellulolytic microorganism producing cellulosome, a multi-enzyme complex assembled on the cell surface [57]; *C. acetobutylicum* barely uses D-xylose as a carbon source with the presence of D-glucose due to the CCR effect, and prefers to use L-arabinose compared to D-xylose [58], while no apparent diauxie growth pattern reflecting glucose-mediated CCR [59-61] has been observed in *C. cellulolyticum*, and D-xylose can be faster consumed than L-arabinose (Figure 3); *C. acetobutylicum* and *C. cellulolyticum* contain different organizations of gene clusters for D-xylose and L-arabinose utilization pathways as well as the DNA recognition motifs of transcriptional regulators [47,62]. The inducing sensitivity of ARAi system in *C. cellulolyticum* is similar to that of AraR-P_{ptk} regulon in *C. acetobutylicum* [47]. Thus, it is feasible to investigate the application of the ARAi system in other mesophilic clostridial strains which uptake L-arabinose into the cell.

Conclusion

This study provided a stringent arabinose-inducible genetic operation system ARAi to support the metabolic engineering of *C. cellulolyticum* and other *Clostridium* strains. The ARAi system includes an effective gene expression platform with an oxygen-independent fluorescent reporter, a sensitive MazF-based counterselection genetic marker, and a precise gene knock-out method based on an inducible ClosTron system, and its inducible effect is specific to L-arabinose. Using the ARAi system, the gene expression was up-regulated over 800-fold with highly inducing stringency, and the target specificity of ClosTron was greatly improved to support precise gene disruption in *C. cellulolyticum*. We suggest that the ARAi system can be widely used for clostridial strains with various natural substrates, and this system will contribute to the efficient metabolic engineering of *C. cellulolyticum* and other mesophilic *Clostridium* strains for lignocellulose bioconversion.

Materials and methods

Bacterial strains and cultivation

The bacterial strains used in this study are listed in Table 1. All *E. coli* strains were grown aerobically at 37°C in Luria-Bertani (LB) medium with shaking at 180 rpm or on solid LB plate with 1.5% agar. Then 100 μg/mL ampicillin was supplemented when necessary. *C. cellulolyticum* strains were routinely cultured anaerobically at 34°C in modified GS-2 medium supplemented with 5.0 g/L cellobiose as a carbon source [14]. Furthermore, 20 μg/mL erythromycin or 500 μg/mL 5-fluoroorotic acid (FOA) dissolved in dimethyl sulfoxide was added when required.

Plasmid construction

All plasmids and primers used in this study are listed in Table 1 and Additional file 5, respectively. Plasmids containing ARAi system were constructed on the basis of pMTC6, an *E. coli* and *C. cellulolyticum* shuttle vector containing an expression cassette of an oxygen-independent fluorescent protein (PpFbFPm) under the control of a thiolase (*thl*) promoter and terminator [14,63]. The ARAi system consisted of a promoter P_{ptk} and a repressor AraR. Then 370 bp from the upstream region of the phosphoketolase (*ptk*) encoding gene in *C. acetobutylicum* (CAC1343), the predicted AraR-binding site [47] was amplified with primer set Pptk-F/R and determined as promoter P_{ptk}. The AraR-expression cassette, including the AraR encoding gene (CAC1340) and its predicted promoter and terminator regions, was amplified with primer set araR-F/R using *C. acetobutylicum* chromosome DNA as a template. The promoter P_{ptk} was ligated into pMTC6 using PstI and MluI sites to replace the *thl* promoter and yield the plasmid pPTK-PpFbFPm. The plasmid pARA-PpFbFPm was further obtained by cloning the AraR cassette into the PstI site of pARA-PpFbFPm for inducible expression of fluorescent protein in *C. cellulolyticum* (Figure 1A). pARA-GusA and pARA-MazE/F were constructed based on pARA-PpFbFPm for GusA and MazF expression, respectively. The *gusA* gene was amplified from plasmid pGusA2-2tetO1 [20] using primer set gusA-F/R and ligated into pARA-PpFbFPm by double digestion with NheI and SalI to take the place of *PpFbFPm* (Figure 1A). To construct pARA-MazE/F, *mazE* and *mazF* genes were cloned from the genomic DNA of *E. coli* BL21(DE3) with primer sets MazE-BL21-F/R and MazF-BL21-F/R and ligated into pARA-PpFbFPm in turn using NarI site and MluI and SalI sites, respectively (Additional file 1).

Targetron plasmids were constructed based on pSY6-*mspI* [14] or pGZ-*cipC* [11]. The entire ARAi system including P_{ptk} promoter and AraR cassette was amplified from pARA-PpFbFPm with primer set int-araR/int-Pptk, and cloned into pSY6-*mspI* or pGZ-*cipC* to replace *ptb* promoter using XmaI and XhoI sites to generate pARA-*mspI* or pARA-*cipC*, respectively. Targetron plasmids pARA-PyrF-*mspI* and pARA-PyrF-*cipC* were finally obtained by cloning the P_{fd}-PyrF expression cassette into the XmaI site of pARA-*mspI* and pARA-*cipC*, respectively, as described by [11] (Additional file 1).

Electroporation of *C. cellulolyticum*

Plasmid methylation with MspI methyltransferase, competent cell preparation, and electroporation of *C. cellulolyticum* were precisely performed as described in previous studies [11,14]. All manipulations were performed under anaerobic conditions. Transformants carrying target plasmids were selected on solid GS-2 medium supplemented with erythromycin.

Fluorescent microscopy

In vivo expression of PpFbFPm was monitored with a BX51TRF fluorescent microscope (Olympus Corporation, Shinjuku-ku, Japan). Five hundred microliters of *C. cellulolyticum* cells grown at mid-log phase were harvested by centrifugation, washed twice and resuspended in 20 μL of distilled water. Two microliters of the resuspended cells were used for imaging of intracellular fluorescence (Ex = 460–490 nm, Em = 520 nm).

β-Glucuronidase assays

For the β-glucuronidase assay, *C. cellulolyticum* strains were cultivated in 100-mL serum bottles until late-log phase with 5 g/L cellobiose as carbon source. Then, 0–10 g/L various sugars (L-arabinose, D-arabinose, D-xylose, D-fructose, D-galactose, or D-mannose) or sugar combinations (L-arabinose + D-glucose, L-arabinose + D-xylose, L-arabinose + D-fructose, L-arabinose + D-galactose, or L-arabinose + D-mannose) were supplemented as inducers. Inducible expression was sustained for 0 to 6 h at 34°C, and the crude enzyme samples were prepared according to a modified protocol [20]. In detail, 30 mL *C. cellulolyticum* cultures were chilled in an ice-water bath for 20 min, and centrifuged at 5000 × *g* for 10 min at 4°C. The cell pellets were washed twice with 30 mL of ice-cold Tris-EDTA buffer (10 mM Tris–HCl, 1 mM EDTA, pH 8.0) and resuspended in 2 mL GusA buffer (50 mM sodium phosphate, 1 mM EDTA, pH 7.0). Cell lysis was achieved by sonication under following conditions: 30% duty cycle, 3 s sonication with a 3 s pause for 10 min (Scientz-IID, Scientz Biotech Company, Ningbo, China). The resulting lysates were centrifuged at 8000 × *g* for 15 min at 4°C, and the obtained supernatants were used as crude enzyme samples for GusA assays. The total protein concentrations of the enzyme samples were determined by the Bradford method [64].

Both chromogenic substrate 5-bromo-4-chloro-3-indolyl-β-D-glucuronic acid (X-gluc) and fluorogenic substrate 4-methylumbelliferyl-β-D-glucuronic acid (MUG) were used to determine GusA activity. For the X-gluc assay, 20 μL of the enzyme samples were mixed with 1 mL of X-gluc assay buffer (GusA Buffer containing 0.086 mM X-gluc) [65], and incubated at 37°C for 2 h until a blue color was observed. When MUG was used as the substrate, the reaction mixture of 0.2 mL diluted enzyme sample and 1.8 mL MUG assay buffer (GusA Buffer containing 4 mM MUG) pre-warmed at 37°C was loaded into a quartz fluorometer cuvette (light path = 1 cm) after adequate mixing, and the fluorescence intensity was measured using a fluorescence spectrophotometer (F-4600, Hitachi Limited, Tokyo, Japan) at 37°C. The fluorescence kinetic curve was recorded with a time-scan mode (Ex = 365 nm, Em = 455 nm, PMT voltage = 700 V) every 30 s for 10 min. The curve slope value and protein concentration of a sample was used to calculate the GusA activity (U/mg) as described [20].

Mutant screening and plasmid curing

The disruption of *mspI* and *cipC* genes in *C. cellulolyticum* H10Δ*pyrF* using targetron plasmids was performed according to a modified protocol [11]. Transformants containing plasmid pARA-PyrF-*mspI* or pARA-PyrF-*cipC* (Table 1) were inoculated and cultivated in liquid GS-2 medium without erythromycin for 24 h. Then, the cultures were supplemented with 0 to 10 g/L L-arabinose as an inducer at 34°C for another 0 to 4 h. The resulting cultures were washed twice with fresh GS-2 medium, and then plated on solid GS-2 medium supplemented with FOA for plasmid curing based on the reported *pyrF*-based screening system [11]. The plasmid-cured strains were verified if no growth was observed with the presence of erythromycin. The *mspI*- and *cipC*-disrupted mutants were finally confirmed by colony PCR using primer sets Ccel2866-F/R or Ccel0728-F/R (Additional file 5) and sequencing (Sangon, Shanghai, China).

Southern hybridization

Southern blotting was performed to identify intron insertion in the genomic DNA of *C. cellulolyticum* mutants as described previously [56]. All positive colonies confirmed by PCR and sequencing were inoculated and cultivated in fresh liquid GS-2 medium until late-log phase. Then 4 mL cultures were centrifuged to obtain cell pellets for *C. cellulolyticum* genomic DNA isolation using a TIANamp Bacteria DNA Extraction Kit (Tiangen Biotech, Beijing, China). Genomic DNA was digested with EcoRI and BamHI at 37°C overnight, purified using a traditional phenol/chloroform extraction and ethanol precipitation method, separated by 0.8% agarose gel electrophoresis, and finally transferred to a Nylon membrane (Hybond-NX, GE Healthcare, Pewaukee, WI, USA). The intron probe was amplified using primer set Probe-F/R (Additional file 5) and labeled with digoxigenin-11-dUTP. Hybridization and immunological detection were performed according to the manufacturer's instructions (DIG-High Prime DNA Labeling and Detection Starter Kit I; Roche, Basel, Switzerland).

Fermentation analysis

Batch fermentation of *C. cellulolyticum* strains was performed anaerobically in GS-2 medium at 34°C in 250-mL serum bottles with 100 mL medium containing various sugars (L-arabinose, D-xylose, D-glucose, cellobiose) or sugar combinations as carbon sources. Two microliters of broth was sampled every 6 or 12 h during cultivation to analyze the optical density at 600 nm (OD_{600}) using a UV-vis spectrophotometer. The supernatants of the samples were obtained by centrifugation and micro-filtration (0.22-μm pore diameter) and were used to quantify the residual sugars directly using high performance liquid chromatography (HPLC) (Rigol L-3000, Rigol Technologies, Oakwood Village, OH, USA) equipped with an Aminex HPX-87H column (Bio-Rad, Hercules, CA, USA) and a refractive index detector (Agilent 1260 infinity RID; Agilent, Santa Clara, CA, USA). Furthermore, 5 mM H_2SO_4 was used as the mobile phase at 55°C with a flow rate of 0.5 mL/min [12].

Additional files

Additional file 1: Schematic representation of the plasmids constructed for the arabinose-inducible genetic operation (ARAi) system. (A) plasmids constructed for inducible gene expression and the counterselection marker; (B) plasmids constructed for the inducible ClosTron system. All plasmids contained an ampicillin resistance gene (*amp*^{R}) and an erythromycin resistance gene (*mls*^{R}). Green dashed boxes indicated the inducible promoter system containing a P_{ptk} promoter and an AraR expression cassette from *C. acetobutylicum*.

Additional file 2: Investigation of inducing activity of the ARAi system in *C. cellulolyticum* H10 in company with L-arabinose consumption. H10::pARA-GusA was firstly cultivated with 5 g/L cellobiose for 24 h and 1 g/L L-arabinose was added to induce the expression of GusA. Interval sampling was performed after induction for 2, 6, 12, 24, 48, and 72 h, and GusA activity (Black bars) and the concentration of residual L-arabinose (Green line) and cellobiose (Blue line) were then determined. Two independent setups were prepared and measured for every time point.

Additional file 3: PCR confirmation of *C. cellulolyticum* H10 mutant strains. The transformants of H10Δ*pyrF* containing pARA-PyrF-*mspI* or pARA-PyrF-*cipC* were cultivated in liquid GS-2 medium without antibiotic and induced with L-arabinose for 4 h. After plasmid curing via FOA screening, 48 colonies of each recombinant strain were selected as templates for PCR with primer sets Ccel2866-F/R or Ccel0728-F/R, respectively. Arrows indicated the PCR products (~1.3 kb) with the full-length intron sequence. The bands of approximately 0.4 or 0.3 kb indicate the PCR products of wild-type strain. For H10ΔpyrF::pARA-PyrF-*mspI*, 3 of 48 colonies were confirmed as mutant H10Δ*pyrF*Δ*mspI* (A); for H10ΔpyrF::pARA-PyrF-*cipC*, 6 of 48 colonies were confirmed as mutant H10Δ*pyrF*Δ*cipC* (B). M, DNA marker (from top to bottom, 5,000, 3,000, 2,000, 1,500, 1,000, 800, 500, and 300 bp).

Additional file 4: Southern hybridization analysis of *C. cellulolyticum H10* mutant strains. H10Δ*pyrF* was used as a negative control (NC). The inducible ClosTron was employed to construct H10Δ*pyrF*Δ*mspI* (lane 2-4) and H10Δ*pyrF*Δ*cipC* (lane 6-8). Two to six colonies verified as mutants by colony PCR (Additional file 3) were analyzed by southern blotting. The L-arabinose induction was performed for 0 h (lane 2,6), 2 h (lane 3,7) or 4 h (lane 4,8). H10::MspI297s (lane 1) and H10Δ*pyrF*::CipC117a (lane 5) constructed by the reported ClosTron method [1] were also analyzed in parallel. Black triangles indicate the additional off-target insertions.

Additional file 5: Primers used in this study.

Abbreviations

ARAi: arabinose-inducible genetic operation system; CBP: consolidated bioprocessing; CCR: carbon catabolize repression; FOA: 5-fluoroorotic acid; GusA: β-glucuronidase; HPLC: high performance liquid chromatography; IEP: intron-encoded protein; IPTG: isopropyl-β-d-thiogalactoside; LB: Luria-Bertani; MUG: 4-methylumbelliferyl-β-D-glucuronic acid; OD_{600}: optical density at 600 nm; PpFbFPm: oxygen-independent fluorescent protein from *Pseudomonas putida*; *ptk*: phosphoketolase gene; *thl*: thiclase gene; X-gluc: 5-bromo-4-chloro-3-indolyl-β-D-glucuronic acid.

Competing interests

The authors declare that they have no competing interests.

Authors' contributions

JZ, YJL, and QC designed the research; JZ and GZC performed experiments; JZ and YJL analyzed data and wrote the paper. All authors have read and approved of the final version of the manuscript.

Acknowledgements

We thank Prof. Yin Li and Dr. Hongjun Dong from Institute of Microbiology, CAS for providing the plasmid pGusA2-2tetO1. We thank Prof. Yingang Feng for critically reading the manuscript. This work was supported by the National Basic Research Program of China (grant 2011CB707404) to QC; the National Natural Science Foundation of China (grant 31300035) and Qingdao Science and Technology Development Plan (grant 13-1-4-196-jch) to YJL; and China Postdoctoral Science Foundation to GZC.

Author details

[1]Shandong Provincial Key Laboratory of Energy Genetics, Qingdao Institute of Bioenergy and Bioprocess Technology, Chinese Academy of Sciences, 189, Songling Road, Qingdao 266101, People's Republic of China. [2]Key Laboratory of Biofuels, Qingdao Institute of Bioenergy and Bioprocess Technology, Chinese Academy of Sciences, 189, Songling Road, Qingdao 266101, People's Republic of China. [3]Qingdao Engineering Laboratory of Single Cell Oil, Qingdao Institute of Bioenergy and Bioprocess Technology, Chinese Academy of Sciences, 189, Songling Road, Qingdao 266101, People's Republic of China. [4]University of Chinese Academy of Sciences, Chinese Academy of Sciences, 19, Yuquan Road, Beijing 100049, People's Republic of China.

References

1. Demain AL. Biosolutions to the energy problem. J Ind Microbiol Biotechnol. 2009;36:319–32.
2. Antoni D, Zverlov W, Schwarz WH. Biofuels from microbes. Appl Microbiol Biotechnol. 2007;77:23–35.
3. Gronenberg LS, Marcheschi RJ, Liao JC. Next generation biofuel engineering in prokaryotes. Curr Opin Chem Biol. 2013;17:462–71.
4. Hoffert MI. Farewell to fossil fuels? Science. 2010;329:1292–4.
5. Hasunuma T, Okazaki F, Okai N, Hara KY, Ishii J, Kondo A. A review of enzymes and microbes for lignocellulosic biorefinery and the possibility of their application to consolidated bioprocessing technology. Bioresour Technol. 2013;135:513–22.
6. Olson DG, McBride JE, Shaw AJ, Lynd LR. Recent progress in consolidated bioprocessing. Curr Opin Biotechnol. 2012;23:396–405.
7. Xu Q, Singh A, Himmel ME. Perspectives and new directions for the production of bioethanol using consolidated bioprocessing of lignocellulose. Curr Opin Biotechnol. 2009;20:364–71.
8. Lynd LR, van Zyl WH, McBride JE, Laser M. Consolidated bioprocessing of cellulosic biomass: an update. Curr Opin Biotechnol. 2005;16:577–83.
9. Tracy BP, Jones SW, Fast AG, Indurthi DC, Papoutsakis ET. Clostridia: the importance of their exceptional substrate and metabolite diversity for biofuel and biorefinery applications. Curr Opin Biotechnol. 2012;23:364–81.
10. Demain AL, Newcomb M, Wu JH. Cellulase, clostridia, and ethanol. Microbiol Mol Biol Rev. 2005;69:124–54.
11. Cui GZ, Zhang J, Hong W, Xu C, Feng Y, Cui Q, et al. Improvement of ClosTron for successive gene disruption in *Clostridium cellulolyticum* using a *pyrF*-based screening system. Appl Microbiol Biotechnol. 2014;98:313–23.
12. Mohr G, Hong W, Zhang J, Cui G-Z, Yang Y, Cui Q, et al. A targetron system for gene targeting in thermophiles and its application in *Clostridium thermocellum*. PLoS One. 2013;8:e69032.
13. Guedon E, Desvaux M, Petitdemange H. Improvement of cellulolytic properties of *Clostridium cellulolyticum* by metabolic engineering. Appl Environ Microbiol. 2002;68:53–8.
14. Cui GZ, Hong W, Zhang J, Li WL, Feng Y, Liu YJ, et al. Targeted gene engineering in *Clostridium cellulolyticum* H10 without methylation. J Microbiol Methods. 2012;89:201–8.
15. Olson DG, Giannone RJ, Hettich RL, Lynd LR. Role of the CipA scaffoldin protein in cellulose solubilization, as determined by targeted gene deletion and complementation in *Clostridium thermocellum*. J Bacteriol. 2013;195:733–9.
16. Harris LM, Welker NE, Papoutsakis ET. Northern, morphological, and fermentation analysis of *spo0A* inactivation and overexpression in *Clostridium acetobutylicum* ATCC 824. J Bacteriol. 2002;184:3586–97.
17. Li H, Liao JC. A synthetic anhydrotetracycline-controllable gene expression system in *Ralstonia eutropha* H16. ACS Synthetic Biol. 2014. in press.
18. Banerjee A, Leang C, Ueki T, Nevin KP, Lovley DR. Lactose-inducible system for metabolic engineering of *Clostridium ljungdahlii*. Appl Environ Microbiol. 2014;80:2410–6.
19. Al-Hinai MA, Fast AG, Papoutsakis ET. Novel system for efficient isolation of *Clostridium* double-crossover allelic exchange mutants enabling markerless chromosomal gene deletions and DNA integration. Appl Environ Microbiol. 2012;78:8112–21.
20. Dong H, Tao W, Zhang Y, Li Y. Development of an anhydrotetracycline-inducible gene expression system for solvent-producing *Clostridium acetobutylicum*: a useful tool for strain engineering. Metab Eng. 2012;14:59–67.
21. Heap JT, Pennington OJ, Cartman ST, Carter GP, Minton NP. The ClosTron: a universal gene knock-out system for the genus *Clostridium*. J Microbiol Methods. 2007;70:452–64.
22. Karberg M, Guo H, Zhong J, Coon R, Perutka J, Lambowitz AM. Group II introns as controllable gene targeting vectors for genetic manipulation of bacteria. Nat Biotechnol. 2001;19:1162–7.
23. Lambowitz AM, Zimmerly S. Group II Introns: mobile ribozymes that invade DNA. Cold Spring Harb Perspect Biol. 2011;3:a003616.
24. Chen Y, McClane BA, Fisher DJ, Rood JI, Gupta P. Construction of an alpha toxin gene knockout mutant of *Clostridium perfringens* type A by use of a mobile group II intron. Appl Environ Microbiol. 2005;71:7542–7.
25. Shao L, Hu S, Yang Y, Gu Y, Chen J, Jiang W, et al. Targeted gene disruption by use of a group II intron (targetron) vector in *Clostridium acetobutylicum*. Cell Res. 2007;17:963–5.
26. Heap JT, Kuehne SA, Ehsaan M, Cartman ST, Cooksley CM, Scott JC, et al. The ClosTron: mutagenesis in *Clostridium* refined and streamlined. J Microbiol Methods. 2010;80:49–55.
27. Cai G, Jin B, Saint C, Monis P. Genetic manipulation of butyrate formation pathways in *Clostridium butyricum*. J Biotech. 2011;155:269–74.
28. Li Y, Tschaplinski TJ, Engle NL, Hamilton CY, Rodriguez Jr M, Liao JC, et al. Combined inactivation of the *Clostridium cellulolyticum* lactate and malate dehydrogenase genes substantially increases ethanol yield from cellulose and switchgrass fermentations. Biotechnol Biofuels. 2012;5:2.
29. Cooksley CM, Zhang Y, Wang H, Redl S, Winzer K, Minton NP. Targeted mutagenesis of the *Clostridium acetobutylicum* acetone-butanol-ethanol fermentation pathway. Metab Eng. 2012;14:630–41.
30. Matsuura M, Saldanha R, Ma HW, Wank H, Yang J, Mohr G, et al. A bacterial group II intron encoding reverse transcriptase, maturase, DNA endonuclease activities: biochemical demonstration of maturase activity

and insertion of new genetic information within the intron. Genes Dev. 1997;11:2910–24.

31. Mohr G, Smith D, Belfort M, Lambowitz AM. Rules for DNA target-site recognition by a lactococcal group II intron enable retargeting of the intron to specific DNA sequences. Genes Dev. 2000;14:559–73.
32. Wang T, Ma X, Du G, Chen J. Overview of regulatory strategies and molecular elements in metabolic engineering of bacteria. Mol Biotechnol. 2012;52:300–8.
33. Guzman LM, Belin D, Carson MJ, Beckwith J. Tight regulation, modulation, and high-level expression by vectors containing the arabinose P-BAD promoter. J Bacteriol. 1995;177:4121–30.
34. Bertram R, Hillen W. The application of Tet repressor in prokaryotic gene regulation and expression. Microbial Biotechnol. 2008;1:2–16.
35. Jensen PR, Westerhoff HV, Michelsen O. The use of lac-type promoters in control analysis. Eur J Biochem. 1993;211:181–91.
36. Pan SH, Malcolm BA. Reduced background expression and improved plasmid stability with pET vectors in BL21 (DE3). BioTechniques. 2000;29:1234–8.
37. Studier FW, Rosenberg AH, Dunn JJ, Dubendorff JW. Use of T7 RNA polymerase to direct expression of cloned genes. Methods Enzymol. 1990;185:60–89.
38. Hartman AH, Liu H, Melville SB. Construction and characterization of a lactose-inducible promoter system for controlled gene expression in *Clostridium perfringens*. Appl Environ Microbiol. 2011;77:471–8.
39. Pyne ME, Bruder M, Moo-Young M, Chung DA, Chou CP. Technical guide for genetic advancement of underdeveloped and intractable *Clostridium*. Biotechnol Adv. 2014;32:623–41.
40. Nuyts S, Van Mellaert L, Theys J, Landuyt W, Bosmans E, Anne J, et al. Radio-responsive *recA* promoter significantly increases TNFalpha production in recombinant clostridia after 2 Gy irradiation. Gene Ther. 2001;8:1197–201.
41. Girbal L, Mortier-Barriere I, Raynaud F, Rouanet C, Croux C, Soucaille P. Development of a sensitive gene expression reporter system and an inducible promoter-repressor system for *Clostridium acetobutylicum*. Appl Environ Microbiol. 2003;69:4985–8.
42. Nariya H, Miyata S, Kuwahara T, Okabe A. Development and characterization of a xylose-inducible gene expression system for *Clostridium perfringens*. Appl Environ Microbiol. 2011;77:8439–41.
43. Newman JR, Fuqua C. Broad-host-range expression vectors that carry the L-arabinose-inducible *Escherichia coli araBAD* promoter and the *araC* regulator. Gene. 1999;227:197–203.
44. Sukchawalit R, Vattanaviboon P, Sallabhan R, Mongkolsuk S. Construction and characterization of regulated L-arabinose-inducible broad host range expression vectors in *Xanthomonas*. FEMS Microbiol Lett. 1999;181:217–23.
45. Carroll P, Brown AC, Hartridge AR, Parish T. Expression of *Mycobacterium tuberculosis* Rv1991c using an arabinose-inducible promoter demonstrates its role as a toxin. FEMS Microbiol Lett. 2007;274:73–82.
46. Zhang Y, Shang X, Lai S, Zhang G, Liang Y, Wen T. Development and application of an arabinose-inducible expression system by facilitating inducer uptake in *Corynebacterium glutamicum*. Appl Environ Microbiol. 2012;78:5831–8.
47. Zhang L, Leyn SA, Gu Y, Jiang W, Rodionov DA, Yang C. Ribulokinase and transcriptional regulation of arabinose metabolism in *Clostridium acetobutylicum*. J Bacteriol. 2012;194:1055–64.
48. Servinsky MD, Kiel JT, Dupuy NF, Sund CJ. Transcriptional analysis of differential carbohydrate utilization by *Clostridium acetobutylicum*. Microbiology. 2010;156:3478–91.
49. Sa-Nogueira I, Mota LJ. Negative regulation of L-arabinose metabolism in *Bacillus subtilis*: characterization of the *araR* (*araC*) gene. J Bacteriol. 1997;179:1598–608.
50. Yamaguchi Y, Inouye M. mRNA interferases, sequence-specific endoribonucleases from the toxin-antitoxin systems. Prog Mol Biol Transl Sci. 2009;85:467–500.
51. Yang J, Jiang W, Yang S. mazF as a counter-selectable marker for unmarked genetic modification of Pichia pastoris. FEMS Yeast Res. 2009;9:600–9.
52. Conrad B, Savchenko RS, Breves R, Hofemeister J. A T7 promoter-specific, inducible protein expression system for *Bacillus subtilis*. Mol Gen Genet. 1996;250:230–6.
53. Suzuki H, Murakami A, Yoshida K. Counterselection system for *Geobacillus kaustophilus* HTA426 through disruption of *pyrF* and *pyrR*. Appl Environ Microbiol. 2012;78:7376–83.
54. Cartman ST, Kelly ML, Heeg D, Heap JT, Minton NP. Precise manipulation of the *Clostridium difficile* chromosome reveals a lack of association between the tcdC genotype and toxin production. Appl Environ Microbiol. 2012;78:4683–90.
55. Jia K, Zhu Y, Zhang Y, Li Y. Group II intron-anchored gene deletion in *Clostridium*. PLoS One. 2011;6:e16693.
56. Hong W, Zhang J, Feng Y, Mohr G, Lambowitz AM, Cui G-Z, et al. The contribution of cellulosomal scaffoldins to cellulose hydrolysis by *Clostridium thermocellum* analyzed by using thermotargetrons. Biotechnol Biofuels. 2014;7:80.
57. Doi RH, Kosugi A, Murashima K, Tamaru Y, Han SO. Cellulosomes from mesophilic bacteria. J Bacteriol. 2003;185:5907–14.
58. Xiao H, Gu Y, Ning Y, Yang Y, Mitchell WJ, Jiang W, et al. Confirmation and elimination of xylose-metabolic bottlenecks in glucose-PTS-deficient *Clostridium acetobutylicum* to realize simultaneous utilization of glucose, xylose and arabinose. Appl Environ Microbiol. 2011;77:7886–95.
59. Goerke B, Stulke J. Carbon catabolite repression in bacteria: many ways to make the most out of nutrients. Nat Rev Microbiol. 2008;6:613–24.
60. Deutscher J. The mechanisms of carbon catabolite repression in bacteria. Curr Opin Microbiol. 2008;11:87–93.
61. Abdou L, Boileau C, de Philip P, Pages S, Fierobe HP, Tardif C. Transcriptional regulation of the *Clostridium cellulolyticum* cip-cel operon: a complex mechanism involving a catabolite-responsive element. J Bacteriol. 2008;190:1499–506.
62. Gu Y, Ding Y, Ren C, Sun Z, Rodionov DA, Zhang W, et al. Reconstruction of xylose utilization pathway and regulons in Firmicutes. BMC Genomics. 2010;11:255.
63. Drepper T, Eggert T, Circolone F, Heck A, Krauss U, Guterl JK, et al. Reporter proteins for *in vivo* fluorescence without oxygen. Nat Biotechnol. 2007;25:443–5.
64. Bradford MM. A rapid and sensitive method for the quantitation of microgram quantities of protein utilizing the principle of protein-dye binding. Anal Biochem. 1976;72:248–54.
65. Jefferson RA, Burgess SM, Hirsh D. β-Glucuronidase from Escherichia coli as a gene-fusion marker. Proc Natl Acad Sci U S A. 1986;83:8447–51.

12

Characterization of a copper responsive promoter and its mediated overexpression of the xylanase regulator 1 results in an induction-independent production of cellulases in *Trichoderma reesei*

Xinxing Lv, Fanglin Zheng, Chunyan Li, Weixin Zhang, Guanjun Chen and Weifeng Liu*

Abstract

Background: *Trichoderma reesei* represents an important workhorse for industrial production of cellulases as well as other proteins. The large-scale production is usually performed in a substrate-inducing manner achieved by a fine-tuned cooperation of a suite of transcription factors. Their production and subsequent analysis are, however, often either difficult to manipulate or complicated by the concomitant production of other inducible proteins. Alternatives to control gene expression independent of the nutritional state are thus preferred in some cases to facilitate not only biochemical studies of proteins but also genetic engineering of the producer.

Results: We identified a copper transporter encoding gene *tcu1* (jgi:Trire2:52315) in *T. reesei*, the transcription of which was highly responsive to copper availability. Whereas excess copper repressed the expression of *tcu1* from *T. reesei*, eliminating copper addition in the medium resulted in a high-level transcription of *tcu1*. The usefulness of the system was further illustrated by the high-level expression of specific cellulases driven by the *tcu1* promoter in *T. reesei* when cultivated on D-glucose or glycerol as the sole carbon source. A recombinant *T. reesei* strain, which overexpressed the main transcription activator of hydrolases (xylanase regulator 1) under the control of *tcu1* promoter, was found to be relieved from the carbon catabolite repression and thus displayed a constitutive cellulase expression. Moreover, the amount and activities of cellulases produced by this strain on glycerol or glucose fully recapitulated those of the parental strain produced on Avicel.

Conclusion: Expression of *T. reesei tcu1* gene was tightly controlled by copper availability, and a homologous protein expression system was developed based on this promoter. Deregulation of XYR1 (xylanase regulator 1) mediated by the *tcu1* promoter not only overcame the carbon catabolite repression of cellulases but also resulted in their full expression even on the non-inducing carbon sources.

Keywords: *Trichoderma reesei*, Copper responsive promoter, XYR1, Cellulase, Carbon catabolite repression

Background

The ascomycete *Trichoderma reesei* (teleomorph *Hypocrea jecorina*), a saprophytic filamentous fungus, is one of the most prolific lignocellulosic enzyme producer in industry. The secreted cellulase mixture synergistically acts upon the insoluble cellulosic materials to achieve the efficient conversion to fermentable sugars [1]. Up to 100 g/l of secreted proteins under optimal culture conditions has been reported for *T. reesei* strains [2]. High-yield production of the bulk of plant cell wall-degrading machinery in *T. reesei* is, however, dependent on the induction by insoluble substrates that include cellulose, hemicellulose, or mixtures of plant polymers. Considering the ease of manipulation and the complication of separating enzymes from insoluble plant cell wall materials, soluble inducing substrates are usually preferred or required. These soluble inducers are, however, either non-economic (for example, sophorose) or with low efficiency (for example, lactose).

* Correspondence: weifliu@sdu.edu.cn
State Key Laboratory of Microbial Technology, School of Life Science, Shandong University, No.27 Shanda South Road, Jinan 250100, Shandong, People's Republic of China

In *T. reesei*, most of the cellulase genes are coordinately controlled by a suite of transcription factors [3]. Whereas CRE1 is the main transcription factor mediating carbon catabolite repression (CCR) of cellulase gene expression, xylanase regulator 1 (XYR1) is absolutely necessary for activating the expression of most of cellulases/hemicellulases. Lack of XYR1 has been shown to eliminate cellulase induction by almost all inducers [4]. Regulation of *xyr1* expression has thus been considered to exert a significant impact on the ability of *T. reesei* to produce the various hydrolytic enzymes though the exact mechanism remains elusive [5,6]. Two seemingly opposing but not necessarily exclusive assumptions exist regarding the mechanism by which XYR1 may govern cellulase expression. Studies by Mach-Aigner *et al.* have indicated that posttranslational modification of XYR1 constitutes a major activating aspect of its activation [5], whereas other studies suggested that a strongly elevated transcription level of *xyr1* is important for hyperproduction of cellulases under inducing conditions [6,7].

Developments over the past nearly 30 years have also established *T. reesei* as an incredibly useful host for large-scale protein production, largely based on the regulatory elements of the gene encoding the most abundant secreted cellulase component, cellobiohydrolase I (CEL7A) [8,9]. On the other hand, the need to further improve the performance of the native cellulase system of *T. reesei* also requires its being augmented by the introduction of heterologous enzymatic activities or improved variants of native enzymes. Problems may arise, however, with these inducible systems due to the unselective expression of endogenous cellulase components leading to contamination of target proteins for further analysis. Alternative promoters with ability to control gene expression by both activation and repression independent of nutritional state are thus desirable in these cases. In *T. reesei*, many constitutive promoters, most of which are from genes participating in glucose metabolism, have been reported to drive gene expression including specific cellulase and xylanase encoding genes on glucose [10]. These promoters are useful, but not ideal, because they are highly influenced by the nutritional state of the culture and their behavior on carbon sources other than glucose has not been systematically studied. More recently, promoters of the high-affinity copper transporter encoding genes have been used successfully in several fungi to drive high-level gene expression [11,12]. Nevertheless, the existence of such copper transporters and the potential of using their promoters to tightly control gene expression in *T. reesei* have not been examined yet.

In this paper, we developed a copper-responsive expression system based on the promoter of a *T. reesei* copper transporter gene *tcu1*. The success of this system was further demonstrated by the tunable expression of proteins including two cellulases (CEL7A and CEL7B) on glucose. Finally, we showed that overexpression of *xyr1* alone driven by P_{tcu1} resulted in the full production of cellulases on glucose up to a level comparable to those produced on cellulose.

Results

T. reesei tcu1 gene expression is controlled by copper availability but independent of carbon sources

Copper transporter genes are sensitive to the environmental copper levels to tightly control the uptake of copper [13,14]. To investigate whether such an ortholog of copper transporter also exists in cellulolytic *T. reesei*, we first searched the genome of *T. reesei* for homologues of the *Neurospora crassa* protein TCU-1 (NCU00830) and a putative copper transporter (jgi:Trire2:52315) was identified. We then examined the expression of *tcu1* in a *T. reesei* strain QM9414 cultured at 12 h in the presence of different concentrations of copper sulfate (0 to 10,000 nM). The results showed that *tcu1* was highly transcribed when less than 200 nM of copper sulfate was added (Figure 1A). The transcription was dramatically decreased by 500 nM of copper sulfate and was almost abolished by copper sulfate above 1 μM (Figure 1A). Importantly, the same response was also observed when the same strain was cultured on glycerol or even cellulose (Figure 1B,C). To determine the kinetics of turnoff of *tcu1* by $CuSO_4$, we analyzed the effect of addition of 10 μM copper to a preculture on 1% glucose on the transcription of *tcu1* over time in strain QM9414. As shown in Figure 1D, turnoff of *tcu1* was rapid and robust with the transcription of *tcu1* being shut down as early as 15 min after addition of $CuSO_4$ and last for up to 4 h. To further investigate whether the repressed *tcu1* could be directly turned on, we applied different concentrations of a copper chelator bathocuproinedisulfonic (BCS) acid to a QM9414 culture in the presence of 1 μM Cu^{2+} (Figure 1E). The result showed that, as compared with the transcription in the absence of copper, BCS at a concentration up to 100 μM only partially activated the transcription of *tcu1*. Taken together, these data indicated that the transcription of *tcu1* from *T. reesei* was very sensitive to the environmental copper and that the promoter of *tcu1* may well be used for the regulated expression of target genes.

Copper-responsive expression of the green fluorescent protein by the promoter of *tcu1*

In order to confirm that the promoter of *tcu1* (P_{tcu1}) could be used to regulate the expression of a heterologous protein, a recombinant QM9414 strain (P_{tcu1}-*gfp*) expressing the green fluorescent protein (GFP) under the control of P_{tcu1} was constructed. The transformant P_{tcu1}-*gfp* was highly fluorescent when cultured in the absence of $CuSO_4$ on 1% glucose for 12 h, whereas no fluorescence was observed when 10 μM of $CuSO_4$ was included (Figure 2A). Mycelia fluorescence intensity was quantitatively analyzed

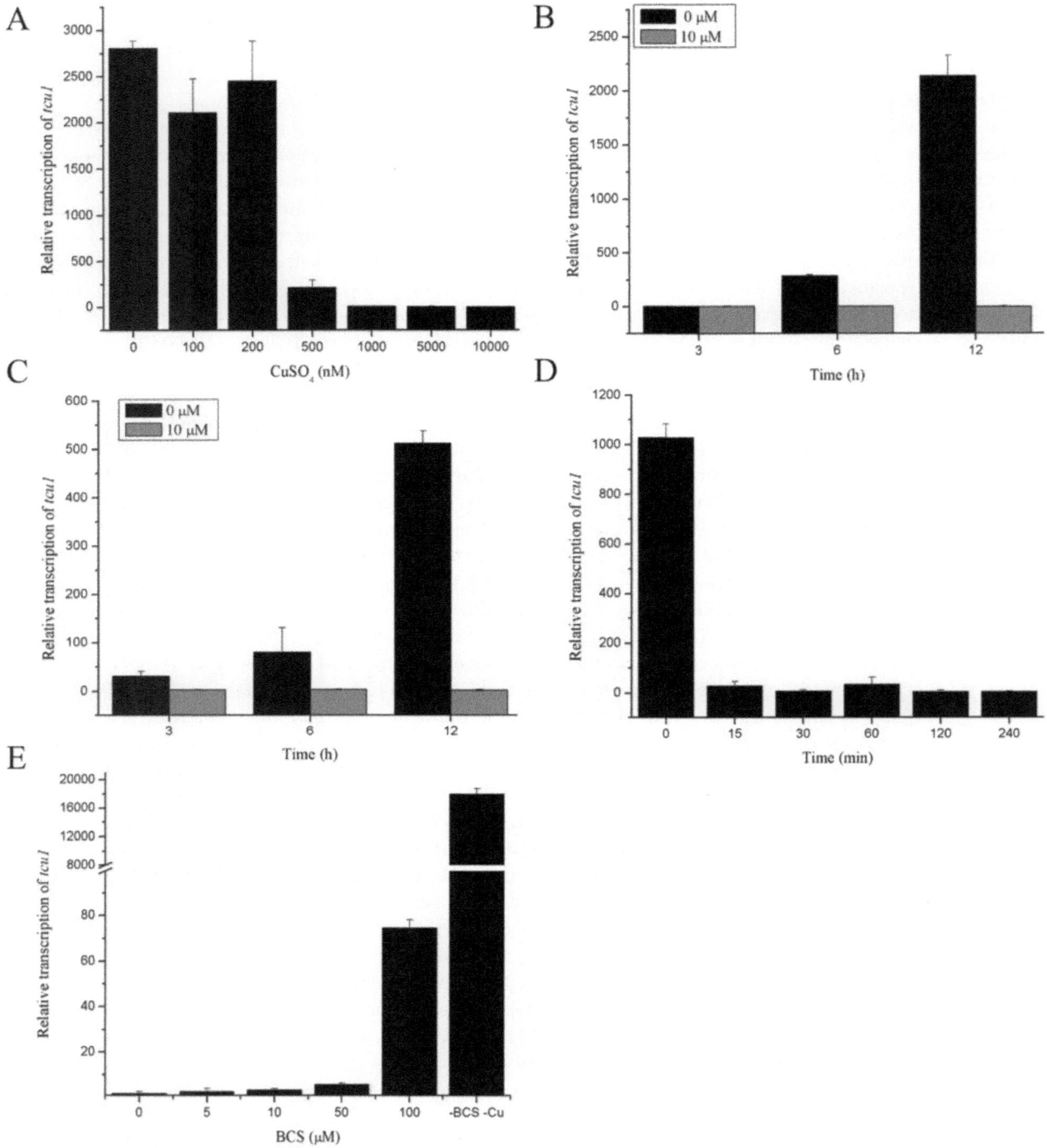

Figure 1 Expression of the *tcu1* gene was influenced by copper availability in *T. reesei*. **(A)** qRT-PCR analysis of the endogenous *tcu1* mRNA after strain QM9414 was cultured for 12 h with 1% glucose as carbon source and with different concentrations of $CuSO_4$. A significant difference ($P < 0.05$) was calculated with a Student's *t* test for the transcription of *tcu1* between concentrations from 0 to 200 nM of copper and those above 500 nM. Levels of endogenous *tcu1* mRNA from strain QM9414 cultured in the presence of 10 μM $CuSO_4$ with 1% glycerol **(B)** and 1% Avicel **(C)** as the sole carbon source, respectively. A significant difference ($P < 0.05$) existed in the transcription of *tcu1* at 12 h. **(D)** The kinetics of turnoff of *tcu1* after 10 μM $CuSO_4$ addition. A significant difference ($P < 0.05$) existed in the transcription of *tcu1* after adding copper for 15 min compared to 0 min. **(E)** qRT-PCR analysis of *tcu1* mRNA after the indicated concentrations of BCS were added to a QM9414 culture for 12 h in the presence of 1 μM $CuSO_4$. A significant difference ($P < 0.05$) existed for the transcription of *tcu1* with and without 100 μM BCS. Error bars are the SD from three biological replicates.

to reflect the expression level of GFP driven by P_{tcu1} at different concentrations of copper sulfate (Figure 2B). In accordance with the results of the expression of the endogenous *tcu1* gene, the fluorescence intensity was the highest when $CuSO_4$ was present at a concentration lower than 50 nM. Increasing concentrations of $CuSO_4$ decreased the fluorescence intensity with a 90% decrease at 500 nM, and the fluorescence was nearly extinguished by copper concentrations above 5 μM. These data demonstrated that the heterologous protein level expressed under the control of P_{tcu1} correlates with copper availability and that P_{tcu1} behaves in a highly responsive manner to exogenous copper.

Homologous expression of two cellulases by P_{tcu1}

To further test the potential of developing a P_{tcu1}-based expression system in *T. reesei*, two cellulase genes including cellobiohydrolase encoding gene (*cel7a*) and endocellulase encoding gene (*cel7b*) were chosen for testing the expression. The respective expression plasmids were transformed into a Δ*cel7a* strain to avoid the potential contamination of endogenous CEL7A, and the extracellular protein production as well as the hydrolytic activity was followed during growth on 1% D-glucose or 1% glycerol. The growth rate was similar with or without $CuSO_4$ (data not shown). As shown in Figure 3A,B,G, a significant amount of CEL7B and CEL7A was detected in the culture supernatant of transformants P_{tcu1}-*cel7b* and P_{tcu1}-*cel7a*. The produced CEL7B and CEL7A was not due to an endogenous expression since no corresponding protein bands could be detected in the parental Δ*cel7a* strain on glucose or glycerol regardless of the presence or absence of copper (data not shown).

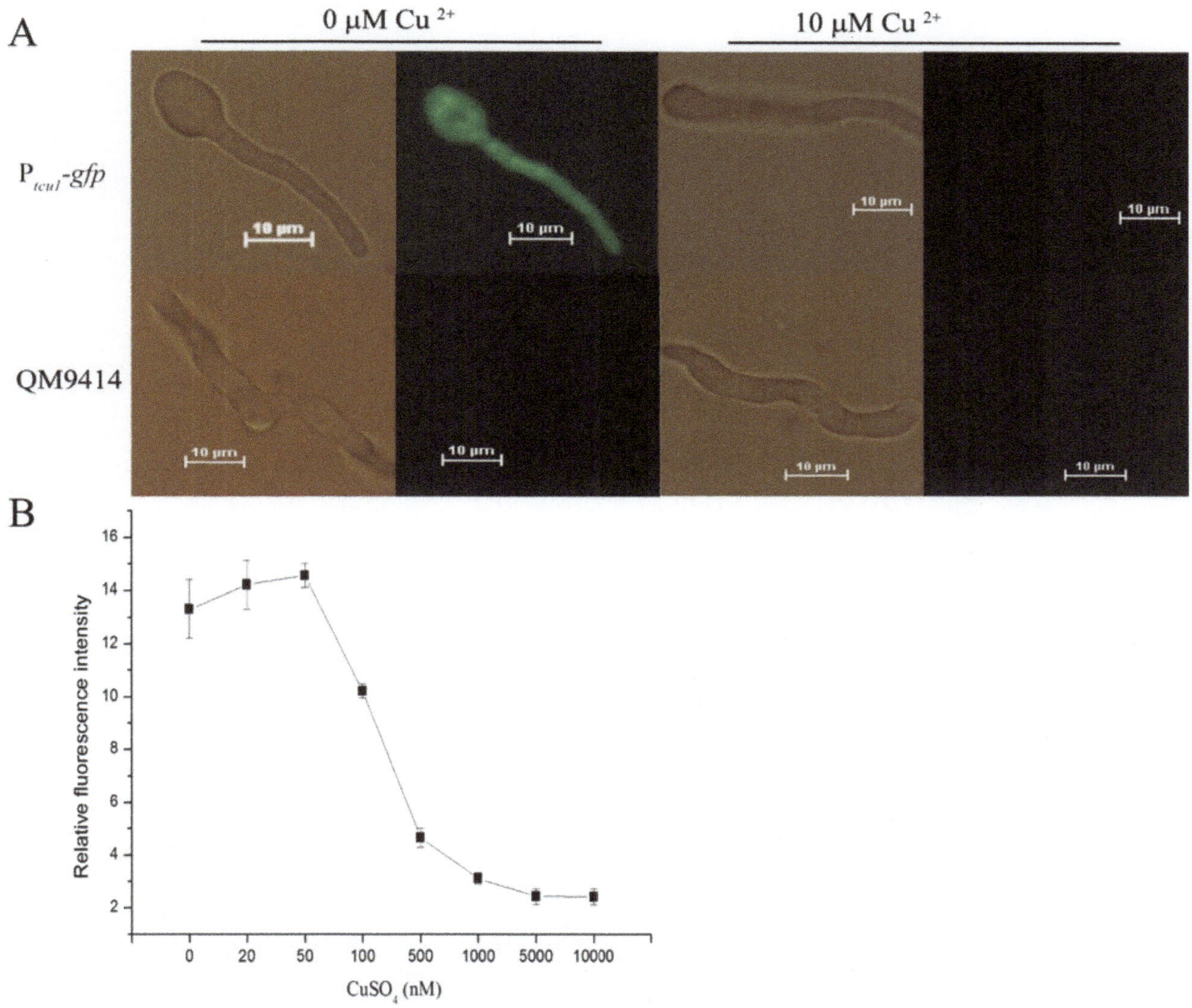

Figure 2 Fluorescent analysis of the recombinant P_{tcu1}-*gfp* strain. **(A)** Microscopic analysis of the GFP signal of the P_{tcu1}-*gfp* and QM9414 strains with and without $CuSO_4$ addition for 16 h. **(B)** Relative fluorescence intensity of the P_{tcu1}-*gfp* mycelia in response to the indicated concentrations of $CuSO_4$. Error bars are the SD from three biological replicates.

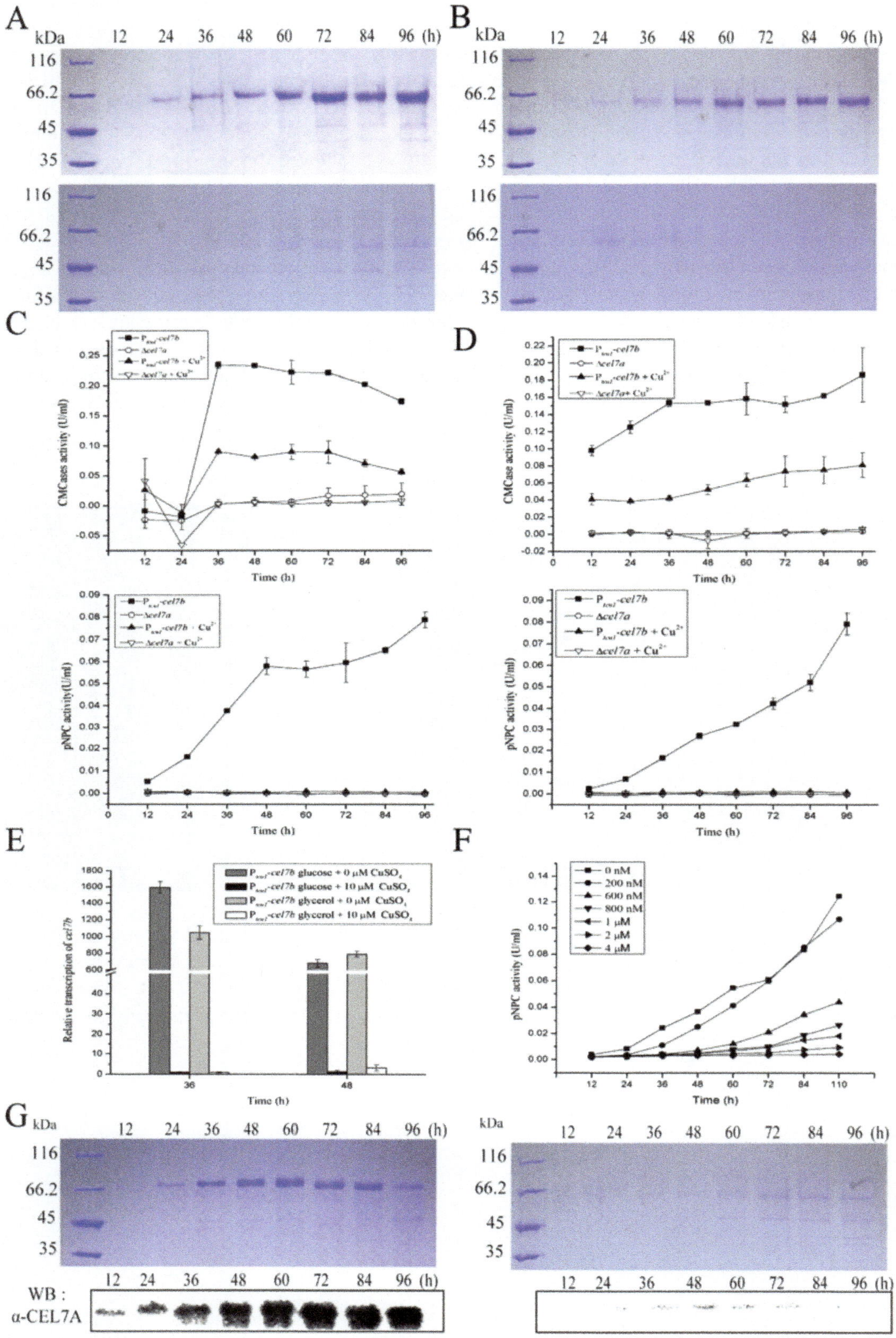

Figure 3 (See legend on next page.)

(See figure on previous page.)
Figure 3 Homologous expression of CEL7A and CEL7B mediated by P_{tcu1}. SDS-PAGE analysis of the extracellular proteins of P_{tcu1}-*cel7b* grown on 1% (wt/vol) glucose **(A)** or glycerol **(B)** with (lower panel) or without (upper panel) addition of CuSO4. The extracellular CMC and pNPC hydrolytic activities of P_{tcu1}-*cel7b* and its parental strain Δ*cel7a* grown on 1% (wt/vol) glucose **(C)** or glycerol **(D)**. Error bars are the SD from two biological replicates. **(E)** qRT-PCR analysis of the *cel7b mRNA* of P_{tcu1}-*cel7b* cultured with 1% (wt/vol) glucose or 1% (wt/vol) glycerol in the presence or absence of 10 μM $CuSO_4$. A significant difference ($P < 0.05$) existed for the transcription of *cel7b* with and without 10 μM Cu^{2+}. Error bars are the SD from three biological replicates. **(F)** The extracellular pNPC hydrolytic activities of P_{tcu1}-*cel7b* in the presence of different concentrations of copper sulfate. **(G)** SDS-PAGE and Western blot analysis of the extracellular production of CEL7A by P_{tcu1}-*cel7a* grown on 1% (wt/vol) glucose with (right panel) or without (left panel) addition of $CuSO_4$. Equal amount of culture supernatant relative to biomass was measured for all the assays.

Moreover, these proteins were absent from the culture supernatant of the respective transformants when 10 μM $CuSO_4$ was included, suggesting the expression was specifically responsive to the copper. Further examination of the *cel7b* transcripts by quantitative real-time PCR (qRT-PCR) revealed that, while the expression of the endogenous *cel7b* was below a detectable level, a significant induced transcription was observed for *cel7b* in the P_{tcu1}-*cel7b* strain in the absence of copper, which was dramatically decreased by 10 μM copper (Figure 3E). The produced CEL7A was verified by Western blot (Figure 3G). Analysis of the hydrolytic activities verified that the produced CEL7B displayed significant carboxymethylcellulose (CMC) and *p*-nitrophenol-D-cellbioside (pNPC) hydrolytic activities which were strictly dependent on copper availability (Figure 3C,D,F), whereas hardly any pNPC hydrolytic activity was detected from the culture supernatant of P_{tcu1}-*cel7a* (data not shown), indicating that CEL7A produced under this condition may not adopt a correct conformation. Thus, the *T. reesei tcu1* promoter can be used for copper-controlled expression of both homologous and heterologous proteins.

P_{tcu1}-mediated overexpression of XYR1 recapitulates fully induced expression of cellulases on non-inducing carbons

XYR1 has been established as the major regulator for the cellulolytic response to inducing substrate including Avicel or lactose in *T. reesei* [4,5,15]. Strongly increased basal expression of *xyr1* has thus been found to correlate with the hyperproduction of cellulases though that still entails the presence of inducers [16,17]. To probe the effect of copper-controlled expression of *xyr1* on cellulase production, we constructed a recombinant strain expressing *xyr1* under the control of P_{tcu1}. After transformation of the Δ*xyr1* strain, several independent transformants were isolated based on their recovered ability to form hydrolytic halos on Avicel plate (data not shown). One transformant with the largest halo (P_{tcu1}-*xyr1*) was chosen for further analysis. To first test whether the P_{tcu1}-driven expression of *xyr1* could recover the cellulolytic phenotype of the Δ*xyr1* strain, the proteins induced by 1% Avicel in the culture filtrate were resolved by sodium dodecyl sulphate-polyacrylamide gel electrophoresis (SDS-PAGE). Whereas cellulases were efficiently induced in the strain QM9414 regardless of the addition of copper or not (Figure 4A), comparable levels and patterns of induced cellulases were only observed for P_{tcu1}-*xyr1* without addition of copper (Figure 4B), indicating that the induced production of cellulases in the Δ*xyr1* strain was absolutely dependent on the copper-controlled expression of *xyr1*. Analysis of the extracellular pNPC and *p*-nitrophenyl-β-D-glucopyranoside (pNPG) hydrolytic activities demonstrated that, in accordance with the SDS-PAGE results, the hydrolytic activities as displayed by the culture supernatant of P_{tcu1}-*xyr1* without addition of $CuSO_4$ were overall comparable to those of QM9414 (Figure 4C). Hardly any hydrolytic activities were detected for P_{tcu1}-*xyr1* in the presence of 10 μM $CuSO_4$. Efficient production of cellulases by P_{tcu1}-*xyr1* was also observed upon incubation with lactose, and the detected hydrolytic activities of the culture filtrate were significantly higher than those of QM9414 under the same culture conditions after 72 h of induction (Figure 5).

To further ask to what extent overexpression of *xyr1* alone would effect on launching the production of cellulases under non-inducing conditions, we then determined the synthesis of cellulases by P_{tcu1}-*xyr1* on various other carbon sources. As shown in Figure 6A,B, the expression of *xyr1* in P_{tcu1}-*xyr1* resulted in a constitutive expression of cellulases on either glucose or glycerol, whereas no cellulases could be detected for QM9414 under these conditions. Consistent with these results, significant extracellular pNPC and pNPG hydrolytic activities were detected for P_{tcu1}-*xyr1* cultured without addition of copper, whereas hardly any such activities were observed when QM9414 was cultured under the same conditions (Figure 6A,B). Analysis of the hydrolytic activity toward filter paper revealed that significant activities were detected for cellulases produced by P_{tcu1}-*xyr1* on glycerol over time, which was almost equivalent to cellulases produced by QM9414 induced with Avicel at 48 h (Figure 6C). Again, the cellulase production was inhibited with increasing amount of $CuSO_4$ added to the culture, and BCS was able to partially recover the production after a complete inhibition was inflicted (Figure 6C,D). Altogether, the above results indicated that P_{tcu1}-*xyr1*-mediated expression of *xyr1* is capable of achieving the full production of cellulases without induction.

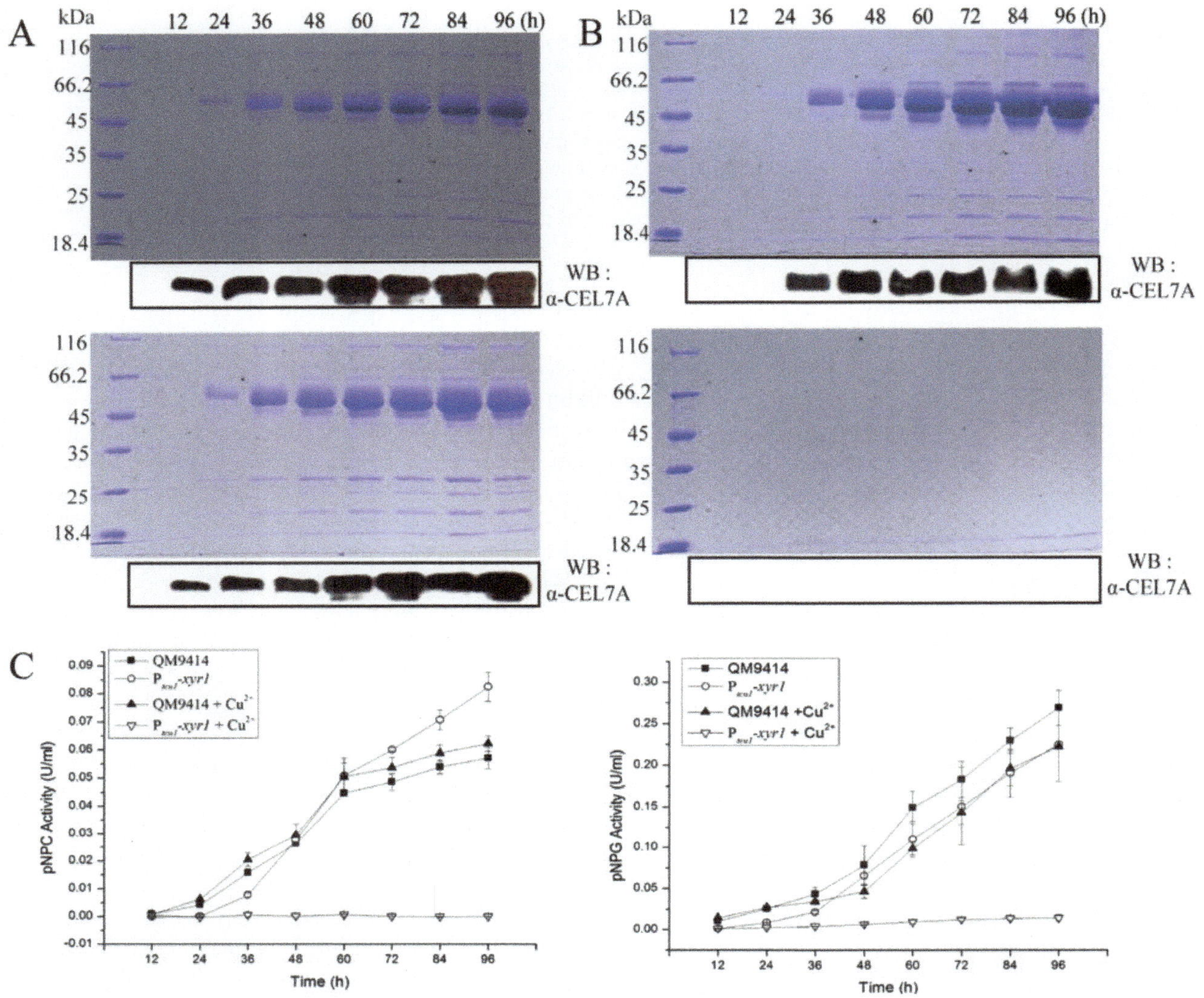

Figure 4 P_{tcu1}-*xyr1* recovered cellulase gene expression on induction by 1% Avicel. SDS-PAGE and Western blot analysis of the secreted proteins in the culture supernatant of QM9414 **(A)** and P_{tcu1}-*xyr1* **(B)** strains grown on 1% (wt/vol) Avicel with (lower panel) or without (upper panel) addition of $CuSO_4$. **(C)** pNPC and pNPG hydrolytic activities of the culture filtrates from QM9414 and P_{tcu1}-*xyr1* strains induced with 1% (wt/vol) Avicel and cultured in the presence or absence of $CuSO_4$. A significant difference ($P < 0.05$) was detected for the culture filtrates of P_{tcu1}-*xyr1* after 60 h with and without 10 μM CuSO4. Equal amount of culture supernatant relative to biomass was measured for all the assays. Error bars are the SD from at least two biological replicates.

To test whether the deregulated expression of cellulases in strain P_{tcu1}-*xyr1* was absolutely due to the overexpression of *xyr1*, the transcription of *xyr1* was analyzed on 1% glucose or glycerol. The results showed that, during cultivation on glucose or glycerol, the abundance of *xyr1* mRNA in P_{tcu1}-*xyr1* remained constantly high after 24 h, which was three- to fivefold higher than that in QM9414 cultured on Avicel (Figure 7A). The relative transcript abundance of *cel7a* in strain P_{tcu1}-*xyr1* was also found to display a high level that was largely comparable to that in QM9414 induced with Avicel though a slight reduction in the *cel7a* abundance was observed at 48 h of culture on glucose (Figure 7B).

Discussion

T. reesei is well known for its high capacity to secrete large amounts of lignocellulosic enzymes. Whereas tremendous efforts have been made to continuously improve the performance of the native cellulase system of *T. reesei* either by optimizing existing enzymatic activities or increasing the production efficiency of cellulases, research has been also directed toward developing it into a heterologous host for large-scale protein production. Although several gene expression systems have been successfully used in *T. reesei* including those based on the inducible *cel7a* promoter as well as the constitutive promoters from glucolytic genes [8-10,18], alternatives

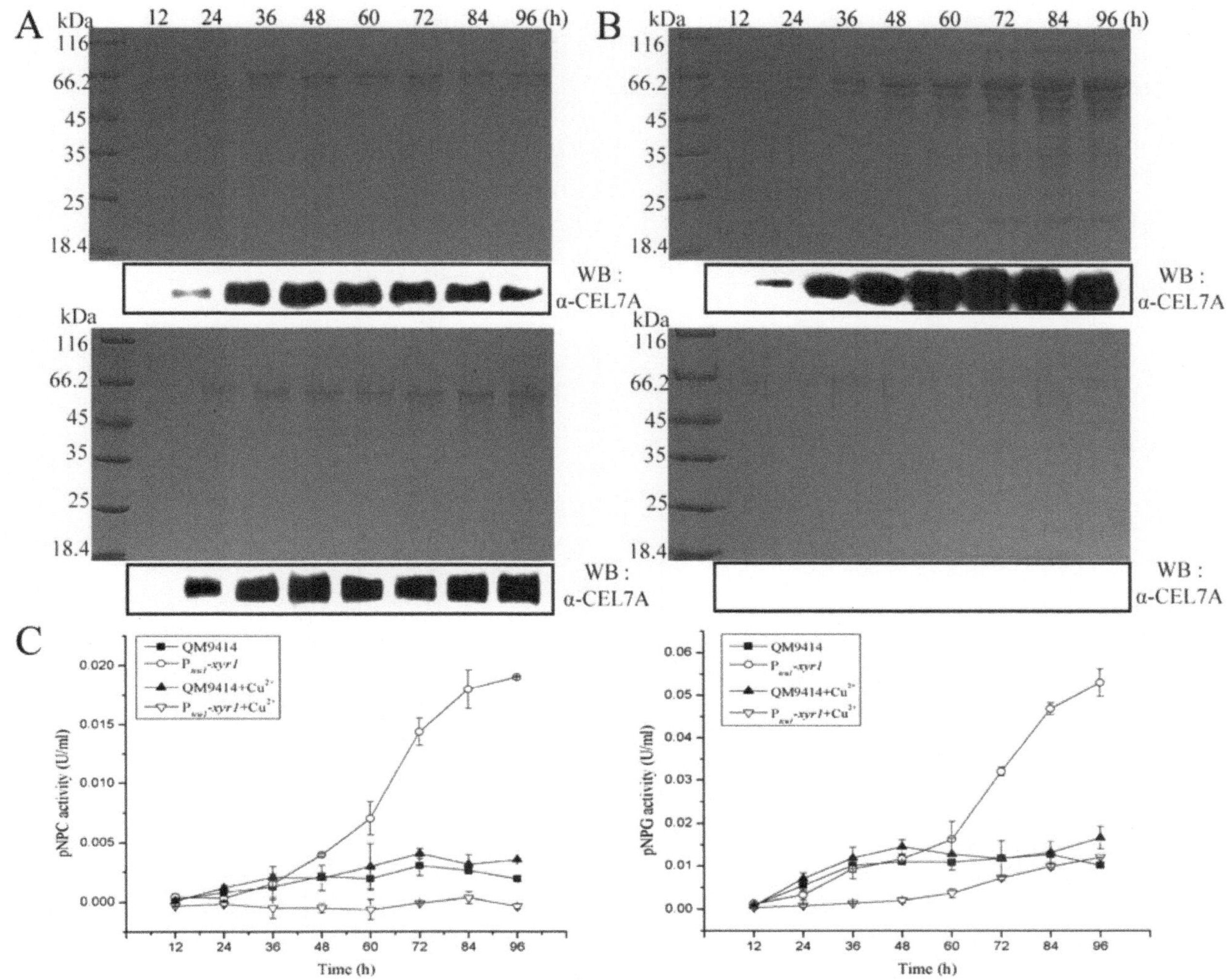

Figure 5 P_{tcu1}-*xyr1* recovered cellulase gene expression on induction by 1% lactose. **(A)** SDS-PAGE and Western blot analysis of the secreted proteins in the culture supernatant of QM9414 **(A)** and P_{tcu1}-*xyr1* **(B)** strains grown on 1% (wt/vol) lactose with (lower panel) or without (upper panel) addition of $CuSO_4$. **(C)** pNPC and pNPG hydrolytic activities of the culture filtrates from QM9414 and P_{tcu1}-*xyr1* strains induced with 1% (wt/vol) lactose and cultured in the presence or absence of $CuSO_4$. A significant difference ($P < 0.05$) was detected for the culture filtrates of P_{tcu1}-*xyr1* after 72 h without copper addition compared with those of QM9414. Error bars are the SD from at least two biological replicates.

to control gene expression by both activation and repression independent of the nutritional state of the culture are still desirable. In the present study, we developed a copper-controllable expression system based on the promoter of a putative copper transporter encoding gene. The system is highly responsive to environmental copper levels and can mediate the high-level expression of heterologous and homologous proteins. Importantly, the controllable expression does not seem to be influenced by the carbon sources used. Moreover, the growth and the cellulases induction are hardly affected no matter whether the exogenous addition of copper sulfate or not. Although we cannot exclude the possibility that the trace amount of copper in the medium even without exogenous addition still inflicts a repressive effect on P_{tcu1}, our observation that the expression of *tcu1* was not significantly decreased by exogenous addition of 200 nM of copper but was almost abolished by concentrations above 500 nM copper sulfate, suggested that there might be a threshold of copper required to shut off the expression of *tcu1* probably through inactivating a homologous transcription activator CUF1p [19,20]. In contrast with a previous report [21], copper chelator BCS only partially relieved the repression of *tcu1* expression compared with that without any exogenous addition of copper. This discrepancy may lie in the differences either in the permeability of BCS in *T. reesei* and *N. crassa* or in the copper binding affinities of CUF1p. Regardless of this, the successful expression of the native cellulases by P_{tcu1} under non-inducing conditions demonstrates that the developed system provides a useful alternative to overexpress selected proteins including cellulases or their variants with relatively high purity and free of other cellulase contaminants. It

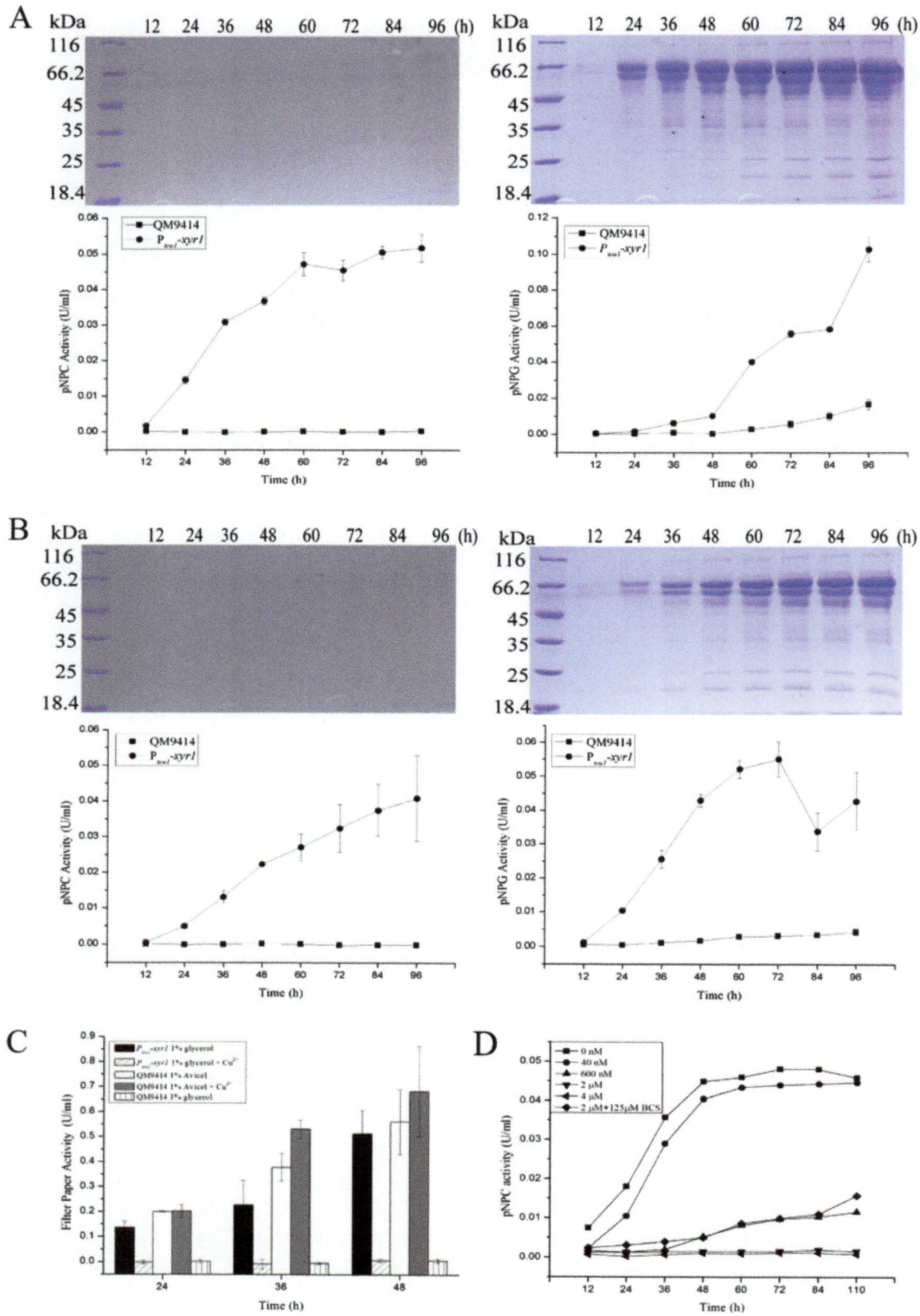

Figure 6 (See legend on next page.)

(See figure on previous page.)

Figure 6 P_{tcu1}-*xyr1* displayed a full production of cellulases on non-inducing carbon sources. **(A)** SDS-PAGE analysis and pNPC hydrolytic activities of the extracellular proteins from QM9414 (left panel) and P_{tcu1}-*xyr1* (right panel) strains grown on 1% (wt/vol) glucose without exogenous addition of $CuSO_4$. **(B)** The same as in **(A)**, but the strains were grown on 1% (wt/vol) glycerol. **(C)** Filter paper hydrolytic activity of the cellulases produced by P_{tcu1}-*xyr1* on 1% (wt/vol) glycerol or those produced by QM9414 strain on 1% (wt/vol) Avicel or glycerol cultured with or without $CuSO_4$. A significant difference ($P < 0.05$) was detected between the produced cellulases of P_{tcu1}-*xyr1* on 1% (wt/vol) glycerol without $CuSO_4$ and those of QM9414 on 1% (wt/vol) glycerol. **(D)** The extracellular pNPC hydrolytic activity of P_{tcu1}-*xyr1* cultured on 1% (wt/vol) glycerol in the presence of the indicated concentrations of $CuSO_4$. Equal amount of culture supernatant relative to biomass was measured for all the assays. Error bars are the SD from at least two biological replicates.

should be noted though that CEL7A expressed under these conditions is almost inactive, suggesting that a proper folding into the active conformation may not be achieved. The precise mechanism warrants further study. As reported previously, the stringent control of gene expression as displayed by P_{tcu1} also provides a powerful tool to elucidate gene function in *T. reesei* [11].

Cellulase production in *T. reesei* is a highly coordinated process regulated by a suite of transcription factors with XYR1 being considered the main activator of cellulase/hemicellulase gene expression [4,5]. Regulation of *xyr1* expression has thus been proposed to have a significant impact on the production of various polysaccharide hydrolytic enzymes. Transcription of *xyr1* is kept at a very low level on D-glucose, whereas its expression is strongly upregulated under inducing conditions [7]. Consistent with the role of XYR1, it has been observed that the constitutive expression of *xyr1* results in a changed hydrolase expression profile and elevated extracellular hydrolytic enzyme activities [5,6,16]. The strongly elevated basal expression of *xyr1* has been also observed in a hyperproducer strain even before induction [7]. Moreover, it has reported that a single point mutation of XYR1, A824V, results in a highly elevated basal level of cellulase expression [22]. Further supporting the above assumption, we found that ectopic expression of *xyr1* driven by P_{tcu1} led to strongly increased production of cellulases on non-inducing carbon sources to a level comparable to that achieved under inducing conditions. Although the possibility cannot be excluded that other genetic traits may have been altered and contributed to the observed phenotype due to the random integrations of the *xyr1* cassette, the success to achieve the efficient cellulase production in the recombinant strain seems to rely on the overexpression of *xyr1* since the observed cellulase production is absolutely responsive to exogenous copper levels. Our results are also in accordance with a recent report that a strong increase in *xyr1* basal expression levels results in very high levels of CMCase activity during growth on glucose although no direct evidence regarding their relationship has been presented [23]. Whereas the exact mechanism of how the overexpressed XYR1 results in the deregulated production of cellulases warrants further study to provide a better understanding of cellulase formation, genetic engineering on the basis of the present strategy targeted other transcription factors holds great potential of further improving cellulase production in a more straightforward way in *T. reesei.*

Conclusion

We have developed a protein expression system based on the promoter of a copper transporter (P_{tcu1}) in *T. reesei*, the usefulness of which was illustrated by the high-level expression of specific cellulases and GFP when cultivated on D-glucose or glycerol as the sole carbon source. Meanwhile, full production of cellulases was achieved by P_{tcu1}-mediated expression of XYR1 even under the non-inducing conditions, which allowed us to better understand the mechanism of cellulase production and further improve the production of cellulases by *T. reesei*.

Methods

Strains, medium, and culture conditions

T. reesei QM9414 (ATCC 26921), a cellulase-enhanced mutant derived from the original strain QM6a, was used as a control strain for analysis of cellulase production. Strain Δ*xyr1* derived from QM9414 was kindly provided by Dr. Robert L. Mach [24]. The Δ*cel7a* strain was constructed as previously described [25]. Strains were grown in 1-l Erlenmeyer flasks on a rotary shaker (250 rpm) at 30° C in the medium described by Mandels and Andreotti with respective carbon sources added at a final concentration of 10 g/l [26]. *Escherichia coli* DH5α was used for routine plasmid construction and amplification.

For transcription and cellulase production analysis, *T. reesei* was precultured on glycerol for 48 h and grown for another 12 h in the same fresh medium. Mycelia were harvested by filtration and washed twice with medium with no carbon source. Equal amounts of mycelia were transferred to a fresh medium containing 10 g/l of corresponding carbon sources without peptone, and the incubation was continued for the indicated time periods.

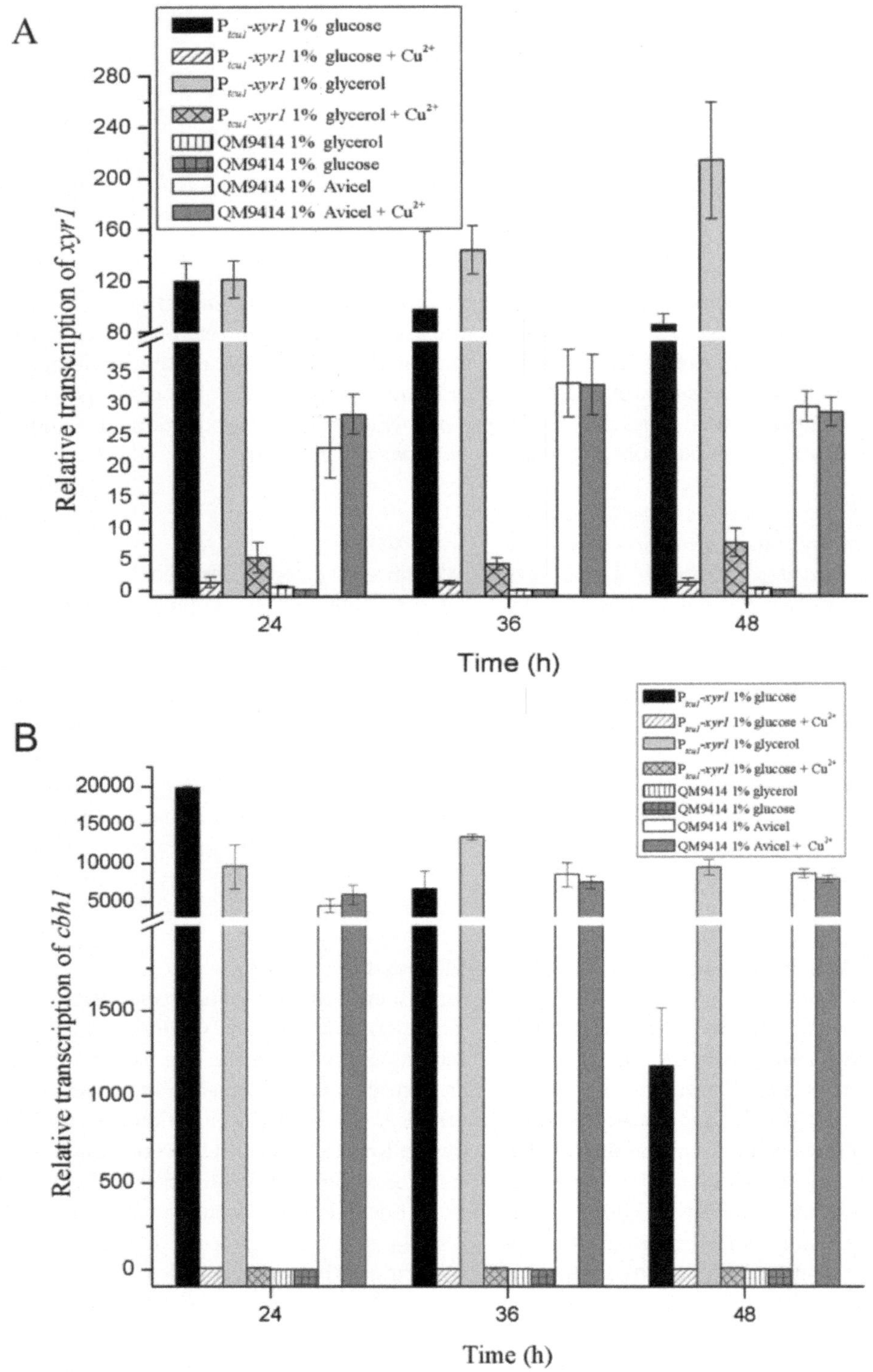

Figure 7 qRT-PCR analysis of the transcription of *xyr1* and *cel7a* in strains P_{tcu1}-*xyr1* and QM9414. **(A)** Transcript abundance of *xyr1 of* P_{tcu1}-*xyr1* on non-inducing carbon sources compared with those of QM9414 on Avicel in the presence or absence of 10 μM $CuSO_4$. **(B)** Levels of *cel7a* mRNA in P_{tcu1}-*xyr1* on non-inducing carbon sources compared with those in QM9414 on Avicel with or without addition of 10 μM $CuSO_4$. Significant differences ($P < 0.05$) were detected for the transcription of *xyr1* between P_{tcu1}-*xyr1* on glucose or glycerol without $CuSO_4$ and QM9414 on Avicel at all three indicated time points. For the transcription of *cel7a*, a significant difference ($P < 0.001$) was detected between P_{tcu1}-*xyr1* at 48 h on glucose and those in QM9414 on Avicel. Error bars are the SD from three biological replicates.

Plasmid construction

The 1715-bp fragment comprising the promoter of the *tcu1* gene (jgi:Trire2:52315) was amplified from the genomic DNA of strain QM9414 and subsequently ligated into pMD19-T to generate pMDP*tcu1*. After digestion with *Not*I and *Hin*dIII, the *trpC* terminator was inserted into $pMDP_{tcu1}$ to obtain $pMDP_{tcu1}$-T_{trpC}. The 2.7-kb *pyr4* gene fragment was amplified from pFG1 [27] and digested with *Hin*dIII and *Sal*I, and was incorporated into $pMDP_{tcu1}$-T_{trpC} to obtain $pMDP_{tcu1}$-T_{trpC}-*pyr4*.

To construct the plasmids for respective expression of *cel7a* and *cel7b* driven by the promoter of *tcu1*, the coding sequences of *cel7a* and *cel7b* were amplified from the genomic DNA of QM9414, respectively, digested with *Asc*I and *Spe*I, and then ligated into $pMDP_{tcu1}$-T_{trpC}-*pyr4* to obtain $pMDP_{tcu1}$-*cel7a*-T_{trpC}-*pyr4* and $pMDP_{tcu1}$-*cel7b*-T_{trpC}-*pyr4*. To construct the expression plasmid for *gfp*, the coding sequence of *gfp* was amplified from pIG1783 [28], digested with *Pme*I and *Spe*I, and subsequently ligated into $pMDP_{tcu1}$-T_{trpC} to obtain $pMDP_{tcu1}$-*gfp*-T_{trpC}. Similarly, the coding sequence of *xyr1* was amplified with cDNA of QM9414 as the template, digested with *Asc*I and *Spe*I, and then incorporated into $pMDP_{tcu1}$-T_{trpC} to generate the expression plasmid for *xyr1*, $pMDP_{tcu1}$-*xyr1*-T_{trpC}.

Construction of recombinant *T. reesei* strains

Fungal transformation was performed as described by Penttila *et al.* [29]. Constructed plasmids of $pMDP_{tcu1}$-*cel7a*-T_{trpC}-*pyr4* and $pMDP_{tcu1}$-*cel7b*-T_{trpC}-*pyr4* were transformed into Δ*cel7a* strain to obtain recombinant strains of P_{tcu1}-*cel7a* and P_{tcu1}-*cel7b*, respectively. The plasmid of $pMDP_{tcu1}$-*gfp-TtrpC* was co-transformed into *T. reesei* QM9414 with pRLMex-30 to obtain the recombinant strain of P_{tcu1}-*gfp*. Similarly, $pMDP_{tcu1}$-*xyr1*-T_{trpC} was co-transformed with pRLMex-30 into strain Δ*xyr1* to obtain the recombinant strain of P_{tcu1}-*xyr1*. Transformants of P_{tcu1}-*cel7a* and P_{tcu1}-*cel7b* were selected on minimal medium for uridine prototroph, and transformants of P_{tcu1}-*gfp* and P_{tcu1}-*xyr1* were selected for resistance to 100 μg/ml of hygromycin. All the recombinant strains were verified for the presence of the entire expression cassette by PCR.

Fluorometric analysis

The fluorescence of the mycelium of P_{tcu1}-*gfp* was detected with a Nikon Eclipse 80i fluorescence microscope (Nikon, Melville, NY, USA), and images were captured and processed with the NIS-ELEMENTS AR software program. For fluorescence quantification of GFP expression driven by the promoter of *tcu1* in the presence of various concentrations of copper in liquid culture, strain P_{tcu1}-*gfp* was precultured on Mandels-Andreotti medium with glycerol, transferred into Mandels-Andreotti medium with 1% glucose plus 0 to 10 μM $CuSO_4$, and harvested for 60 h. After being washed with 0.9% NaCl, the mycelia were suspended 50 mM HEPES buffer at pH 7.4 and grinded with micro glass beads by Precellys homogenizers (Bertin Technologies, Montigny-le-Bretonneux, France). Cell debris was removed by centrifugation at 14,000 × *g* for 10 min, and 200 μl of the supernatant was used to determine the GFP fluorescent intensity with a 96-well spectrofluorometer at an excitation wavelength of 485 nm and an emission wavelength of 535 nm. The protein concentration in the supernatant was determined using the method of the Bradford protein assay with bovine serum albumin (BSA) as the standard, and the relative fluorescent intensity was calculated according to an equation as follows:

The relative fluorescent intensity = [measured fluorescent intensity/protein concentration]/[background fluorescent intensity of QM9414/background protein concentration of QM9414].

Quantitative RT-PCR

Total RNAs were extracted using TRIzol reagent (Invitrogen, Grand Island, NY, USA) and purified using the TURBO DNA-free kit (Ambion, Austin, TX, USA) to remove DNA according to the manufacturer's instructions. Reverse transcription was carried out using the PrimeScript RT reagent Kit (Takara, Tokyo, Japan) according to the instructions. Quantitative PCR was performed on a Bio-Rad myIQ2 thermocycler (Bio-Rad, Richmond, CA, USA). Amplification reactions were performed using the SYBR Green Supermix (Takara, Tokyo, Japan) according to the instructions. Data analysis was performed using the relative quantitation/comparative CT (ΔΔCT) [30]. The endogenous *actin* gene was used as a normalized control gene. For analysis of *tcu1* transcription, *tcu1* mRNA level under the most repressed conditions was set to 1 and other points showed the fold of difference relative to these most repressed conditions. And for cellulase gene expression, mRNA level in uninduced conditions (glucose or glycerol) was set to 1 and other time points showed the fold of difference relative to this uninduced sample. At least three biological replicates were carried out for each experiment.

Enzyme activity measurements

Cellobiohydrolase and β-glucosidase activities were determined with 2 mM *p*-nitrophenol-D-cellbioside (Sigma-Aldrich, St. Louis, MO, USA) plus 0.1% δ-gluconolactone and 5 mM *p*-nitrophenyl-β-D-glucopyranoside (Sigma-Aldrich, St. Louis, MO, USA) as substrates at 45°C, respectively. The assays were carried out in 200 μl of reaction mixture containing 50 μl of culture supernatant and 50 μl of the respective substrate plus 100 μl of 50 mM sodium acetate buffer (pH 4.8) with incubation at 45°C for 30 min. The assay for endoglucanase activity measurement was carried out in 100 μl of reaction mixture containing

50 μl of culture supernatant and 50 μl of 0.5% CMC in 50 mM sodium acetate buffer (pH 4.8) and incubated at 50°C for 30 min. The total cellulase activity assay was carried out in 200 μl of reaction mixture including 50 μl of culture supernatant and 150 μl 50 mM sodium acetate buffer (pH 4.8) with a Whatman No. 1 filter paper strip as the substrate and incubated at 50°C for 1 h. The reducing sugars were determined spectrophotometrically by using 3,5-dinitrosalicylic (DNS) acid method with glucose as standards. At least two biological replicates were carried out for each experiment.

Protein analysis

SDS-PAGE and Western blotting were performed according to standard protocols [31]. The cellobiohydrolase CEL7A was detected by immunoblotting using a polyclonal antibody raised against amino acids (426 to 446) of CEL7A as previously described [32].

Abbreviations

BCS: bathocuproinedisulfonic acid; BSA: bovine serum albumin; CCR: carbon catabolite repression; CMC: carboxymethylcellulose; FPA: filter paper activity; GFP: green fluorescent protein; pNPC: *p*-nitrophenol-D-cellbioside; pNPG: *p*-nitrophenyl-β-D-glucopyranoside; P_{tcu1}: promoter of a copper transporter encoding gene *tcu1* in *T. reesei*; XYR1: xylanase regulator 1.

Competing interests

The authors declare that they have no competing interests.

Authors' contributions

XXL and WFL designed the experiments, and XXL, FLZ and CYL performed the experiments. XXL, FLZ, GJC, and WFL analyzed the results. XXL, WXZ, and WFL wrote the manuscript which was reviewed and approved by all authors.

Acknowledgements

This work is supported by grants from the National Natural Science Foundation of China (31470162, 31270116 and 31300059). Zhang WX would also like to acknowledge the support from the Scientific Research Foundation for Excellent Young and Middle-Aged Scientists of Shandong Province of China (BS2013NJ021) and China Postdoctoral Science Foundation (2012 M521325). We thank Dr. Robert L. Mach for kindly providing the Δ*xyr1* strain.

References

1. Lynd LR, Weimer PJ, van Zyl WH, Pretorius IS. Microbial cellulose utilization: fundamentals and biotechnology. Microbiol Mol Biol Rev. 2002;66:506–77.
2. Peterson R, Nevalainen H. *Trichoderma reesei* RUT-C30--thirty years of strain improvement. Microbiology (Reading, England). 2012;158:58–68.
3. Kubicek CP, Mikus M, Schuster A, Schmoll M, Seiboth B. Metabolic engineering strategies for the improvement of cellulase production by *Hypocrea jecorina*. Biotechnol Biofuels. 2009;2:19.
4. Stricker AR, Grosstessner-Hain K, Wurleitner E, Mach RL. Xyr1 (xylanase regulator 1) regulates both the hydrolytic enzyme system and D-xylose metabolism in *Hypocrea jecorina*. Eukaryotic cell. 2006;5:2128–37.
5. Mach-Aigner AR, Pucher ME, Steiger MG, Bauer GE, Preis SJ, Mach RL. Transcriptional regulation of *xyr1*, encoding the main regulator of the xylanolytic and cellulolytic enzyme system in *Hypocrea jecorina*. Appl Environ Microbiol. 2008;74:6554–62.
6. Seiboth B, Karimi RA, Phatale PA, Linke R, Hartl L, Sauer DG, et al. The putative protein methyltransferase LAE1 controls cellulase gene expression in *Trichoderma reesei*. Mol Microbiol. 2012;84:1150–64.
7. Portnoy T, Margeot A, Seidl-Seiboth V, Le Crom S, Ben Chaabane F, Linke R, et al. Differential regulation of the cellulase transcription factors XYR1, ACE2, and ACE1 in *Trichoderma reesei* strains producing high and low levels of cellulase. Eukaryotic cell. 2011;10:262–71.
8. Nevalainen KM, Te'o VS, Bergquist PL. Heterologous protein expression in filamentous fungi. Trends Biotechnol. 2005;23:468–74.
9. Singh A, Taylor LE, 2nd, Vander Wall TA, Linger J, Himmel ME, Podkaminer K, et al. Heterologous protein expression in Hypocrea jecorina: a historical perspective and new developments. Biotechnology advances. 2014:doi:10.1016/j.biotechadv.2014.1011.1009.
10. Li JX, Wang J, Wang SW, Xing M, Yu SW, Liu G. Achieving efficient protein expression in *Trichoderma reesei* by using strong constitutive promoters. Microb Cell Fact. 2012;11:84.
11. Lamb TM, Vickery J, Bell-Pedersen D. Regulation of gene expression in Neurospora crassa with a copper responsive promoter. G3-Genes Genom Genet. 2013;3:2273–80.
12. Ory JJ, Griffith CL, Doering TL. An efficiently regulated promoter system for Cryptococcus neoformans utilizing the CTR4 promoter. Yeast (Chichester, England). 2004;21:919–26.
13. De Freitas J, Wintz H, Kim JH, Poynton H, Fox T, Vulpe C. Yeast, a model organism for iron and copper metabolism studies. Biometals. 2003;16:185–97.
14. Thiele DJ. Integrating trace element metabolism from the cell to the whole organism. J Nutr. 2003;133:1579s–80s.
15. Stricker AR, Steiger MG, Mach RL. Xyr1 receives the lactose induction signal and regulates lactose metabolism in *Hypocrea jecorina*. FEBS letters. 2007;581:3915–20.
16. Pucher ME, Steiger MG, Mach RL, Mach-Aigner AR. A modified expression of the major hydrolase activator in *Hypocrea jecorina* (*Trichoderma reesei*) changes enzymatic catalysis of biopolymer degradation. Catal Today. 2011;167:122–8.
17. Wang SW, Liu G, Wang J, Yu JT, Huang BQ, Xing M. Enhancing cellulase production in *Trichoderma reesei* RUT C30 through combined manipulation of activating and repressing genes. J Ind Microbiol Biot. 2013;40:633–41.
18. Uzbas F, Sezerman U, Hartl L, Kubicek CP, Seiboth B. A homologous production system for *Trichoderma reesei* secreted proteins in a cellulase-free background. Appl Microbiol Biot. 2012;93:1601–8.
19. Graden JA, Winge DR. Copper-mediated repression of the activation domain in the yeast Mac1p transcription factor. Proc Natl Acad Sci U S A. 1997;94:5550–5.
20. Serpe M, Joshi A, Kosman DJ. Structure-function analysis of the protein-binding domains of Mac1p, a copper-dependent transcriptional activator of copper uptake in *Saccharomyces cerevisiae*. J Biol Chem. 1999;274:29211–9.
21. Jensen LT, Posewitz MC, Srinivasan C, Winge DR. Mapping of the DNA binding domain of the copper-responsive transcription factor Mac1 from *Saccharomyces cerevisiae*. J Biol Chem. 1998;273:23805–11.
22. Derntl C, Gudynaite-Savitch L, Calixte S, White T, Mach RL, Mach-Aigner AR. Mutation of the xylanase regulator 1 causes a glucose blind hydrolase expressing phenotype in industrially used *Trichoderma* strains. Biotechnol Biofuels. 2013;6:62.
23. Gupta VK, Schmoll M, Herrera-Estrella A, Upadhyay RS, Druzhinina I, Tuohy M. Biotechnology and biology of trichoderma. In: Shaowen Wang, Gang Liu. Applications of RNA interference for enhanced cellulase production in trichoderma. Elsevier Science; 2014;201-226.
24. Rauscher R, Wurleitner E, Wacenovsky C, Aro N, Stricker AR, Zeilinger S, et al. Transcriptional regulation of *xyn1*, encoding xylanase I, in *Hypocrea jecorina*. Eukaryotic cell. 2006;5:447–56.
25. Qi FF, Zhang WX, Zhang FJ, Chen GJ, Liu WF. Deciphering the effect of the different N-glycosylation sites on the secretion, activity, and stability of cellobiohydrolase I from *Trichoderma reesei*. Appl Environ Microbiol. 2014;80:3962–71.
26. Mandels M, Andreotti R. Problems and challenges in the cellulose to cellulase fermentation. Proc Biochem. 1978;13:6–13.
27. Gruber F, Visser J, Kubicek C, de Graaff L. Cloning of the *Trichoderma reesei pyrG* gene and its use as a homologous marker for a high-frequency transformation system. Curr Genet. 1990;18:447–51.
28. Poggeler S, Masloff S, Hoff B, Mayrhofer S, Kuck U. Versatile EGFP reporter plasmids for cellular localization of recombinant gene products in filamentous fungi. Curr Genet. 2003;43:54–61.
29. Penttila M, Nevalainen H, Ratto M, Salminen E, Knowles J. A versatile transformation system for the cellulolytic filamentous fungus *Trichoderma reesei*. Gene. 1987;61:155–64.
30. Schmittgen TD, Livak KJ. Analyzing real-time PCR data by the comparative CT method. Nat Protocols. 2008;3:1101–8.

31. Chong L. Molecular cloning - a laboratory manual, 3rd edition. Science. 2001;292:446–6.
32. Zhou QX, Xu JT, Kou YB, Lv XX, Zhang X, Zhao GL, et al. Differential involvement of beta-Glucosidases from *Hypocrea jecorina* in rapid induction of cellulase genes by cellulose and cellobiose. Eukaryotic cell. 2012;11:1371–81.

Cellobiohydrolase and endoglucanase respond differently to surfactants during the hydrolysis of cellulose

Chia-wen C Hsieh[1*], David Cannella[1], Henning Jørgensen[2], Claus Felby[1] and Lisbeth G Thygesen[1]

Abstract

Background: Non-ionic surfactants such as polyethylene glycol (PEG) can increase the glucose yield obtained from enzymatic saccharification of lignocellulosic substrates. Various explanations behind this effect include the ability of PEG to increase the stability of the cellulases, decrease non-productive cellulase adsorption to the substrate, and increase the desorption of enzymes from the substrate. Here, using lignin-free model substrates, we propose that PEG also alters the solvent properties, for example, water, leading the cellulases to increase hydrolysis yields.

Results: The effect of PEG differs for the individual cellulases. During hydrolysis of Avicel and PASC with a processive monocomponent exo-cellulase cellobiohydrolase (CBH) I, the presence of PEG leads to an increase in the final glucose concentration, while PEG caused no change in glucose production with a non-processive endoglucanase (EG). Also, no effect of PEG was seen on the activity of β-glucosidases. While PEG has a small effect on the thermostability of both cellulases, only the activity of CBH I increases with PEG. Using commercial enzyme mixtures, the hydrolysis yields increased with the addition of PEG. In parallel, we observed that the relaxation time of the hydrolysis liquid phase, as measured by LF-NMR, directly correlated with the final glucose yield. PEG was able to boost the glucose production even in highly concentrated solutions of up to 150 g/L of glucose.

Conclusions: The hydrolysis boosting effect of PEG appears to be specific for CBH I. The mechanism could be due to an increase in the apparent activity of the enzyme on the substrate surface. The addition of PEG increases the relaxation time of the liquid-phase water, which from the data presented points towards a mechanism related to PEG-water interactions rather than PEG-protein or PEG-substrate interactions.

Keywords: PEG, Surfactants, Enzymatic saccharification of cellulose, Monocomponent cellulase hydrolysis, Avicel hydrolysis, PASC hydrolysis, Water constraint

Background

Surfactants have long been added to the hydrolysis step within bioconversion to improve process yields [1-3]. The two most widely used surfactants are polyethylene glycol (PEG) [4,5] and Tween [2,6,7]. When present in concentrations of 0.5 to 5 wt% biomass, these surfactants can increase hydrolysis yields for a given enzyme loading. For example, the addition of PEG improves wheat straw hydrolysis conversion rates by up to 45% [8]. For steam-pretreated spruce, another lignocellulosic substrate, the hydrolysis yield increased 20% with PEG [9]. Experimental evidence has confirmed that PEG increases lignocellulose hydrolysis yields not only via adsorption to lignin to prevent unproductive binding of cellulases, but also by interacting with cellulases to increase their thermal stability [4,5,10]. Other observed mechanisms have been decreased precipitation [11] and subsequent inactivation of the enzymes [12] and by changing the adsorption parameters of cellulases to facilitate enzyme desorption and reduce enzyme loss through irreversible binding [13-15]. Of these mechanisms, the interaction with lignin seems to be the most important from a practical point of view.

Comparing Avicel, corn stover, and bagasse, PEG improved hydrolysis yield mostly for lignin-containing

* Correspondence: chsi@ign.ku.dk
[1]Department of Geosciences and Natural Resource Management, Faculty of Science, University of Copenhagen, Rolighedsvej 23, DK-1958 Frederiksberg C, Denmark
Full list of author information is available at the end of the article

substrates [16]. While there is no doubt that PEG improves hydrolysis of lignin-containing substrates, it has also been shown to boost the hydrolysis of a pure cellulosic substrate, Avicel [9,14,16,17]. Therefore, simple homogeneous model systems consisting of Avicel and PEG can be used as a tool to study the sensitivity of different enzymes to non-substrate-related factors affecting the rate of hydrolysis. In this study, we use such model systems to look into whether individual cellulases are differently affected by addition of PEG and whether the properties of the hydrolysis liquid phase play a role in this effect.

The aim of this work is to present new results on cellulose hydrolysis in the presence of PEG, which we tested using two different monocomponent enzymes, cellobiohydrolase (CBH) I and endoglucanase (EG), as well as two different commercial enzyme preparations. We investigate whether PEG introduces hydrophobic interactions that 'free up' water so that it is more available for hydrolysis. Low-field nuclear magnetic resonance (LF-NMR) relaxometry can be used to measure the strength of water-water interactions, and in this paper, we apply LF-NMR to monitor water constraint in the presence or absence of PEG. It has previously been shown that the hydrolysis liquid phase affects hydrolytic enzyme activity [18,19], and in this work, we study whether the effect of PEG on hydrolysis of a pure cellulosic substrate can be linked to a change in water constraint during saccharification.

Results and discussion

We determined the impact of PEG addition to Avicel hydrolysis using both monocomponent enzymes and commercially available cellulase mixtures. Monocomponent enzyme hydrolysis indicates which cellulases are most affected by the presence of PEG. We also measured the spin-spin (T_2) relaxation times of the hydrolysis liquid phase at the onset of hydrolysis using LF-NMR for a series of experiments designed to change the water constraint via the presence of PEG and monosaccharides.

Effect of PEG on different cellulases

Previous studies in the literature on the effect of PEG on lignocellulosic biomass used commercially available cellulase preparations [5,9]. The major components of commercial cellulase mixtures are cellobiohydrolases (Cel7A, also known as CBH I) and endoglucanases (EG) [20]. In this study, a monocomponent CBH I and an endoglucanase were tested on Avicel or phosphoric acid swollen cellulose (PASC) in the presence of PEG as shown in Figure 1A,B. The results in Figure 1A show that PEG improved Avicel hydrolysis with CBH I by 45%, while the increase in hydrolysis yield for EG

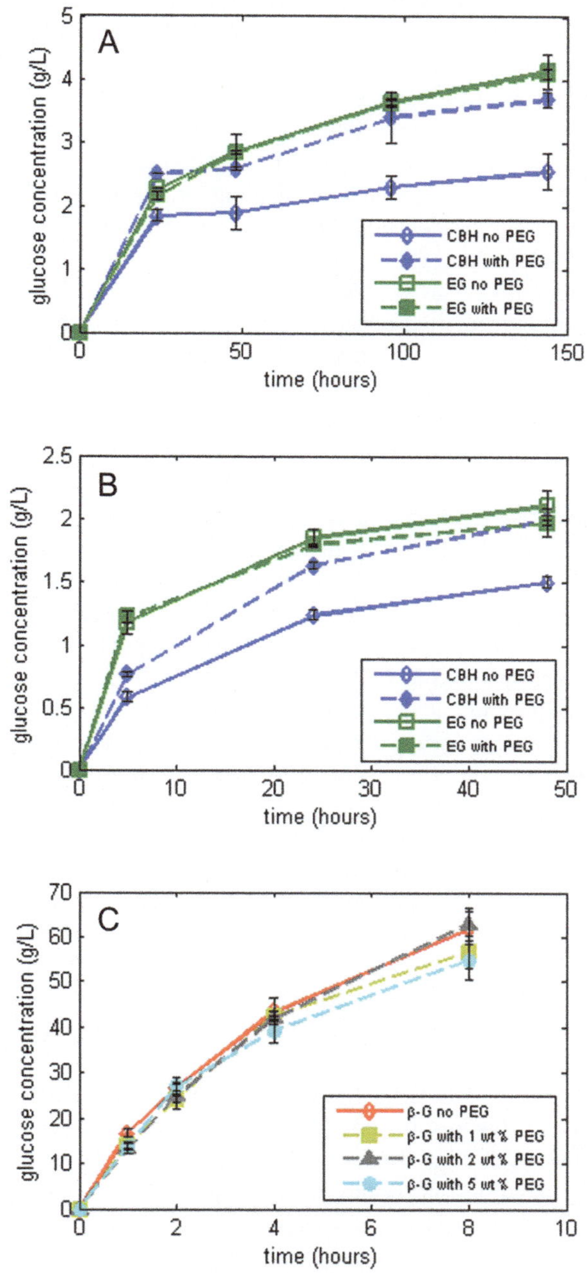

Figure 1 Effect of PEG on the glucose yields obtained using different substrates. (A) Hydrolysis of 5% Avicel using monocomponent CBH I and EG with β-glucosidase supplementation in the absence (open symbols) and presence (solid symbols) of PEG 3000, **(B)** hydrolysis of 3% PASC using the same monocomponent enzymes with β-glucosidase supplementation, and **(C)** Hydrolysis of 100-g/L cellobiose using Cellic CTec2 with different concentrations of PEG 3000. CBH, cellobiohydrolase; EG, endoglucanase; PEG, polyethylene glycol.

was only 1%. Figure 1B shows that qualitatively similar results were obtained for PASC, confirming that the difference seen is not substrate specific. That is, as a similar difference between hydrolysis with and without PEG is seen for a highly accessible substrate like PASC, the effect of PEG is not to increase substrate accessibility of Avicel for CBH I. While this effect is observed for one specific CBH I and EG obtained from Megazyme (see the 'Materials and methods' section), it nevertheless sheds some light on the impact of PEG on cellulases having different hydrolytic mechanisms.

In Figure 1A, at an enzyme loading of 1.2 mg protein/g Avicel, the reaction rate without and with PEG was 77 and 105 mg/L/h, respectively, from time 0 to 24 h, amounting to a 36% difference for CBH I. Towards the end of hydrolysis, that is, from 96 to 144 h, the rate of glucose production without and with PEG is calculated to be 5.2 and 6 mg/L/h, respectively, corresponding to a 15% difference. The main difference, therefore, is observed within the first hours of hydrolysis where PEG plays a larger role in the kinetics of the reaction. We speculate that PEG either increases the activity (k_{cat}) [21,22] of the enzyme or decreases the k_{off} of the CBH I [23-26]. We have also tested different CBH I dosages and observed that the increase in activity was due to the presence of PEG regardless of the dosage of enzyme (data not shown).

The experimental setup shown in Figure 1A,B contained an excess dosage of β-glucosidase in order to avoid end-product inhibition from cellobiose and short-chain oligomers. Thus, the effect of PEG on β-glucosidase activity was also tested by hydrolysis of pure cellobiose with commercially available CTec2. Figure 1C shows little difference between the glucose production with 1, 2, and 5 wt% PEG addition to cellobiose and the control (no PEG). Also, cellobiose was not detected at the final time point in any of the experiments reported in Figure 1, proving that the β-glucosidase dosage was not the limiting factor of the reaction.

That PEG affects CBH I activity only is in agreement with hydrolysis results obtained using another surfactant, Tween 20. Ooshima *et al.* [1] showed that the specific activities of β-glucosidase and endoglucanase were not influenced by 0.05 wt% Tween 20, which was presumed to only enhance the catalytic function of CBH. However, the endoglucanase adsorption to Avicel decreased in the presence of Tween, and its activity in solution increased. Hence, Tween was able to make the endoglucanases desorb more easily from the substrate (increase the k_{off}) and retain their activity in the solution longer. The mechanism of surfactant addition was then believed to balance the adsorption profiles of exo- and endo-cellulases on the substrate and, hence, increase hydrolysis. Park *et al.* [13] also reported non-ionic surfactants in general to aid in the desorption of cellulases tightly adsorbed to the substrate surface, increasing both the amount of free cellulases in solution and conversion yields. Our experiment was only designed to look at enzymatic activity based on hydrolysis yields, and for this parameter, no significant effect of PEG addition was found for the endoglucanase tested. From these experiments, we can only speculate on the influence of PEG on k_{off} as the adsorption profiles of the enzymes were not monitored over time.

Effect of PEG on enzyme thermal stability

Another way in which PEG may enhance the overall activity of cellulases is by increasing the stability of the enzyme. Reese [12] showed increased stability of cellulases against changes in pH, temperature, shaking, and other environmental factors with the addition of PEG, and Chylenski *et al.* [11] have shown that PEG can prevent enzyme precipitation in solution. We tested the thermostability of both the CBH I and the EG by measuring its retention of activity after incubating the enzymes (without substrate) in buffer at 50°C for 24 h [27], and the results are shown in Figure 2A for the CBH I and Figure 2B for the EG. Figure 2A shows that there was no significant change in the glucose production

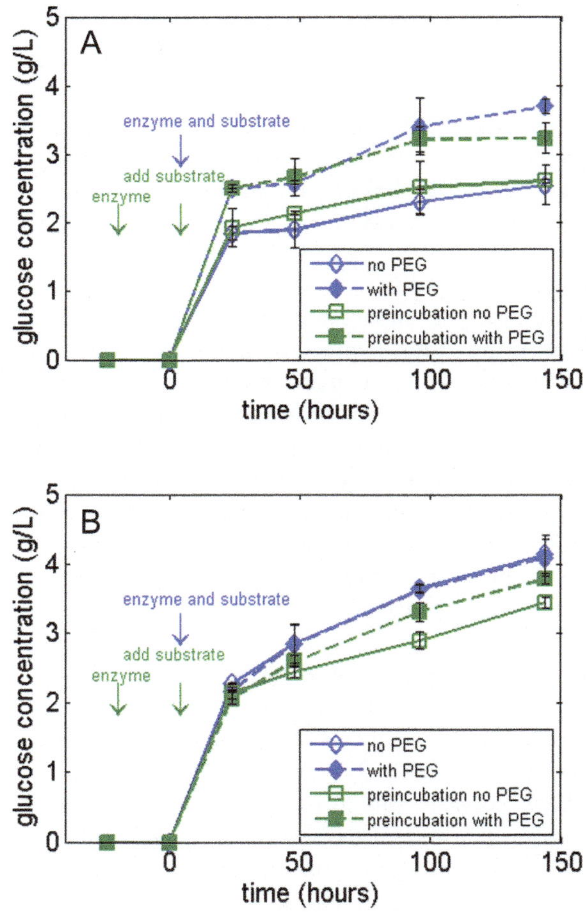

Figure 2 Thermal stability of monocomponent cellulases. Hydrolysis of 5% Avicel at 50°C using **(A)** CBH I and **(B)** EG with β-glucosidase supplementation in the absence (open symbols) and presence (solid symbols) of PEG 3000. PEG, polyethylene glycol.

after CBH I pre-incubation at 50°C. With the presence of PEG, the enzyme activity increased as expected, but the effect of heat treatment on the enzyme activity can be seen only in the later stages of hydrolysis (after 96 h) where yields drop by 14% if the CBH I was subject to a pre-incubation period. For the initial stages of hydrolysis, however, the pre-incubated CBH I yields similar, if not better, glucose yields than those without heat treatment. Thus, it is inconclusive whether PEG has a true effect on the stability of CBH I. As for the EG activity shown in Figure 2B, PEG is shown to stabilize the enzyme after 24 h of hydrolysis after undergoing a 24-h pre-incubation period, increasing the yield 10%. Thus, PEG was able to stabilize the EG against denaturation, verifying that the substantial increase in hydrolysis yield seen for CBH I but not for EG when PEG was added (Figure 1A,B) does not have to do with better thermal stability. CBH I acting on Avicel is generally less stable than endoglucanase under standard hydrolysis conditions [28], but an increase in stability does not necessarily lead to increased hydrolysis when PEG is added as the enzymes are still active and the total concentration of the final product (glucose) in the hydrolysis reaction increases over time even without PEG (Figure 1A,B).

Effect of PEG on hydrolysis of Avicel via reduced water constraint

To explore the possible link between yields and a PEG-induced reduction in water constraint/increased relaxation time, we added 10 to 150 g/L of galactose (a stereoisomer of glucose which does not inhibit cellulases and is not part of the hydrolysis reaction) during hydrolysis to introduce severe water constraint and tested the effect of PEG in these situations. These concentrations of galactose are based on lignocellulosic industrial production conditions, which can result in up to 150 g/L of glucose equivalent [8]. In Figure 3, we observe the following: as the concentration of galactose increases in the hydrolysis liquid phase, the T_2 relaxation time of the solution goes down, consistent with an increase in water constraint; moreover, when PEG is added to highly constrained solutions of galactose, the T_2 relaxation time goes up corresponding to a decrease in the water constraint; additionally, glucose production by the CBH I decreases when higher concentrations of galactose are present in the hydrolysis liquid phase, consistent with our earlier report that hydrolysis yields decrease when more solutes are present in hydrolysis [19]; lastly, the presence of PEG increases the hydrolysis yields for solutions with up to 270 mM galactose, but for even higher concentrations, there is no difference in yield with or without PEG, indicating a probable threshold for the boosting effect of PEG at this enzyme dosage. For hydrolysis using EG, the glucose production with and without PEG remained the same at all galactose concentrations (data not shown). These experiments had no cellobiose end-products; thus, β-glucosidase was not inhibited in the process even at >150-g/L galactose concentrations.

To better understand the role of CBH I-water interactions in the presence of PEG in more relevant industrial scenarios, we used a commercial cellulase mix, Cellic CTec2 (Figure 4). Here, we studied the correlation between the cellulose hydrolysis yield and the T_2 relaxation time for a setup containing high concentrations of

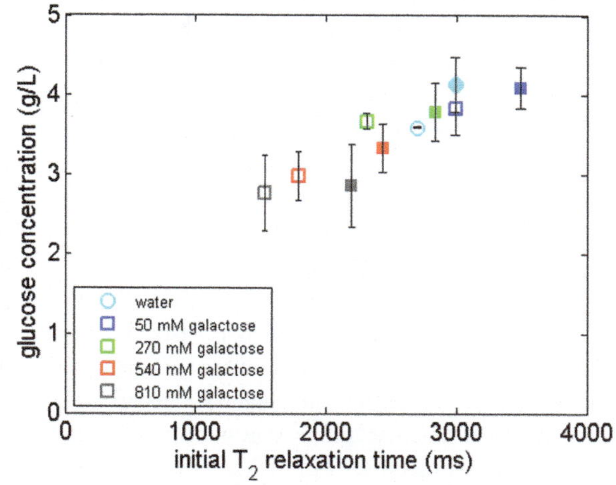

Figure 3 Hydrolysis of 5% Avicel with monocomponent CBH I in galactose solutions at different concentrations without (open symbols) and with (solid symbols) PEG 3000. The figure shows the relationship between the overall glucose production and the T_2 relaxation time of the galactose solutions in which the hydrolysis was carried out. T_2, spin-spin relaxation time.

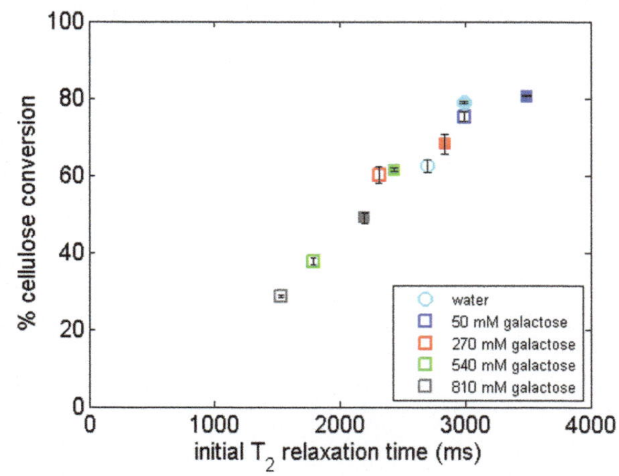

Figure 4 Hydrolysis of 5% Avicel with Cellic CTec2 in galactose solutions at different concentrations without (open symbols) and with (solid symbols) PEG 3000. The figure shows the relationship between the overall cellulose conversion and the T_2 relaxation time of the galactose solutions in which the hydrolysis was carried out. T_2, spin-spin relaxation time.

galactose (equivalent to a high-solids loading) which also constrains water in to an equivalent extent as glucose. As can be seen, there is a direct correlation between the T_2 relaxation time of the hydrolysis liquid phase at the onset of hydrolysis and the cellulose conversion yield in the presence of galactose. For relaxation times higher than pure water, the overall correlation is less pronounced, maybe because water constraint is not limiting hydrolysis under these conditions.

Increasing the galactose concentration in hydrolysis decreased the overall conversion, which is also observed when additional glucose is added instead of galactose (Figure 5). Yields with additional galactose tend to be higher than glucose because glucose not only acts as a water-constraining molecule, but is also a powerful cellulase inhibitor. During the hydrolysis experiments shown in Figure 5, we did not observe any cellobiose at the end of the hydrolysis, showing that the β-glucosidases were still active in the presence of high concentrations of sugars. While PEG increases the cellulose hydrolysis rate and final glucose production, the cellobiose produced by CBH I is consumed regardless of whether or not PEG is present, consistent with the fact that the cellulases most inhibited by water constraint are those acting on the insoluble substrate and not the β-glucosidases [19]. This phenomenon is supported by the results of cellobiose hydrolysis using Cellic CTec2, where in the presence or absence of PEG, the cellobiose hydrolysis yield did not change (Figure 1C).

Looking back at the hydrolysis results obtained using monocomponent CBH I and EG in light of the correlation observed between PEG addition, increased yields and decreased water constraint, we suggest that the increased hydrolysis yield seen for CBH I in the presence of PEG has to do with the lower water constraint brought on by the surfactant. Thus, since kinetic and mechanistic studies reported in the literature suggest that the saccharification rate for CBH-type cellulases depends upon enzyme desorption rate [21-23,26], and since the presence of PEG increases the saccharification rate, then a link between enzyme desorption rate and the presence of PEG seems likely. Determining the influence of PEG on the kinetics of CBH I would be a subject of future work.

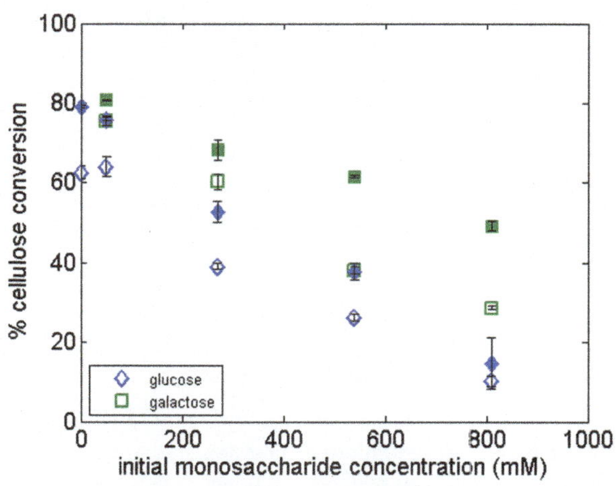

Figure 5 Hydrolysis of 5% Avicel with Cellic CTec2, 1 wt% PEG 3000 and additional glucose or galactose. The control samples (without PEG) are shown using open symbols.

Effect of different PEG molecular weights and concentrations on commercial cellulase mixtures

As PEG increases the activity of CBH I, it can also boost the glucose production of commercial cellulase preparations acting on Avicel. Figure 6 shows the hydrolysis of 5% Avicel in the presence of 1 wt% PEG using Cellic CTec2 or Celluclast 1.5 L with additional Novozym 188. The enzyme activity in the liquid phase was the same for both enzyme mixes, that is, the protein content was higher for Celluclast than for CTec2. With this setup, the addition of PEG gives approximately the same increase in conversion for CTec2 as for Celluclast with additional Novozym 188 supplementation, which suggests that PEG has a universal Avicel hydrolysis-boosting effect, as also shown by Ouyang *et al.* [17] and Zhang *et al.* [16]. Li *et al.* [14] have also shown a PEG-boosting effect on Avicel using Accellerase 1000 (Genencor, Palo Alto, CA, USA) with different enzyme loadings and PEG concentrations. Given that CBHs make up more than 70% of the *Trichoderma reesei* secretome [29] from which most cellulolytic enzyme mixtures are produced industrially [30], it is no surprise

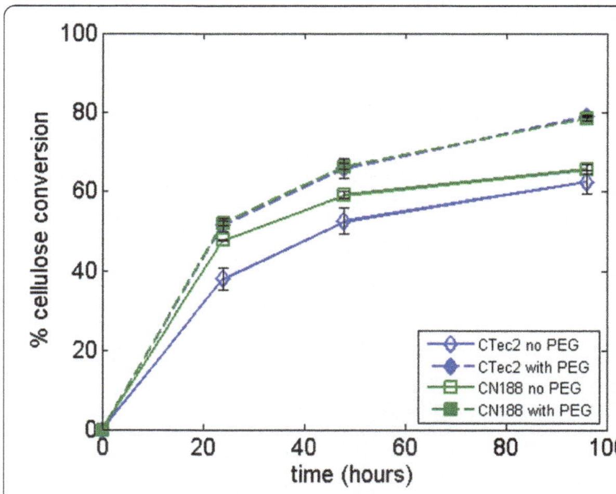

Figure 6 Comparison of 5% Avicel hydrolysis using PEG 3000 between Cellic CTec2 and Celluclast 1.5 L with Novozym 188 supplementation (CN188) based on the same FPU activity (10 FPU per gram of Avicel). The control samples (without PEG) are shown using open symbols. PEG, polyethylene glycol.

that the addition of PEG increases the hydrolysis yield for a range of enzyme preparations, which can be attributed to the PEG effect on CBH.

We also tested different molecular weights of PEG (Figure 7A) and different PEG dosages (Figure 7B) on Avicel hydrolysis with CTec2. As seen in Figure 7A, if the same dosage is used, there is no significant difference in the glucose production by using solid PEG with different molecular weights in the 1,500 to 8,000 g/mol range, while liquid PEG with a molecular weight of 400 g/mol has a smaller impact. The hydrolysis results are in agreement with those reported by Ouyang *et al.* [17], who observed very small changes in the hydrolysis yield of Avicel with Celluclast 1.5 L supplemented with Novozym 188 in citrate buffer using PEG with different molecular weights (2,000 to 8,000 g/mol), while in a later study, Zhang *et al.* [16] observed a yield difference of 2% to 3% using different molecular weights of PEG (2,000 to 10,000 g/mol) under the same conditions, albeit using higher PEG loadings. However, other studies have found a more significant effect of PEG molecular weight on hydrolysis yields, most likely due to other factors including the different enzyme mix, PEG concentration, and the substrate being hydrolyzed. For example, when using Celluclast 1.5 L with Novozym 188 supplementation for the hydrolysis of either steam-pretreated spruce [4] or wheat straw [5], higher cellulose conversions were reported when PEG with higher chain lengths were used.

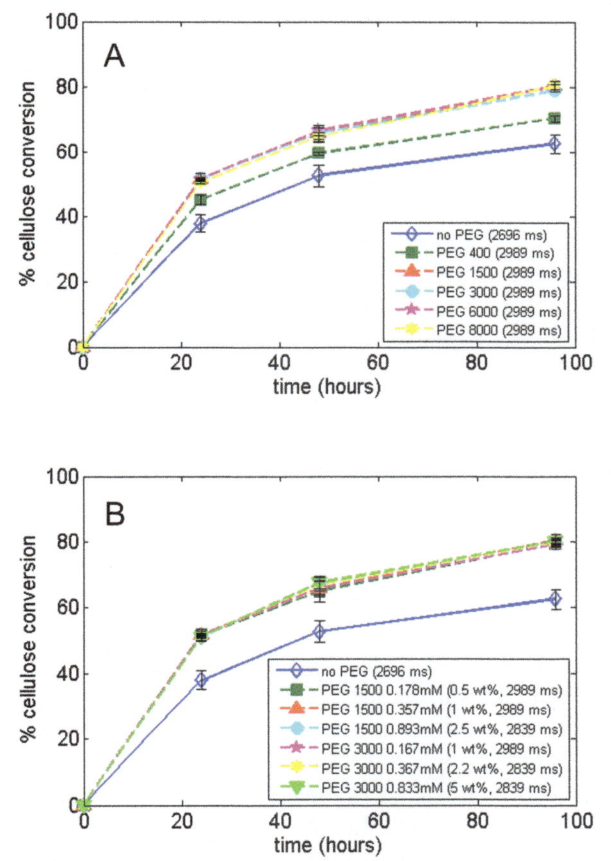

Figure 7 Hydrolysis of Avicel using CTec2 at 5% dry matter at 24, 48, and 96 h with (A) different molecular weight of PEG (1% *w/w*) in water and (B) different concentrations (molar concentration and percentage weight per gram of cellulose) of PEG. The control hydrolysis contained no PEG and was conducted in water (open symbols), and the T_2 relaxation times of the hydrolysis liquid fraction at the onset of hydrolysis are shown in parenthesis next to the legend. PEG, polyethylene glycol.

The reason why the molecular weight of PEG does not affect the boost in hydrolysis of Avicel can to a large extent be understood as an effect of water constraint. As shown in Figure 7A (legend), T_2 relaxation time data of the hydrolysis liquid phase shows no difference between PEG molecular weights. Thus, an increase in PEG chain length does not affect water constraint. That yields are, however, different for liquid *vs.* solid PEG (400 g/mol *vs.* 1,500 to 8,000 g/mol) while their relaxation time is the same shows that LF-NMR spin-spin relaxation times depend on many factors and can only be used as an indicator of expected hydrolysis yields between sets of strictly comparable samples.

Regarding PEG dosage used, Figure 7B shows that PEG boosts Avicel hydrolysis with CTec2 compared to the control, but from 0.17 to 0.89 M (0.5 to 5 wt% of cellulose) of PEG in solution, the difference is minimal with respect to the increase in yield. LF-NMR data given in the figure also confirm the hypothesis that, as there is no difference in the hydrolysis liquid phase relaxation time between different concentrations of PEG, the glucose production remains the same. It is important to note, however, that the concentrations of PEG used in these studies are within a 1 to 5 wt% cellulose range, and results would probably differ if higher concentrations were used, as shown in Table 1. The relaxation time measurements show that PEG does not have an additive effect, that is, relaxation times do not increase

Table 1 T_2 relaxation times of PEG 3000 solutions

PEG 3000 concentration		Avicel 5 wt% equivalent (wt%)	T_2 relaxation time (ms)
(wt% solution)	(mM)		
0	0	0	2,696
0.0025	0.00833	0.05	2,818
0.005	0.0167	0.1	2,989
0.025	0.0833	0.5	2,989
0.05	0.167	1	2,989
0.125	0.417	2.5	2,839
0.25	0.833	5	2,839
0.5	1.667	10	2,839
1	3.333	20	2,696

with higher concentrations of PEG and eventually decrease when PEG concentrations are above 20% in solution. Such high-PEG concentrations, however, would be unrealistic for an industrial hydrolysis setting.

Conclusions

The addition of PEG to hydrolysis of Avicel has a positive effect on the hydrolysis yield. When present during hydrolysis using a CBH I, PEG can increase the overall hydrolysis yield up to 45%, whereas there is no significant increase in hydrolysis yields when using an EG with PEG. PEG does not seem to affect the performance of β-glucosidase, as yields remain the same with and without PEG during hydrolysis. Increased thermal stability and higher substrate accessibility in the presence of PEG have been ruled out as factors which could explain the higher activity of CBH I. We hypothesize that part of the increase in hydrolysis yield using a pure cellulosic substrate in the presence of PEG was due to the increased water availability upon addition of the surfactant. Based on LF-NMR data, the increase in hydrolysis yield was directly correlated to the increase in the relaxation time of the hydrolysis liquid phase.

This study stresses the fact that the properties of the hydrolysis liquid phase play a role in the overall hydrolysis yield aside from properties of the substrate and that the extent of water constraint in the hydrolysis liquid phase may either be increased or decreased compared to pure water depending on the solutes present. In a broader sense, this work illustrates that water constraint can be a mirror through which not only factors impeding cellulose saccharification can be studied, but also those boosting this process.

Materials and methods

Materials

PEG 400 (weight average molecular weight (M_w) 380 to 420 g/mol), 1500 (M_w 1,400 g/mol), 3000 (M_w 3,000 g/mol), and 6000 (M_w 5,400 g/mol), were obtained from Merck, Darmstadt, Germany. Avicel, PEG 8000 (M_w 8,000 g/mol), D-glucose, D-galactose, and D-cellobiose standards were purchased from Sigma-Aldrich, St. Louis, MO, USA. Commercial cellulase mixtures Celluclast 1.5 L and Cellic CTec2, as well as the β-glucosidase mixture Novozym 188, were obtained from Novozymes A/S, Bagsværd, Denmark. Celluclast 1.5 L had a protein content of 127 mg/g, filter paper activity of 62 FPU/g, and a β-glucosidase activity of 15 U/g based on the following assays: filter paper activity was determined according to Ghose [31] and the β-glucosidase activity was measured using 5 mM *p*-nitrophenyl-β-D-glucopyranoside as substrate [32]. The protein content was measured using the Ninhydrin assay with BSA as a protein standard [33]. Cellic CTec2 had a protein content of 161 mg/g (120 FPU/g) and a β-glucosidase activity of 2,731 U/g. Novozym 188 had a protein content of 220 mg/g and a β-glucosidase activity of 231 U/g. The purified monocomponent enzymes CBH I from *Trichoderma longibrachiatum* (0.05 U/mg on *p*NP-lactoside), endo-β-glucanase from *Talaromyces emersonii* (64 U/mg on carboxymethylcellulose), and β-glucosidase from *Aspergillus niger* (52 U/mg on *p*NP-β-glucoside) were purchased from Megazyme, Wicklow, Ireland. All reagents and enzymes were used as received without further purification.

Avicel hydrolysis with monocomponent cellulases and β-glucosidase

Fifty milligrams of Avicel were weighed in 2-mL screw-cap tubes and filled with diluted monocomponent enzyme solution in 50 mM citrate buffer at pH 4.8 for a total water-insoluble solids loading of 5% *w/w* (1-mL working volume). The first set of experiments consisted of CBH I (6 μL, or 1.2 mg protein/g Avicel) and β-glucosidase (6 μL, or 0.1 mg protein/g Avicel) (1:1 *v/v*), while in the second set, the CBH I was replaced by EG (6 μL, or 1.2 mg protein/g Avicel). The PEG 3000 concentration was 1 wt% (10 mg/g Avicel) and added to the diluted enzyme solution before adding to the substrate. Control samples without PEG were run in parallel. Hydrolysis was carried out in triplicate at 50°C in an Eppendorf Thermomixer (Hamburg, Germany) with shaking at 800 rpm. Time points were taken at 24, 48, 96, and 144 h. The hydrolyses were terminated by boiling at 100°C for 10 min, centrifuging at 13,200 × *g* and 5°C for 10 min, and filtering through a 0.45-μm syringe filter (Millipore, Billerica, MA, USA). The filtrates were stored at −20°C before HPAEC sugar quantification.

PASC hydrolysis with monocomponent cellulases and β-glucosidase

The preparation of PASC was based on the procedure by Walseth [34] with a few modifications: 4 g of Avicel was suspended in 100 mL of phosphoric acid (85% *w/v*) at 1°C and magnetically stirred for 1 h. The mixture was then poured into 1,900 mL of ice cold water and kept at 1°C with further stirring for 1 h. The suspension was left stationary to allow the fibers to sink to the bottom of the flask while the supernatant was decanted. The suspension was washed four times with 2 L of MilliQ, two times 1% $NaHCO_3$ solution to increase the pH to 5, and a further three times with water. The PASC then underwent a solvent exchange with 50 mM citrate buffer at pH 4.8 and stored at 1°C until further use. A full 24-h hydrolysis of PASC with excess CTec2 confirmed the cellulose content at 3.5% *w/v*.

PASC was hydrolyzed using the same monocomponent enzymes as described in the previous section using Avicel. PASC (0.5 mL) was pipetted into 2-mL screw-cap tubes with 10 μL of 1:1 cellulase:β-glucosidase mix (2 mg cellulase protein/g Avicel, 0.14 mg β-glucosidase protein/g Avicel) and a further 2.5 wt% PEG 3000. Control samples without PEG were also run in parallel. Hydrolysis was carried out in triplicates at 50°C in an Eppendorf Thermomixer with shaking at 800 rpm, where samples were analyzed for glucose yields after 5, 24, and 48 h. The samples were processed as described in the previous section with Avicel.

HPAEC hydrolysis products quantification

Avicel and PASC hydrolysis products were quantified using an ICS5000 HPAEC system equipped with a PAD detector (Dionex, Sunnyvale, CA, USA). The separation was performed using a Dionex CarboPac PA1 anion exchange column, a 2 mm × 50 mm guard column and 2 mm × 250 mm analytical column. The column operated at a flow of 0.25 mL/min and maintained at 30°C. Peak separation for glucose and cellobiose was obtained by applying the following elution gradient and using fucose as an internal standard: water for 28 min, gradient to 0.1 M NaOH for 9 min and 0.1 M NaOH for 8 min, gradient to 0.2 M NaOH for 2 min and 0.2 M NaOH for 5 min, followed by column reconditioning with water for 8 min for a total run time of 60 min. Column elution was followed by a post-column addition with 0.2 M NaOH at 0.12 mL/min. For the analysis of oligosaccharides, the method was modified from Westereng *et al.* [35]. The elution gradient consisted of 0.1 M NaOH for the initial 10 min, a linear gradient to 0.1 M NaOH with 0.22 M NaOAc for 12 min, gradient to 0.1 M NaOH with 1 M NaOAc for 3 min, followed by a column reconditioning with 0.1 M NaOH for 15 min, for a total run time of 40 min.

Cellobiose hydrolysis with Novozymes Cellic CTec2

D-cellobiose was dissolved in 100 mM sodium acetate buffer at pH 4.8 to make a concentrated solution of 100 g/L. The cellobiose was hydrolyzed using CTec2 with a protein loading of 0.064 mg/g of cellobiose. PEG 3000 was added to the reaction at 1, 2, and 5 wt% of cellobiose. The samples were mixed in an Eppendorf Thermomixer at 50°C and 800 rpm. Samples were prepared in triplicates, and time points were taken at 1, 2, 4, and 8 h. The hydrolysis was terminated by adding 72% H_2SO_4 and diluting to a final concentration of 3 mM H_2SO_4, which then the samples were centrifuged at 13,200 × *g* and 5°C for 10 min, filtered through a 0.45-μm syringe filter and analyzed using high-performance liquid chromatography (HPLC).

Stability of monocomponent enzymes

Monocomponent enzymes (CBH I and EG) were diluted (6 μL, or 1.2 mg protein/g Avicel, in 1-mL 50 mM sodium citrate buffer at pH 4.8) with or without PEG 3000 (0.05% solution). The solutions were incubated at 50°C for 24 h in a rotating incubator, to which a monocomponent enzyme equivalent volume of β-glucosidase (6 μL, or 0.1 mg protein/g Avicel) was then added before adding to Avicel and hydrolyzing the substrate as described previously.

To test the monocomponent enzyme stability towards high-solute concentrations, four sets of galactose concentrations were added to the hydrolysis mix (5% Avicel) with and without PEG 3000 (1 wt% Avicel) using CBH I (1.2 mg protein/g Avicel) and EG (1.2 mg protein/g Avicel) with additional β-glucosidase supplementation (0.1 mg protein/g Avicel). PEG 3000 was first diluted in solution with the cellulases before adding to the substrate. The galactose concentrations were 50, 270, 540, and 810 mM. Hydrolysis was carried out as described previously, with control samples also run in parallel.

Enzymatic hydrolysis of Avicel with Novozymes Cellic CTec2

Fifty milligrams of Avicel were weighed in 2-mL screw-cap tubes and filled with diluted enzyme solution for a total water-insoluble solids loading of 5% *w*/*w* (1-mL working volume). The enzymatic loading was 13 mg protein/g dry matter (approximately 10 FPU/g of Avicel). For hydrolysis with PEG, the loading was 1 wt% of the water-insoluble solids (10-mg/g Avicel), unless otherwise indicated. In a separate set of experiments, glucose and galactose were each used to spike the hydrolysates at 50, 270, 540, and 810 mM with and without PEG. Glucose was chosen as it is a known inhibitor of cellulases, while galactose, though not an inhibitor, is a stereoisomer of glucose. Hydrolysis was carried out as described above. Control samples (without enzyme) were also run in parallel.

Enzymatic hydrolysis of Avicel with Celluclast 1.5 L/Novozym 188

A 5% *w*/*w* hydrolysate was prepared by measuring 50-mg Avicel in 2-mL screw-cap tubes and filled with diluted enzyme and PEG 3000 solution (1% weight per gram of dry matter) for a 1-mL working volume. A 5:1 weight mix of Celluclast and Novozym 188 at 10-FPU/g dry matter (19 mg protein/g) enzyme solution was used. The combined enzyme mixture had a filter paper activity of 76 FPU/g. Hydrolysis was carried out in triplicates at 50°C in an incubator mounted with an orbital shaker at 150 rpm. Time points were taken at 24, 48, and 96 h. The hydrolyses were terminated by centrifuging each sample at 13,200 × *g* and 5°C for 10 min, filtering through a 0.45-μm syringe filter, and analyzed using HPLC.

HPLC sugar quantification

D-cellobiose, D-glucose, and D-galactose (standards obtained from Sigma-Aldrich, St. Louis, MO, USA) were quantified with a Dionex UltiMate 3000 (Dionex, Germering, Germany) equipped with a refractive index detector (Shodex, Minato, Tokyo, Japan). The separation was performed in a Phenomenex Resex ROA column at 80°C with 5 mM H_2SO_4 as eluent at a flow rate of 0.6 mL/min for 15 min. The results were analyzed using the Chromeleon software program from Dionex.

LF-NMR T_2 relaxation

LF-NMR measurements were performed on a Bruker mq20 minispec with a 0.47 T permanent magnet (equivalent to 20-MHz proton resonance frequency). The internal magnet temperature was 40°C. Samples were prepared by mixing 50, 270, 540, and 810 mM of galactose in separate vials with and without PEG. Then, 1 mL of sample was placed in 15 dia glass tubes and heated to 40°C, where the T_2 relaxation times of the solutions were then measured using the CPMG pulse sequence. Each sample was run with 32 scans and a 5-s recycle delay containing 8,000 points and a pulse separation of 1.2 ms. The obtained relaxation curves were analyzed using the inverse Laplace transformation method CONTIN [36], where all the solutions gave a single T_2 relaxation peak. The T_2 reported are the peak maxima positions of this peak.

Abbreviations

CBH: cellobiohydrolase; EG: endoglucanase; LF-NMR: low-feld nuclear magnetic resonance; M_w: weight average molecular weight; PASC: phosphoric acid swollen cellulose; PEG: polyethylene glycol; T_2: spin-spin relaxation time.

Competing interests

The authors declare that they have no competing interests.

Authors' contributions

CCH and DC designed and planned the study with contributions from HJ, CF, and LGT. CCH and DC conducted the experiments, and CCH wrote the manuscript. DC, HJ, CF, and LGT critically reviewed and edited the manuscript. All authors read and approved the final manuscript.

Acknowledgements

DC was financially supported by the Nordic Top-Level Research Initiative project TFI-PK-bio 02 'High gravity hydrolysis and fermentation of lignocellulosic material for production of bio-fuels.' We would like to thank Novozymes A/S, Denmark, for the provision of Celluclast 1.5 L, Novozym 188, and Cellic CTec2. We also acknowledge Dr. Katja S. Johansen for helpful comments and discussion on the manuscript.

Author details

[1]Department of Geosciences and Natural Resource Management, Faculty of Science, University of Copenhagen, Rolighedsvej 23, DK-1958 Frederiksberg C, Denmark. [2]Present address: Center for Bioprocess Engineering, Department of Chemical and Biochemical Engineering, Technical University of Denmark, Søltofts Plads, Building 229, DK-2800 Kgs. Lyngby, Denmark.

References

1. Ooshima H, Sakata M, Harano Y. Enhancement of enzymatic hydrolysis of cellulose by surfactant. Biotechnol Bioeng. 1986;28:1727–34.
2. Duff SJB, Murray WD. Bioconversion of forest products industry waste cellulosics to fuel ethanol: a review. Bioresour Technol. 1996;55:1–33.
3. Sun Y, Cheng J. Hydrolysis of lignocellulosic materials for ethanol production: a review. Bioresour Technol. 2002;83:1–11.
4. Börjesson J, Peterson R, Tjerneld F. Enhanced enzymatic conversion of softwood lignocellulose by poly(ethylene glycol) addition. Enzyme Microb Technol. 2007;40:754–62.
5. Kristensen JB, Börjesson J, Bruun MH, Tjerneld F, Jørgensen H. Use of surface active additives in enzymatic hydrolysis of wheat straw lignocellulose. Enzyme Microb Technol. 2007;40:888–95.
6. Kaar WE, Holtzapple MT. Benefits from Tween during enzymic hydrolysis of corn stover. Biotechnol Bioeng. 1998;59:419–27.
7. Tu M, Zhang X, Paice M, McFarlane P, Saddler JN. Effect of surfactants on separate hydrolysis fermentation and simultaneous saccharification fermentation of pretreated lodgepole pine. Biotechnol Prog. 2009;25:1122–9.
8. Cannella D, Jørgensen H. Do new cellulolytic enzyme preparations affect the industrial strategies for high solids lignocellulosic ethanol production? Biotechnol Bioeng. 2014;111:59–68.
9. Börjesson J, Engqvist M, Sipos B, Tjerneld F. Effect of poly(ethylene glycol) on enzymatic hydrolysis and adsorption of cellulase enzymes to pretreated lignocellulose. Enzyme Microb Technol. 2007;41:186–95.
10. Eriksson T, Börjesson J, Tjerneld F. Mechanism of surfactant effect in enzymatic hydrolysis of lignocellulose. Enzyme Microb Technol. 2002;31:353–64.
11. Chylenski P, Felby C, Haven MØ, Gama M, Selig MJ. Precipitation of Trichoderma reesei commercial cellulase preparations under standard enzymatic hydrolysis conditions for lignocelluloses. Biotechnol Lett. 2012;34:1475–82.
12. Reese ET. Inactivation of cellulase by shaking and its prevention by surfactants. J Appl Biochem. 1980;2:36–9.
13. Park JW, Takahata Y, Kajiuchi T, Akehata T. Effects of nonionic surfactant on enzymatic hydrolysis of used newspaper. Biotechnol Bioeng. 1992;39:117–20.
14. Li J, Li S, Fan C, Yan Z. The mechanism of poly(ethylene glycol) 4000 effect on enzymatic hydrolysis of lignocellulose. Colloids Surf B Biointerfaces. 2012;89:203–10.
15. Helle SS, Duff SJB, Cooper DG. Effect of surfactants on cellulose hydrolysis. Biotechnol Bioeng. 1993;42:611–7.
16. Zhang M, Ouyang J, Liu B, Yu H, Jiang T, Cai C, et al. Comparison of hydrolysis efficiency and enzyme adsorption of three different cellulosic materials in the presence of poly(ethylene glycol). BioEnergy Res. 2013;6:1252–9.
17. Ouyang J, Dong Z, Song X, Lee X, Chen M, Yong Q. Improved enzymatic hydrolysis of microcrystalline cellulose (Avicel PH101) by polyethylene glycol addition. Bioresour Technol. 2010;101:6685–91.
18. Selig MJ, Hsieh C-WC, Thygesen LG, Himmel ME, Felby C, Decker SR. Considering water availability and the effect of solute concentration on high solids saccharification of lignocellulosic biomass. Biotechnol Prog. 2012;28:1478–90.
19. Hsieh CC, Cannella D, Jørgensen H, Felby C, Thygesen LG. Cellulase inhibition by high concentrations of monosaccharides. J Agric Food Chem. 2014;62:3800–5.
20. Chundawat SPS, Lipton MS, Purvine SO, Uppugundla N, Gao D, Balan V, et al. Proteomics-based Compositional analysis of complex Cellulase–Hemicellulase mixtures. J Proteome Res. 2012;10:4365–72.
21. Horn SJ, Sikorski P, Cederkvist JB, Vaaje-Kolstad G, Sørlie M, Synstad B, et al. Costs and benefits of processivity in enzymatic degradation of recalcitrant polysaccharides. Proc Natl Acad Sci. 2006;103:18089–94.
22. Kurašin M, Väljamäe P. Processivity of cellobiohydrolases is limited by the substrate. J Biol Chem. 2011;286:169–77.
23. Igarashi K, Uchihashi T, Koivula A, Wada M, Kimura S, Okamoto T, et al. Traffic jams reduce hydrolytic efficiency of cellulase on cellulose surface. Science. 2011;333:1279–82.
24. Várnai A, Viikari L, Marjamaa K, Siika-aho M. Adsorption of monocomponent enzymes in enzyme mixture analyzed quantitatively during hydrolysis of lignocellulose substrates. Bioresour Technol. 2011;102:1220–7.
25. Pribowo A, Arantes V, Saddler JN. The adsorption and enzyme activity profiles of specific Trichoderma reesei cellulase/xylanase components when hydrolyzing steam pretreated corn stover. Enzyme Microb Technol. 2012;50:195–203.

26. Bu L, Nimlos MR, Shirts MR, Ståhlberg J, Himmel ME, Crowley MF, et al. Product binding varies dramatically between processive and nonprocessive cellulase enzymes. J Biol Chem. 2012;287:24807–13.
27. Skovgaard PA, Jørgensen H. Influence of high temperature and ethanol on thermostable lignocellulolytic enzymes. J Ind Microbiol Biotechnol. 2013;40:447–56.
28. Reese ET, Mandels M. Stability of the cellulase of Trichoderma reesei under use conditions. Biotechnol Bioeng. 1980;22:323–35.
29. Merino ST, Cherry J. Progress and challenges in enzyme development for biomass utilization. In: Biofuels, vol. 108. Heidelberg: Springer Berlin; 2007. p. 95–120.
30. Allen AL, Mortensen RE. Production of cellulase from trichoderma reesei in fed-batch fermentation from soluble carbon sources. Biotechnol Bioeng. 1981;23:2641–5.
31. Ghose TK. Measurement of cellulase activities. Pure Appl Chem. 1987;59:257–68.
32. Saha BC, Bothast RJ. Production, purification, and characterization of a highly glucose-tolerant novel beta-glucosidase from *Candida peltata*. Appl Environ Microbiol. 1996;62:3165–70.
33. Starcher B. A ninhydrin-based assay to quantitate the total protein content of tissue samples. Anal Biochem. 2001;292:125–9.
34. Walseth CS. Occurrence of cellulases in enzyme preparations from microorganisms. Tappi. 1952;35:228–33.
35. Westereng B, Agger JW, Horn SJ, Vaaje-Kolstad G, Aachmann FL, Stenstrøm YH, et al. Efficient separation of oxidized cello-oligosaccharides generated by cellulose degrading lytic polysaccharide monooxygenases. J Chromatogr A. 2013;1271:144–52.
36. Provencher SW. CONTIN: a general purpose constrained regularization program for inverting noisy linear algebraic and integral equations. Comput Phys Commun. 1982;27:229–42.

The difference in effective light penetration may explain the superiority in photosynthetic efficiency of attached cultivation over the conventional open pond for microalgae

Junfeng Wang[1], Jinli Liu[2] and Tianzhong Liu[1*]

Abstract

Background: The 'attached cultivation' technique for microalgae production, combining the immobilized biofilm technology with proper light dilution strategies, has shown improved biomass production and photosynthetic efficiency over conventional open-pond suspended cultures. However, how light is transferred and distributed inside the biofilm has not been clearly cefined yet.

Results: In this research, the growth, photosynthetic oxygen evolution, and specific growth rate for microalgal cells in both open-pond and attached cultivation were studied to determine the effective light penetration at different phases of the cultivation. As a result, the light conditions inside the culture broth as well as the biofilm were revealed for the first time. Results showed that outdoor, in a conventional 20-cm deep open pond, all of the algal cells were fully illuminated in the first 3 days of cultivation. As the biomass concentration increased from day 4 to day 10, the light could only effectively penetrate 45.5% of the open-pond depth, and then effective light penetration gradually decreased to 31.1% at day 31, when the biomass density reached a maximum value of 0.45 g L^{-1} or 90 g m^{-2}. In the attached cultivation system, under nitrogen-replete condition, almost 100% of the immobilized algal cells inside the biofilm were effectively illuminated from day 0 through day 10 when the biomass density increased from 8.8 g m^{-2} to 107.6 g m^{-2}.

Conclusion: Higher light penetration efficiency might be the reason why, using attached cultivation, observed values for photosynthetic efficiency were higher than those recorded in conventional open-pond suspended cultures.

Keywords: Microalgae, Photosynthetic efficiency, Open pond, Attached cultivation, Effective light penetration

Background

Microalgae are a group of photosynthetic microorganisms represented by more than 40,000 species, many of which can produce high-value bioactive compounds [1]. In recent years, due to the price increase of fossil fuel as well as the concerns on environment deterioration, the global R&D investment on microalgae biofuel vastly increased. The reasons for this are that microalgae have high photosynthetic efficiency and high oil content and are believed to be the most promising feedstock for environment-friendly renewable liquid biofuel thanks to the fact that they do not compete with food crops for arable land and water supply [2-7]. However, until now, the success of microalgae cultures only happened in some small-scale tests and no success in biofuel production or CO_2 mitigation has yet been achieved on a commercial scale anywhere in the world with microalgae cultivation [8,9]. The primary reason for this is that even when state-of-the-art techniques and designs are adopted on a large scale for open ponds or closed photobioreactors (PBRs), microalgal cell growth rate values currently cannot reach those obtained in highly controlled lab environments. The highest recorded biomass productivities for open ponds and PBRs are around 40 g dry mass m^{-2} day^{-1} and long-term averaged productivity is 10~20 g dry mass m^{-2} day^{-1} [3,10,11], which are far less than the theoretical maximum

* Correspondence: liutz@qibebt.ac.cn
[1]Key Laboratory of Biofuels, Qingdao Institute of Bioenergy and Bioprocess Technology, Chinese Academy of Sciences, Qingdao, Shandong 266101, People's Republic of China
Full list of author information is available at the end of the article

of 120~150 g m^{-2} day^{-1}, or photosynthetic efficiency of 12.4% (based on total solar radiation spectrum) and 28% (based on visible light spectrum, 400~700 nm) [12,13].

The biofilm cultivation of microalgae, in which the algal cells are generally immobilized and fixed onto/into supporting materials in high density and fed with nutrient solutions, is a different cultivation method than conventional aqueous suspended cultures. Open ponds and PBRs have existed for a long time now; however, they often exhibited relatively low biomass productivity [14-21]. Recently, our research group proposed an improved cultivation method based on a reactor where both the immobilized biofilm technology and the light dilution theory work together in a patented design called 'attached cultivation' [22]. The basic principles of this 'attached cultivation' technique as well as the related photobioreactors include the following: (i) a highly dense wet algal paste attached onto artificial supporting materials to form a thin layer of the algal biofilm and (ii) many layers of these biofilms arranged in an array fashion to dilute the high light so that the light intensity impinging the algal biofilm is much lower than the light that reaches the photobioreactor footprint area (Figure 1A). The biomass productivity potential of this 'attached cultivation' method was evaluated both under indoor and outdoor conditions. Results showed that with indoor light of 700 μmol m^{-2} s^{-1} and a light dilution rate of 10, the maximum biomass productivity for *Scenedesmus* was close to 120 g m^{-2} day^{-1}, with photosynthetic efficiency of *ca.* 18% (based on visible light). Under outdoor conditions, the maximum biomass productivity reached *ca.* 80 g m^{-2} day^{-1} corresponding to a photosynthetic efficiency of 17.3% (based on visible light) and 8.3% (based on total solar irradiation), which were both seven times higher than the data reported for a conventional open pond under the same environment conditions [22]. Similar biomass productivity and photosynthetic efficiency were also achieved

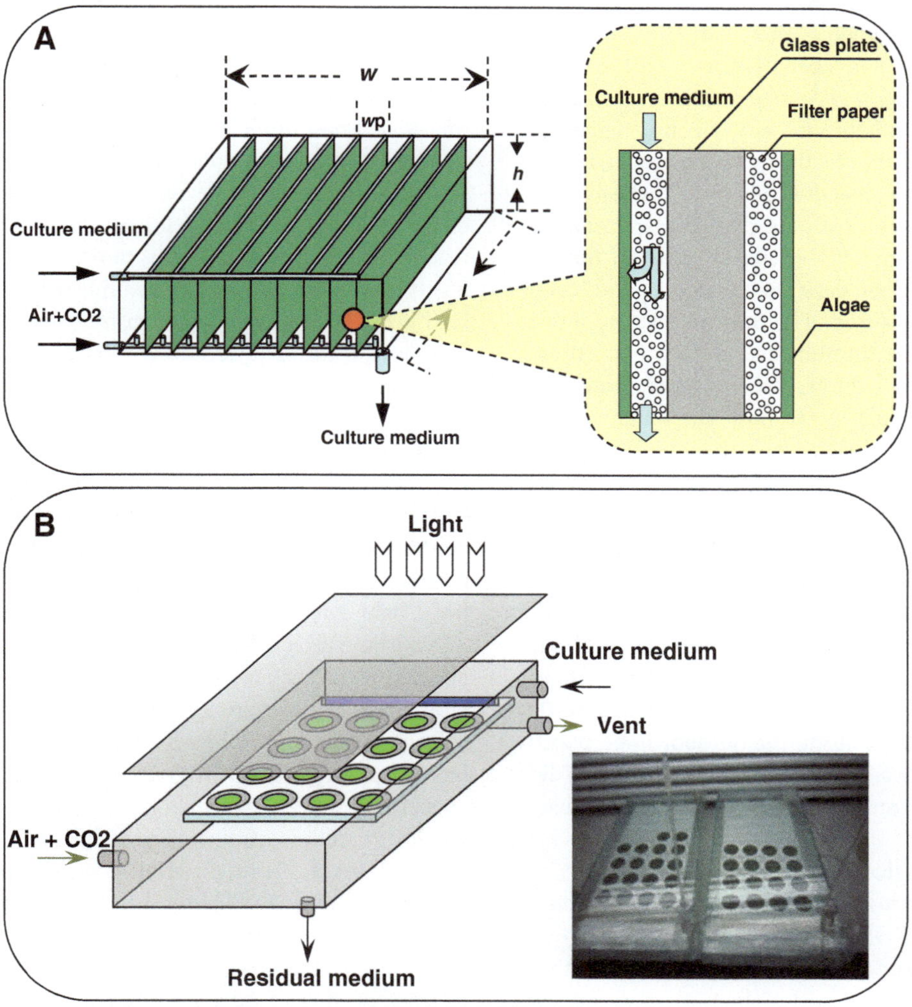

Figure 1 The schematic diagrams of the photobioreactors for attached cultivation. (A) The schematic diagram for the multiple-layer photobioreactor for attached cultivation (adapted from Liu *et al.* [22]). **(B)** The schematic diagram and the actual photograph for the single-layer photobioreactor for attached cultivation used in the research.

with *Botryococcus braunii* which grows slowly with aqueous suspended cultivation but exhibits a high biomass productivity of *ca.* 50 g m^{-2} day^{-1} with attached cultivation, corresponding to a photosynthetic efficiency of *ca.* 15% (based on visible light) [23].

As shown in these results, the 'attached cultivation' technique greatly improves the biomass productivity as well as photosynthetic efficiency of microalgae. However, the reason(s) and mechanism(s) for the superiorities of this method over suspended cultivation have not been fully understood yet. Maybe the secret of its superior performances over traditional suspended cultures lies in the differences in delivery of light and nutrients and in CO_2 transfer between the two systems [24,25]. In this research, we especially focused on how the illumination worked, because light is always considered as the most important environmental factor that determines the behaviors of the microalgal cells [12,26]. A basic but important aspect to consider is how deep light penetrates during the microalgal cultivation. For open-pond suspended cultures, it is obvious that with the increases of biomass density, the light penetration depth decreases [12] and results in less and less percentage of algal cells being illuminated with light intensity which is higher than the compensation point (effectively illuminated) where the biomass accumulation rate due to photosynthesis equilibrates with the biomass loss rate due to respiration at a specific biomass density [11]. However, until to now, the dynamics of the 'light' *vs.* 'dark' for the open pond has not been quantitatively assessed. The information on illumination properties of the immobilized biofilm cultivation is scarce. How does the light transfer and penetrate inside the immobilized microalgal biofilm? Does light transfer and penetrate differently in the microalgal biofilm and in a suspended culture? How is the difference(s) related with the performances in photosynthetic efficiency? To answer these questions, the growth, photosynthetic oxygen evolution property, and specific growth rate for microalgal cells in open-pond and 'attached cultivation' systems were studied to determine the dynamics of effective light penetration (d_E, depth from the surface where the light intensity is equal to the light compensation point) during cultivation. The light condition inside the outdoor open pond as well as the biofilm was revealed, and the effect of light distribution on the photosynthetic efficiency was discussed.

Results and discussion

The light distribution inside the suspended cultivation system

The light intensity inside the open pond declined with the increase of broth depth as well as biomass concentration (Figure 2B). Similar results had been reported by Tredici [12]. According to the oxygen evolution properties, light intensity of 12.5 ± 3.9 μmol m^{-2} s^{-1} was considered as the light compensation point (LCP) for *Scenedesmus dimorphus*; under these circumstances, the oxygen consumption due to respiration already surpassed the oxygen evolution due to photosynthesis (Figure 2C). It should be noticed that the photosynthesis-light intensity (PI) curve of Figure 2C was measured on healthy vegetative algal cells that had been cultivated in the outdoor open pond for 5 days. According to our pilot experiment, from day 0 to day 30, there were non-significant differences among LCPs of algal cells from the outdoor open pond supplied with a non-nitrogen-deficient nutrient solution. The d_E for any given biomass density can be estimated from the results of light attenuation and LCP (Figure 2D). Accordingly, the d_E was only 0.57 cm at a biomass density of 5.0 g L^{-1}, which means that only 40.4% of cells were effectively illuminated for a glass column having a diameter of 5 cm. In these conditions, the maximum biomass density would be 5.0 g L^{-1} after 10 days with continuous illumination of 100 μmol m^{-2} s^{-1} under indoor conditions. This d_E value might also be affected by light conditions, for example, the d_E might decrease in cloudy weather. To avoid this uncertainty, in this experiment, d_E measurement was performed under outdoor conditions with natural sunlight intensity of 1,500~1,600 μmol m^{-2} s^{-1} so that this equation could be directly applied to outdoor cultivation.

In this study, we proved that for a conventional 20-cm deep open pond, the maximum biomass density would reach 0.45 g L^{-1} in 10 days and remain stable in the following days (Figure 3A). This maximum biomass density of 0.45 g L^{-1} in the open pond, corresponding to 90 g m^{-2}, was similar with the results of Chisti [11] where the biomass density for open pond could not exceed 0.5 g L^{-1}. The pH during the cultivation slightly fluctuated in the range of 6.7 ~ 7.5, indicating that the carbon source was not limited during the experiment (Figure 3A). According to the equation in Figure 2D, thanks to the low biomass concentration, during the first 3 days of cultivation, light could easily penetrate the suspended culture delivering optimal light intensity to each algal cell. After 31 days, only 31.1% of the algal cells were effectively illuminated, with the effective d_E being only 6.2 cm. In this research, the fastest biomass accumulation for the open pond was recorded from day 3 to day 7, corresponding to an increase of the biomass concentration from 0.1 to 0.3 g L^{-1} or areal biomass density from 20 to 60 g m^{-2}. During the same time window, biomass productivity reached *ca.* 10 g m^{-2} day^{-1}, and the effective light penetration continuously decreased from 100% to 45.5% (Figure 3B). The biomass productivity did not decrease with the reduced effective illumination depth during these times but remained at 10 g m^{-2} day^{-1}. In microalgal suspended cultures, light distribution is not uniform at all. Light intensity decreases exponentially as we move farther away from the illuminated surface; for this reason, a thin layer on the water surface usually receives

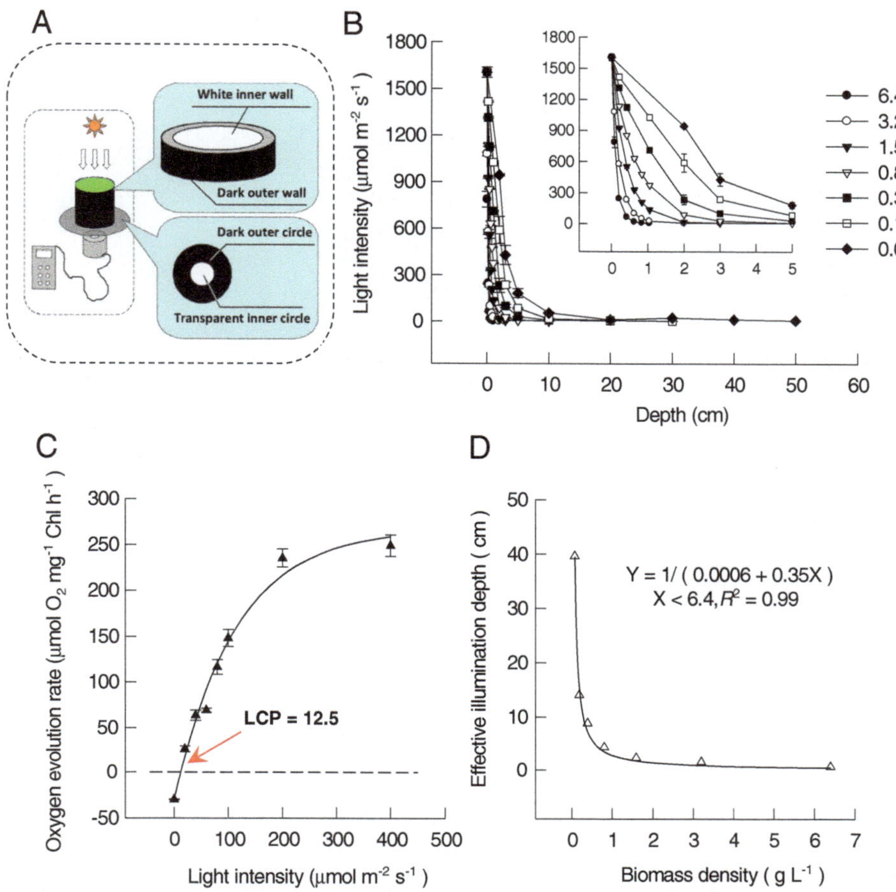

Figure 2 The light attenuation inside the culture broth of suspended cultivation of *Scenedesmus dimorphus*. (A) The schematic diagrams of light intensity measurements under different depths of the culture broth. **(B)** The dependence of light attenuation on culture density as well as depth. Seven different culture densities were tested, *viz.* 6.4 (black circle), 3.2 (white circle), 1.59 (black down-pointing triangle), 0.81 (white down-pointing triangle), 0.39 (black square), 0.18 (white square), and 0.07 (black diamond). The measurements were carried out outdoors with natural light. Data were mean ± standard deviation of three measurements. **(C)** The relationship of oxygen evolution rate *versus* light intensity (black triangle). Data were mean ± standard deviation of three measurements. The light compensation point (LCP) was indicated by arrows. **(D)** The effective illumination depth of aqueous suspended *S. dimorphus* culture broth at different biomass densities (white triangle).

oversaturating light intensity while the bottom of the pond, as the algae cells reproduce and the biomass concentration increases, lies in total darkness. The algal cells traveling through the dark portion of the reactor consume biomass by dark respiration, if they spend a long enough time in this kind of environment. An appropriate depth of the water layer in an open pond is required to avoid biomass loss by cellular respiration. The optimal depth of any suspended culture of algae cells should be equal to the maximum depth at which a sufficient amount of solar light can penetrate. The aforementioned values for how 'light' and 'dark' regions expand and contract as the culture grows thicker are very likely to change depending on the different geographic locations considered other than Qingdao, China, where this experiment was conducted and the particular algal strain considered or the reactor design adopted. Identifying the upper limit of biomass density and the maximum depth at which algae cells are still effectively illuminated depending on the specific design of the reactor (different thicknesses of the water layer) has been an interesting task for our research team. Obtaining this information will definitely help to better understand how light travels through the water layer of an open pond, and it will also be useful knowledge for enhancing biomass productivity.

The light distribution inside the biofilm of the attached cultivation system

One of the methods to determine the d_E of the attached microalgal biofilm is to directly measure the light intensity from underneath the biofilm. If the light intensity was equal to or higher than the LCP, then the corresponding depth could be effectively illuminated. However, we must also consider that the photosynthetically active wavelengths (red

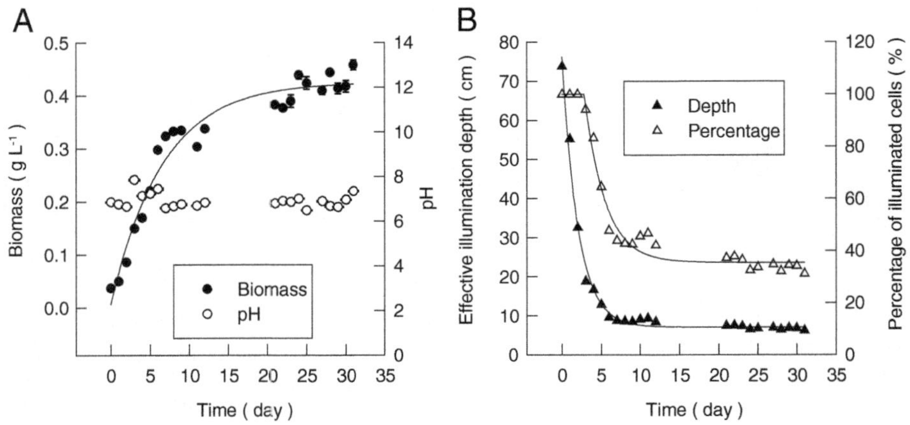

Figure 3 The biomass increase, pH, and effective illumination depth of conventional aqueous suspended open ponds outdoors. **(A)** The biomass density (black circle) and pH (white circle) changes of the open pond outdoors. Data for biomass were mean ± standard deviation of six measurements for two independent open ponds (three measurements for each). Data for pH were mean ± standard deviation of two measurements for two independent open ponds (one measurement for each). **(B)** The effective illumination depth (black triangle) and percentage of effective illuminated algal cells (white triangle) at different days of cultivation in outdoor open ponds.

light) are more prone to be absorbed by chlorophyll than other less desired wavelengths (green light). The method we just described to measure light intensity might overestimate the d_E if only the quantum flux intensity is measured because the undesired wavelengths travel farther away through layers of cells than the desired ones. To avoid the aforementioned problem in this research, we used a different method to estimate the d_E by monitoring the specific growth rate of a thin layer of microalgal cells, which was called 'marker' layer, inserted just beneath the re-constructed biomass layer. An optimum 'marker' layer should have two main characteristics, *viz.* 1) thin enough to avoid the formation of light gradient inside the layer and 2) sensitive enough to the changes in light intensity. Algal cells from this thin layer should experience the same conditions as the upper layer. However, in pilot experiments, we found that such thin marker layers failed to meet requirement #2. In other words, the biomass changes for this kind of thin layers were too slight to be detected by regular gravimetric method, especially under lower light conditions. As an alternative, the algal cells used to inoculate the attached cultivation were chosen as the 'marker' layer because they were photosynthetically active even at low biomass densities. However, the risk of over- or underestimating the d_E with this method was still present if the LCP of the inoculum algal cells was lower or higher than the LCPs of daughter cells that only experienced attached cultivation conditions. The LCP for the newly formed microalgal cells was firstly measured at different days. Typical light curves for oxygen evolution rate were obtained for all of these detached and re-suspended samples (Figure 4A). The LCPs of the daughter cells from day 0 to day 10 were in the range of 12.6 ± 2.2 ~ 19.8 ± 3.2 μmol m^{-2} s^{-1}, without significant differences ($P > 0.05$) (Figure 4B). These results indicated that it is safe to determine the d_E of the attached algal biofilm with this double-layer method under nutrient-replete condition without worrying about over- or underestimation. We also tested the LCP changes for the attached algal biofilm under nitrogen-depleted conditions and found that it increased dramatically to *ca.* 60 μmol m^{-2} s^{-1} at day 2. Because of this, a 'marker' layer consisting of the cells used to inoculate the attached cultivation surface cannot be used to estimate the d_E of the nitrogen-starved attached cultivation.

The results of the double-layer experiments for attached cultivation of *S. dimorphus* under nitrogen-replete condition are shown in Figure 5. For the biofilm layer that has been attached cultivated for several days, the specific growth rate of the 'marker' layer (μ) decreased as the biomass of the upper layer increased, mainly because the light traveling through the upper layer became weaker and weaker (Figure 5A). Meanwhile, the rate of this decrease for μ gradually slowed down as the days went by, resulting in the increase of the thicknesses for the effectively illuminated upper layers (Figure 5A). According to the results in Figure 5A, the d_E for attached cultivation under nitrogen-replete condition increased from 33.9 ± 6.5 μm for day 0 to 237.3 ± 11.8 μm for day 10, corresponding respectively to biomass densities of 14.6 ± 1.6 g m^{-2} and 111.5 ± 4.1 g m^{-2}, whereas the actual biomass density increased from 8.8 ± 1.3 g m^{-2} to 107.6 ± 2.6 g m^{-2} in 10 days, corresponding to thicknesses of 21.7 ± 5.9 μm and 229.1 ± 8.6 μm (Figure 5B), respectively. The ratios of d_E *vs.* actual biomass were almost equal to 1 throughout the experiments except for day 0 and day 2 which were 1.66 and 0.68, respectively. The fact that d_E was similar to the

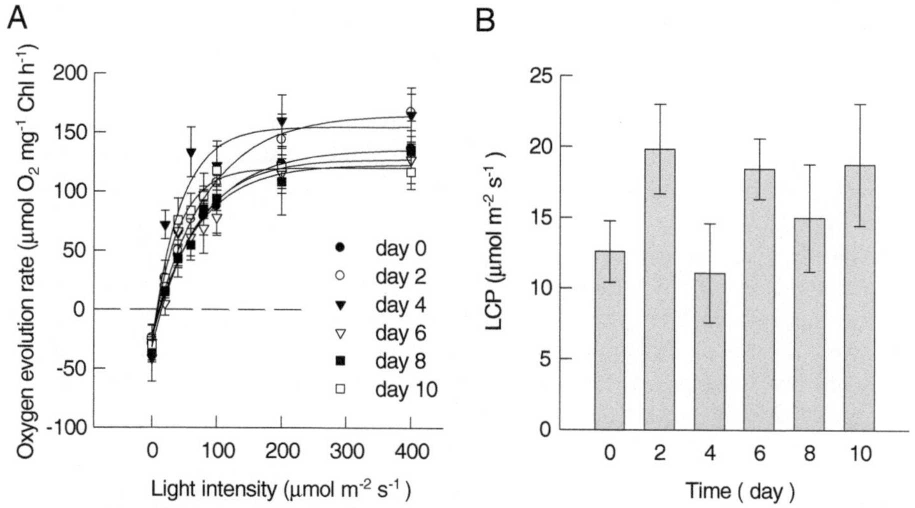

Figure 4 The oxygen evolution characters of *Scenedesmus dimorphus* under attached cultivation for different days of cultivation. **(A)** The oxygen evolution rate versus light intensity (light curve) of *S. dimorphus* under attached cultivation for 0 (black circle), 2 (white circle), 4 (black down-pointing triangle), 6 (white down-pointing triangle), 8 (black square), and 10 (white square) days of cultivation. Data were mean ± standard deviation of six measurements for three independent experiments (two measurements for each). **(B)** The light compensation point of *S. dimorphus* under attached cultivation for different days of cultivation. The ANOVA (SPSS, Chicago, IL, USA) result indicated that there were no differences for different days.

actual biomass density indicated that nearly 100% of the algal cells inside the immobilized biofilm were effectively illuminated under attached cultivation.

Chlorophyll is the main light-harvesting molecule for photosynthetic organism, but some of the carotenoid molecules can also capture light and pass the excited energy to chlorophyll [27]. In this research, the content of chlorophyll as well as carotenoid was measured and values have been expressed as both a function of the dry biomass and the cultivated surface area. From day 0 to day 10, the dry mass-based chlorophyll content decreased from 3.3% to 0.4% and the carotenoid content decreased from 0.5% to 0.2%. The areal chlorophyll content increased from 0.29 to 0.70 g m^{-2} during the first 8 days of attached cultivation and then decreased to 0.45 g m^{-2} at day 10. The areal carotenoid content increased from 0.05 to 0.26 g m^{-2} from day 0 to day 8 and then remained constant. In general, during the attached

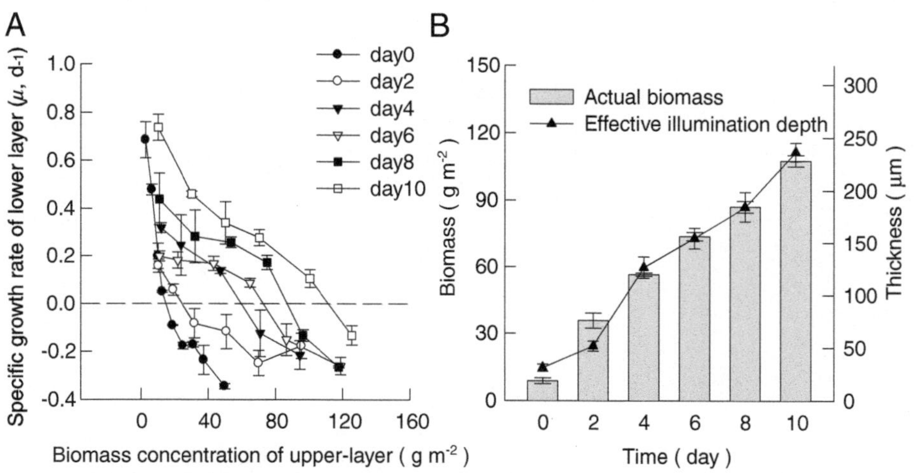

Figure 5 The estimation of effective illumination depth for the attached cultivation of *S. dimorphus* with twin-layer method. **(A)** The specific growth rate of lower layer *versus* the upper layer biomass density. Data were mean ± standard deviation of nine measurements for three independent experiments (three measurements for each). The upper layers were consisted with biomass grown with attached cultivated technique for 0 (black circle), 2 (white circle), 4 (black down-pointing triangle), 6 (white down-pointing triangle), 8 (black square), and 10 (white square) days. **(B)** The actual biomass density for the attached cultivation from day 0 to day 10 (closed bar), and the effective illumination depth for the attached cultivation from day 0 to day 10 (black triangle).

cultivation, the increase of the d_E was accompanied by the decrease in biomass-based chlorophyll and carotenoid contents as well as the increase in areal chlorophyll and carotenoid contents (Figure 6). Figure 7 suggests how the light transfer and distribution inside the immobilized microalgal biofilm are thought to be happening. In the earlier phase of attached cultivation, the pigment content in every single cell is relatively high, so that the light intensity decreased sharply when passing through each algal layer and can penetrate only a small depth of the algal biofilm. With the increase of the cultivation days, the algal layers as well as the pigment content along the light path increased but the pigment content in each single cell decreased, and only a moderate decrease in light intensity happens when passing through each algal layer. In the end, light can penetrate a bigger depth of the algal biofilm (Figure 7).

Comparison of the illumination aspects of the suspended and attached cultivation

The obvious difference between the two cultivation systems was that full illumination only happened for a short period of time (day 0 to day 3) in the open-pond system, when biomass concentration was very dilute (<20 g m^{-2}), whereas in the attached cultivation system, the full illumination happened during a prolonged interval of 10 days as well as a wider range of biomass density (up to 100 g m^{-2} for a single layer). This high percentage of photosynthetic active biomass might explain the high biomass productivity and photosynthetic efficiency of attached cultivation under high light conditions. For example, in our previous research with a multiple-plate attached cultivation bioreactor with a light dilution rate of 10 (refer to Figure 1A), the land areal biomass density increased from 163.7 to 484.5 g m^{-2} in 5 days [22] under outdoor high light conditions (1,500 ~ 2,000 μmol m^{-2} s^{-1}). Since the averaged light intensity impinging the cultivated surface was only about 100 ~ 150 μmol m^{-2} s^{-1}, we can assume that nearly 100% of the biomass in this case was effectively illuminated. Similar results were also achieved under indoor conditions. The high photosynthetic efficiency of *ca.* 18% (based on visible light) was reached for *Scenedesmus* indoors with a light intensity of 700 μmol m^{-2} s^{-1} in the land areal biomass density range of 192.5 to 828.1 g m^{-2} [22]. Photosynthetic efficiency of *ca.* 15% (based on visible light) was also reached for *Botryococcus braunii* indoors with a light intensity of 500 μmol m^{-2} s^{-1} in the standing areal biomass density range of *ca.* 70 to *ca.* 580 g m^{-2} [23]. It is hard to imagine an open pond working with such high percentage of photosynthetic active cells and consequently such a condensed biomass concentration range. In general, for the attached cultivation, the high light intensity is diluted and absorbed by a cultivated area hosting an extremely dense biomass, where there is virtually no space left between adjacent cells, which is totally photosynthetically active, whereas in conventional open ponds, the high light intensity is only absorbed by a much smaller amount of algal cells so that the average light energy received by each photosynthetically active cell is higher in the open pond and a large portion of light is dissipated in a non-photosynthetic pathway before reaching other cells.

Conclusions

In a conventional 20-cm deep open pond, all of the algal cells were fully illuminated for the first 3 days of outdoor cultivation. With the increase of the biomass concentration, the effective illuminated portion of the water layer decreased steeply to 45.5% from day 4 to day 10 and

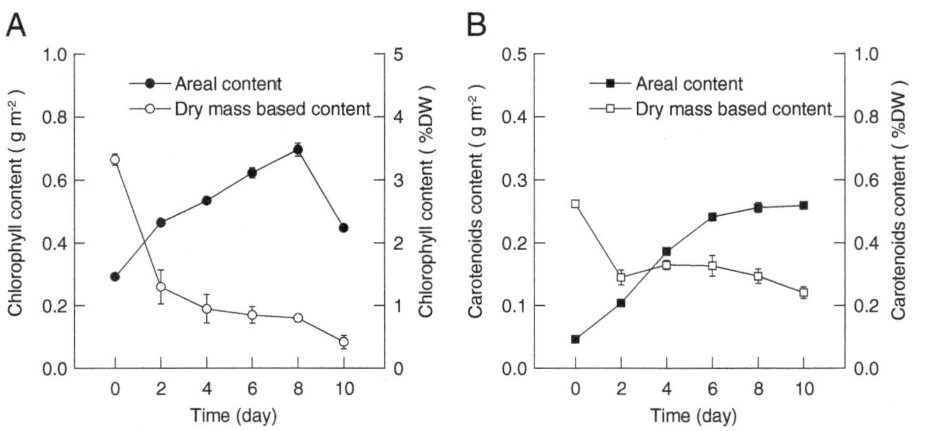

Figure 6 The chlorophyll and carotenoid contents for the attached cultivation of *S. dimorphus*. (A) The areal (black circle) and dry mass-based (white circle) chlorophyll contents for attached cultivation of *S. dimorphus* for different days of cultivation. Data were mean ± standard deviation of six measurements for three independent experiments (two measurements for each). **(B)** The areal (black square) and dry mass-based (white square) carotenoid contents for attached cultivation of *S. dimorphus* for different days. Data were mean ± standard deviation of six measurements for three independent experiments (two measurements for each).

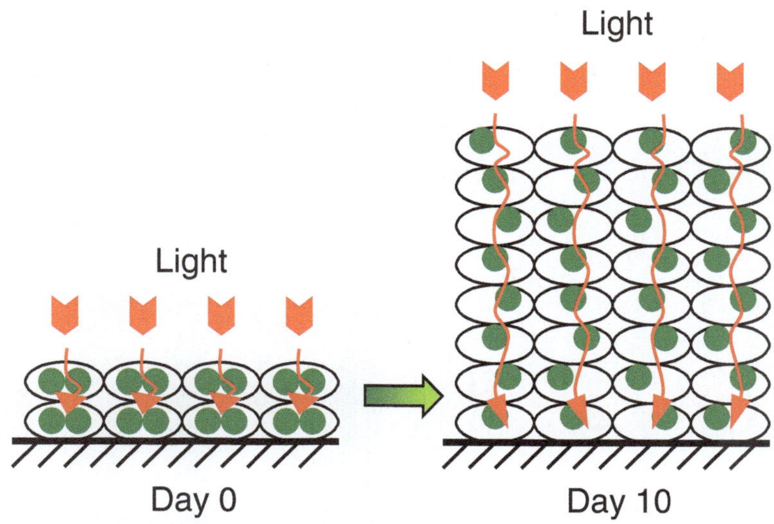

Figure 7 The schematic model for the chlorophyll and light distribution inside the attached cultivation biofilm of *S. dimorphus*. The small oval represents the algal cells inside the biofilm. The green dots indicate the chlorophyll content of algal cells. The suggested light penetration path inside the biofilm is indicated by red arrows.

then gradually decreased to 31.1% at day 31, when the biomass density reached 0.45 g L^{-1} or 90 g m^{-2}. In the attached cultivation system under nitrogen-replete condition, almost 100% of the immobilized algal cells inside the biofilm were effectively illuminated from day 0 through day 10, with biomass density increasing from 8.8 to 107.6 g m^{-2}. Attached cultivation facilitates super high biomass density (up to *ca.* 800 g m^{-2}) and super high efficient illumination (100%) when the reactor consists of multiple panels hosting the biofilm surface which might be the key to the high biomass productivity as well as high photosynthetic efficiency.

Methods

Microalgae strain and medium

The freshwater microalgal species *S. dimorphus* was locally screened in Qingdao, China. The algal cells were maintained in BG11 medium [28], each liter of which contains 1.5 g $NaNO_3$, 0.075 g $MgSO_4{\cdot}7H_2O$, 0.036 g $CaCl_2{\cdot}2H_2O$, 0.04 g $KH_2PO_4{\cdot}H_2O$, 0.02 g Na_2CO_3, 6.0 × 10^{-3} g citric acid, 1.0 × 10^{-3} g Na_2EDTA, 6.0 × 10^{-3} g ferric ammonium citrate, 2.22 × 10^{-4} g $ZnSO_4{\cdot}7H_2O$, 6.9 × 10^{-5} g $CuSO_4{\cdot}5H_2O$, 1.81 × 10^{-3} g $MnCl_2{\cdot}4H_2O$, 3.9 × 10^{-4} g $Na_2MoO_4{\cdot}2H_2O$, 4.94 × 10^{-5} g $Co(NO_3)_2{\cdot}6H_2O$ and 2.86 × 10^{-3} g H_3BO_3.

Inoculum preparation

The algal inoculum was prepared in glass columns (diameter = 0.05 m, working volume = 0.7 L) under indoor conditions of continuous illumination (100 μmol m^{-2} s^{-1}; cold fluorescent lamp, white light; FSL T8 36W, Foshan Electrical Lighting Co., Ltd., Foshan, China) and continuous aeration (1.0 vvm; 98% air + 2% CO_2, *v*/*v*). The temperature of the culture broth was 25°C ± 1°C. The algal biomass at late exponential phase (5~7 days after inoculation, 2~3 g L^{-1}) was used to inoculate the experimental systems.

Experimental design

The d_E for the algal biofilm of the attached cultivation and conventional aqueous suspended open pond were determined under indoor and outdoor conditions, respectively. For the attached cultivation, the d_E was measured with a single-layer attached cultivation system (Figure 1B) following the flow chart of Figure 8. Firstly, the algal cells were attached cultivated for 10 days with BG-11 (step 1; refer to the 'Cultivation systems' section for details). Every other day, some of the algal disks were sampled and the attached cells on it were detached and washed three times with de-ionized water (step 2). A 'marker' layer was built by filtering some fresh inoculum onto new cellulose acetate/nitrate membranes at the biomass density of *s* (g m^{-2}). Then, some aliquots of the re-suspended algal culture from step 2 were gently filtered onto the 'marker' layers to form 'double-layer' attached algal disks. The corresponding 'single-layer' attached algal 'disks' were also prepared at the same time as a control using identical aliquots of re-suspended algal culture but without the 'marker' layer (step 3). Both of the newly prepared 'double-layer' and 'single-layer' algal disks were cultivated for 24 h under the same environmental conditions using the PBR depicted in Figure 1B. The biomass increase for the 'double-layer' algal disk was denoted by *a* (g m^{-2}) and the biomass

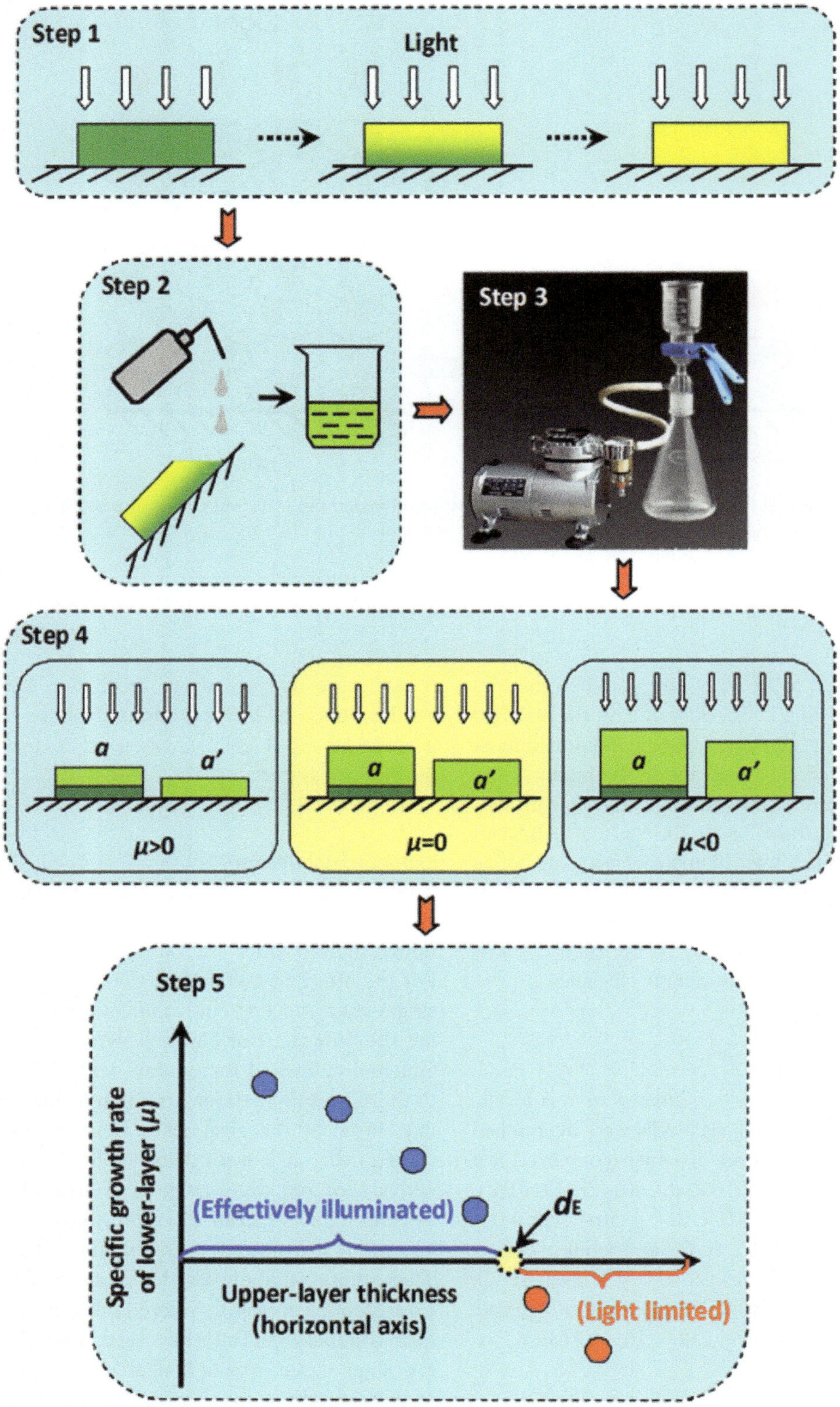

Figure 8 The schematic flow chart of the experiment design to determine the effective illumination depth for the attached cultivation.

increase for the corresponding 'single-layer' algal disk was a' (g m^{-2}), so that the net biomass gain for the 'marker' layer was calculated as $\delta = a - a'$ (g m^{-2}), and μ was calculated as:

$$\mu = \ln(s + \delta) - \ln(s) \tag{1}$$

If $\mu > 0$, it means that the 'marker' layer is receiving a light intensity higher than the LCP. In other words, the upper layer is not thick enough to fully attenuate the impinging light and the depth of the upper layer (d_A) was only a fraction of the d_E, $d_A < d_E$.

If $\mu < 0$, it means that the light intensity reaching the 'marker' layer is lower than the LCP and the upper layer is too thick to allow the 'marker' layer to accumulate biomass, $d_A > d_E$.

If $\mu = 0$, the light intensity at the 'marker' layer is equal to the LCP and the biomass increase of the biofilm due to photosynthesis is equal to the biomass loss due to cellular respiration. The thickness of the upper layer could be considered as the maximum distance the light can effectively penetrate, $d_A = d_E$ (Figure 8, step 4). In actual experiments, however, considering the difficulties in precisely manipulating the amount of the upper-layer biomass to make $d_A = d_E$, the d_E was generally approximately estimated as the intercept of the x-axis (Figure 8, step 5).

The percentage of the effectively illuminated cells (P_{light}, %) for attached cultivation was calculated as:

$$P_{light} = d_E \times 100/d_G \tag{2}$$

where d_G represents the actual thickness of the attached algal 'disk' at the considered sampling day.

The d_E in the suspended cultivation system was determined with a conventional open-pond system and a set of homemade equipment with different light paths (Figure 2A; refer to the sections 'Equipments for determining the effective illumination depth for suspended cultures' and 'Cultivation systems' for detailed specifications and operations). Firstly, the relationships of biomass concentrations *vs.* penetrated light intensities were measured (Figure 2B). Secondly, according to the LCP determined by oxygen evolution rates at different light intensities (Figure 2C; refer to the 'Determination of LCP' section for details), the d_E were calculated for each tested biomass densities. An equation was proposed to calculate the d_E at any given biomass density in the range of 0.07 ~ 6.4 g L^{-1} (Figure 2D). Finally, the changes in biomass density of *S. dimorphus* in open ponds were studied outdoors (Figure 3A), and the d_E was estimated according to the proposed equation. The percentage of the effectively illuminated cells (P_{light}, %; Figure 3B) was calculated as:

$$P_{light} = d_E \times 100/d_P \tag{3}$$

where the d_P represents the depth of the open pond, 0.2 m in this research.

Equipments for determining the effective illumination depth for suspended cultures

A series of white polyvinyl chloride (PVC) tubes (inner diameter = 0.03 m) with different lengths (2, 3, 5, 10, 20, 30, 40, and 50 cm; three tubes for each length) were prepared, and one of the open ends of these PVC tubes was covered with a piece of highly transparent polystyrene plate (thickness = 0.8 mm; >98% light transmittance). The outer wall of the PVC tubes and the outer circles of the polystyrene sheets were painted black to prevent the unwanted penetration of outside irradiance, whereas the inner wall and the inner circles (diameter = 1 cm) were not painted but thoroughly cleaned with water (Figure 2A). Algal suspensions with different biomass densities, that is, 0.07, 0.18, 0.39, 0.81, 1.59, 3.20 and 6.4 g L^{-1}, were prepared in beakers and then used to fill the PVC tubes, so that the heights of the algal broth, *viz.* light paths, were equal to the lengths of the PVC tubes. The PVC pipes were placed vertically in open air, under direct sunlight with an intensity of approximately 1,500 ~ 1,600 μmol m^{-2} s^{-1}. A photosynthetic quantum sensor (Li-250A, Li-Cor, Lincoln, NE, USA) was placed underneath each tube and tightly attached to the polystyrene sheets to measure the penetrated photon flux density.

When low water layer thicknesses were to be tested, that is, 0.084 ~ 1.05 cm, the algal suspension was poured into a Petri dish made of highly transparent polystyrene (thickness = 0.8 mm; >98% light transmittance) and then the penetrated light intensity was measured. The thickness of the water layer in the Petri dish was calculated according to the volume of algal suspension poured into and the diameter of the Petri dish used. The external surface of the Petri dish was also painted with black paint with the exception of a small circular window (diameter = 1 cm) where the quantum sensor was to be attached.

Cultivation systems

Two different cultivation systems were adopted in this research to evaluate the active proportions for photosynthesis. The first one was a mini open-pond system which measured 2 m in length, 1 m in width, and 0.2 m in culture depth. The culture broth was propelled by a paddle wheel at a flow speed of 0.3 m s^{-1}. A gas stream of pure CO_2 (>99%, v/v) was injected into the bottom of the pond through a ventilation stone at an aeration speed of 4 L min^{-1} (0.01 vvm) from 8 am to 4 pm to supply the carbon source for the algal cells and maintain a pH of 7.5 ± 0.5. The measured maximum light intensity at the water surface was *ca.* 2,000 μmol m^{-2} s^{-1} during the daytime. The temperature of the culture broth was 30°C~35°C from 12 pm to 2 pm in the afternoon on a sunny day and 10°C~25°C for the rest of the time. Two identical ponds were operated synchronously in outdoor conditions at Qingdao, China (35°35′ N, 119°30′ E) from 17 September to 17 October 2012, with the same inoculated biomass density of 0.04 g L^{-1} (8 g m^{-2}).

The second adopted cultivation system was the single-layer attached biofilm cultivation system (Figure 1B) which has been introduced in detail in our previous research as a 'type 1' photobioreactor [22]. In brief, algal cells were evenly filtered onto a cellulose acetate/nitrate membrane (pore size 0.45 μm) to form algal disks with 10 ± 0.5 cm^2 of footprint at an inoculated areal biomass density of 8 g m^{-2}. These algal disks were then put onto a layer of regular filter paper and finally onto a 0.2 m × 0.4 m glass plate (3 mm thick). The glass plate as well as the filter paper and algal disks on it were sealed in a 0.5 m × 0.3 m × 0.05 m chamber made out of 5-mm-thick transparent glass plates. During the cultivation, the algal disks were continuously illuminated with 100 ± 10 μmol m^{-2} s^{-1} of white light from cold fluorescent lamps. A continuous air stream that contained 1.5% (v/v) of CO_2 was injected into the chamber with a flow rate of 0.75 L min^{-1} (0.1 vvm). The BG11 medium was dripped onto the filter paper with a speed of 0.5 mL min^{-1} to keep the algal disks attached to the filter paper and provide nutrients for the growth of the algal cells. Only fresh medium was used, and any excess was immediately discarded. The temperature inside the glass chamber was kept at 25°C ± 2°C during the experiment.

Analysis

Growth analysis

In this research, the growth of the algal culture was investigated by measuring the changes in dry biomass concentration (DW). Some aliquots of the culture broth (open pond) or detached biomass from the 'algal disk' (attached cultivation) were first washed with 50 mL of de-ionized water and then filtered onto a pre-weighted 0.45-μm GF/C filter membrane (Whatman, Little Chalfont, England; DW_0). The membrane was oven dried at 105°C for 24 h and then weighted (DW_1).

The DW for open pond (g L^{-1}) was calculated as:

$$DW = (DW_1 - DW_0) \times 1,000/v \quad (4)$$

where v represents the sample volume expressed in milliliters.

The DW for attached cultivation was calculated as:

$$DW = (DW_1 - DW_0)/0.001 \quad (5)$$

where 0.001 represents the footprint area (m^2) of attached algal cells.

Determination of LCP

The LCP was determined according to the photosynthetic oxygen evolution rate *vs.* light intensity curve (PI curve); this was measured with a Chlorolab-2 liquid-phase Clark-type oxygen electrode (Hansatech, Norfolk, England). The newly prepared inoculum for the algal disks (refer to the 'Inoculum preparation' section) was collected by centrifugation at 5,000 × g for 30 s (Allegra X-22R, Beckman Coulter, Brea, CA, USA). The obtained pellet was then re-suspended with BG-11 medium containing 100 mM $NaHCO_3$ and adjusted to a proper biomass concentration so that the chlorophyll content was as close as possible to 10 μg per milliliter. One milliliter of the sample was added to the reaction cuvette and then bubbled with nitrogen gas for 30 s to expel the dissolved oxygen gas. Light intensities of 400, 200, 100, 80, 60, 40, 20 and 0 μmol m^{-2} s^{-1} were applied to drive the photosynthetic oxygen evolution. The measurement was repeated three times for each light intensity. The LCP was obtained as the intercept on the horizontal axis of the PI curve [29].

Determination of the thickness of the attached algal disk

Prior to the experiments, the thickness of the algal disks at different cultivation days was measured by means of a surface profiler (Veeco Dektak150, Veeco Instruments Inc., Plainview, NY, USA) and the corresponding areal biomass density was also measured (refer to the 'Growth analysis' section). Accordingly, the correlation of the thickness (d, expressed in μm) of the algal film with the areal biomass density (DW, in g m^{-2}) was obtained as follows:

$$d = 2.1 \times DW + 3.2 \quad (6)$$

For later experiments, the thickness (μm) of the attached algal disks during cultivation was estimated by this equation. According to the pilot experiments, this equation could be applied in the biomass range of 0 ~ 120 g m^{-2} for *S. dimorphus* grown with the attached cultivation technique for up to 10 days under both nitrogen-replete or nitrogen-depleted conditions ($R^2 = 0.99$).

Determination of the chlorophyll and carotenoid contents

The chlorophyll and carotenoid contents were determined according to the methods described by Wellburn [30]. The algal biomass was harvested and washed with plenty of de-ionized water and then mixed with methanol under 60°C for 12 h in darkness until the algal cells whitened completely. The optical densities of the extraction were measured with a spectrophotometer at 666, 653, and 470 nm (Cary 50, Varian Inc., Palo Alto, CA, USA). The chlorophyll a (Chl_a), chlorophyll b (Chl_b), and carotenoid (Car) concentrations in the extraction were calculated as follows (mg L^{-1}):

$$Chl_a = 15.65 \times OD_{666} - 7.34 \times OD_{653} \quad (7)$$

$$Chl_b = 27.05 \times OD_{653} - 11.21 \times OD_{666} \quad (8)$$

$$Car = (1,000 \times OD_{470} - 2.86 \times Chl_a - 129.2 \times Chl_b)/221 \quad (9)$$

Abbreviations

a: biomass increase for the 'twin-layer' algal disk; *a'*: biomass increase for the corresponding 'single-layer' algal disk; Car: concentration of carotenoid; Chl_a: concentration of chlorophyll a; Chl_b: concentration of chlorophyll b; *D*: thickness of the algal film; d_A: the depth of the upper layer; d_E: effective illumination depth; d_G: actual thickness of the attached algal 'disk' at different sampling days; DW: dry biomass concentration; DW_0: dry weight of the filter membrane; DW_1: dry weight of the filter membrane after filtering the algal cells; LCP: light compensation point; PBR: photobioreactor; PI curve: curve of photosynthetic oxygen evolution rate *vs*. light intensity; P_{light}: percentage of the effectively illuminated cells; PVC: polyvinyl chloride; *V*: volume of the sample for biomass measurement; δ: net biomass gain for the 'seed' layer; μ: specific growth rate.

Competing interests

The authors declare that they have no competing interests.

Authors' contributions

JW and TL proposed the idea and hypothesis. JW carried out the experiment design and outdoor open-pond experiment and drafted the manuscript. JL carried out the biofilm experiment. TL performed the statistical analysis and helped to draft and revise the manuscript. All authors read and approved the final manuscript for publication.

Acknowledgements

This work was supported by the National Natural Science Foundation of China (41276144), the Director Innovation Foundation of Qingdao Institute of Bioenergy and Bioprocess Technology, CAS, and the Science and Technology Development Planning of Shandong Province (2013GHY11520).

Author details

[1]Key Laboratory of Biofuels, Qingdao Institute of Bioenergy and Bioprocess Technology, Chinese Academy of Sciences, Qingdao, Shandong 266101, People's Republic of China. [2]Low-Carbon Energy Conversion Science and Engineering Center, Shanghai Advanced Research Institute, Chinese Academy of Sciences, Shanghai 201203, People's Republic of China.

References

1. Hu Q, Sommerfeld M, Jarvis E, Ghirardi M, Posewitz M, Seibert M, et al. Microalgal triacylglycerols as feedstocks for biofuel production: perspectives and advances. Plant J. 2008;54:621–39.
2. Chisti Y. Biodiesel from microalgae. Biotechnol Adv. 2007;25:294–306.
3. Mata TM, Martins AA, Caetano NS. Microalgae for biodiesel production and other applications: a review. Renew Sustain Energ Rev. 2010;14:217–32.
4. Sostaric M, Klinar D, Bricelj M, Golob J, Berovic M, Likozar B. Growth, lipid extraction and thermal degradation of the microalga *Chlorella vulgaris*. New Biotechnol. 2012;29:325–31.
5. Lam MK, Lee KT. Microalgae biofuels: a critical review of issues, problems and the way forward. Biotechnol Adv. 2012;30:673–90.
6. Larkum AWD, Ross IL, Kruse O, Hankamer B. Selection, breeding and engineering of microalgae for bioenergy and biofuel production. Trends Biotechnol. 2012;30:198–205.
7. Weisz PB. Basic choices and constraints on long-term energy supplies. Phys Today. 2004;57:47–52.
8. Stephens E, Ross IL, Mussgnug JH, Wagner LD, Borowitzka MA, Posten C, et al. Future prospects of microalgal biofuel production systems. Trends Plant Sci. 2010;15:554–64.
9. Wijffels R, Barbosa M. An outlook on microalgal biofuels. Science. 2010;329:796–9.
10. Brennan L, Owende P. Biofuels from microalgae - a review of technologies for production, processing, and extractions of biofuels and co-products. Renew Sustain Energ Rev. 2010;14:557–77.
11. Chisti Y. Raceways-based production of algal crude oil. In: Posten C, Walter C, editors. Microalgal biotechnology: potential and production. Berlin: de Gruyter; 2012. p. 113–46.
12. Tredici MR. Photobiology of microalgae mass cultures: understanding the tools for the next green revolution. Biofuels. 2010;1:143–62.
13. Zhu XG, Long SP, Ort DR. What is the maximum efficiency with which photosynthesis can convert solar energy into biomass? Curr Opin Biotechnol. 2008;19:153–9.
14. Shi J, Podola B, Melkonian M. Removal of nitrogen and phosphorus from wastewater using microalgae immobilized on twin layers: an experimental study. J Appl Phycol. 2007;19:417–23.
15. Mulbry W, Kondrad S, Pizarro C, Kebede-Westhead E. Treatment of dairy manure effluent using freshwater algae: algal productivity and recovery of manure nutrients using pilot-scale algal turf scrubbers. Bioresour Technol. 2008;99:8137–42.
16. Boelee NC, Temmink H, Janssen M, Buisman CJN, Wijffels RH. Nitrogen and phosphorus removal from municipal wastewater effluent using microalgal biofilms. Water Res. 2011;45:5925–33.
17. Naumann T, Çebi Z, Podola B, Melkonian M. Growing microalgae as aquaculture feeds on twin-layers: a novel solid-state photobioreactor. J Appl Phycol. 2013;25:1413–20.
18. Zamalloa C, Boon N, Verstraete W. Decentralized two-stage sewage treatment by chemical–biological flocculation combined with microalgae biofilm for nutrient immobilization in a roof installed parallel plate reactor. Bioresour Technol. 2013;130:152–60.
19. Bayless DJ, Kremer G, Vis M, Stuart B, Shi L. Photosynthetic CO_2 mitigation using a novel membrane-based photobioreactor. J Environ Eng Manag. 2006;16:209–15.
20. Johnson MB, Wen Z. Development of an attached microalgal growth system for biofuel production. Biotechnol Prod Process Eng. 2010;85:525–34.
21. Ozkan A, Kinney K, Katz L, Berberoglu H. Reduction of water and energy requirement of algae cultivation using an algae biofilm photobioreactor. Bioresource Technol. 2012;114:542–8.
22. Liu T, Wang J, Hu Q, Cheng P, Ji B, Liu J, et al. Attached cultivation technology of microalgae for efficient biomass feedstock production. Bioresour Technol. 2013;127:216–22.
23. Cheng P, Ji B, Gao L, Zhang W, Wang J, Liu T. The growth, lipid and hydrocarbon production of Botryococcus braunii with attached cultivation. Bioresour Technol. 2013;138:95–100.
24. Ji C, Wang J, Zhang W, Liu J, Wang H, Gao L, et al. An applicable nitrogen supply strategy for attached cultivation of *Aucutodesmus obliquus*. J Appl Phycol. 2014;26:173–80.
25. Chisti Y. Constraints to commercialization of algal fuels. J Biotechnol. 2013;167:201–14.
26. Wang J, Han D, Sommerfeld M, Lu C, Hu Q. Effect of initial biomass density on growth and astaxanthin production of *Haematococcus pluvialis* in an outdoor photobioreactor. J Appl Phycol. 2013;25:253–60.
27. Green BR, Durnford DG. The chlorophyll-carotenoid proteins of oxygenic photosynthesis. Annu Rev Plant Physiol Plant Mol Biol. 1996;47:685–714.
28. Stanier RY, Kunisawa R, Mandel M, Cohenbaz G. Purification and properties of unicellular blue-green algae (order Chroococcales). Bacteriol Rev. 1971;35:171–205.
29. Ye Z. A new model for relationship between light intensity and the rate of photosynthesis in *Oryza sativa*. Photosynthetica. 2007;45:637–40.
30. Wellburn AR. The spectral determination of chlorophylls *a* and *b*, as well as total carotenoids, using various solvents with spectrophotometers of different resolution. J Plant Physiol. 1994;144:307–13.

Uniconazole-induced starch accumulation in the bioenergy crop duckweed (*Landoltia punctata*) II: transcriptome alterations of pathways involved in carbohydrate metabolism and endogenous hormone crosstalk

Yang Liu[1,2,3,4†], Yang Fang[1,3,4†], Mengjun Huang[1,2,3,4], Yanling Jin[1,3,4], Jiaolong Sun[1,3,4], Xiang Tao[1,3,4], Guohua Zhang[1,3,4], Kaize He[1,3,4], Yun Zhao[5] and Hai Zhao[1,3,4*]

Abstract

Background: *Landoltia punctata* is a widely distributed duckweed species with great potential to accumulate enormous amounts of starch for bioethanol production. We found that *L. punctata* can accumulate starch rapidly accompanied by alterations in endogenous hormone levels after uniconazole application, but the relationship between endogenous hormones and starch accumulation is still unclear.

Results: After spraying fronds with 800 mg/L uniconazole, *L. punctata* can accumulate starch quickly, with a dry weight starch content of up to 48% after 240 h of growth compared to 15.7% in the control group. Electron microscopy showed that the starch granule content was elevated after uniconazole application. The activities of key enzymes involved in starch synthesis were also significantly increased. Moreover, the expression of regulatory elements of the cytokinin (CK), abscisic acid (ABA) and gibberellin (GA) signaling pathways that are involved in chlorophyll and starch metabolism also changed correspondingly. Importantly, the expression levels of key enzymes involved in starch biosynthesis were up-regulated, while transcript-encoding enzymes involved in starch degradation and other carbohydrate metabolic branches were down-regulated.

Conclusion: The increase of endogenous ABA and CK levels positively promoted the activity of ADP-glucose pyrophosphorylase (AGPase) and chlorophyll content, while the decrease in endogenous GA levels inactivated α-amylase. Thus, the alterations of endogenous hormone levels resulted in starch accumulation due to regulation of the expression of genes involved in the starch metabolism pathway.

Keywords: Bioethanol, Starch accumulation, Endogenous hormones, Uniconazole, Crosstalk, Pathway

* Correspondence: zhaohai@cib.ac.cn
[†]Equal contributors
[1]Chengdu Institute of Biology, Chinese Academy of Sciences, No.9 Section 4, Renmin Nan Road, 610041 Chengdu, China
[3]Key Laboratory of Environmental and Applied Microbiology, Chinese Academy of Sciences, No.9 Section 4, Renmin Nan Road, 610041 Chengdu, China
Full list of author information is available at the end of the article

Background

Environmental pollution, global warming, and energy shortages are urgent problems for sustainable development. To gradually decrease our excessive dependence on oil and reduce greenhouse gas emissions, many countries are looking for alternative energy sources. Renewable and clean bioethanol is a promising alternative to oil. However, most feedstocks for bioethanol production are terrestrial crops, such as corn, cassava, and sweet potato, which may compete with food or feed crops for agricultural land and may lead to other environmental problems [1-3]. Therefore, it is necessary to explore novel feedstocks to make the development of the bioethanol industry more sustainable and environmentally friendly.

Duckweed, the smallest and fastest-growing aquatic plant on earth [4], has become a novel potential alternative for bioethanol production in recent years [5]. Duckweed can double its biomass in 16 h to 2 days [6] and hence grows much faster than most other higher plants [7]. The growth rate of duckweed can reach 12.4 g/m^2/day dry weight in warm regions [8], and its yield has been documented up to 26.50 tons/ha/year dry weight [9]. The dry weight of starch content can reach 75% under ideal growth conditions [10]. Moreover, duckweed can grow on eutrophic wastewater to recover pollution nutrients, and it has been widely applied for wastewater treatment, including industrial wastewater and domestic sewage [11,12]. Importantly, duckweed biomass exhibits good characteristics for bioethanol production due to its relatively high starch and low lignin percentages [13], and it has been successfully converted to bioethanol in recent years [14,15]. Therefore, duckweed could be an ideal candidate for renewable bioenergy sources. Duckweed has great potential to accumulate high starch, and a high starch percentage is the key to energy utilization for duckweed. The starch content of duckweed can be considerably increased by manipulating growing conditions, such as phosphate concentration, nutrient starvation [16], and plant growth regulators [17-19]. Plant growth regulators are common and efficient synthetic compounds that are widely used to regulate plant growth and development [20,21]. To obtain high quality and quantity duckweed for bioethanol utilization, we systematically screened more than 20 plant growth regulators, including auxin, cytokinins (CKs), abscisic acid (ABA), and gibberellins (GAs), to improve the starch and biomass yield of duckweed. Screening results showed that uniconazole can be used as an effective candidate for starch and biomass accumulation of duckweed in Hoagland nutrient solution. Uniconazole (S3307) is a potent and active member of the triazole family that was developed as plant growth retardants [22]. It can enhance plant photosynthetic rates, increase soluble protein and total sugar content, elevate yield components in various crop plants [23,24], and change the endogenous hormone content [25]. However, there have been few in-depth studies into the responsive mechanism of plant growth regulators. There has thus far been little research linking uniconazole with expression changes in hormone biosynthesis enzymes and on the roles of certain hormone variations that cause high starch accumulation. Starch is the major storage form of sugar and energy in plants. The synthesis of starch in plant cells begins with the enzyme ADP-glucose pyrophosphorylase (AGPase), which catalyzes the reaction of glucose-1-phosphate with ATP to form ADP-glucose. The ADP-glucose is then used a substrate by starch synthase (SS) enzymes to build up a starch molecule. Branches in the chain are introduced by starch-branching enzymes (SBEs), which hydrolyse 1, 4-glycosidic bonds, and in their place, create 1, 6 bonds with other glucose units. [26,27].

Next-generation sequencing (NGS) technology is a new development of sequencing technology, and it can provide a novel method to uncover transcriptomics data. It is difficult to research metabolic pathways using conventional biological techniques in non-model plants. However, NGS technologies are not limited to detecting transcripts that correspond to existing genomic sequences, it is particularly attractive for non-model plants with genomic sequences that are yet to be determined [28,29]. This technology has been applied to investigations in some non-model plants and was successfully used to study metabolic pathways in duckweed last year [30]. In the accompanying report, we showed that uniconazole elevated chlorophyll content, enhanced the net photosynthetic rate, and altered the endogenous hormone levels of duckweed (data not shown). However, the relationship between the alteration of endogenous hormone levels and starch accumulation is still unclear. In this study, we constructed a comprehensive transcriptome using NGS technology in combination with physiological and biochemical analyses to investigate the process of starch accumulation mediated by endogenous hormones in *Landoltia punctata*.

Results

Impact of uniconazole on starch accumulation of *L. punctata*

L. punctata 0202, originally collected from Sichuan, China, is a widely distributed duckweed species with great potential for starch accumulation. In this study, frond samples were collected at 13 time points after treatment with uniconazole for measurement of starch percentages and the activity of enzymes related to starch metabolism.

As shown in Figure 1, the starch content was 3.16% (DW) at 0 h, but reached 10.31% at 3 h post-treatment. The starch content reached 19.46% (DW) at 12 h and finally reached 48.01% (DW) at 240 h following uniconazole

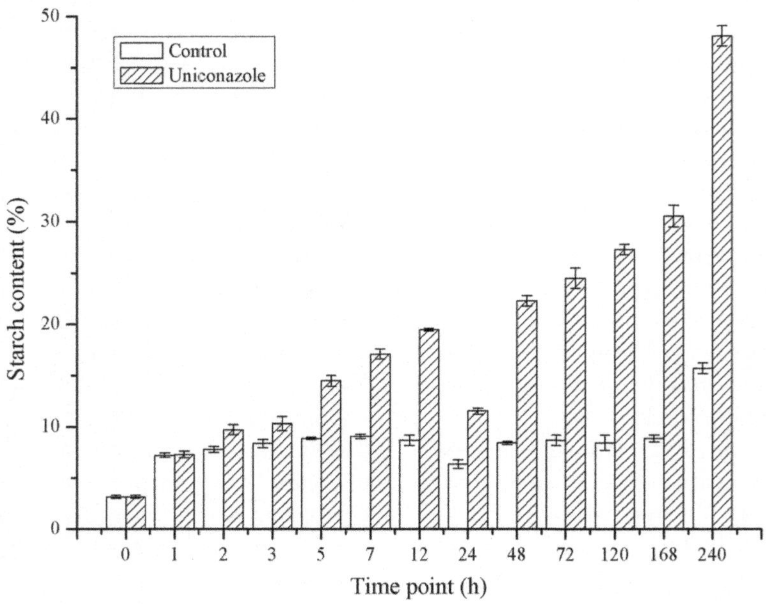

Figure 1 Starch percentage of uniconazole-treated *L. punctata*. Fronds were collected at different time points and used for starch percentage analysis. The starch percentage was calculated basing on dry weight. Each data point represents the mean of triplicate values; error bars indicate the standard deviation.

treatment. The starch content in the control fronds over the same time course remained mostly steady, reaching 8.7% (DW) at 12 h and 15.68% (DW) at 240 h post-treatment. Next, frond samples were examined by electron microcopy (Figure 2). The control frond cell contained several chloroplasts with a few small starch granules. In the uniconazole treatment group, several huge starch granules were found in the chloroplasts.

Effect of uniconazole treatment on the activities of enzymes involved in starch metabolism

To gain insight into the rapid accumulation of high starch following uniconazole application, the activity of enzymes involved in starch metabolism was analyzed. The activities of two of the most important key enzymes involved in starch synthesis (starch biosynthesis related enzymes AGPase and soluble starch synthase (SSS)) were measured (Figure 3). The activity of AGPase increased significantly from 8.20 to 27.59 U/mg protein at 5 h. After 24 h, the activity of AGPase increased slightly until 48 h when it reached a nearly stable level, while the activities of AGPase remained mostly constant in the control sample. SSS activity increased from the initial 8.03 to 25.69 U/mg protein at 2 h, and then decreased to 7.24 U/mg protein at 240 h. SSS activity did not change in the control sample.

The activities of starch degradation related enzymes in *L. punctata* were also investigated (Figure 3). The activity of alpha-amylase (α-amylase) was very low and changed very little in both in the control and treatment samples. The activity of α-amylase was 0.0025 U/mg protein at 0 h and reached to 0.037 and 0.0051 U/mg protein in the treated and control samples at 240 h, respectively. However, following treatment, beta-amylase (β-amylase) activity increased gradually from 0.0319 to 0.0747 U/mg protein at 1 h, and finally increased to the 0.3446 U/mg protein at 240 h in the treated sample, with little change in the control sample.

Sequencing, *de novo* assembly, and functional annotation of the *L. punctata* transcriptome

To investigate the genome-wide expression patterns of uniconazole treated *L. punctata*, samples collected at the 0, 2, 5, 72, and 240 h time points were used for RNA-Seq analysis. Results indicated that most of the contigs were protein-encoding transcripts. For more details on assembly statistics of the *L. punctata* transcriptome, please see the accompanying report (data not shown). To analyze temporal expression patterns of each transcript following uniconazole treatment, all RNA-Seq reads from each *L. punctata* sample were used for mapping analysis. The expression value of each transcript was calculated and normalized according to the RESM-based algorithm. We identified 70,090, 71,268, 71,170, 75,092, and 95,367 transcripts expressed at 0, 2, 5, 72, and 240 h, respectively (Figure 4). According to edgeR [31], compared with 0 h, there were 9,722, 11,139, 15,261, and 25,323 transcripts that were

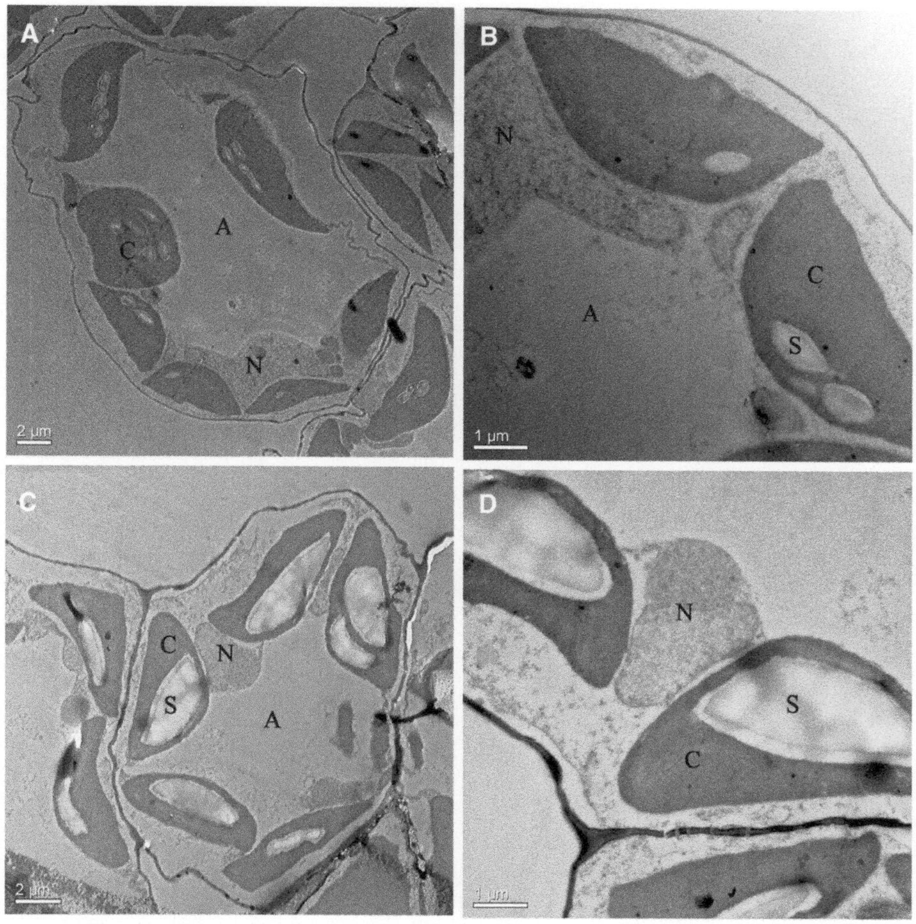

Figure 2 Transmission electron micrographs (TEM) study of *L. punctata*. (**A**) TEM picture of frond cells under lower magnification without treatment, Bars = 2 µm; (**B**) TEM picture of a section of a frond cell under higher magnification without treatment, Bar = 1 µm; (**C**) TEM picture of frond cells under lower magnification treated by uniconazole, Bar = 2 µm; (**D**) TEM picture of a section of a frond cell under higher magnification treated with uniconazole, Bar = 1 µm; Abbreviations are chloroplast (C), starch granule (S), intercellular air space (A), nucleus (N).

significantly differentially expressed at 2, 5, 72, and 240 h, respectively (Additional file 1: Table S1). Among these four differentially expressed transcripts (DETs: 2 *vs* 0 h, 5 *vs* 0 h, 72 *vs* 0 h, 240 *vs* 0 h), 2,929 transcripts were shared between these four DET sets. Compared with each other in these four DET sets, 5,503 transcripts were shared between 2 *vs* 0 h and 5 *vs* 0 h, 6,281 were shared between 5 *vs* 0 h and 72 *vs* 0 h, 5,339 transcripts were shared between 2 *vs* 0 h and 72 *vs* 0 h, 5,738 transcripts were shared between 2 *vs* 0 h and 240 *vs* 0 h, 6,641 transcripts were shared between 5 *vs* 0 h and 24 *vs* 0 h, and 10,621 were shared between 72 *vs* 0 h and 240 *vs* 0 h.

Expression analysis of transcript-encoding regulatory proteins and transcription factors involved in the CK, ABA, and GA signaling pathways

The transcripts of regulatory proteins and transcription factors involved in the CK, ABA, and GA signaling pathways changed significantly in response to uniconazole treatment (Figure 5). Cytokinins are degraded by cytokinin oxidase/dehydrogenase (CKXs) which is thought to play a key role in controlling cytokinin levels in plants. The expression of CKX was down-regulated from 11.97 fragments per kilobase of transcripts per million mapped fragments (FPKM) at 0 h to 4.22 FPKM at 240 h. In the signaling pathway of cytokinin, the cytokinin receptors histidine kinases (HKs) were up-regulated. For example, HK3 expression increased from 27.22 to 49.06 FPKM at 72 h (Additional file 2: Table S2). Their downstream elements histidine phosphotransfer proteins (AHPs) carry conserved amino acids required for phosphotransfer via a conserved histidine residue. There was no significant change in expression of AHPs. Type-B *Arabidopsis thaliana* response regulator (type-B ARR) interacts with the promoter of STAY-GREEN2 (SGR2) and interrupts transcription of the chlorophyll degradation pathway. SGR expression was up-regulated significantly from an initial value of 8.46 to 130.77 FPKM.

The core ABA signaling components have been well described in recent years. The family of START proteins

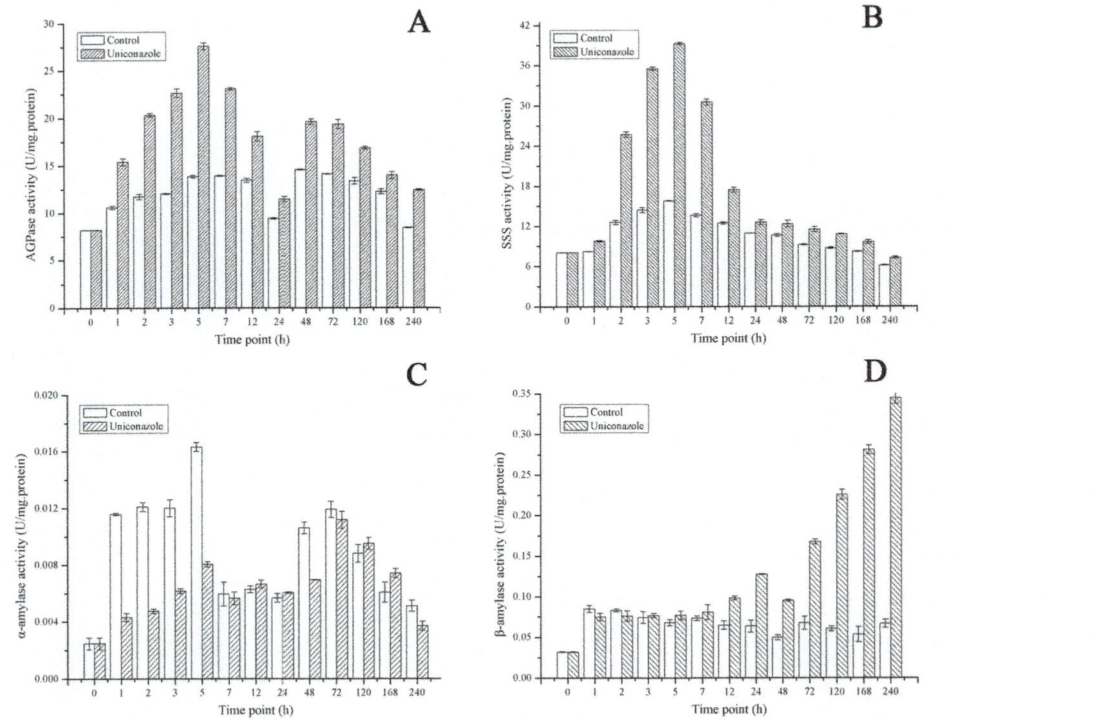

Figure 3 AGPase, SSS, α-amylase, and β-amylase activity. Fronds were collected at different time points and used for starch metabolism-related enzymatic activity assay after uniconazole treatment. (**A**) The activity of ADP-glucose pyrophosphorylase (AGPase); (**B**) The activity of soluble starch synthase (**C**) The activity of α-amylase; (**D**) The activity of β-amylase. All data are presented as the mean of triplicate measurements ± standard deviation.

(PYLs) act as ABA receptors, and 13 of 14 members of the Arabidopsis PYL family have been identified. PYL1 and PYL8 were identified in duckweed, and the expression levels of PYL1 increased from 36.88 FPKM at 0 h to 45.27 FPKM at 240 h. The expression of the PP2C negative regulator (comp31119_c0_seq1) decreased from 308.55 FPKM at 0 h to 122.48 FPKM at 240 h. There were no significant changes observed for the expression of transcript-encoding abscisic acid insensitive 4 (ABI4) (comp31717_c0_seq1). The expression was 0.14, 0.22, 0.05, 0.09 and 0.14 FPKM, respectively. The expression of two identical large subunits of AGPase (AGP-LS comp37255_c0_seq1) was up-regulated from the initial value of 122.13 to 192.48 FPKM at 240 h (Additional file 2: Table S2).

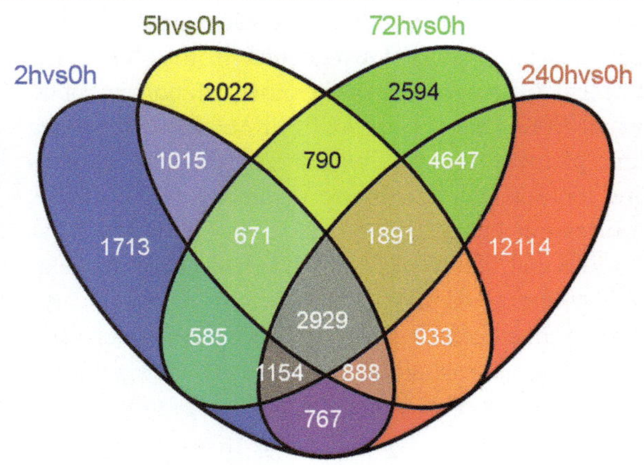

Figure 4 Differential expression between each pair of samples. Venn diagram showing unique and shared genes between time points. Overlapping examinations were performed based on the resulting gene lists from four comparisons by VENNY [74]. Overlap among four groups, 2 *vs* 0 h (blue), 5 *vs* 0 h (yellow), 7 *vs* 0 h (green), and 240 *vs* 0 h (red) are shown.

In the GAs signal pathway, transcriptomics data showed that the expressions of GA receptor GA insensitive dwarf 1 (GID1 comp41567_c0_seq1) were down-regulated from 44.03 to 23.93, 28.28, and 33.5 FPKM at different time points. The expressions of transcript-encoding DELLA proteins (comp33138_c1_seq1) were down-regulated from 6.3 to 4.55, 3.05, 3.61, and 2.12 FPKM, respectively. Moreover, the expression levels of transcript-encoding α-amylase were down-regulated from the initial value of 8.03 to 6 FPKM.

Expression analysis of transcript-encoding key enzymes involved in starch accumulation

Starch is the major storage carbohydrate in plants. To investigate the mechanisms by which uniconazole treatment resulted in starch accumulation, the expression patterns of transcript-encoding key enzymes were analyzed

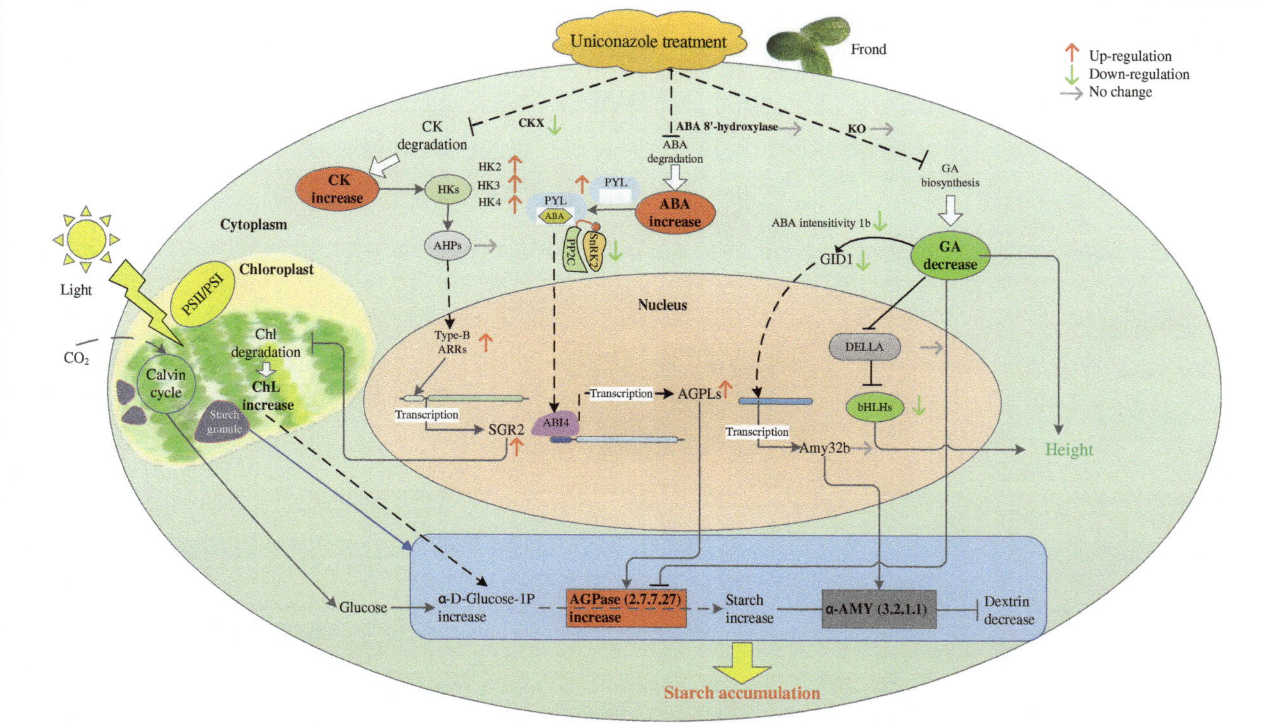

Figure 5 A hypothetical model of cytokinin, abscisic acid, and gibberellin signal pathways related to carbohydrate metabolism. Red indicates up-regulated expression, green down-regulated gene expression, gray means no significant difference was observed, and white means this enzyme was not found in this study. The major signaling pathways are indicated by black lines and arrows. Dotted arrowed lines indicate indirect or unconfirmed connections. Blue arrow indicates enlarged image. Cytokinin is perceived by the cytokinin receptor HKs. Cytokinin binding to HKs activates autophosphorylation (*P*) via AHPs (histidine phosphotransfer proteins) in the cytoplasm. Then type-B *Arabidopsis thaliana* response regulator (type-B ARR) interacts with the promoter of STAY-GREEN2 (SGR2). The family of START proteins (PYLs) act as ABA receptors. ABA combines with intracellular PYL and type 2C protein phosphatase (PP2C) to form an ABA-PYL-PP2C complex. This complex inhibits the activity of PP2C in an ABA-dependent manner and activates SNF1-related protein kinase 2 families (SnRK2s). Abscisic acid insensitive 4 (ABI4) induces ADP-glucose pyrophosphorylase subunit AGPLs (*ApL3*) gene expression. The main components of the GA signal pathway include GA receptor (GID1) and DELLA growth inhibitors. The GA-GID1-DELLA complex stimulates the degradation of DELLAs to regulate plant growth. GID1 regulated the transcription of amylase by a number of transcriptional regulatory.

(Figure 6). Gene expression profiling results showed that the expression of transcript-encoding AGPase were up-regulated from 24 to 95 FPKM at 240 h (comp37852_c0_seq1). Transcript-encoding granule-bound starch synthase (GBSS) exhibited an expression level of 106 FPKM at 0 h and increased to 311 FPKM at 240 h (comp31254_c0_seq1). There were no significant changes observed for the expression of transcript-encoding SSS and SBE (Additional file 3: Table S3).

Transcript-encoding enzymes involved in starch degradation and other carbohydrate metabolic branches were also analyzed. The expression level of transcript-encoding trehalose-6-phosphate synthase (EC: 2.4.1.15; TPS), which catalyze the biosynthesis of trehalose using UDPGlucose as substrate, was down-regulated from 78.1 to 18.7 FPKM (comp36386_c1_seq6) at 240 h after uniconazole treatment. Carbohydrate metabolic branches that compete with the synthesis of starch were also measured, including hexokinase (EC: 2.7.1.1), beta-glucosidase (EC: 3.2.1.21), phosphoglucomutase (EC: 5.4.2.2), sucrose-phosphate synthase (EC: 2.4.1.14), and sucrose synthase (SuSy EC: 2.4.1.13). Specifically, hexokinase (comp24929_c1_seq2) was significantly down-regulated from 59 to 18 FPKM at 240 h. No significant increase was observed for the expression of transcript-encoding α-amylase. However, transcript-encoding β-amylase exhibited an expression level of 15 FPKM at 0 h and increased to 309 FPKM at 240 h (comp16912_c0_seq1).

Discussion

The relationship between endogenous hormones induced by uniconazole and starch accumulation in *L. punctata*

Numerous studies have investigated different types of plant growth regulators that regulate growth and development in plants. Some articles focus their investigations on certain stress responses mediated by one or two types of plant hormones [32-34]. These studies often analyze phenotypic, biochemical, and physiological data. Some studies research the function of regulatory elements on hormones of signaling pathways [35,36]. Other studies investigate the relationship

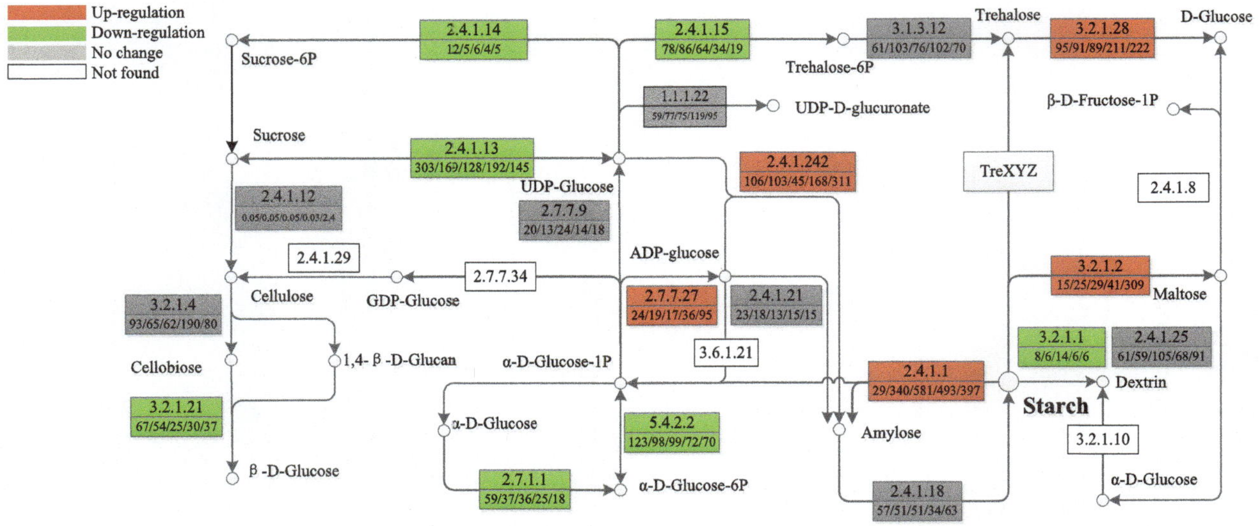

Figure 6 Expression patterns of carbohydrate metabolism-related transcripts. Expression variations of some carbon metabolism-related transcripts are displayed in the simplified starch and sucrose metabolism pathway. Red boxes indicate the up-regulated enzymes involved in carbohydrate metabolism, green boxes indicate the down-regulated enzymes, gray boxes mean no significant difference was observed, and white boxes mean this enzyme was not found in this study. The numbers in the upper half of the boxes correspond to the EC numbers and the numbers in the lower half, separated by slashes, correspond to the expression levels of these enzymes shown in FPKM at 0, 2, 5, 72, and 240 h, respectively. 1.1.1.22: UDP-glucose 6-dehydrogenase; 2.4.1.1: glycogen phosphorylase; 2.4.1.13: sucrose synthase; 2.4.1.14: sucrose phosphate synthase; 2.4.1.21: soluble starch synthase; 2.4.1.15: trehalose-6-phosphate synthase; 2.4.1.18: starch-branching enzyme; 2.4.1.12: cellulose synthase; 2.4.1.242: granule bound starch synthase; 2.7.7.27: ADP-glucose pyrophosphorylase; 2.7.7.9: UDP-glucose pyrophosphorylase; 2.7.1.1: hexokinase; 3.2.1.1: alpha-amylase; 3.2.1.2: beta-amylase; 3.1.3.12: trehalose 6-phosphate phosphatase; 3.2.1.4: endoglucanase; 3.2.1.28: trehalase; 5.4.2.2: phosphoglucomutase.

between different types of hormones [37-39]. Regulatory proteins are the main focus in these articles, which utilize molecular biology techniques with little to no focus on metabolic pathways. Additionally, some articles used transcriptome analyses to study the response of plants treated with hormones [40-42]; however, these articles emphasized the up- or down-regulation of genes or discovery of new genes, but did not consider metabolic pathways. Li [43] used RNA sequencing technology to understand the mechanisms of parthenocarpy and predicted 14 genes as putative parthenocarpic genes. The transcription analyses of these candidate genes revealed that auxin, cytokinin, and gibberellin crosstalk at the transcriptional level during parthenocarpic fruit set, but the metabolic pathways of these hormones were not mentioned. In this study, we analyzed metabolic pathways using NGS technology; this data, combined with physiological and biochemical analyses and crosstalk of different plant hormones, allowed us to elucidate the process of starch accumulation in *L. punctata*.

The alteration of the endogenous CK, ABA, and GAs co-regulates starch metabolism in *L. punctata*. The endogenous hormone content changed dramatically following uniconazole application. ABA content increased from 61.47 to 166.53 ng/g (FW), ZR increased from 7.73 to 11.87 ng/g (FW), and the level of GA_{1+3} decreased from 9.25 to 5.57 ng/g (FW) following treatment (accompanying report). CKs elevated the chlorophyll content by controlling regulatory proteins involved in the chlorophyll biosynthesis signaling pathway. CKs are a class of plant growth substances that promote cell division, chloroplast synthesis, and amyloplast formation [44,45]. Furthermore, the increase of CKs plays an important role in regulating grain filling pattern and consequently elevated starch accumulation [46]. In the CK mediated chlorophyll synthesis signaling pathway [47,48], CKs are degraded by CKXs. The transcriptomics data suggested that HKs and SGR were up-regulated. The expression of SGR was up-regulated significantly from the initial value of 8.46 to 130.77 FPKM. The interaction of type-B ARR with SGR2 was assayed to determine whether the cytokinin signaling pathway interacted with a key step in chlorophyll degradation within the chloroplast [49]. The increase in SGR can suppress the degradation of chlorophyll, thereby improving chlorophyll content. Furthermore, the expression pattern of the regulatory proteins described above coincided with the increase in chlorophyll content. The chlorophyll a content increased from the initial value of 0.998 to 1.239 mg/g (FW), and chlorophyll b increased from the initial value of 0.426 to 0.488 mg/g (FW). The chlorophyll a and chlorophyll b content increased by 25.6% and 27%, respectively, compared to the control sample. Importantly, the improvement of the net photosynthetic rate was consistent with the increase of starch and biomass accumulation. The net photosynthetic rate increased from the initial value of 8.83 μmol $CO_2/m^2/s$ to 22.05 and 25.6 μmol $CO_2/m^2/s$ in the control and treatment groups, respectively (accompanying report). Thus, the improvement of chlorophyll content and net photosynthetic rate may lead to starch and biomass accumulation in *L. punctata*.

ABA up-regulated the expression of AGPase large subunit gene transcription by controlling the expression of regulatory elements of the ABA signal pathway. Studies have shown that ABA can up-regulate AGPase gene transcription in rice suspension cells [50] and suppress the expression of gene-encoding amylases and proteases [51]. Reports also indicated that the rates of starch accumulation are positively correlated with ABA levels in wheat grains. As shown in Figure 5, the expression of transcript-encoding ABA receptors of PYLs was up-regulated from 36.88 to 45.27 FPKM at 240 h. ABA combines with intracellular PYL and negative regulator PP2C (type 2C protein phosphatase) to form an ABA-PYL-PP2C complex. ABI4 induces ADP-glucose pyrophosphorylase subunit (*ApL3*) gene expression [52-55]. Moreover, the expression of the AGPase large subunit gene (*ApL3*) also increased. Importantly, the up-regulated expression of the AGPase large subunit gene strongly supported the improved activity of AGPase, and the increased activity of AGPase promoted starch accumulation in duckweed.

GAs suppressed the expression of the amylase gene by controlling expression of regulatory factors. A study of GA regulating the growth and carbohydrate metabolism of potatoes showed that GA_3 could substantially reduce the activity of AGPase in the growing tubers of potatoes [56]; therefore, uniconazole treatment might eliminate the obstacle by blocking GA synthesis and enhancing starch accumulation. GAs can induce or activate α-amylase and other hydrolases, which is not conducive to the synthesis and accumulation of starch [57,58]. Moreover, reports showed that during grain filling, the ratio of endogenous GAs and ABA changes greatly in rice. The ABA content was significantly increased and GA content dramatically decreased, which enhanced the remobilization of prestored carbon to the grains and accelerated the grain filling rate [59]. In this study, the expression of α-amylase was down-regulated from the initial 8.03 to 6 FPKM. The decrease of GA levels suppressed α-amylase expression, and the reduced activities of α-amylase prevented starch degradation in duckweed. These findings also support starch accumulation.

The improvement of starch accumulation following uniconazole treatment was closely associated with the elevated level of endogenous ABA and CK and reduced GA content in duckweed. In this study, high levels of CKs accelerated the biosynthesis of chlorophyll and the net photosynthetic rate, the increased ABA content promoted the activity of AGPase, and the low levels of GAs inactivated amylase. Overall, the alterations in endogenous hormone levels following uniconazole treatment improved starch accumulation in duckweed by influencing the related enzymes involved in carbohydrate metabolism and processes.

Starch accumulation of *L. punctata* under uniconazole treatment

The rapid starch accumulation in fronds of *L. punctata* after uniconazole treatment displayed some similarities to grain filling which is a major process of starch biosynthesis and accumulation in seeds. For instance, both processes are rapid and show similar alteration of endogenous hormone levels such as a decrease in GAs and an increase in ABA. In addition, some key enzymes (AGPase and SSS) involved in starch biosynthesis are regulated in a similar way in both processes.

In this study, the transcriptomics analyses, enzymatic assays, and starch percentages were integrated to uncover the process of rapid accumulation of high starch after uniconazole application. The data from three lines of evidence were analyzed and compared. Investigation of starch composition showed that the starch content and biomass yield in *L. punctata* accumulated rapidly. After culturing for 240 h, the starch content reached 48% in the treated samples and 15.7% in the control samples from an initial yield of 3.2% (Figure 1). The biomass of treated samples (dry weight) improved 10% over the control samples (data not shown). As a result, the total starch that accumulated in the treated samples was 3.4 times higher than that in the control samples. Meanwhile, the enzyme activities involved in starch synthesis, such as AGPase and SSS, were improved dramatically by uniconazole treatment (Figure 3). The activity of AGPase increased significantly from 8.20 to 27.59 U/mg protein, representing a 3.4-fold increase. Importantly, the expression patterns of transcript-encoding key enzymes involved in starch biosynthesis and degradation further supported the physiological and biochemical results described above. Transcriptome analysis showed that the expression of GBSS transcripts were up-regulated significantly.

Starch phosphorylation and glucan hydrolysis are two necessary steps in the degradation process. Glucan water dikinase (GWD) and phosphoglucan water dikinase (PWD) are responsible for starch phosphorylation. B-amylases catalyze the hydrolysis of a-1, 4-glycosidic linkages and release maltose from the exposed nonreducing ends of glucan chains. A-amylases hydrolyze α-1, 4 linkages within polymers exposed on the surface or in channels within granules, releasing soluble glucans that are the substrate for further degradation [60,61]. In this study, there no significant changes were observed for the expression of transcript-encoding GWD. The expression level of GWD (comp38348_c1_seq1) was 40.24, 39.54, 63.45, 39.05, and 47.73 FPKM in different time points, respectively. The starch degradation enzyme activities of α-amylase changed little between the control and treated samples. Though the expression of β-amylase was increased in a manner contradictory with starch accumulation, physiological data showed that the activity of β-amylase was too low to compare with the improved enzyme activity of starch biosynthesis. The expression of enzymes involved in competitive starch metabolic branches, including hexokinase sucrose-phosphate synthase, phosphoglucomutase (EC: 5.4.2.2) and others were also down-regulated (Figure 6). Coupled with the up-regulation of starch biosynthesis related key enzyme-encoding transcripts, the down-regulation of transcripts finally redirected alpha-D-glucose-1P and UDP-glucose to the starch biosynthesis branch.

In this study, up-regulation of key enzymes in starch biosynthesis, in combination with down-regulation of transcripts of key enzymes related to starch degradation and other carbohydrate metabolic branches that compete with the synthesis of starch, eventually led to the accumulation of starch in *L. punctata*.

Uniconazole has a similar chemical structure to paclobutrazol. It reduces plant growth more than paclobutrazol when applied as a soil drench in equal amounts. On average, the amount of paclobutrazol required is four to ten times that of uniconazole, to obtain a similar effect on plant size [62]. Early research indicated that uniconazole can be very persistent in retarding plant growth without causing phytotoxicity [63]. Half-lives of paclobutrazol and uniconazole in water were 24.4 and 5.2 days, respectively. Uniconazole-p is non-toxic to birds, bees, and earthworms, but slightly toxic to fish and aquatic invertebrates. Therefore, it can be applied to high starch accumulation of duckweed in large-scale cultivation.

Conclusions

In this study, high starch accumulation in *L. punctata* 0202 was achieved after uniconazole application in a nutrient-rich environment. The process of starch accumulation was investigated at physiological, biochemical, and transcriptome levels. The increase in endogenous ABA and CK levels further enhanced the activity of AGPase and chlorophyll biosynthesis, while decreased endogenous GA levels significantly correlated with the inactivation of α-amylase. Moreover, uniconazole increased the levels of substrates of starch synthesis and regulated transcriptional expression of enzymes by changing the biosynthesis of endogenous hormones, resulting in starch accumulation in duckweed. Because of the complex interaction among different hormones, the alteration of endogenous hormone levels can provide further insight into the relationship of endogenous hormones to starch accumulation. In this study, an operable process for high starch accumulation in duckweed was developed, paving the way for large-scale treatment of wastewater and the application of duckweed to bioenergy.

Materials and methods

Duckweed cultivation and uniconazole treatments

L. punctata 0202 was originally collected from Sichuan province, China. It was cultivated in standard 1/6Hoagland E+ solution (Total $N = 58.3$ mg/L, $P = 25.8$ mg/L) [64] culture for 3 days under a 16/8 h day/night photoperiod, with a light intensity of 130 μmol/m^2/s and a temperature of 25°C/15°C at day/night. Then, 6 g of fronds were transferred into 1,000 mL 1/6 Hoagland E+ culture plastic containers (23 × 14 × 4.5 cm) for further cultivation over a period of 10 days. Uniconazole powder was produced in Japan and purchased from Aoke Biotech Corp (Beijing, China). The concentration of uniconazole used in this study was 800 $mg \cdot L^{-1}$. To investigate the effect of uniconazole treatment on *L. punctata*, a 5-mL solution of 800 $mg \cdot L^{-1}$ uniconazole was sprayed evenly on the surface of fronds. Controls were sprayed with 5 mL water containing 10% methanol. The experiments were carried out with three replicates. Thirteen different time points, including 0, 1, 2, 3, 5, 7, 12, 24, 48, 72, 120, 168, and 240 h after fronds were cultured in solution and were chosen for composition and enzymatic activity assays. For each time point, fronds were collected from three culture plastic containers. Samples collected at 0, 2, 5, 72, and 240 h were frozen in liquid nitrogen immediately for the RNA-Seq study.

Material composition

The starch content was described as glucose content in total sugar by HPLC (Thermo 2795, Thermo Corp, Waltham, USA)-ELSD (All-Tech ELSD 2000, All-tech, Corp, Nicholasville, USA) using the following method. The starch content was determined using the total sugar content (starch content = glucose content × 0.909). Dry duckweed powder was hydrolyzed with 1.2 M HCl in a boiling water bath. After adjusting the pH to 7 with 10 M NaOH, PbAc was added to precipitate protein. After the solution was measured, filtered, and treated with a C18 extraction column, the hydrolyzate was analyzed by HPLC (Thermo 2795, Thermo Corp.) with an Evaporative Lightscattering Detector (All-Tech ELSD 2000, All-tech., Corp.) [65].

Microscopic analysis of fronds

Fronds in the uniconazole treatment group and control group were fixed, embedded, and dehydrated as described [66,67]. Samples were fixed in 5% glutaraldehyde in 0.1 M PBS (pH 7.4) containing 2% Suc in a 2 mL tube at 4°C overnight followed by 3 h at room temperature. Samples were rinsed with 0.1 M PBS (pH 7.4) and postfixed in buffered 1% osmium tetroxide at 4°C overnight, followed by dehydration in a graded series of acetone washes. The dehydrated samples were then embedded in epon resin. The 1 mm-thick sections were picked up on a glass slide, stained with methylene blue, and scoped with a light microscope. Ultrathin sections were cut with an ultramicrotome (Leica EM UC6, Wetzlar, Germany) and observed with transmission electron microscopy (TEM; Tecnai G^2 F20S-Twin, FEI, Hillsboro, USA) at 200 kV after staining with uranyl acetate and lead citrate.

Carbohydrate metabolism enzyme activity assay

To investigate enzyme activity, 1 g fresh weight duckweed was homogenized with a ceramic pestle in an ice-cold mortar in 5 mL of 50 mmol/L HEPES-NaOH (pH = 7.6), 5 mmol/L DL-Dithiothreitol, 8 mmol/L $MgCl_2$, 2 mmol/L EDTA, 2% (*w*/*v*) polyvinylpyrrolidone-40, and 12.5% (*w*/*v*) glycerol. The homogenate was centrifuged at 10,000 × *g* for 5 min. The supernatant extract was used as a crude enzyme solution stored at −20°C. All procedures were carried out at 0 to 4°C. The activities of α-amylase (1, 4, d-glucan glucanohydrolase) and β-amylase (1, 4, d-glucan maltohydrolase) were estimated following the method of Tarrago and Nicolas [68,69]. The enzymatic activities of SSS and AGPase were assayed according to Nakamura *et al.* [69].

RNA extraction and cDNA fragment library construction

Five *L. punctata* samples were collected at the 0, 2, 5, 72, and 240 h time points after treatment with uniconazole. For each sample, total RNA was extracted from 200 mg fronds using the OMEGATM Plant DNA/RNA kit (OMEGA, Norcross, USA) and genomic DNA was digested by DNase I (Fermentas, Waltham, USA) according to the manufacturer's instructions. RNA concentration, OD260/280, OD260/230, 28S/18S and RNA integrity number (RIN) were measured with the Agilent 2100 Bioanalyzer or NanoDrop (Agilent, Santa Clara, USA). Qualified total RNA extracted from each sample was submitted to the Beijing Genomics Institute (BGI)-Shenzhen, Shenzhen, China, (http://www.genomics.cn) for RNA sequencing by Illumina HiSeq 2000 (Illumina, San Diego, USA). cDNA fragment libraries were constructed according to the manufacturer's instructions using the TruSeq RNA Sample Prep kit. Library quality control analysis was performed using the Agilent 2100 Bio-analyzer.

RNA sequencing and paired-end reads assembly

The validated 200 bp fragment cDNA libraries were submitted to the Illumina HiSeq 2000 platform for paired-end (PE) RNA sequencing. PE read sequencing quality was assessed by fastqc (http://www.bioinformatics.bbsrc.ac.uk/projects/fastqc/) and then *de novo* assembled using Trinity (v2012-06-08) [70] under default parameter choices. All PE reads were used to align back to these assembled sequences using the Bowtie2 (v2.0.0-beta5) program [71]. Accordingly, the read align rate was calculated. Length distribution analysis was performed with Perl scripts (Additional file 4) to calculate the N50 number, average length, and max length. The best candidate open reading frame (ORF) was predicted using Perl scripts in the Trinity package (v2012-06-08) [70].

Functional annotation and cluster

All contigs assembled by Trinity (v2012-06-08) [70] were submitted to Blast2GO [72,73] for functional annotation. A BLASTX similarity search was performed against the NR database (http://www.ncbi.nlm.nih.gov/) by Blast2GO with a threshold of *E* value $<10^3$. Enzyme codes were extracted, and Kyoto Encyclopedia of Genes and Genomes (KEGG) pathways were retrieved from the KEGG web server (http://www.genome.jp/kegg/).

Expression pattern analysis

To analyze the express levels of each transcript at different time points following uniconazole treatment, all PE reads for each sample were used for mapping analysis with Perl scripts in the Trinity package (v2012-06-08) [70] under default parameter choices. The expression value of each transcript was calculated and normalized according to the RESM-based algorithm using the Perl scripts in the Trinity (v2012-06-08) package to obtain FPKM values. *P* values and log_2 fold change (log_2FC) were calculated, and significantly DETs between each sample set were identified with *P* value ≤0.05 and log_2FC ≥1. Hypergeometric tests based on the KEGG annotation were performed for each DET group identified between each sample set using R scripts (Additional file 4) to extract the enriched KEGG pathway. Additionally, we depicted differences and commonalities in the number of DEGs using the VENNY under default parameter choices (http://bioinfogp.cnb.csic.es/tools/venny/index.html) [74].

Calculations and statistics

Each data point represents the results of three sample experiments; the results are provided as means ± standard error in the figures.

Additional files

Additional file 1: Table S1. Sequence annotations of *L. punctata* transcripts and the gene expression profiling of five samples.

Additional file 2: Table S2. Expression levels of some regulatory proteins and transcription factors involved in CK, ABA, and GA signaling pathways.

Additional file 3: Table S3. Expression levels of some carbon metabolism related genes.

Additional file 4: Common scripts. The common scripts including Perl scripts and R scripts for assembly statistic and pathway enrich.

Abbreviations
ABA: abscisic acid; ABI4: abscisic acid insensitive 4; AHPs: histidine phosphotransfer proteins; CKs: cytokinins; DET: differentially expressed transcript; DW: dry weight; EC: enzyme codes; FPKM: fragments per kilobase of transcripts per million mapped fragments; FW: fresh weight; GAs: gibberellins; HKs: histidine kinases; KEGG: Kyoto Encyclopedia of Genes and Genomes; log2FC: log2 fold change; NGS: next-generation sequencing; PE: paired-end; PP2C: type 2C protein phosphatase; PYLs: family of START proteins; SGR2: STAY-GREEN2 protein; type-B ARR: type-B *Arabidopsis thaliana* response regulator.

Competing interests
The authors declare that they have no competing interests.

Authors' contributions
YL carried out biochemical assays, the data analysis, drafted and revised the manuscript. YF participated in the design of the study, the data analysis, conceived the study and revised the manuscript. Y-IJ and YZ participated in the design of the study and revised the manuscript. M-jH and XT participated in the data analysis and revised the manuscript. J-IS participated in the data analysis. G-hZ participated in the part of data analysis and interpretation and revised the manuscript. K-zH participated in the design of the study. HZ conceived the study and revised the manuscript. All authors read and approved the final manuscript.

Acknowledgements
The authors acknowledge financial support received from the National Key Technology R&D Program of China (No. 2015BAD15B01), the Projects of International Cooperation of Ministry of Science and Technology of China (2014DFA30680), Key Laboratory of Environmental and Applied Microbiology, Chengdu Institute of Biology, Chinese Academy of Sciences (No. KLCAS-2014-02), and West Light Foundation of The Chinese Academy of Sciences (Y2C5021100). We thank Dr. Wei-zao Huang, Prof Song-hu Wang, and Prof Wan Xiong for revising the manuscript. We thank Miss Wen Zheng, Dr. Zhen Liu, and Dr. Ying-hong Gu for writing some scripts.

Author details
[1]Chengdu Institute of Biology, Chinese Academy of Sciences, No.9 Section 4, Renmin Nan Road, 610041 Chengdu, China. [2]University of Chinese Academy of Sciences, No.19A Yuquan Road, 100049 Beijing, China. [3]Key Laboratory of Environmental and Applied Microbiology, Chinese Academy of Sciences, No.9 Section 4, Renmin Nan Road, 610041 Chengdu, China. [4]Environmental Microbiology Key Laboratory of Sichuan Province, No.9 Section 4, Renmin Nan Road, 610041 Chengdu, China. [5]Key Laboratory of Bio-Resources and Eco-Environment, Ministry of Education, College of Life Sciences, Sichuan University, N0.24 South Section 1, Yihuan Road, 610064 Chengdu, China.

References
1. Ge L, Wang P, Mou H. Study on saccharification techniques of seaweed wastes for the transformation of ethanol. Renew Energ. 2011;36(1):84–9.
2. Crutzen PJ, Mosier AR, Smith KA, Winiwarter W. N2O release from agro-biofuel production negates global warming reduction by replacing fossil fuels. Atmos Chem Phys. 2008;8(2):389–95.
3. Searchinger T, Heimlich R, Houghton RA, Dong F, Elobeid A, Fabiosa J, et al. Use of US croplands for biofuels increases greenhouse gases through emissions from land-use change. Science. 2008;319(5867):1238–40.
4. Landolt E. Biosystematic investigations in the family of duckweed (Lemnaceae), vol. 2. Geobotanischen Inst ETH: Stiftung Rubel, Zúrich, Suiza; 1986.
5. Ge X, Zhang N, Phillips GC, Xu J. Growing Lemna minor in agricultural wastewater and converting the duckweed biomass to ethanol. Bioresour Technol. 2012;124:485–8.
6. Leng R, Stambolie J, Bell R. Duckweed-a potential high-protein feed resource for domestic animals and fish. Livest Res Rural Dev. 1995;7(1):36.
7. Hillman WS, Culley Jr DD. The uses of duckweed. Am Sci. 1978;66:442–51.
8. Xu J, Cui W, Cheng JJ, Stomp A-M. Production of high-starch duckweed and its conversion to bioethanol. Biosyst Eng. 2011;110(2):67–72.
9. Zhao Y, Fang Y, Jin Y, Huang J, Bao S, Fu T, et al. Potential of duckweed in the conversion of wastewater nutrients to valuable biomass: a pilot-scale comparison with water hyacinth. Bioresour Technol. 2014;163:82–91.
10. Reid M, Bieleski R. Response of Spirodela oligorrhiza to phosphorus deficiency. Plant Physiol. 1970;46(4):609–13.
11. Bayrakci AG, Kocar G. Second-generation bioethanol production from water hyacinth and duckweed in Izmir: a case study. Renew Sust Energ Rev. 2014;30:306–16.
12. El-Shafai SA, El-Gohary FA, Nasr FA, Peter Van Der Steen N, Gijzen HJ. Nutrient recovery from domestic wastewater using a UASB-duckweed ponds system. Bioresour Technol. 2007;98(4):798–807.
13. Blazey EB, McClure JW. The distribution and taxonomic significance of lignin in the Lemnaceae. Amer J Bot. 1968;55:1240–5.
14. Cui W, Xu J, Cheng J, Stomp A. Starch accumulation in duckweed for bioethanol production. Biol Eng. 2011;3:187–97.
15. Chen Q, Jin Y, Zhang G, Fang Y, Xiao Y, Zhao H. Improving production of bioethanol from Duckweed (Landoltia punctata) by pectinase pretreatment. Energies. 2012;5(8):3019–32.
16. Xiao Y, Fang Y, Jin Y, Zhang G, Zhao H. Culturing duckweed in the field for starch accumulation. Ind Crop Prod. 2013;48:183–90.
17. McLAREN JS, Smith H. The effect of abscisic acid on growth, photosynthetic rate and carbohydrate metabolism in Lemna minor L. New Phytol. 1976;76(1):11–20.
18. McCombs P, Ralph R. Protein, nucleic acid and starch metabolism in the duckweed *Spirodela oligorrhiza* treated with cytokinins. Biochem J. 1972;129:403–17.
19. Wang W, Messing J. Analysis of ADP-glucose pyrophosphorylase expression during turion formation induced by abscisic acid in Spirodela polyrhiza (greater duckweed). BMC Plant Biol. 2012;12:5.
20. Pavlista AD. Growth regulators increased yield of Atlantic potato. Am J Potato Res. 2011;88(6):479–84.
21. Fletcher RA, Hofstra G. Improvement of uniconazole-induced protection in wheat seedlings. J Plant Growth Regul. 1990;9(4):207–12.
22. Fletcher R, Hofstra G, Gao JG. Comparative fungitoxic and plant growth regulating properties of triazole derivatives. Plant Cell Physiol. 1986;27(2):367–71.
23. Zhou W, Ye Q. Physiological and yield effects of uniconazole on winter rape (Brassica napus L.). J Plant Growth Regul. 1996;2:69–73.
24. Wan-zhuo G, Zheng-yi Z, Weng-yu Y, Wen-zhu L. Effect of uniconazloe for dry seed treatment on morphological characteristics and yield of soybean. Soybean Science. 2007;26(3):369.
25. Zhang M, Duan L, Tian X, He Z, Li J, Wang B, et al. Uniconazole-induced tolerance of soybean to water deficit stress in relation to changes in photosynthesis, hormones and antioxidant system. J Plant Physiol. 2007;164(6):709–17.
26. Smith AM. The biosynthesis of starch granules. Biomacromolecules. 2001;2(2):335–41.
27. Martin C, Smith AM. Starch biosynthesis. Plant Cell. 1995;7(7):971–85.
28. Wu Y, Wei W, Pang X, Wang X, Zhang H, Dong B, et al. Comparative transcriptome profiling of a desert evergreen shrub, Ammopiptanthus mongolicus, in response to drought and cold stresses. BMC Genomics. 2014;15:671.
29. Wang X, Zhou G, Xu X, Geng R, Zhou J, Yang Y, et al. Transcriptome profile analysis of adipose tissues from fat and short-tailed sheep. Gene. 2014;549(2):252–7.
30. Tao X, Fang Y, Xiao Y, Jin YL, Ma XR, Zhao Y, et al. Comparative transcriptome analysis to investigate the high starch accumulation of duckweed (Landoltia punctata) under nutrient starvation. Biotechnol Biofuels. 2013;6(1):72.
31. Robinson MD, McCarthy DJ, Smyth GK. edgeR: a Bioconductor package for differential expression analysis of digital gene expression data. Bioinformatics. 2010;26(1):139–40.
32. Guan C, Wang X, Feng J, Hong S, Liang Y, Ren B, et al. Cytokinin antagonizes abscisic acid-mediated inhibition of cotyledon greening by promoting the degradation of abscisic acid insensitive5 protein in Arabidopsis. Plant Physiol. 2014;164(3):1515–26.

33. Ramireddy E, Chang L, Schmulling T. Cytokinin as a mediator for regulating root system architecture in response to environmental cues. Plant Signal Behav. 2014;9:e27771-1.
34. Reguera M, Peleg Z, Abdel-Tawab YM, Tumimbang EB, Delatorre CA, Blumwald E. Stress-induced Cytokinin synthesis increases drought tolerance through the coordinated regulation of carbon and nitrogen assimilation in rice. Plant Physiol. 2013;163(4):1609–22.
35. Bastias A, Yanez M, Osorio S, Arbona V, Gomez-Cadenas A, Fernie AR, et al. The transcription factor AREB1 regulates primary metabolic pathways in tomato fruits. J Exp Bot. 2014;65(9):2351–63.
36. Kim HJ, Chiang Y-H, Kieber JJ, Schaller GE. SCFKMD controls cytokinin signaling by regulating the degradation of type-B response regulators. Proc Natl Acad Sci U S A. 2013;110(24):10028–33.
37. Yang D-L, Yao J, Mei C-S, Tong X-H, Zeng L-J, Li Q, et al. Plant hormone jasmonate prioritizes defense over growth by interfering with gibberellin signaling cascade. Proc Natl Acad Sci U S A. 2012;109(19):E1192–200.
38. Giulia E, Alessandro B, Mariano D, Andrea B, Benedetto R, Angelo R. Early induction of apple fruitlet abscission is characterized by an increase of both isoprene emission and abscisic acid content. Plant Physiol. 2013;161(4):1952–69.
39. Wang Y, Li L, Ye T, Zhao S, Liu Z, Feng Y-Q, et al. Cytokinin antagonizes ABA suppression to seed germination of Arabidopsis by downregulating ABI5 expression. Plant J. 2011;68(2):249–61.
40. Wang Y, Tao X, Tang X-M, Xiao L, Sun J-l, Yan X-F, et al. Comparative transcriptome analysis of tomato (Solanum lycopersicum) in response to exogenous abscisic acid. BMC Genomics. 2013;14:841.
41. Chandrasekaran U, Xu W, Liu A: Transcriptome profiling identifies ABA mediated regulatory changes towards storage filling in developing seeds of castor bean (Ricinus communis L.). Cell Biosci 2014, 4:33.
42. Dugas DV, Monaco MK, Olsen A, Klein RR, Kumari S, Ware D, et al. Functional annotation of the transcriptome of Sorghum bicolor in response to osmotic stress and abscisic acid. BMC Genomics. 2011;12:514.
43. Li J, Wu Z, Cui L, Zhang T, Guo Q, Xu J, et al. Transcriptome comparison of global distinctive features between pollination and parthenocarpic fruit set reveals transcriptional phytohormone cross-talk in cucumber (Cucumis sativus L.). Plant Cell Physiol. 2014;55(7):1325–42.
44. Sakai A, Yashiro K, Kawano S, Kuroiwa T. Amyloplast formation in cultured tobacco cells; effects of plant hormones on multiplication, size, and starch content. Plant Cell Rep. 1996;15(8):601–5.
45. Fletcher RA, McCullag D. Cytokinin-induced chlorophyll formation in cucumber cotyledons. Planta. 1971;101(1):88.
46. Yang J, Peng S, Visperas RM, Sanico AL, Zhu Q, Gu S. Grain filling pattern and cytokinin content in the grains and roots of rice plants. Plant Growth Regul. 2000;30(3):261–70.
47. To JPC, Kieber JJ. Cytokinin signaling: two-components and more. Trends Plant Sci. 2008;13(2):85–92.
48. Werner T, Schmuelling T. Cytokinin action in plant development. Curr Opin Plant Biol. 2009;12(5):527–38.
49. Pilkington SM, Montefiori M, Galer AL, Emery RJN, Allan AC, Jameson PE. Endogenous cytokinin in developing kiwifruit is implicated in maintaining fruit flesh chlorophyll levels. Ann Bot. 2013;112(1):57–68.
50. Akihiro T, Mizuno K, Fujimura T. Gene expression of ADP-glucose pyrophosphorylase and starch contents in rice cultured cells are cooperatively regulated by sucrose and ABA. Plant Cell Physiol. 2005;46(6):937–46.
51. Gomez-Cadenas A, Zentella R, Walker-Simmons MK, Ho THD. Gibberellin/abscisic acid antagonism in barley aleurone cells: Site of action of the protein kinase PKABA1 in relation to gibberellin signaling molecules. Plant Cell. 2001;13(3):667–79.
52. Hubbard KE, Nishimura N, Hitomi K, Getzoff ED, Schroeder JI. Early abscisic acid signal transduction mechanisms: newly discovered components and newly emerging questions. Gene Dev. 2010;24(16):1695–708.
53. Park S-Y, Fung P, Nishimura N, Jensen DR, Fujii H, Zhao Y, et al. Abscisic acid inhibits type 2C protein phosphatases via the PYR/PYL family of START proteins. Science. 2009;324(5930):1068–71.
54. Umezawa T, Nakashima K, Miyakawa T, Kuromori T, Tanokura M, Shinozaki K, et al. Molecular basis of the core regulatory network in ABA responses: sensing. Signaling Trans Plant Cell Physiol. 2010;51(11):1821–39.
55. Rook F, Corke F, Card R, Munz G, Smith C, Bevan MW. Impaired sucrose-induction mutants reveal the modulation of sugar-induced starch biosynthetic gene expression by abscisic acid signalling. Plant J. 2001;26(4):421–33.
56. Mares DJ, Marschner H, Krauss A. Effect of gibberellic-acid on growth and carbohydrate-metabolism of developing tubers of potato (solanum-tuberosum). Physiol Plant. 1981;52(2):267–74.
57. Kaur S, Gupta AK, Kaur N. Gibberellin A3 reverses the effect of salt stress in chickpea (Cicer arietinum L.) seedlings by enhancing amylase activity and mobilization of starch in cotyledons. Plant Growth Regul. 1998;26(2):85–90.
58. Rentzsch S, Podzimska D, Voegele A, Imbeck M, Mueller K, Linkies A, et al. Dose- and tissue-specific interaction of monoterpenes with the gibberellin-mediated release of potato tuber bud dormancy, sprout growth and induction of alpha-amylases and beta-amylases. Planta. 2012;235(1):137–51.
59. Yang J, Zhang J, Wang Z, Zhu Q, Wang W. Hormonal changes in the grains of rice subjected to water stress during grain filling. Plant Physiol. 2001;127(1):315–23.
60. Smith AM, Zeeman SC, Smith SM. Starch degradation. Annu Rev Plant Biol. 2005;56:73–98.
61. Ghiena C, Schulz M, Schnabl H. Starch degradation and distribution of the starch-degrading enzymes in Vicia faba leaves. Plant Physiol. 1993;101(1):73–9.
62. Barrett JE, Nell TA. Irrigation interval and growth retardants affect poinsettia development. Proceed Florida State Horticultural Soc. 1982;95:167–9.
63. Davis TD, Steffens GL, Sankhla N. Triazole plant growth regulators. Hortic Rev. 1988;63:105.
64. Hoagland DR, Arnon DI: The water-culture method for growing plants without soil. Circular California Agricultural Experiment Station,1950;347:2nd edit pp.32 pp.
65. Zhang L, Zhao H, Gan M, Jin Y, Gao X, Chen Q, et al. Application of simultaneous saccharification and fermentation (SSF) from viscosity reducing of raw sweet potato for bioethanol production at laboratory, pilot and industrial scales. Bioresour Technol. 2011;102(6):4573–9.
66. Wu Y, Messing J. RNA interference-mediated change in protein body morphology and seed opacity through loss of different Zein Proteins1 C W OA. Plant Physiol. 2010;153(1):337–47.
67. Ji X, Gai Y, Zheng C, Mu Z. Comparative proteomic analysis provides new insights into mulberry dwarf responses in mulberry (Morus alba L.). Proteomics. 2009;23:5328–39.
68. Tárrago JF, Nicolás G. Starch degradation in the cotyledons of germinating lentils. Plant Physiol. 1976;58(5):618–21.
69. Nakamura Y, Yuki K, Park S-Y, Ohya T. Carbohydrate metabolism in the developing endosperm of rice grains. Plant Cell Physiol. 1989;30(6):833–9.
70. Grabherr MG, Haas BJ, Yassour M, Levin JZ, Thompson DA, Amit I, et al. Full-length transcriptome assembly from RNA-Seq data without a reference genome. Nat Biotechnol. 2011;29(7):644–52.
71. Langmead B, Trapnell C, Pop M, Salzberg SL. Ultrafast and memory-efficient alignment of short DNA sequences to the human genome. Genome Biol. 2009;10(3):R25.
72. Conesa A, Götz S. Blast2GO: A comprehensive suite for functional analysis in plant genomics. Int J Plant Genomics. 2008;2008:1–13.
73. Conesa A, Götz S, García-Gómez JM, Terol J, Talón M, Robles M. Blast2GO: a universal tool for annotation, visualization and analysis in functional genomics research. Bioinformatics. 2005;21(18):3674–6.
74. Oliveros JC. VENNY: An interactive tool for comparing lists with Venn Diagrams. 2007.

Semi-continuous anaerobic digestion for biogas production: influence of ammonium acetate supplement and structure of the microbial community

Haijia Su*, Luo Liu, Shaojie Wang, Qingfeng Wang, Yixin Jiang, Xiaocong Hou and Tianwei Tan

Abstract

Background: As an efficient disposal method of food waste, anaerobic digestion (AD) for biogas production is widely used. In order to understand the enhanced efficiency and stability of AD by appropriate amounts of ammonia and volatile fatty acids (NH_4^+/VFAs), the characteristics of the corresponding microbial community with ammonium acetate supplement were investigated by denatured gradient gel electrophoresis (DGGE) and pyrosequencing analyses of samples, with or without supplement of NH_4^+/VFAs.

Results: In this study, four different supplement strategies of adding ammonium acetate were investigated, including a blank group (without supplement of ammonium acetate), a low group (L group, 0.7 g/L/d), a moderate group (M group, 1.0 g/L/d) and a high group (H group, 1.3 g/L/d), respectively. The average daily gas production was 1,839 mL/d, 1,655 mL/d, 1,448 mL/d and 1,488 mL/d for L, M, H and blank groups, respectively. The results reveal that the absence or overload of NH_4^+/VFAs leads to the inhibition or failure of the AD operation. The blank and H groups were selected for further investigation of the microbial community by DGGE and pyrosequencing analyses. A significant difference of the microbial communities at different AD stages was observed between the blank and H groups.

Conclusions: Ammonium acetate, as an efficient supplement, significantly influences the characteristics of a semi-continuous AD operation. The DGGE and pyrosequencing analyses indicated that the different bacterial and archaeal communities occurred in the blank and H groups at different AD stages. Thus, an appropriate ammonium acetate supplement may maintain the balance of the microbial community and could be applied to adjust the AD operation and microbial composition towards optimal biogas production.

Keywords: Anaerobic digestion, Biogas production, Ammonium acetate, Glucose, Microbial community

Background

As an efficient disposal method in the treatment of food, and fruit or vegetable wastes [1-3], anaerobic digestion (AD) for biogas production is widely used. As previous research has reported [4], the AD process includes steps of hydrolysis, acidogenesis, acetogenesis and methanogenesis. In the first three steps, the bacterial community at phylum level is commonly dominated by *Proteobacteria*, *Firmicutes* and *Bacteroidetes* [5-7], which generate sufficient volatile fatty acids (VFAs) for the methane production. The archaeal community is, however, often dominated by the *Euryarchaeota phylum*, which utilizes acetate, H_2, CO_2 and methyl compounds for methane production. The stability of AD operation was affected significantly by the microbial communities, whose diversity was directly affected by varying digestion conditions [8].

The operation mode of AD and its parameters [9], the pretreatment of the substrate [10] and the added C- and N-resources [11-13] were studied in view of AD operational stability. Whereas the addition of metal-ions could

* Correspondence: suhj@mail.buct.edu.cn
Beijing Key Laboratory of Bioprocess, Beijing University of Chemical Technology, No.15, Beisanhuan East Road, Beijing 100029, People's Republic of China

maintain a stable and efficient digestion [12], the effect of adding other additives, especially ammonia (NH_4^+), has not been investigated in detail, although an overload of ammonia can hamper the aerogenesis in AD and reduce the activity of methano-bacteria [14,15]. Some previous research has indicated that NH_4^+/VFAs could form a weak buffer system at the beginning of the digestion operation, thus enhancing the tolerance of AD, whilst the failure of the buffer action resulted in a direct inhibition [16].

Wang *et al.* reported that the formed buffer of NH_4^+/VFAs during the semi-continuous AD process boosted the biogas production and efficiently maintained the AD stability in the treatment of food waste. When ammonia acetate was added to the AD, the liquid environment changed and caused a shift of microbial communities [17]. Previous studies focused upon the AD microbial community analysis, either during the domestication or the inhibition period [18-20].

The relationship of the exogenous buffer impact and the microbial communities was not previously investigated; thus it is the objective of the present work to investigate the effect of different ammonium acetate concentrations on the AD characteristics and the change in microbial communities. In order to eliminate the complex influence of other factors, a solution of glucose with ammonia acetate was used to simulate the food waste.

The key objectives of this work involve: (i) studying the effects of different ammonium acetate supplement strategies on the biogas and CH_4 production, and on the liquid characteristics during the AD process, whilst evaluating the stability and inhibition of AD when exogenous buffer is added; and (ii) assessing the effect of the supplement onto the microbial communities to identify the microbial communities' change under different operating modes.

Results and discussion

Effect of different ammonium acetate concentrations on the characteristics of anaerobic digestion

Biogas production trends for the blank and supplement groups with different addition strategies (low (L), moderate (M) or high (H)) were assessed. The average daily gas production was 1,839 mL/d, 1,655 mL/d, 1,448 mL/d and 1,488 mL/d for L, M, H and blank groups, respectively. The H and blank groups were gently inhibited after about 25 days, with average daily biogas production of 1,266.9 mL/d and 1,256.8 mL/d at 51 days as listed in Table 1, respectively. The ammonium acetate supplement moreover caused an accelerated startup in the groups with ammonium acetate supplement. Due to the VFAs inhibition, the daily average biogas production of the blank group during approximately 26 to 51 days decreased by 15% in comparison with the period 0 to 25 days (approximately). The results clearly stress the importance of adding a stable N-resource supplement as the key factor to the stability and high efficiency of the AD system.

The groups with the supplement maintained a significantly higher methane production during the whole process. The methane production of L, M and H groups was 44.5%, 26.5% and 4.1% higher than that of the blank group, respectively. However, after 25 days, the NH_4^+ inhibition clearly occurred in the H group. The average methane production of the H group at the end of the AD operation (about 51 days) was only 581.0 mL/d; 96.3% and 57.0% lower than that of L and M groups, respectively. Similarly, the VFAs inhibition of the blank group also limited the methanogen activity during the end period of the digestion (approximately 26 to 51 days) (see Table 1).

Whereas the VFAs inhibition occurred in the blank group, a weaker buffer in the L and M groups was formed from the ammonium acetate as supplement, thus

Table 1 Effect of ammonium acetate concentration on characteristics of anaerobic digestion

Group	Different period (days)	Average daily total biogas production (mL)	Average daily CH_4 concentration (%)	NH_4^+ (g/L)	Volatile fatty acids (g/L)
L	10	1,837.0 ± 46.4	61.66 ± 1.56	1.47	2.44
	25	1,767.5 ± 48.3	61.14 ± 2.38	1.86	4.88
	51	1,880.4 ± 37.9	59.11 ± 1.62	1.97	8.00
M	10	1,757.3 ± 54.1	63.38 ± 1.52	1.27	2.83
	25	1,708.5 ± 37.1	61.80 ± 2.17	2.73	5.69
	51	1,586.4 ± 40.8	54.53 ± 2.08	3.22	8.70
H	10	1,710.5 ± 56.3	64.39 ± 1.81	1.48	3.80
	25	1,586.5 ± 43.6	61.39 ± 1.69	3.35	9.69
	51	1,266.9 ± 49.6	45.86 ± 1.96	5.97	15.44
Blank	10	1,678.0 ± 42.4	52.42 ± 1.66	0.15	3.63
	25	1,778.3 ± 53.1	56.65 ± 2.20	0.12	6.44
	51	1,256.8 ± 51.6	47.08 ± 2.12	0.27	9.55

defusing the impact of the VFAs inhibition and resulting in a higher methane production. However, an overload of ammonium acetate also caused NH_4^+-inhibition directly, which was more obvious in later periods of the M and H groups, especially in the H group.

As shown in Figure 1, the supplement of ammonium acetate affected the pH value differently. An instable pH change was obvious in the blank group, without the N-resource supplement; below 7.0 during the whole process. However, the pH value of L and M groups were relatively stable during the whole process, because a weaker buffer system of NH_4^+/VFAs was formed due to the supplement of ammonium acetate, which weakened the VFAs impact. Nevertheless, the pH of the H group showed a severe increase to 8.0 after about 25 to 30 days, which can be attributed to a continuous NH_4^+ accumulation, followed by NH_4^+ inhibition.

The effect of different supplement strategies on the NH_4^+ concentration from different stages was shown in Table 1, where the blank group maintained the lowest NH_4^+ concentration during the whole process; the highest NH_4^+ concentration in the blank group was only 0.27 g/L during the whole AD process. With a continuous VFAs accumulation, VFAs inhibition resulted in a poor biogas and methane performance of the blank group during the whole process (see Table 1).

The fastest NH_4^+ accumulation rate was obtained in the H group, but the AD operation would be hampered at NH_4^+ concentration in excess of 3 g/L after a period of about 20 to 25 days. The average biogas production decreased from 1,586.5 to 1,266.9 mL/d and the average methane concentration decreased from 61.4 to 45.9%, comparing day 25 with 51, respectively. The present study indicates that the critical point of NH_4^+ concentration was about 2.5 to 3.0 g/L, so it is remarkable that the M group was also slightly inhibited when the NH_4^+ concentration reached 2.7 g/L, at around 25 days. However there was no obvious difference in the biogas and methane production of the L group during the whole AD process, maintaining the optimum productive capacity.

Table 1 also clearly shows the variation of VFAs accumulation in the different groups during the whole AD process. The VFAs concentration in the blank group increased significantly from 3.6 to 9.6 g/L. Similarly, Table 1 illustrates that the VFAs concentration in the L and M groups were slightly lower than that of the blank group, but these two groups maintained a stable biogas production without inhibition. However, the higher ammonium acetate supplement caused a severe VFAs increase in the H group, reaching 15.4 g/L at the end.

A weak buffer system between VFAs and NH_4^+ in the AD operation was necessary to maintain its high efficiency and stability. Both absence and overload of NH_4^+/VFAs will directly lead to the inhibition or failure of the AD operation.

Denatured gradient gel electrophoresis analysis of bacterial and archaeal communities at different anaerobic digestion stages

Throughout the AD, adding exogenous buffer would generate a significant effect on the microbial communities, as

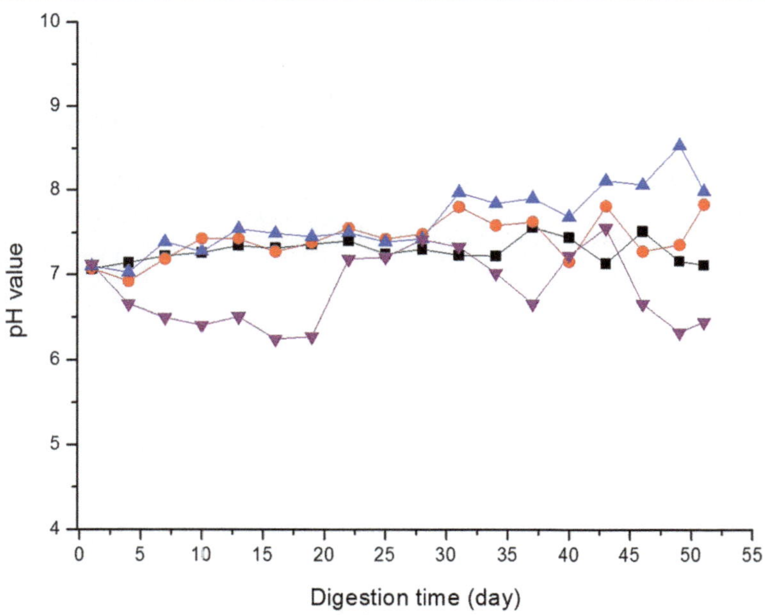

Figure 1 Effect of different ammonium acetate concentrations on the pH value of the anaerobic digestion (AD). The different amount of ammonium acetate was supplemented at L group (black square), M group (red circle symbol), H group (blue triangle symbol) and blank sample (violet triangle symbol). The experiments were operated in a 1 L fermentor at a mesophilic temperature of 37 ± 1°C for the semi-continuous AD.

shown in the denatured gradient gel electrophoresis (DGGE) results. The microbial structure of the community in each group throughout the AD process changed at different stages, but the dominant bacteria did not change too much, as shown in Figure 2. With the extension of the digestive time, some bacteria in the blank and H groups were inhibited, as reflected to the DGGE bands, reducing in brightness or even disappearing.

A bacterial community change was also observed from Figure 2A. At the beginning stage of AD, the DGGE

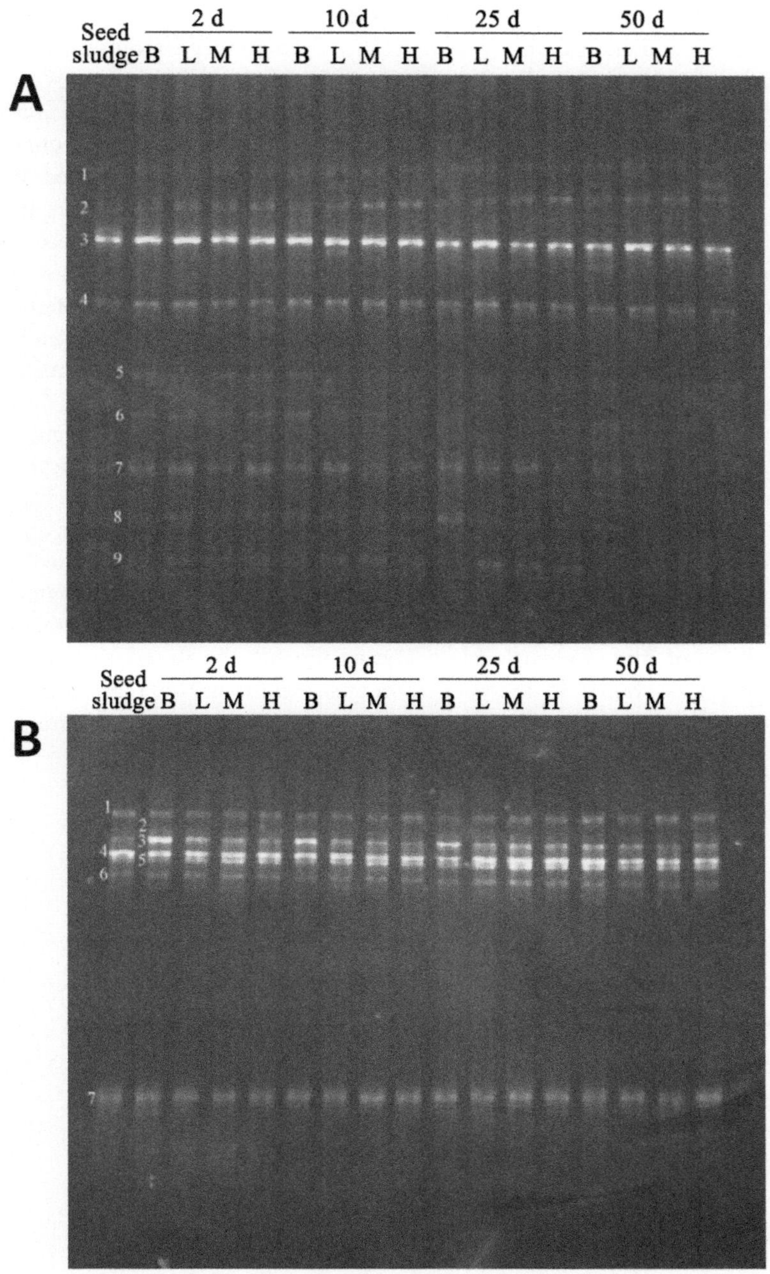

Figure 2 The DGGE profile for the microbial communities analyses at different AD stages. A, Bacterial community at different AD stages; **B**, Archaeal community at different AD stages. AD, anaerobic digestion; B, blank group; D, days; DGGE, denatured gradient gel electrophoresis; H, high group; L, low group; M, moderate group. The bands in Figure 2**A** indicate the presence of different microorganisms: 1.*Thermotogae* bacterium JX473505.1; 2.Uncultured bacterium DQ490120.1; 3.Uncultured *Planctomycetales* bacterium AM406812.1; 4.*Mesotoga* sp.KC800693.1; 5.*Thermotogae* bacterium JX473505.1; 6.*Trichococcus* sp.FJ374769.1; 7.Uncultured *Spirochaetes* bacterium JN998192.1; 8.*Trichococcus* sp.AM933652.1; 9.*Synergistetes* bacterium JX473564.1.The bands in Figure 2**B** indicate the presence of different microorganisms: 1.*Methanobacterium formicicum* JX042445.1; 2. *Methanobacteriaceae* archaeon JN836400.1; 3.Uncultured *Methanobacterium* sp.HQ231787.1; 4.Uncultured methanogenic archaeon DQ682552.1; 5.Uncultured *crenarchaeote* HQ141818.1; 6.Uncultured methanogenic archaeon DQ682560.1; 7.Unkonw.

bands were all similar, with a slight change due to the seed sludge. A difference of the DGGE bands between different groups was observed after 10 days, and the difference increased with the progress of AD. Regarding the biogas production, before 25 days all groups exhibited a stable biogas production, while the biogas production began to decrease both in the blank and H groups in the last stage (Table 1). This indicated that the initial microbial communities had the capability of adaptation and self-regulation, with a slight effect on the overall biogas production. However, after a certain change in microbial communities, a reduced biogas production was observed. It is likely that in this case the ability of the bacterial community to self-repair may be insufficient.

For further understanding of the effect of adding an exogenous buffer system on the bacterial community between the groups showing a significant difference, we selected samples of the blank and H groups at 10, 25 and 50 days for pyrosequencing analysis in depth (Figure 3).

The DGGE analysis was carried out for the archaeal community as well. No significant change was observed throughout the entire AD for all groups (Figure 2B). In order to study the effect of ammonia acetate on the archaeal community, we choose the H group as the research object since it had the highest ammonia concentration. Further analysis of the archaeal *16S rRNA* gene by pyrosequencing was shown in the follow text.

Bacterial *16S rRNA* variation under different inhibitory conditions

The above research showed that certain operation factors could cause inhibition effects in the AD process, with effect on the biogas production, but could also be expected to influence the diversity of the microbial communities. In order to correlate the ammonium acetate supplement and AD microbial communities, especially under the inhibitory conditions, the microbial communities of the blank and H groups were selected for further investigation by pyrosequencing analysis.

Discarding the unquantified sequences, a total of 23,703 effective sequences were obtained. Effective sequences of 2,700 were randomly picked from each sample which guaranteed the same analyzed sequencing depth. At the species level, observed numbers of operational taxonomic units (OTUs) were quite different from the estimated (abundance-based coverage estimator (ACE) and Chao 1 (species richness estimator) number of OTUs (Table 2). Regarding bacterial community, the blank group showed higher OTUs than that of the H group in the early periods (about 0 to 25 days) at species level; however, at the end of the digestion, the two groups exhibited similar OTUs. The high number of OTUs, ACE and Chao 1 reveals that, without the supplement of ammonium acetate, no inhibition effect occurred and the abundance of diversity was high in the early periods. While the VFAs accumulated during the AD, the abundance decreased, as reflected by a reduced number of OTUs, ACE and Chao 1. Contrary to the blank group, the H group was supplied with

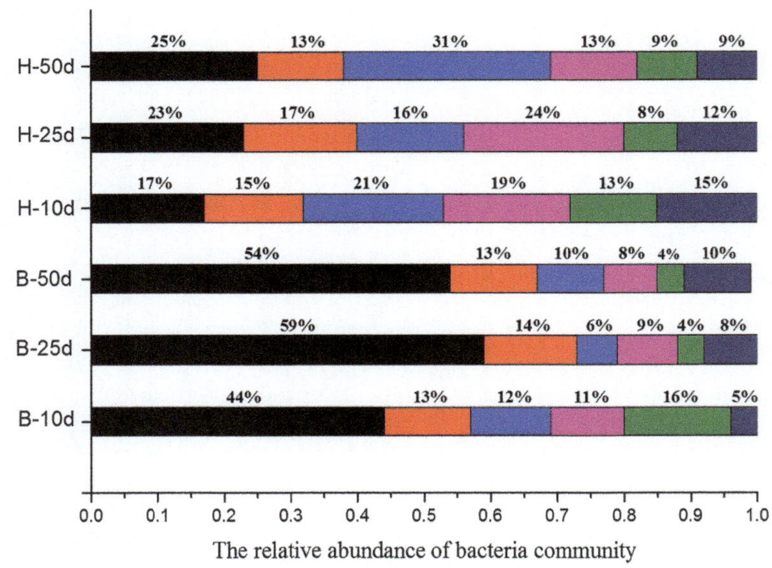

Figure 3 Taxonomic composition of the bacterial community at phyla level in each sample. The relative taxonomic abundances of the bacteria community at the phylum level. Phyla *Firmicutes* (black rectangle), WWE1 (green rectangle symbol), *Bacteroidetes* (red rectangle symbol), *Proteobacteria* (blue rectangle symbol), *Spirochetes* (pink rectangle symbol) and other (dark blue rectangle symbol) dominated in both blank and H groups.

Table 2 Effective DNA sequences, OTUs (Operational taxonomic units), ACE (abundance-based coverage estimator), Chao 1 (species richness estimator), Shannon (diversity index) and GOOD's coverage (percent of total species represented in the samples) of samples

Sample code		Effective reads	2,700 effective reads/sample (3% cut-off)				
			OTUs	ACE	Chao 1	Shannon	GOOD's coverage
Blank group	10 days	3,975	794	3,239.3	1,964.1	5.49	81%
	25 days	5,392	780	2,898.1	1,722.2	5.61	82%
	50 days	3,469	611	1,488.0	1,152.0	5.43	88%
H group	10 days	2,941	461	902.7	722.4	5.01	92%
	25 days	3,074	499	1,247.8	878.5	4.83	91%
	50 days	4,852	617	1,990.8	1,209.8	5.15	87%

ammonium acetate at the beginning (10 days), and this supplement exhibited strong inhibition on the abundance (low number of OTUs, ACE and Chao 1 in Table 2). Although the VFAs accumulated during the AD, the increased concentration of ammonium formed a buffer system with VFAs, which facilitate growth of certain species in the late phase (50 days), and led to an increase of abundance of diversity (a high number of OTUs, ACE and Chao 1). In addition, the blank group maintained a higher Shannon index compared with the H group during the whole process, which also indicates that the ammonium acetate supplement caused a simpler community structure than that without ammonium acetate supplement.

Figure 3 illustrates the relative taxonomic abundances of the bacterial community at the phylum level. *Phyla Firmicutes*, WWE1, *Bacteroidetes*, *Proteobacteria*, and *Spirochetes* (with relative abundance of higher than 10% in one sample) dominated in both the blank and H groups, and these five phyla are the common bacteria in AD [5,6]. However, ammonium acetate as supplement caused a change of phyla abundance. Although the alike responses of the low biogas production in both blank and H groups were observed, the inhibition could be attributed to different factors.

In the case of the blank group, the *Firmicutes phylum* was the dominant bacteria during the whole process, with a relative abundance of about 44 to 59%, while the relative abundance of *Bacteroidetes* at about 13 to 14% was the second largest phylum. The reason for this was probably that the lack of N-resource resulted in the overgrowth of *Firmicutes* in AD. However, the relative abundance of several dominant phyla in the H group did not show a significant difference (Figure 3); the most dominant bacteria were *Proteobacteria*, whose relative abundance was only about 16 to 31%. The ammonium acetate supplement influenced the growth of particular microbes, which resulted in the different patterns of phyla distributions between the blank and H groups. Furthermore, the gradual changes of the bacterial community during the process is worthy of attention.

Figure 4 illustrates more details about the change of the blank and H groups at genus level. For the blank group, from 10 to 25 days, *Firmicutes* (as the most abundant phylum) increased from 44 to 59%, which could be attributed to the significant increase of the proportion of *Caldanaerocella* and *Sedimentibacter* from 0.2 to 12.3% and 0.2 to 10.3%, respectively. As a fermentative bacteria, *Sedimentibacter* sp. consumes pyruvate and amino acids for acetate and butyrate production [21]. The relative abundance of *Proteobacteria* and WWE1 decreased from 10 to 25 days, because *Syntrophobacter* and BHB21 decreased from 8.0 to 2.4% and 14.8 to 3.8%, respectively. During this period of AD, the inhibition on methane production of the blank group was not obvious. However, the acid-generating species such as *Sedimentibacter* sp., *Clostridium* sp. reduced the pH value in the blank group. As a consequence, the acidic environment further affected the bacterial community, leading to an increase of the adaptive species and a decrease of the non-adaptive species.

From 25 to 50 days, the community in the blank group changed slightly compared with the early period. The abundance of *Caldanaerocella* decreased from 12.3 to 6.7%, while the abundance of *Sedimentibacter* increased from 10.3 to 13.1%. In the same period, however, the daily average biogas and methane production met a severe decrease, and fluctuation of pH was also observed. This illustrated that the inhibition of biogas production was not as sensitive as the microorganisms' change to the operating conditions from 25 to 50 days.

The abundance of each species in the H group also shifted during the whole operating process, although a far greater evenness of species was observed in the sample of the H group (Figure 4). The change of bacterial community in the H group was not as significant as that in the blank group from 10 to 25 days. At species level, the abundance of *Treponema* decreased from 11.9 to 7.5%, while BHB21 increased from 18.2 to 23.7%. This indicated that the microbial communities in the H group were relatively stable at the beginning of AD. As described above, according to the experiment data of biogas production

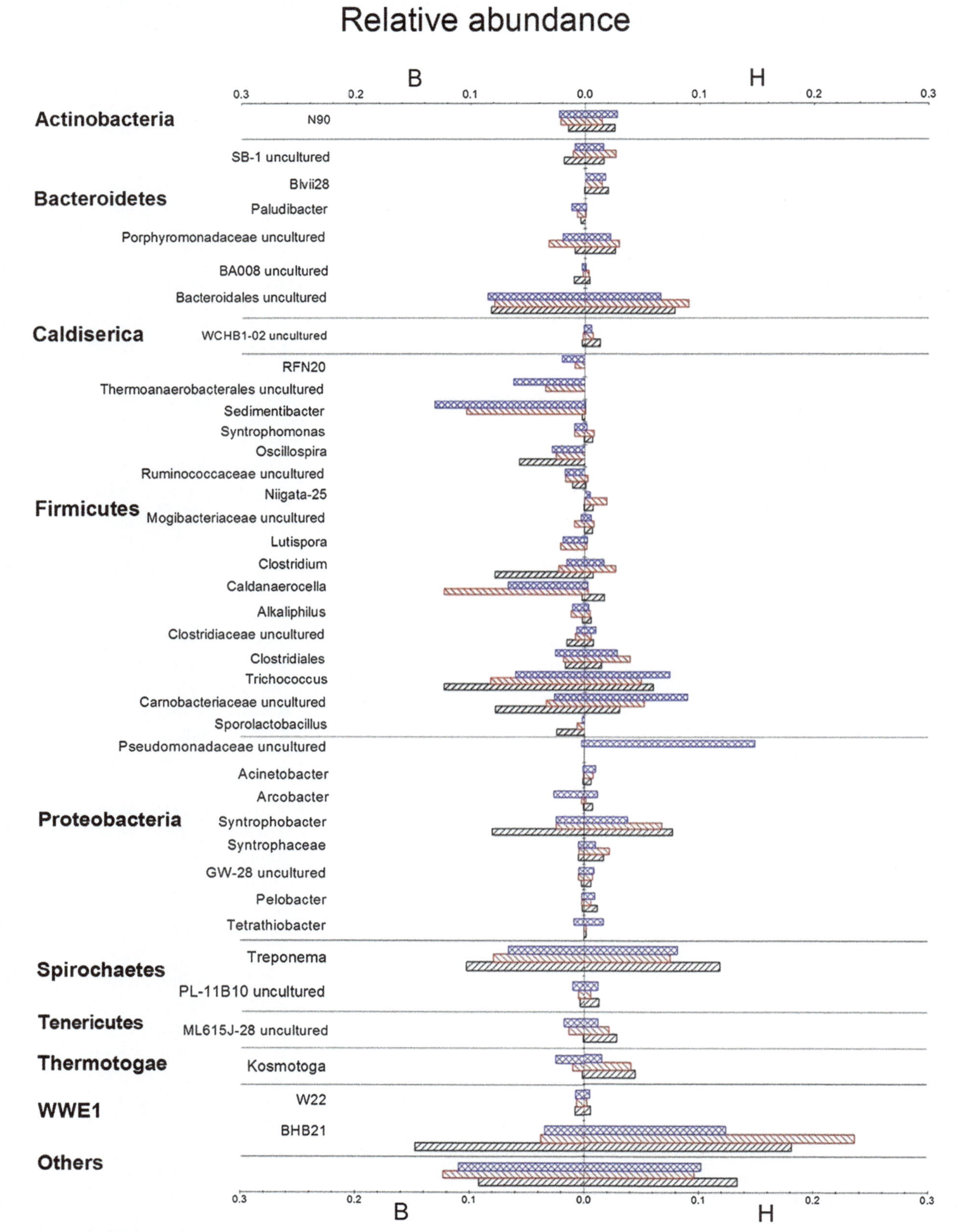

Figure 4 (See legend on next page.)

(See figure on previous page.)
Figure 4 Taxonomic composition of bacterial community at the genus level from the sequencing. Taxonomic compositions of bacterial community at the genus level from the sequencing, the left side of the axis shows the results of blank group, the right side of the axis shows the results of H group (brown rectangle symbol: 10 days; red rectangle symbol: 25 days; blue rectangle symbol: 50 days). The length of each bar indicates the relative abundance.

and liquid characteristics, the H group was not inhibited in this period, in which biogas production was similar to L and M. This reveals that an appropriate ammonium acetate concentration maintained the balance of microbial ecology.

Compared with 25 days, the microbial communities' obvious change at 50 days was shown from Table 3. *Pseudomonadaceae* uncultured bacterium was not detected at 25 days, however, its abundance reached 14.9% in 50 days. In addition, BHB21 decreased from 23.7 to 12.4%. The biogas production of the H group was lower than that of the L and M groups since the buffer system was broken down during the same period. However, the change of bacterial population lagged behind the shift of AD characters. On the other hand, along with increased NH_4^+ concentration, it was shown that the ecological balance was gradually destroyed from the excess ammonium acetate supplement.

Although the blank and H groups were inhibited and resulted in a similar poor biogas production, there was an essential distinction regarding to the microbial community structure. In the blank group, along with the VFAs accumulation, the microbial communities changed immediately, and thereafter the biogas production was inhibited. However, the system maintained a relatively constant microbial community in the inhibitory period. On the contrary, in the H group, the buffer formed from the supplement offset the impact of the VFAs and stabilized the balance of the microbial colony. Further, the microbial communities in the H group were sensitive when the NH_4^+ concentration increased, and drastic species changes were detected when the H group was inhibited, which means that the bacterial community in the blank and H groups was different because of ammonium acetate supplement, especially in the similar VFAs concentration range within 20 days.

The microbial community structure was not the unique factor for the biogas production. The pyrosequencing analysis indicated that an appropriate ammonium acetate concentration not only optimized the community structure, but also promoted the AD activity. Therefore, finely controlled ammonium acetate supplement will facilitate biogas production.

Table 3 Taxonomic composition of the archaeal community at the genus level for the sequences

	Relative abundance of H group		
Genus	**10 days**	**25 days**	**50 days**
Methanosaeta	82.19%	82.52%	63.56%
Methanoculleus	9.85%	10.30%	28.41%
Methanolinea	4.04%	3.15%	3.15%
Methanomicrobiales uncultured	2.22%	1.96%	3.44%
Coverage	98.30%	97.93%	98.56%
Other	1.70%	2.07%	1.44%

Archaeal *16S rRNA* gene analysis by pyrosequencing

As described above, when the NH_4^+ concentration was beyond the tolerance of the system, the balance of the bacterial community was broken with the change of community structure. According to the low biogas production in the H group, the reason for this may be that the formation and the breakdown of the buffer system also resulted in the change of the aerogenes community, especially methanogen. In the following section, the archaeal community structure was investigated in order to find the relationship of the supplement strategies and the change of the archaeal community.

The relative taxonomic abundances of the archaeal community at genus level were showed in Table 3. It was clear that, the predominant archea included four genera (>1%) before 10 days, *Methanosaeta, Methanoculleus, Methanolinea* and uncultured Archaea of the order *Methanomicrobiales*, all of which were in the phylum *Euryarchaeota*. The relative abundance of *Methanosaeta,* which only consumes acetate for methane production [22] and shows a much lower minimum threshold for acetate utilization, reached up to 82.2% [23]. As reported [8,24], because of the comparison with other aceticlastic methanoarchaea, *Methanosarcina* showed a lower metabolic activity in the high acetate concentrations condition [25]. In this study, seed sludge had been used for the food waste treatment for over one year, and the archaeal community had acclimated at a low acetate concentration during the operation process, which caused *Methanosaeta* to dominate in the community at the beginning of the digestion, before 25 days. At this period, the buffer system formed by ammonium acetate maintained a constant archaeal community in the reactor.

However, when the system was blocked at about 50 days, the proportion of *Methanosaeta* decreased from 82.5 to 63.6%, and *Methanoculleus* increased from 10.3 to 28.4%. It indicated that, with the release of free VFAs, the archaeal community shifted from acetoclastic methanogens to hydrogenotrophic methanogens, as in the research of Kim *et al.* [26], and the free NH_4^+ also enhanced the inhibition of acetoclastic methanogens [27]. However, the formation of the buffer delayed the change of archaeal community. When the NH_4^+ concentration reached 1.7 g/L, the hydrogen-consuming methanoarchaea maintained more activity than the aceticlastic methanoarchaea [28]. With the breakdown of the buffer (NH_4^+/VFAs), the abundance of *Methanosaeta* decreased and *Methanoculleus* became a significant genus, which means that the appropriate ammonia acetate supplement protected both bacteria and archaeal communities. However, excessive addition affected the community of bacteria and archaea negatively, and decreased both biogas and methane production.

Conclusions

The research of ammonium acetate acting as an efficient supplement demonstrated that the characteristics of a semi-continuous AD operation were significantly influenced. The results showed that an ammonium acetate concentration of 0.7 g/L/d could maintain a stable digestion process, with an average daily methane production of 1,122 mL/d; 44.5% higher than that of the blank group. The present study indicated that the optimal ammonium acetate concentration was about 2.5 to 3.0 g/L. The absence or overload of NH_4^+/VFAs led to the inhibition or failure of the AD operation. The DGGE and pyrosequencing analyses of bacterial and archaeal communities indicated the different inhibition in the blank and H groups. The ammonium acetate supplement could cause a simpler community structure than that without ammonium acetate supplement, finally resulting in the different patterns of phyla distributions between the blank and the H groups. The results demonstrate that an appropriate ammonium acetate supplement strategy can be applied to adjust the AD operation and microbial ecology towards the optimal biogas production.

Methods

The anaerobic digestion of food waste

The seed sludge used in this study was collected from the Shunyi wastewater treatment plant in Beijing, China, and had been anaerobically adapted to food waste treatment over one year in the laboratory. The characteristics of the seed sludge were assessed in Wang *et al.* [17] and are listed in Table 4. Seed sludge was separated into eight reactors (Shengda Glassware Factory, Shandong province, China) with a working volume of 1 L, and total solids (TS) of 19 wt%, operated at a mesospheric temperature of 37 ± 1°C for the semi-continuous AD. Four groups were examined in duplicate: the blank group with only a glucose feed, and three test groups, where glucose was supplemented by ammonium acetate (Xilong Chemical Co., Ltd, Guangdong province, China) in increasing concentrations of 0.7 g/L/d (L group), 1.0 g/L/d (M group) and 1.3 g/L/d (H group). The feedstock was diluted by the supernatant of the reactor, and then added to the reactor semi-continuously every 24 hours. All systems maintained the same total organic carbon (TOC) load (1.2 g TOC/d) with different N-resource concentrations. The reactors were rubber-sealed. Two sampling tubes were installed: one tube connected to a gas bag for biogas collection, and the other tube (under the liquid surface) for the collection of liquid samples. Gas samples were analyzed every 24 hours, while liquid samples were analyzed every 72 hours.

Table 4 Characteristics of seed sludge [17] (TS, Total solids; VS, volatile solids; VFAs, Volatile fatty acids)

Parameter	Value
TS (% wet weight)	19
VS (% wet weight)	17
pH	7.3
NH_4^+ (g/L)	0.8
VFAs (g/L)	0

Analytical methods

The TS and volatile solids (VS) content were measured according to Standard Method 2540 G (Shanghai Jing Hong Laboratory Instrument Co., Lid, Shanghai, China) [17]. The pH was measured using a pH probe (MP5C1, Shanghai San-Xin Instrumentation, Inc, Shanghai, China). The NH_4^+ was measured according to the Nessler's reagent colorimetric method of the China State Environmental Protection Administration (GB7479-87). The VFAs and the composition of biogas were measured by gas chromatograph (GC-2014C, SHIMADZU, Kyoto, Japan) [17].

Microbial structure analyses by denatured gradient gel electrophoresis and pyrosequencing

The extraction of total genomic DNA was carried out by Soil DNA extraction kit (OMEGA, Georgia, United States). For DGGE analysis, the extracted DNA was used for PCR with the primers 341 F-GC/518R (Table 5). To ensure the PCR amplification for archaeal DGGE analysis, a nest PCR

Table 5 The primer sequences for PCR analysis

Primer	Sequence (5′-3′)
341 F-GC	CCTACGGGAGGCAGCAG
518R	ATTACCGCGGCTGCTGGC
ARCH46F	YTAAGCCATGCRAGT
ARCH1017R	GGCCATGCACCWCCTCT
PARCH344F-GC	CGCCCGCCGCGCGCGGCGGGCGGGGCG
	GGGGCACGGGGGGHGCAGCAGGCGCGA

was carried out using the primers ARCH46F/ARCH1017R in the first round, and using the primers PARCH344F-GC/UNIV 522R in the second round. The PCR products were analyzed using Bio-Rad Dcode TM Universal Mutation Detection System (Bio-Rad, California, United States). Two polyacrylamide (Amresco, Ohio, United States) gel solutions with concentration of 8% were prepared, and urea (Amresco, Ohio, United States) and deionized formamide (Amresco, Ohio, United States) (containing 7 mol/L urea and 40% formamide, defined as the denaturant concentration 100%) were added in one of the two solutions. The gel was prepared by using the Bio-Rad gradient mixing device (Bio-Rad, California, United States), to make the bacterial denaturant concentration range of approximately 40 to 65%, wherein the concentration of denaturant was decreasing from bottom to top. The sheet was put into a completely solidified electrophoresis tank containing 1 × Tris-Acetate-EDTA buffer (Biotopped, Beijing, China), and the temperature was maintained at 60°C during the whole electrophoresis. Pre-run was performed for 20 minutes under conditions of 180 V, in order to remove impurities in the gel. The 30 μL samples were loaded to the inlet. After electrophoresis at 180 V for 5.5 hours, the gel was stained for 30 minutes using a 3 × GelRed (Biotium, California, United States) for further analysis.

For pyrosequencing, the DNA extracted by the above mentioned method was used for amplification of the bacterial hypervariable region with the primers 341 F-GC/518R and the archaeal hypervariable region with the primers ARC344/ARC915 of the *16S rRNA* gene (Table 5). Specific 10-bases long barcode sequences were attached to primers as a tag for identification. The PCR products were sent to Hanyu biotech Co. Ltd. (Shanghai, China) for pyrosequencing using a 454 GS-FLX sequencer (Roche Diagnostics Co., Indiana, United States) using a Titanium Sequencing Kit (Roche Diagnostics Co., Indiana, United States). The method for use of the DNA extraction kit, and the comparison by extraction of DNA from archaeal sludge, was illustrated previously [29].

In the obtained pyrosequencing data, ambiguous and short sequences with a length less than 200 and quality score of less than 20 nucleotides were removed. In addition, qualified sequences were clustered into OTUs defined by a 3% distance level using complete-linkage clustering. Distance matrices were generated and sequences were grouped to generate rarefaction curves by a 3% distance level [30]. OTU-based diversity analyses and richness estimates of Chao 1 and ACE, the Shannon index and Simpson index calculations were performed using the mothur package (www.mothur.org/wiki/Download_mothur) [31].

Abbreviations

ACE: Abundance-based coverage estimator; AD: Anaerobic digestion; DGGE: Denatured gradient gel electrophoresis; OTUs: Operational taxonomic units; PCR: Polymerase chain reaction; TAE buffer: Tris-Acetate-EDTA buffer; TOC: Total organic carbon; TS: Total solids; VFAs: Volatile fatty acids; VS: Volatile solids.

Competing interests

The authors declare that they have no competing interests.

Authors' contributions

HS and TT participated in the design of the study. QW and YJ executed the experimental study. HS, QW and YJ analyzed the data. HS, LL, SW and XH wrote and revised the manuscript. All authors read and approved the final manuscript.

Acknowledgements

The authors want to express their thanks for the support from the National Basic Research Program (973 Program) of China (2014CB745103), the (863) High Technology Project (2012AA021402), and the project sponsored by the Scientific Research Foundation for the Returned Overseas Chinese Scholars, State Education Ministry (LXJJ2012-001) and the Chinese Universities Scientific Fund (JD1417).

References

1. Bayr S, Rintala J. Thermophilic anaerobic digestion of pulp and paper mill primary sludge and co-digestion of primary and secondary sludge. Water Res. 2012;46:4713–20.
2. Ghanimeh S, El Fadel M, Saikaly P. Mixing effect on thermophilic anaerobic digestion of source-sorted organic fraction of municipal solid waste. Bioresour Technol. 2012;117:63–71.
3. Zhang C, Su H, Baeyens J, Tan T. Reviewing the anaerobic digestion of food waste for biogas production. Renew Sustain Energy Rev. 2014;38:383–92.
4. Appels L, Baeyens J, Degrève J, Dewil R. Principles and potential of the anaerobic digestion of waste-activated sludge. Prog Energy Combustion Sci. 2008;34:755–81.
5. Cardinali-Rezende J, Colturato LFDB, Colturato TDB, Chartone-Souza E, Nascimento AMA, Sanz JL. Prokaryotic diversity and dynamics in a full-scale municipal solid waste anaerobic reactor from start-up to steady-state conditions. Bioresour Technol. 2012;119:373–83.
6. Nelson MC, Morrison M, Yu Z. A meta-analysis of the microbial diversity observed in anaerobic digesters. Bioresour Technol. 2011;102:3730–9.
7. Tang Y, Shigematsu T, Morimura S, Kida K. Microbial community analysis of mesophilic anaerobic protein degradation process using bovine serum albumin (BSA)-fed continuous cultivation. J Biosci Bioeng. 2005;99:150–64.
8. Ike M, Inoue D, Miyano T, Liu TT, Sei K, Soda S, et al. Microbial population dynamics during startup of a full-scale anaerobic digester treating industrial food waste in Kyoto eco-energy project. Bioresour Technol. 2010;101:3952–7.
9. Jard G, Jackowiak D, Carrère H, Delgenes JP, Torrijos M, Steyer JP, et al. Batch and semi-continuous anaerobic digestion of *Palmariapalmata*: comparison with *Saccharinalatissima* and inhibition studies. Chem Eng J. 2012;209:513–9.
10. Ferrer I, Ponsá S, Vázquez F, Font X. Increasing biogas production by thermal (70°C) sludge pre-treatment prior to thermophilic anaerobic digestion. Biochem Eng J. 2008;42:186–92.
11. Bayr S, Pakarinen O, Korppoo A, Liuksia S, Väisänen A, Kaparaju P, et al. Effect of additives on process stability of mesophilic anaerobic monodigestion of pig slaughterhouse waste. Bioresour Technol. 2012;120:106–13.
12. Guo J, Ostermann A, Siemens J, Dong R, Clemens J. Short term effects of copper, sulfadiazine and difloxacin on the anaerobic digestion of pig manure at low organic loading rates. Waste Manag. 2012;32:131–6.
13. Tada C, Yang Y, Hanaoka T, Sonoda A, Ooi K, Sawayama S. Effect of natural zeolite on methane production for anaerobic digestion of ammonium rich organic sludge. Bioresour Technol. 2005;96:459–64.
14. El-Mashad HM, Zhang R. Biogas production from co-digestion of dairy manure and food waste. Bioresour Technol. 2010;101:4021–8.
15. Hansen KH, Angelidaki I, Ahring BK. Anaerobic digestion of swine manure: inhibition by ammonia. Water Res. 1998;32:5–12.

16. Nielsen HB, Angelidaki I. Strategies for optimizing recovery of the biogas process following ammonia inhibition. Bioresour Technol. 2008;99:7995–8001.
17. Wang Q, Peng L, Su H. The effect of a buffer function on the semi-continuous anaerobic digestion. Bioresour Technol. 2013;139:43–9.
18. Jang HM, Park SK, Ha JH, Park JM. Microbial community structure in a thermophilic aerobic digester used as a sludge pretreatment process for the mesophilic anaerobic digestion and the enhancement of methane production. Bioresour Technol. 2013;145:80–9.
19. Lin J, Zuo J, Ji R, Chen X, Liu F, Wang K, et al. Methanogenic community dynamics in anaerobic co-digestion of fruit and vegetable waste and food waste. J Environ Sci. 2012;24:1288–94.
20. Yagi H, Ninomiya F, Funabashi M, Kunioka M. RNA analysis of anaerobic sludge during anaerobic biodegradation of cellulose and poly(lactic acid) by RT-PCR–DGGE. Polym Degrad Stab. 2011;96:547–52.
21. Breitenstein A, Wiegel J, Haertig C, Weiss N, Andreesen JR, Lechner U. Reclassification of Clostridium hydroxybenzoicum as Sedimentibacter hydroxybenzoicus gen. nov., comb. nov., and description of Sedimentibactersaalensis sp. nov. Int J Syst Evol Microbiol. 2002;52:801–7.
22. Smith KS, Ingram-Smith C. *Methanosaeta*, the forgotten methanogen? Trends Microbiol. 2007;15:150–5.
23. Jetten MSM, Stams AJM, Zehnder AJB. Methanogenesis from acetate: a comparison of the acetate metabolism in *Methanothrix soehngenii* and *Methanosarcina* spp. FEMS Microbiol Lett. 1992;88:181–97.
24. Lim JW, Chen CL, Ho IJR, Wang JY. Study of microbial community and biodegradation efficiency for single- and two-phase anaerobic co-digestion of brown water and food waste. Bioresour Technol. 2013;147:193–201.
25. Janssen PH. Selective enrichment and purification of cultures of *Methanosaeta* spp. J Microbiol Methods. 2003;52:239–44.
26. Kim S, Bae J, Choi O, Ju D, Lee J, Sung H, et al. A pilot scale two-stage anaerobic digester treating food waste leachate (FWL): Performance and microbial structure analysis using pyrosequencing. Process Biochem. 2014;49:301–8.
27. Kampmann K, Ratering S, Baumann R, Schmidt M, Zerr W, Schnell S. Hydrogenotrophic methanogens dominate in biogas reactors fed with defined substrates. Syst Appl Microbiol. 2012;35:404–13
28. Koster IW, Lettinga G. The influence of ammonium-nitrogen on the specific activity of pelletized methanogenic sludge. Agric Wastes. 1984;9:205–16.
29. Guo F, Zhang T. Biases during DNA extraction of activated sludge samples revealed by high throughput sequencing. Appl Microbiol Biotechnol. 2013;97:4607–16.
30. Chao A, Lee S-M. Estimating the number of classes via sample coverage. J Am Stat Assoc. 1992;87:210–7.
31. Schloss PD, Westcott SL, Ryabin T, Hall JR, Hartmann M, Hollister EB, et al. Introducing mothur: open-source, platform-independent, community-supported software for describing and comparing microbial communities. Appl Environ Microbiol. 2009;75:7537–41.

Anaerobic detoxification of acetic acid in a thermophilic ethanologen

A Joe Shaw[1,2], Bethany B Miller[1], Stephen R Rogers[1], William R Kenealy[1,3], Alex Meola[1], Ashwini Bhandiwad[4,5], W Ryan Sillers[1,6], Indraneel Shikhare[1], David A Hogsett[1,7] and Christopher D Herring[1,4*]

Abstract

Background: The liberation of acetate from hemicellulose negatively impacts fermentations of cellulosic biomass, limiting the concentrations of substrate that can be effectively processed. Solvent-producing bacteria have the capacity to convert acetate to the less toxic product acetone, but to the best of our knowledge, this trait has not been transferred to an organism that produces ethanol at high yield.

Results: We have engineered a five-step metabolic pathway to convert acetic acid to acetone in the thermophilic anaerobe *Thermoanaerobacterium saccharolyticum*. The first steps of the pathway, a reversible conversion of acetate to acetyl-CoA, are catalyzed by the native *T. saccharolyticum* enzymes acetate kinase and phosphotransacetylase. *ack* and *pta* normally divert 30% of catabolic carbon flux to acetic acid; however, their re-introduction in evolved ethanologen strains resulted in virtually no acetic acid production. Conversion between acetic acid and acetyl-CoA remained active, as evidenced by rapid ^{13}C label transfer from exogenous acetate to ethanol. Genomic re-sequencing of six independently evolved ethanologen strains showed convergent mutations in the *hfs* hydrogenase gene cluster, which when transferred to wildtype *T. saccharolyticum* conferred a low acid production phenotype. Thus, the mutated *hfs* genes effectively separate acetic acid production and consumption from central metabolism, despite their intersecting at the common intermediate acetyl-CoA. To drive acetic acid conversion to a less inhibitory product, the enzymes thiolase, acetoacetate:acetate CoA-transferase, and acetoacetate decarboxylase were assembled in *T. saccharolyticum* with genes from thermophilic donor organisms that do not natively produce acetone. The resultant strain converted acetic acid to acetone and ethanol while maintaining a metabolic yield of 0.50 g ethanol per gram carbohydrate.

Conclusions: Conversion of acetic acid to acetone results in improved ethanol productivity and titer and is an attractive low-cost solution to acetic acid inhibition.

Keywords: Acetate detoxification, Cellulosic ethanol, Metabolic engineering, Inhibitors

Background

Biological conversion of low-cost carbohydrate-based renewable resources [1] is often hindered by inhibitors present in these heterogeneous materials. Acetic acid has consistently been reported as one of the most significant inhibitors, followed by furfural, hydroxymethylfurfural, and phenolic compounds [2,3]. Unlike other inhibitors, which are by-products of specific chemical and physical pretreatment processes [4,5], acetic acid can result directly from acetylated carbohydrates, and its production is unavoidable during carbohydrate fermentation.

* Correspondence: chrisherringfish@gmail.com
[1]Mascoma Corporation, Lebanon, NH 03766, USA
[4]Thayer School of Engineering, Dartmouth College, Hanover, NH 03755, USA
Full list of author information is available at the end of the article

Acetic acid is a particularly potent inhibitor at pH values under 6, a range common for industrial fermentation [6-8]. Undissociated acetic acid (pKa 4.75) freely passes through cellular membranes, where it can then reach a new equilibrium with dissociated acetic acid at the intracellular pH. The net result is both uncoupling of the transmembrane pH gradient and an accumulation of acetate anion in the cytoplasm [9,10].

Several chemical and physical methods to separate or detoxify acetic acid prior to fermentation have been proposed [11], but they require additional expense to implement, which may hinder commercial application. Biological removal of acetate has also been proposed via aerobic respiration of acetate followed by fermentation of carbohydrates to ethanol [12,13]. However, this requires a separate oxygen-dependent incubation, which raises overall process costs. Solventogenic clostridia, typified by *Clostridium acetobutylicum*, have the native ability to convert acetic acid and butyric acid into the solvents acetone and butanol [14]. The conversion is not easily transferrable to the context of ethanol production though, as it normally occurs in two phases with acetate first produced in an acidogenic phase, followed by conversion to acetone in a solventogenic phase [15].

Acetate detoxification can occur under normal fermentation conditions, enabled by an advanced organism without requiring additional capital equipment. We report conversion of acetic acid to acetone while maintaining a high yield of carbohydrates to ethanol in *Thermoanaerobacterium saccharolyticum*, thereby demonstrating a low-cost solution to acetic acid inhibition.

Results

Re-introduction of *ack* and *pta* do not restore acetate production in evolved ethanologen strains

T. saccharolyticum has previously been engineered to produce ethanol at high yield by deletion of *ldh* encoding L-lactate dehydrogenase and *ack* and *pta* encoding acetate kinase and phosphotransacetylase [16,17]. From a carbon-based perspective, these two deletions eliminate organic acid production and force central metabolic flux towards ethanol. From an electron-based perspective, nicotinamide adenine dinucleotide (NADH) generated during glycolysis and reduced ferredoxin generated during pyruvate cleavage must be re-oxidized through reduction of acetyl-CoA to ethanol, rather than through formation of lactic acid or molecular hydrogen (Figure 1). In attempts to engineer ethanologen strains with improved properties for industrial fermentation, we developed multiple strain lineages derived from strain M0355, a Δldh Δack Δpta mutant [16]. Several of these lineages underwent evolutionary selection, either via continuous culture or serial transfer, for improved fermentation fitness (Figure 2). Interestingly, we found on several occasions that re-introduction of phosphotransacetylase and acetate kinase into evolved strains resulted in greatly reduced acetic acid production as compared to the same genes re-introduced into strain M0355 or the wildtype strain.

Spontaneous mutations occurring during evolutionary selection direct electron flux to ethanol

One possible explanation for this phenomenon was that mutations which occurred during the evolutionary process direct metabolic flux to ethanol even in the presence of a functional pathway from acetyl-CoA to acetic acid. We re-sequenced the genomes of six evolved strains (Table 1), revealing 12 to 28 mutations per strain. In strains sharing a common lineage, up to 10 of the mutations were present in both strains. However regardless of lineage, we found recurrent mutations in two loci relevant to acetic acid and ethanol production; in subunits of the *hfs* hydrogenase gene operon and in *adhE*, the bifunctional acetaldehyde alcohol dehydrogenase (Table 2). We thought either of these mutant loci might contribute to the acetate minus phenomenon; entire deletion of the *hfs* hydrogenase gene operon has previously been shown to reduce acetic acid production [18], while mutations in *adhE* could result in a higher affinity for acetyl-CoA or NAD(P)H, effectively out-competing acetic acid or hydrogen formation, respectively. To test these possibilities, we introduced the *hfs* and *adhE* mutations that had occurred in evolved strain M0863 into the wildtype background via homologous recombination with selection based on insertion of a kanamycin resistance marker 3′ to the genomic region of interest. Depending on the exact site of recombination crossover, the mutations were not always carried along with the marker. Strains with the desired mutation were identified by DNA sequencing, and strains without mutations were also evaluated to determine effects that introduction of the resistance marker may have introduced. Results are shown in Table 3. While the mutant *adhE* gene had an impact on end product ratios, only the strain carrying the *hfs* spontaneous mutations produced significantly less acetic acid, with a combined ethanol and lactate to acetate ratio 10-fold higher than wildtype. Introduction of the M0863-acquired *hfs* mutations did not appear detrimental to growth rate, which was in contrast to a complete deletion of the *hfs* operon, which decreased growth rate and the amount of consumed carbohydrate [18].

Acetic acid and acetyl-CoA freely interconvert in evolved ethanologen strains harboring *ack* and *pta*

Figure 3 shows $^{13}C_2$-labeled acetate spiked at 18 hours in a batch carbohydrate fermentation with strain M1667 (an evolved ethanologen from the M0863 lineage with re-introduced *ack* and *pta* genes) in the presence of 60 mM externally added acetic acid. Based on nuclear magnetic resonance (NMR) spectra, the $^{13}C_2$ label is almost completely transferred from acetic acid to ethanol within 7 hours, despite minimal change in the bulk acetic acid concentration. This data indicates a dilution of the $^{13}C_2$ acetate pool by approximately 50% in an hour and a corresponding increase in $^{13}C_2$ ethanol. No other metabolite showed enrichment of ^{13}C beyond that expected from natural abundance indicating that the acetate was metabolized intact. In similar experiments with strain M1291, the direct parent strain of M1667 without *ack* and *pta*, there was no evidence of ^{13}C label transfer from acetic acid for up to 48 hours after spike addition.

The reaction acetic acid + ATP + CoA $\leftrightarrow$ acetyl-CoA + ADP + P_i is near thermodynamic equilibrium (+2 kJ/mol at pH 7) and is used by microorganisms for both acetate production and consumption in various environmental

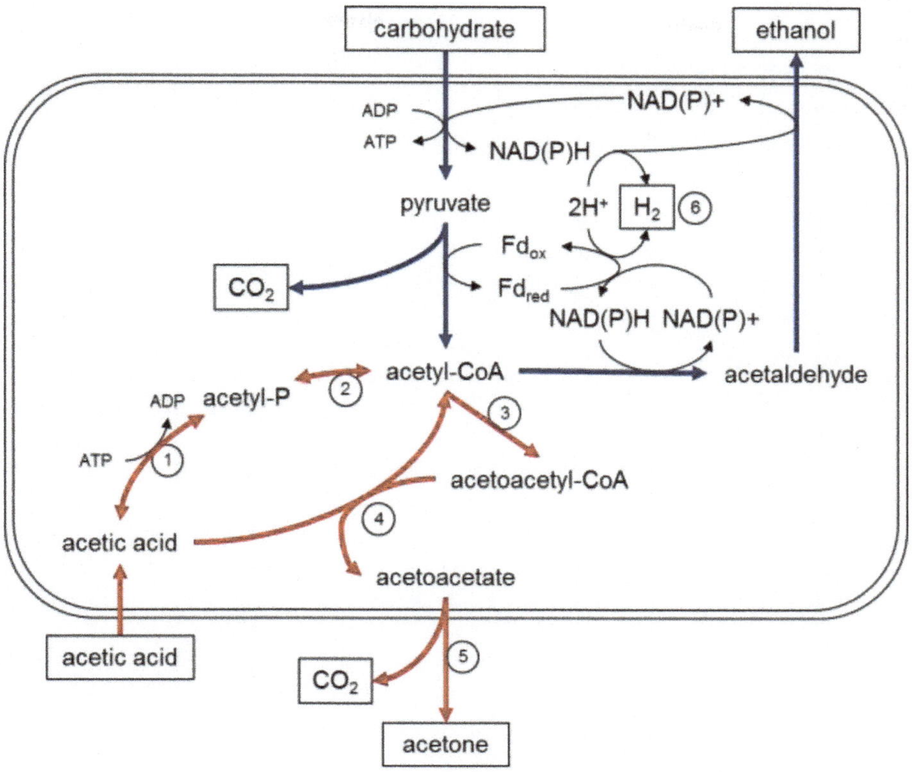

Figure 1 Synthetic pathway to convert acetic acid to acetone. Acetic acid diffuses freely into the cell following hydrolysis of acetylated polysaccharides and is then activated to acetyl-CoA by acetate kinase (1), phosphotransacetylase (2), and a half-reaction of CoA-transferase (4). Two acetyl-CoA molecules are then converted to acetoacetyl-CoA by thiolase (3), acetoacetate by the other half-reaction of CoA-transferase, and finally to acetone and CO_2 by acetoacetate decarboxylase (5). Although the synthetic pathway shares a common intermediate with the ethanol production pathway, carbohydrate to ethanol production remains highly coupled due to the requirement to balance NAD(P)+/NAD(P)H generation. Hydrogenases (6) act to uncouple electron acceptor regeneration and ethanol formation, resulting in production of acetic acid through the reversible acetate kinase and phosphotransacetylase pathway.

conditions [19-22]. Although the bulk concentration of acetic acid remained constant in evolved ethanologen strains containing *pta* and *ack*, we hypothesize that labeled acetate is reversibly converted into labeled acetyl-CoA, a fraction of which is then irreversibly converted into labeled ethanol. The intracellular acetyl-CoA pool is also fed by unlabeled carbohydrate metabolism and, assuming that ^{13}C label is not differentially recognized, can be equally used to produce acetate or ethanol. Carbohydrate-derived acetyl-CoA then acts to dilute the labeled acetate/acetyl-CoA pool and allows the creation of labeled ethanol. Due to stoichiometric coupling with hydrogen formation,

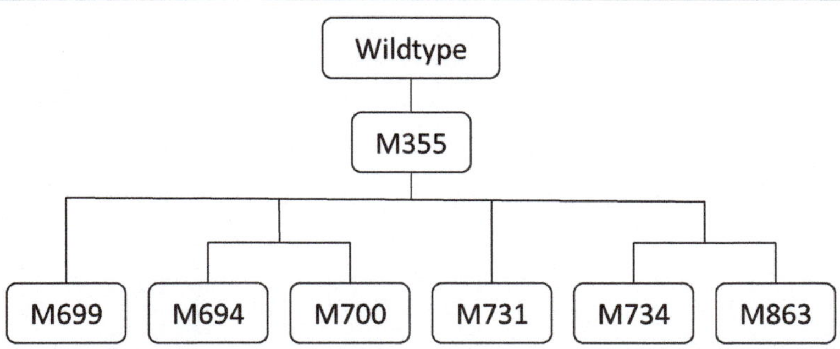

Figure 2 Lineage of strains selected for improved growth rate and ethanol production in this study. M0355 carries deletions of the *ldh*, *pta*, and *ack* but was not selected for improved growth rate. The lineages resulting in M0694, M0699, and M0700 were cultivated in continuous culture to select for improved growth properties. The lineages resulting in M0731, M0734, and M0863 were exposed to chemical mutagenesis and serial transfer for improved growth properties; the lineage resulting in M0863 was subsequently cultured in continuous culture. All strains were isolated as single colonies prior to genomic DNA isolation and re-sequencing.

Table 1 Strains used in this study

Strain	Description	Source
M0010	*T. saccharolyticum* JW/SL-YS485 DSM 8691	DSMZ
M0355	*T. saccharolyticum Δpta Δack Δldh*	(16)
M1291	*T. saccharolyticum Δpta Δack Δldh Cth ureABCDEFG*	This study
M1442	*T. saccharolyticum Δpta Δack Δldh Cth ureABCDEFG*	This study
M1667	*T. saccharolyticum Δpta Δack Δldh::Tsa pta ack kanR Cth ureABCDEFG*	This study
M2030	*T. saccharolyticum Δpta Δack Δldh::Tsa pta ack Cac thl Tme thl Tte thl Cac ctfAB Tme ctfAB Cac adc Bam adc kanR Cth ureABCDEFG*	This study
M2202	*T. saccharolyticum adhE*::kanR*	This study
M2203	*T. saccharolyticum adhE::kanR*	This study
M2204	*T. saccharolyticum hfs*::kanR*	This study
M2205	*T. saccharolyticum hfs::kanR*	This study
M2212	*T. saccharolyticum Δpta Δack Δldh::Tsa pta ack Tte thl Tme ctfAB Bam adc kanR Cth ureABCDEFG*	This study
Top10	*E. coli* cloning strain	Invitrogen, Madison, WI
FY2	*S. cerevisiae* cloning strain	(37)
	Re-sequenced strains with improved growth rate	
M0694	*T. saccharolyticum Δpta Δack Δldh* continuous cultured adapted	This study
M0699	*T. saccharolyticum Δpta Δack Δldh* continuous cultured adapted	This study
M0700	*T. saccharolyticum Δpta Δack Δldh* continuous cultured adapted	This study
M0731	*T. saccharolyticum Δpta Δack Δldh* mutation and serial transfer	This study
M0734	*T. saccharolyticum Δpta Δack Δldh* mutation and serial transfer	This study
M0863	*T. saccharolyticum Δpta Δack Δldh* mutation and serial transfer	This study

DSMZ, Deutsche Sammlung von Mikroorganismen und Zellkulturen.

net acetate formation is arrested via the evolutionarily acquired *hfs* mutations while the capacity to convert acetate to acetyl-CoA is retained.

Construction of a synthetic pathway for the conversion of acetic acid to acetone

Exogenously added acetone is substantially less inhibitory than acetic acid to *T. saccharolyticum* in the pH range of 5.0 to 5.5 relevant to industrial fermentation (Figure 4). Additionally, conversion of two molecules of acetic acid results in one molecule of acetone, further lowering the potential for acetone inhibition. Solventogenic Clostridia such as *C. acetobutylicum* have native biochemical pathways to convert acetic acid into acetone, although they are intricately coupled to carbohydrate metabolism. We began by importing the *C. acetobutylicum* acetone pathway, inclusive of thiolase, acetate:acetoacetyl-CoA-transferase, and acetoacetate decarboxylase (*thl ctfAB adc*) as well as the *T. saccharolyticum* acetate kinase and phosphotransacetylase into the *T. saccharolyticum* expression plasmid pMU1299 that integrates at the *ldh* locus. Despite conferring acetone production in *E. coli* (Table 4),

Table 2 Mutations identified by re-sequencing

Gene	Locus tag	ORF nt	DNA mutation	Coding change	Occurrence in strain
adhE	Tsac_416	1630	G→A	G→D	M0734, M0863
adhE	Tsac_416	1804	C→T	S→L	M0699
adhE	Tsac_416	1808	A→G	E→G	M0694, M0700
hfsA	Tsac_1550	159	A→-	Frameshift	M0700
hfsB	Tsac_1551	434	A→-	Frameshift	M0734, M0863
hfsB	Tsac_1551	652	A→-	Frameshift	M0731
hfsC	Tsac_1552	255	A→-	Frameshift	M0699
hfsD	Tsac_1553	193	G→A	A→T	M0699
hfsD	Tsac_1553	321	A→T	R→S	M0863
hfsD	Tsac_1553	787	G→A	E→K	M0734

Table 3 Re-introduction of spontaneous *adhE* and *hfs* hydrogenase mutations that occurred during growth adaptation of the Δ*ldh*, Δ*pta* Δ*ack* ethanologen strain M0863 into wildtype (WT) *T. saccharolyticum*

		Units in mM				
Strain	Genotype	Cellobiose consumed	Lactate	Acetate	Ethanol	Carbon recovery (%)
M0010	WT	28.2 ± 0.6	13.8 ± 0.5	34.7 ± 0.7	63.5 ± 2.3	99
M2202	*adhE**::*kan*^{R}	30.2 ± 0.0	2.4 ± 0.1	42.3 ± 0.3	76.5 ± 0.6	100
M2203	*adhE*::*kan*^{R}	27.7 ± 2.6	1.4 ± 0.2	29.0 ± 1.6	78.4 ± 8.6	98
M2204	*hfs**::*kan*^{R}	29.9 ± 0.1	20.3 ± 0.5	4.2 ± 0.1	91.7 ± 0.6	97
M2205	*hfs*::*kan*^{R}	28.1 ± 0.7	9.9 ± 2.2	29.8 ± 1.1	67.9 ± 2.7	96

*mutations found in evolved strain M0863 introduced by homologous recombination. Data are average of four replicates, anaerobic TSC7 medium, 55°C, 24-h fermentation of 30.4 ± 0.7 mM cellobiose. Mutations were created via homologous recombination of a kanamycin resistance marker downstream of the *hfs* operon that contained mutations in the upstream flanking region. M2202 - kan^R inserted 3' of the *adhE* gene, with the M0863 mutant *adhE* sequence. M2203 - kan^R inserted 3' of the *adhE* gene, with the WT *adhE* sequence. M2204 - kan^R inserted 3' of the *hfs* gene operon, with the M0863 mutant *hfs* sequence. M2205 - kan^R inserted 3' of the *hfs* gene operon, with the WT *hfs* sequence.

we could find no evidence of acetone formation upon integration of the plasmid in *T. saccharolyticum*. While RT-PCR experiments (not shown) indicated transcriptional activity of the acetone pathway genes, a likely issue was temperature incompatibility of the donor and host organism; *C. acetobutylicum* has grown optimally between 30°C and 37°C, while *T. saccharolyticum*'s growth rate drops precipitously below 48°C, the lowest temperature where we assayed for acetone formation.

We next looked for thermophilic acetone producers but could not find a native acetone producer with an optimal growth temperature higher than 43°C [23]. We could, however, find homologs to each of the three enzymes in the acetone pathway residing in the sequenced genomes of thermophilic bacteria (Table 5). None of these thermophiles were reported to produce acetone, and the entire pathway was not present in a single organism. To explore the ability of these genes to produce acetone, we opted to progressively incorporate expression vectors harboring thermophilic acetone pathway homologs into *T. saccharolyticum* and assay the resulting strains for acetone production. Qualitative evidence of ketone body formation was first identified in strain M2030 (Figure 5), which contained, in addition to the *T. saccharolyticum ack* and *pta* genes, three *thl* genes, two *ctfAB* gene operons, and two *adc* genes. We then constructed strains with minimal sets of acetone pathway genes while continuing to assay for acetone formation; the most efficient acetone producer, M2212, was created by expression of a putative thiolase from *Thermoanaerobacterium thermosaccharolyticum* DSM 571, a putative CoA-transferase from *Thermosipho melanesiensis* DSM 12029, and a putative acetoacetate

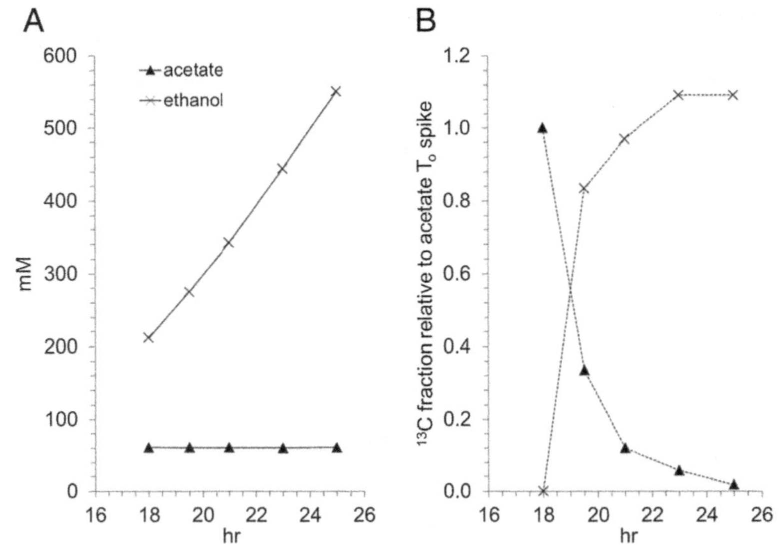

Figure 3 Fate of $^{13}C_2$-labeled acetate during strain M1667 carbohydrate fermentation. Strain M1667 was grown in batch fermentation in a 3:2 weight mixture of maltodextrin and cellobiose at a concentration of 900 mM glucose equivalents and 45 mM acetic acid at an initial pH of 6.0. At 18 hours, the culture was spiked with 15 mM sodium acetate-$^{13}C_2$. Bulk metabolite measurements **(A)** and ^{13}C label **(B)** were monitored for extracellular acetic acid (filled triangles) and ethanol (crosses). No appreciable label was found in metabolites other than acetic acid and ethanol. Data is from one representative experiment.

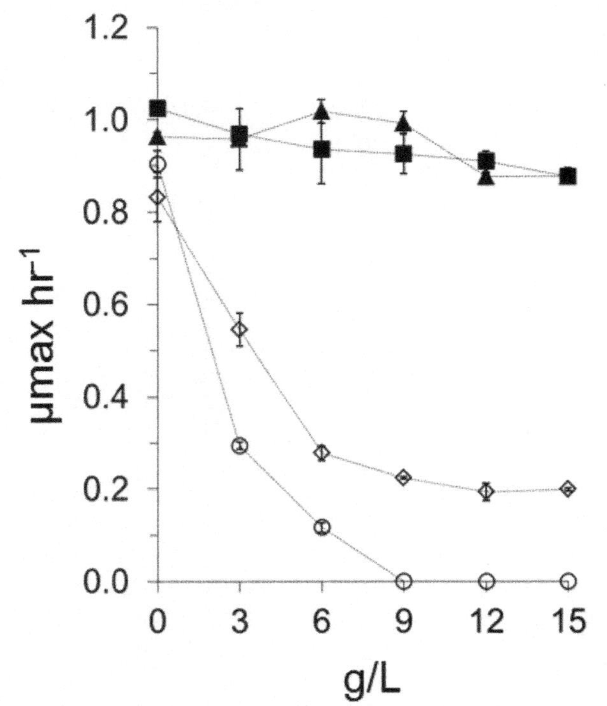

Figure 4 Maximum growth rate of *T. saccharolyticum* in the presence of exogenously added acetic acid or acetone at pH 5.0 and 5.5 in TSC7 medium with 10 g/L cellobiose at 55°C. Open circles - acetic acid pH 5.0, open diamonds - acetic acid pH 5.5, filled squares - acetone pH 5.0, filled triangles - acetone pH 5.5. Data represent the mean of triplicate cultures +/− one standard deviation.

decarboxylase from *Bacillus amyloliquefaciens* FZB42. The *T. thermosaccharolyticum thl* is part of a gene operon with several homologs involved in butanol production, and the organism has been reported to produce butanol [24]. *T. melanesiensis ctfAB* is adjacent to a thiolase in its native genetic organization, which could potentially be involved in acetoacetate degradation. The *B. amyloliquefaciens adc* appears to be independently transcribed; however, the native function of either enzyme is not obvious based on genetic organization and the known physiology of these organisms.

Strain M2212 was also assayed for the *in vitro* ability to catalyze conversions along the pathway of acetyl-CoA to acetone. Table 6 shows data collected from cell-free extract measurements of thiolase, CoA-transferase, and acetoacetate decarboxylase from strain M2212 and M1442, the direct parent ethanologen. While activities for these enzymes were at or below the limit of detection with strain M1442, detectable levels were found for each step with strain M2212.

Table 4 Acetone production in *E.coli*

	Glucose	Lactic	Acetic	Ethanol	Acetone
Media	24.5	0.0	4.2	0.0	0.0
TOP10 + pMU1299	16.1	5.9	3.1	0.5	1.8
TOP10 + pACYC177	19.5	3.6	5.1	0.4	0.0

Fermentation end point at 170 hours, in LB media supplemented with 25 g/L glucose, 4 g/L acetic acid, and 50 μg/mL kanamycin sulfate. Microaerobic conditions were utilized by inoculation into sealed 125 mL pressure bottles with atmospheric pressure air. Incubation was performed without agitation at 30°C, with a $1/10^{th}$ v/v inoculation from aerobically grown overnight LB media supplemented with kanamycin.

Expression of the synthetic acetone pathway improves ethanol productivity and titer in the presence of acetic acid

Strain M2212 and strain M1442 were grown in batch fermentation in the presence of acetic acid (Figure 6). M2212 reduced the overall acetic acid concentration by 2.8 g/L, which in combination with an increase in the pH, reduced the undissociated acetic acid concentration from an initial 1.3 g/L to below 0.6 g/L. 0.78 g/L acetone was produced, a value lower than that expected from the theoretical 0.5 g/g conversion of acetic acid to acetone. However, this measurement may also reflect some acetone evaporation during fermentation gas release, as the boiling point of acetone, 54°C, is slightly below the 55°C fermentation temperature. In contrast, fermentation with strain M1442 resulted in a small 0.4 g/L increase in acetic acid, while the undissociated acetic acid concentration rose to over 4.3 g/L as a result of declining pH during the fermentation. From Figure 4, a 50% decrease in μmax occurs at an undissociated acetic acid concentration of 0.61 to 0.81 g/L. The decline in pH with strain M1442 is likely a result of assimilation of ammonium sulfate and carbonic acid generated during production of carbon dioxide. The conversion of acetic acid to acetone completely reversed this pH trend in strain M2212.

Strain M2212 has favorable ethanol fermentation metrics as well; compared to M1442, the final ethanol titer was 33.1 g/L higher, an increase of more than 300%. Volumetric productivity at 48 hours rose from 0.22 g/L hr^{-1} to 0.92 g/L hr^{-1}, also an increase of over 300%. Final metabolic yields, as a measure of gram ethanol produced per gram glucose equivalent consumed, were 0.48 g/g for M1442 and 0.50 g/g for M2212.

Discussion

The observation that re-introduction of *ack* and *pta* in evolved ethanologen strains resulted only in minute amounts of acetic acid production was surprising; the genes had initially been removed to eliminate acetic acid production, which can account for up to 30% of total organic end products on a molar basis in wildtype cells [17,25]. Predicting that evolutionary selection had imparted mutations resulting in this phenotype, we examined the genome sequences of several strains descended from strain M0355 that had been serially or continuously cultured for several generations. We identified two mutational hotspots, in the *hfs* hydrogenase gene operon and the *adhE* bifunctional acetaldehyde/alcohol dehydrogenase, that could be rationally predicted to impact acetic acid production.

Table 5 Genes tested for acetone production in *T.saccharolyticum*

Source strain	T_{opt}(°C)	Gene	Genbank protein
Thermoanaerobacterium saccharolyticum DSM 8691	60	*pta*	ACA51668
Thermoanaerobacterium saccharolyticum DSM 8691	60	*ack*	ACA51669
Clostridium acetobutylicum ATCC 824	35	*thl*	NP_349476
Thermosipho melanesiensis DSM 12029	70	*thl*	YP_001306374
Kosmotoga olearia DSM 21960	65	*thl*	YP_002940320
Thermoanaerobacterium thermosaccharolyticum DSM 571	60	*thl*	YP_003852249
Thermoanaerobacterium thermosaccharolyticum DSM 571	60	*actA*	CAA93155
Clostridium acetobutylicum ATCC 824	35	*ctfA*	NP_149326
Thermosipho melanesiensis DSM 12029	70	*ctfA*	YP_001306376
Kosmotoga olearia DSM 21960	65	*ctfA*	YP_002940319
Clostridium acetobutylicum ATCC 824	35	*ctfB*	NP_149327
Thermosipho melanesiensis DSM 12029	70	*ctfB*	YP_001306375
Kosmotoga olearia DSM 21960	65	*ctfB*	YP_002940318
Clostridium acetobutylicum ATCC 824	35	*adc*	NP_149328
Acidothermus cellulolyticus B11 ATCC 43068	55	*adc*	YP_872855
Bacillus amyloliquefaciens FZB42 BGSC 10A6	50	*adc*	YP_001422565

T_{opt} is the optimal temperature for growth of the source strain.

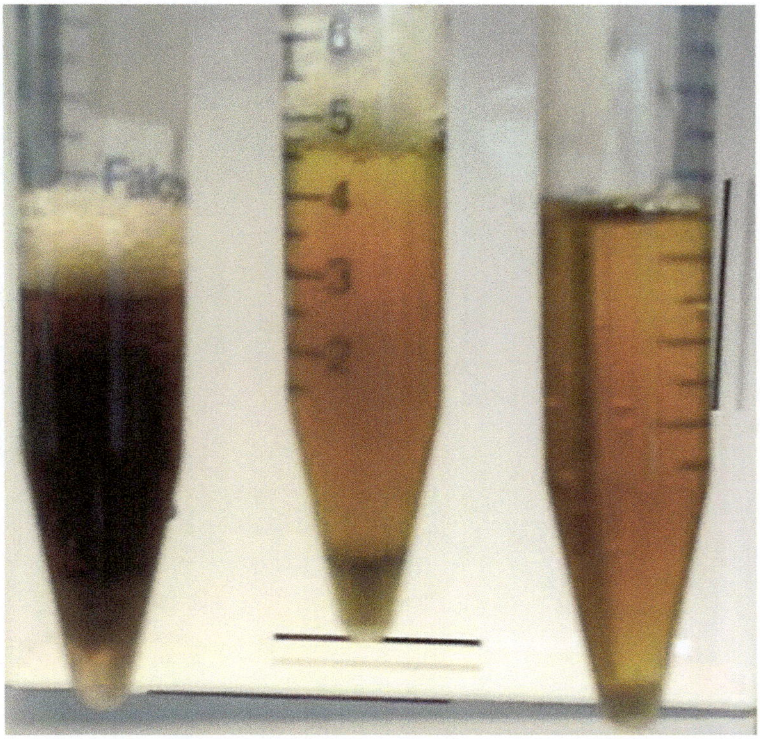

Figure 5 Rothera's test (1) for acetone and acetoacetic acid formation performed on spent fermentation medium, initially containing 50 g/L cellobiose and 5 g/L acetic acid. From left to right, M2030, M1675 + pMU131 (the immediate parent strain of M2030), and uninoculated medium. Acetone and acetoacetic acid are characterized by a purple/red coloremetric reaction with sodium nitroprusside in the presence of saturating ammonia. To perform the test, 4-mL samples were saturated with ammonium sulfate, and 5% sodium nitroprusside was added followed by an overlay of 1 mL 25% ammonium hydroxide. Photograph was taken 18 hours after addition of sodium nitroprusside.

Table 6 Enzymatic activities in cell extracts

	μmol min^{-1} mg^{-1} protein					
	Thiolase	**SD**	**CoA-transferase**	**SD**	**Acetoacetate decarboxylase**	**SD**
M1442	0.06	0.13	0.02	0.02	0.17	0.23
M2212	10.21	0.55	3.57	0.09	1.66	0.26

Data represent the mean of triplicate determinations +/− one standard deviation. SD, standard deviation.

Introducing these mutations to the wildtype strain indicated that the *hfs* mutation is responsible for a sharp decrease in acetic acid production.

An entire deletion of the *hfs* operon has previously been shown to reduce acetic acid production in wildtype and Δ*ldh T. saccharolyticum* strains [18]. However, it also results in strains with poor fermentation kinetics, which are not readily improved after several serial transfers. Deletion of the entire *hfs* operon in evolved Δ*ldh* Δ*ack* Δ*pta* ethanologens also resulted in strains with poor fermentation kinetics, suggesting that the point mutations acquired through evolution have a more nuanced effect than an entire loss of function mutation. Any fitness advantage the *hfs* mutations impart to ethanologen strains is unknown, although it is tempting to speculate that cellular redox regulation may play a role. Of six evolved strains, five acquired frameshift mutations in the genes *hfsA*, *hfsB*, or *hfsC*, encoding a putative

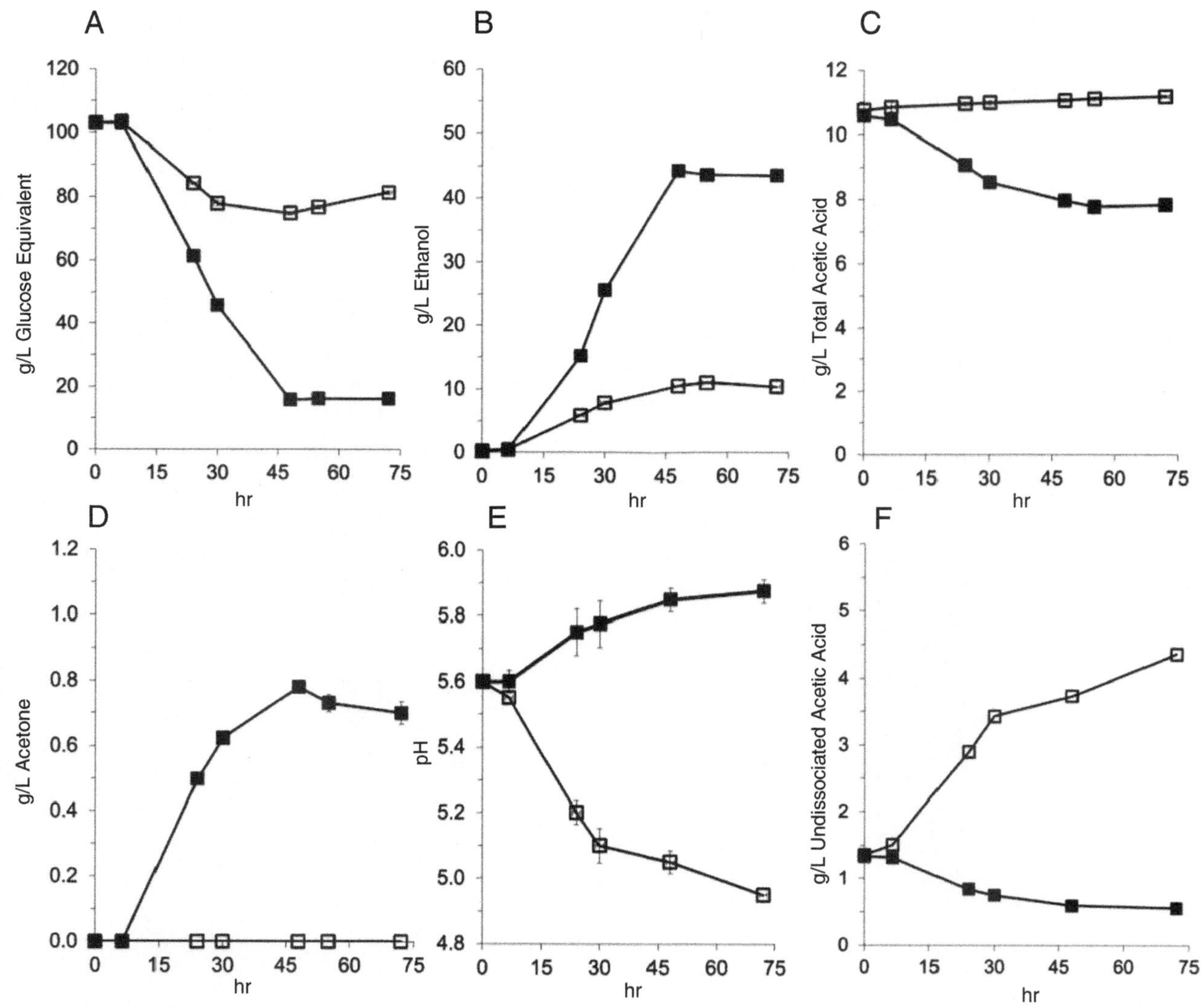

Figure 6 Fermentation of maltodextrin, plotted as glucose equivalents, in the presence of 10 g/L acetic acid in TSC7 medium at 55°C. **(A-F)** show the levels of different substrates, products or pH, as indicated by the vertical axis label. Filled squares - M2212, containing the acetic acid to acetone pathway. Open squares - parental ethanologen strain M1442. Strain M1442 rapidly encounters inhibition as the undissociated acetic acid concentration rises, while M2212 decreases the undissociated acetic acid to below 0.6 g/L **(F)**. Data represent the mean of triplicate cultures +/− one standard deviation.

ferredoxin, a hydrogenase and [PAS] sensory domain fusion and a histidine kinase, respectively. Identification and classification of [FeFe] hydrogenases with possible sensory functions is an area of current investigation [26-28]. These frameshift mutations may be beneficial by loss of function of the affected gene but may also have beneficial effects through polar down-regulation of the gene *hfsD* downstream. This gene, encoding a putative monomeric group B [FeFe] hydrogenase [28], only acquired point mutations, resulting in amino acid changes in three of six strains. A small amount of hydrogen production can be predicted to be important for ethanologen strains to maintain redox balance with cell mass and associated carbon dioxide generation from carbohydrates, in a role parallel to that of glycerol formation in *S. cerevisiae* [29-31].

The *adhE* mutations, although not further investigated here, may be involved in co-factor specificity and increased ethanol tolerance, as previously reported for *C. thermocellum* [32]. Another evolved *T. saccharolyticum* ethanologen, strain ALK2, was shown to have an increase in nicotinamide adenine dinucleotide phosphate (NADPH) specificity and a decrease in NADH specificity for whole cell extract alcohol dehydrogenase activity [17].

The combination of *hfs* gene mutations and *ack* and *pta* re-introduction allows for inter-conversion of acetic acid and acetyl-CoA independent of central metabolism. To detoxify acetic acid with concurrent high yield fermentation of carbohydrates to ethanol, the solvent acetone was an attractive end product of acetic acid metabolism. It is less inhibitory than acetic acid, can be easily recovered through distillation, and can be sold as a commodity chemical. The native solventogenic clostridial pathway takes one acetyl-CoA and one acetic acid and converts them to one CoA, one CO_2, and one acetone. This, combined with action of phosphotransacetylase and acetate kinase, allows the redox neutral conversion of two acetic acid and one ATP to one acetone, one CO_2, one H_2O, and one ADP + P_i. The standard Gibbs free energy change for this reaction is −0.8 kJ/rxn, a fairly small driving force. However, if acetic acid accumulates intracellularly due to a transmembrane pH gradient, the overall reaction will become increasingly thermodynamically favorable. Assuming acetic acid is at intracellular and extracellular equilibrium, an extracellular change of 1.0 pH unit more acidic results in a $\Delta G'$ change of −12.4 kJ/rxn. Thus, the pathway becomes more thermodynamically feasible as acetic acid becomes more inhibitory.

The inability to produce acetone in *T. saccharolyticum* with heterologously expressed enzymes from *C. acetobutylicum* and the lack of natively thermophilic acetone producers led us to test homologs of thiolase, CoA-transferase, and acetoacetate decarboxylase from thermophiles that presumably participate in other metabolic pathways. The final pathway reported here, utilizing genes from *T. saccharolyticum*, *T. thermosaccharolyticum*, *T. melanesiensis*, and *B. amyloliquefaciens*, is assembled entirely from organisms that do not produce acetone from acetic acid.

Despite being carbon and electron independent, production of acetone from acetic acid does require chemical energy in the form of ATP, which must be supplied from central catabolism. Based on a stoichiometric evaluation of the fermentation presented in Figure 6, 5% of the total ATP produced by carbohydrate catabolism is used for acetic acid conversion to acetone during the maximum rate of acetone formation. The rapid and complete conversion of ^{13}C-labeled acetate to ethanol indicates there is sufficient capacity for acetate entry into the cell. Further conversion could be limited by thermodynamic feasibility, as the majority of the residual acetic acid is in the dissociated species when the rate of acetone production decreases.

From economic and process engineering perspectives, biological conversion of acetic acid to acetone provides several distinct advantages over other detoxification methods. In the case of lignocellulose to ethanol conversion, it is likely to increase ethanol productivity and titer, while maintaining the carbohydrate to ethanol yield essential for efficient biofuel production. Acetone is readily recovered by distillation equipment already installed for ethanol recovery, and it has a selling price around $1.15 per kg [33] with a North American market size of 1.6 MM metric tons per year [34]. The conversion of acetic acid to a solvent reduces the demand for pH neutralization during fermentation and lowers the chemical or biological oxygen demand for wastewater treatment during water recycle, as acetic acid composes half or more of soluble, unrecovered organic material in a typical lignocellulose process [35]. Finally, unlike chemical or physical methods to remove acetic acid, conversion to acetone does not require significant additional capital or operating expenditures, as new unit operations are not required beyond separation of acetone from ethanol. With these factors taken together, this technology has the potential to lower costs and improve the feasibility of large scale, anaerobic production of biofuels and biochemicals from acetate containing feedstocks.

Conclusions

Spontaneous mutations acquired in the *hfs* hydrogenase gene cluster prevent net acetate production in strains in which the *pta* and *ack* genes are re-introduced. Surprisingly, uptake and incorporation of exogenous acetate occurs in those strains even when overall acetate levels remain constant, as shown by ^{13}C-labeling. Genes from *C. acetobutylicum* did not confer a functional acetone production pathway, but genes assembled from a variety of thermophilic species did. The best performing strain converted exogenous acetate to acetone, while markedly improving ethanol production. These findings improve our understanding of acetate and ethanol metabolism and demonstrate a novel strategy for converting a potential inhibitor into a valuable co-product.

Methods

Strains and DNA vectors

Strains used in this study are listed in Table 1, and plasmids and primers are given in Additional file 1. Plasmids used to transform *T. saccharolyticum* were constructed with *S. cerevisiae* gap repair cloning techniques and amplified in *E. coli* [36]. *T. saccharolyticum* was transformed using a natural competence based system [37], and transformants were selected by resistance to the antibiotic kanamycin. Chromosomal integrants were evaluated by colony PCR with primers which hybridized to regions external and internal to the integration region.

Fermentation conditions

T. saccharolyticum was primarily grown in modified TSC7 medium, containing per liter 1.0 g sodium citrate tribasic dihydrate, 4.0 g $(NH_4)_2SO_4$, 0.2 g $FeSO_4{\cdot}7H_2O$, 2.0 g $MgSO_4{\cdot}7H_2O$, 1.0 g KH_2PO_4, 0.2 g $CaCl_2{\cdot}2H_2O$, 0.12 g L-methionine, 0.5 g L-cystiene HCl, and 8.5 g yeast extract. For C^{13} acetate labeling experiments, TSC4 base medium was used, with the following differences from TSC7: 5 g urea replacing $(NH_4)_2SO_4$, 0.5 g $MgSO_4{\cdot}7H_2O$, 0.5 g KH_2PO_4, and 2 g sodium citrate tribasic dihydrate. Carbohydrates and acetic acid were added as indicated in the text and figure legends. Chemicals, including $^{13}C_2$ acetate, were acquired from Sigma-Aldrich (St Louis, MO, USA). Fermentations were begun by addition of 10% *v*/*v* from an exponentially growing pre-culture and were performed in sealed pressure vessels (120-mL bottles with a 20-mL working volume for high carbohydrate concentrations and 20-mL tubes with a 5-mL working volume for low carbohydrate concentrations. Vessels were purged with an anaerobic gas mixture containing 95% N_2 and 5% CO_2 and vented to atmospheric pressure at each sampling. Fermentations were performed at 55°C and shaking at 150 to 200 rpm.

Analytical methods

Fermentation metabolite concentrations were determined by high-performance liquid chromatography (HPLC) using an Aminex HPX-87H column (Bio-Rad Laboratories, Hercules, CA, USA) with a refractive index detector. Acetone was additionally identified with a UV detector at 260 nm. Proton-decoupled one-dimensional ^{13}C spectra were obtained from fermentation samples mixed with D_2O (20%) with a Varian Inova 500MHz NMR spectrometer (Bloomington, IN, USA) operating at 125 MHz using the WALTZ-16 sequence for decoupling. The sweep width was 18 kHz, and acquisition time was 1.5 s. A pulse angle of 80° and 0.8 s delay were used. Data (65,000 pts) was Fourier transformed with a line broadening of 0.5 Hz.

Ketone body formation was qualitatively assayed via the Rothera method, as described in Figure 5. Maltodextrin was hydrolyzed with *A. niger* amyloglucosidase according to the starch assay kit protocol (Sigma-Aldrich, St. Louis, MO, USA) prior to detection of glucose by HPLC. Undissociated acetic acid levels were calculated from acetate and pH measurements.

Genome re-sequencing

Descendants of the ethanologen strain M0355 evolved for improved growth rate and ethanol production via continuous culturing and chemical mutagenesis were sequenced at the National Center for Genome Resources on the Illumina platform, with each strain generating 5 to 8 million 36 bp reads. The reads were assembled onto a draft genome of wildtype *T. saccharolyticum* reference sequence using SeqMan NGen from DNAStar (Madison, WI, USA). SNPs, and other small mutations were manually verified by examination of the sequence assemblies and distinguished from errors in the draft genome sequence, evident as sequence differences in all strains.

Enzymatic assays

Cell-free extracts were prepared from cultures grown in 200 mL TSC7 medium with 10 g/L cellobiose at 55°C to an optical density of 1.0. Cells were chilled to 4°C, harvested aerobically by centrifugation at 8,000×*g* for 5 min, washed twice with 50-mL cold water, and pelleted to approximately 250-uL wet volume. Cell lysis was performed by addition of B-PER bacterial cell lysis reagent (Thermo Scientific, Rockford, IL, USA) with 0.1 mg lysozyme, 2.5 U DNAse I, and 1 mM phenylmethanesulfonylfluoride per mL of B-PER reagent. Cell extracts used for acetoacetate decarboxylase assays additionally contained 50 mM dithiothreitol. After cell re-suspension and a 30-min room temperature incubation, cell debris was removed by centrifugation at 15,000×*g* for 10 min. Protein concentration was determined via the Bradford method with Bio-Rad Protein Assay Reagent (Bio-Rad, Hercules, CA, USA). All chemicals and enzymes were obtained from Sigma-Aldrich, and reported data are the average of three replicate assays.

Thiolase was assayed in the direction of acetoacetyl-CoA cleavage, which was monitored at 303 nm with an extinction coefficient of 14.00 mM^{-1} cm^{-1} at room temperature in a Spectramax M2 plate reader (Molecular Devices, Sunnyvale, CA, USA). The assay contained, in a final volume of 200 uL, 100 mM Tris-HCl (pH 8.0), 1 mM DTT, 10 mM $MgCl_2$, 0.05 mM acetoacetyl-CoA, 0.2 mM CoA and 4 μg cell-free extract. Addition of CoA started the reaction, and activity depended on both the presence of acetoacetyl-CoA and cell-free extract.

CoA-transferase was assayed in the direction of acetate and acetoacetyl-CoA conversion to acetoacetate and acetyl-CoA. To measure activity, acetyl-CoA was subsequently converted to citrate and CoA via citrate synthase, and CoA formation was monitored with 5,5′-dithiobis-(2-nitrobenzoic acid) (DTNB) at 412 nm with an extinction coefficient of 14.15 mM^{-1} cm^{-1} at room temperature in a Spectramax M2 plate reader. The assay contained, in a final volume of 200 uL, 67 mM Tris-HCl (pH 8.0), 10 mM potassium acetate, 1 mM oxaloacetate, 0.1 mM DTNB, 2 U porcine heart citrate synthase, 0.2 mM acetoacetyl-CoA, 0.2 mM CoA, and 4 μg cell-free extract. Addition of acetoacetyl-CoA started the reaction, and activity depended on the presence of acetoacetyl-CoA, acetate, and cell-free extract. Control assays without added cell extract were used to subtract baseline signal.

Acetoacetate decarboxylase was assayed in the direction of acetone formation via HPLC with a Fast Acid Analysis column (Bio-Rad, Hercules, CA, USA) with refractive index and UV detection at 260 nm. The assay contained, in a final volume of 1.8 mL, 100 mM KPO_4 buffer at pH 6.1, 25 mM lithium acetoacetate and 180 μg cell-free extract. The assay began with addition of cell-free extract, and samples were taken at 70 min and frozen at −80°C until acidification to pH 2 and HPLC analysis. Control assays without added cell extract were used to subtract baseline signal.

Additional file

Additional file 1: Plasmids and primers used in this study.

Abbreviations

ADP: adenosine diphosphate; ATP: adenosine triphosphate; CoA: coenzyme A; NADH: nicotinamide adenine dinucleotide; NADPH: nicotinamide adenine dinucleotide phosphate; NMR: nuclear magnetic resonance; RT-PCR: reverse transcriptase-polymerase chain reaction.

Competing interests

Authors were former employees of Mascoma Corporation, which held a financial interest in intellectual property related to this work.

Authors' contributions

AJS, WRK, AB, WRS, DAH, and CDH designed experiments and analyzed data. AJS, BBM, SRR, AM, and IS performed experiments. AJS, WRK, and CDH wrote and reviewed the manuscript. All authors have read and approved the final version of this manuscript.

Acknowledgements

We would like to thank Wayne Casey at the Dartmouth College Department of Chemistry for assistance with NMR analysis. This work was supported by the Mascoma Corporation and the Department of Energy under Award Number DE-FC36-07G017057. Manuscript preparation was supported in part by the BioEnergy Science Center, a U.S. Department of Energy Bioenergy Research Center supported by the Office of Biological and Environmental Research in the DOE Office of Science. Disclaimer 'This report was prepared as an account of work sponsored by an agency of the United States Government. Neither the United States Government nor any agency thereof, nor any of their employees, makes any warranty, express or implied, or assumes any legal liability or responsibility for the accuracy, completeness, or usefulness of any information, apparatus, product, or process disclosed, or represents that its use would not infringe privately owned rights. Reference herein to any specific commercial product, process, or service by trade name, trademark, manufacturer, or otherwise does not necessarily constitute or imply its endorsement, recommendation, or favoring by the United States Government or any agency thereof. The views and opinions of authors expressed herein do not necessarily state or reflect those of the United States Government or any agency thereof.'

Author details

[1]Mascoma Corporation, Lebanon, NH 03766, USA. [2]Novogy Inc., 85 Bolton St, Cambridge, MA 02140, USA. [3]Verdezyne Inc., 2715 Loker Avenue West, Carlsbad, CA 92010, USA. [4]Thayer School of Engineering, Dartmouth College, Hanover, NH 03755, USA. [5]Energy Biosciences Institute, 2151 Berkeley Way, Berkeley, CA 94704, USA. [6]Myriant Corporation, 66 Cummings Park, Woburn, MA 01801, USA. [7]OPX Biotechnologies Inc., 2425 55th Street, Boulder, CO 80301, USA.

References

1. Perlack RD, Wright LL, Turhollow AF, Graham RL, Stoke BJ, Erbach DC. Biomass as feedstock for a bioenergy and bioproducts industry: the technical feasibility of a billion-ton annual supply. Technical Report. 2005; ORNL/TM-205/66. http://www.osti.gov/scitech/biblio/885984.
2. Palmqvist E, Hahn-Hägerdal B. Fermentation of lignocellulosic hydrolysates. II: inhibitors and mechanisms of inhibition. Bioresource Technol. 2000;74:25–33.
3. Ingram LO, Aldrich HC, Borges AC, Causey TB, Martinez A, Morales F, et al. Enteric bacterial catalysts for fuel ethanol production. Biotechnol Progr. 2009;15:855–66.
4. Weil JR, Dien B, Bothast R, Hendrickson R, Mosier NS, Ladisch MR. Removal of fermentation inhibitors formed during pretreatment of biomass by polymeric adsorbents. Ind Eng Chem Res. 2002;41:6132–8.
5. Larsson S, Palmqvist E, Hahn-Hägerdal B, Tengborg C, Stenberg K, Zacchi G, et al. The generation of fermentation inhibitors during dilute acid hydrolysis of softwood. Enzyme Microb Tech. 1999;24:151–9.
6. Ranatunga T, Jervis J, Helm R, McMillan J, Hatzis C. Identification of inhibitory components toxic toward *Zymomonas mobilis* CP4(pZB5) xylose fermentation. Appl Biochem Biotech. 1997;67:185–98.
7. Sreenath HK, Jeffries TW. Production of ethanol from wood hydrolyzate by yeasts. Bioresource Technol. 2000;72:253–60.
8. Taherzadeh MJ, Niklasson C, Lidén G. Acetic acid - friend or foe in anaerobic batch conversion of glucose to ethanol by *Saccharomyces cerevisiae*? Chem Eng Sci. 1997;52:2653–9.
9. Lawford H, Rousseau J. Improving fermentation performance of recombinant *Zymomonas* in acetic acid-containing media. Appl Biochem Biotechnol. 1998;70-72:161–72.
10. Booth IR. Regulation of cytoplasmic pH in bacteria. Microbiol Rev. 1985;49:359–78.
11. Mussatto SI, Roberto IC. Alternatives for detoxification of diluted-acid lignocellulosic hydrolyzates for use in fermentative processes: a review. Bioresource Technol. 2004;93:1–10.
12. Schneider H. Selective removal of acetic acid from hardwood-spent sulfite liquor using a mutant yeast. Enzyme Microb Tech. 1996;19:94–8.
13. Lakshmanaswamy A, Rajaraman E, Eiteman MA, Altman E. Microbial removal of acetate selectively from sugar mixtures. J Ind Microbiol Biot. 2011;38:1477–84.
14. Jones DT, Woods DR. Acetone-butanol fermentation revisited. Microbiol Rev. 1986;50:484–524.
15. Lütke-Eversloh T, Bahl H. Metabolic engineering of *Clostridium acetobutylicum*: recent advances to improve butanol production. Curr Opin Biotech. 2011;22:634–47.
16. Shaw AJ, Covalla SF, Hogsett DA, Herring CD. Marker removal system for *Thermoanaerobacterium saccharolyticum* and development of a markerless ethanologen. Appl Environ Microbiol. 2011;77:2534–6.
17. Shaw AJ, Podkaminer KK, Desai SG, Bardsley JS, Rogers SR, Thorne PG, et al. Metabolic engineering of a thermophilic bacterium to produce ethanol at high yield. Proc Natl Acad Sci U S A. 2008;105:13769–74.
18. Shaw AJ, Hogsett DA, Lynd LR. Identification of the [FeFe]-hydrogenase responsible for hydrogen generation in *Thermoanaerobacterium saccharolyticum* and demonstration of increased ethanol yield via hydrogenase knockout. J Bacteriol. 2009;191:6457–64.

19. Wolfe AJ. The acetate switch. Microbiol Mol Biol Rev. 2005;69:12–50.
20. Kumari S, Tishel R, Eisenbach M, Wolfe AJ. Cloning, characterization, and functional expression of *acs*, the gene which encodes acetyl coenzyme A synthetase in *Escherichia coli*. J Bacteriol. 1995;177:2878–86.
21. Fournier GP, Gogarten JP. Evolution of acetoclastic methanogenesis in *Methanosarcina* via horizontal gene transfer from cellulolytic clostridia. J Bacteriol. 2008;190:1124–7.
22. Ingram-Smith C, Martin SR, Smith KS. Acetate kinase: not just a bacterial enzyme. Trends Microbiol. 2006;14:249–53.
23. Weimer PJ. Control of product formation during glucose fermentation by *Bacillus macerans*. J Gen Microbiol. 1984;130:103–11.
24. Freier-Schröder D, Wiegel J, Gottschalk G. Butanol formation by *Clostridium thermosaccharolyticum* at neutral pH. Biotechnol Lett. 1989;11:831–6.
25. Desai SG, Guerinot ML, Lynd LR. Cloning of L-lactate dehydrogenase and elimination of lactic acid production via gene knockout in *Thermoanaerobacterium saccharolyticum* JW/SL-YS485. Appl Microbiol Biotechnol. 2004;65:600–5.
26. Ballor N, Paulsen I, Leadbetter J. Genomic analysis reveals multiple [FeFe] hydrogenases and hydrogen sensors encoded by treponemes from the H_2-rich termite gut. Microb Ecol. 2012;63:282–94.
27. Posewitz MC, Mulder DW, Peters JW. New frontiers in hydrogenase structure and biosynthesis. Curr Chem Biol. 2008;2:178–99.
28. Calusinska M, Happe T, Joris B, Wilmotte A. The surprising diversity of clostridial hydrogenases: a comparative genomic perspective. Microbiol. 2010;156:1575–88.
29. Björkqvist S, Ansell R, Adler L, Lidén G. Physiological response to anaerobicity of glycerol-3-phosphate dehydrogenase mutants of *Saccharomyces cerevisiae*. Appl Environ Microbiol. 1997;63:128–32.
30. Van Dijken JP, Scheffers WA. Redox balances in the metabolism of sugars by yeasts. FEMS Microbiol Lett. 1986;32:199–224.
31. Ansell R, Granath K, Hohmann S, Thevelein JM, Adler L. The two isoenzymes for yeast NAD + -dependent glycerol 3-phosphate dehydrogenase encoded by GPD1 and GPD2 have distinct roles in osmoadaptation and redox regulation. EMBO J. 1997;16:2179–87.
32. Brown SD, Guss AM, Karpinets TV, Parks JM, Smolin N, Yang S, et al. Mutant alcohol dehydrogenase leads to improved ethanol tolerance in *Clostridium thermocellum*. Proc Natl Acad Sci U S A. 2011;108:13752–7.
33. Mariano AP, Dias MOS, Junqueira TL, Cunha MP, Bonomi A, Filho RM. Butanol production in a first-generation Brazilian sugarcane biorefinery: technical aspects and economics of greenfield projects. Bioresource Technol. 2013;135:316–23.
34. Acetone uses and market data. Technical Report. Independent Commodity Information Services; ICIS News. 2007 09:34, updated 2010.
35. Aden A, Ruth M, Ibsen K, Jechura J, Neeves K, Sheehan J, et al. Lignocellulosic biomass to ethanol process design and economics utilizing co-current dilute acid prehydrolysis and enzymatic hydrolysis for corn stover. Technical Report. 2002; NREL/TP-510-32438, NREL, Golden, CO. http://www.osti.gov/scitech/biblio/15001119.
36. Shanks RMQ, Kadouri DE, MacEachran DP, O'Toole GA. New yeast recombineering tools for bacteria. Plasmid. 2009;62:88–97.
37. Shaw AJ, Hogsett DA, Lynd LR. Natural competence in *Thermoanaerobacter* and *Thermoanaerobacterium* species. Appl Environ Microbiol. 2010;76:4713–9.

Rational design of polymer-based absorbents: application to the fermentation inhibitor furfural

Ikechukwu C Nwaneshiudu and Daniel T Schwartz*

Abstract

Background: Reducing the amount of water-soluble fermentation inhibitors like furfural is critical for downstream bio-processing steps to biofuels. A theoretical approach for tailoring absorption polymers to reduce these pretreatment contaminants would be useful for optimal bioprocess design.

Results: Experiments were performed to measure aqueous furfural partitioning into polymer resins of 5 bisphenol A diglycidyl ether (epoxy) and polydimethylsiloxane (PDMS). Experimentally measured partitioning of furfural between water and PDMS, the more hydrophobic polymer, showed poor performance, with the logarithm of PDMS-to-water partition coefficient falling between −0.62 and −0.24 (95% confidence). In contrast, the fast setting epoxy was found to effectively partition furfural with the logarithm of the epoxy-to-water partition coefficient falling between 0.41 and 0.81 (95% confidence). Flory-Huggins theory is used to predict the partitioning of furfural into diverse polymer absorbents and is useful for predicting these results.

Conclusion: We show that Flory-Huggins theory can be adapted to guide the selection of polymer adsorbents for the separation of low molecular weight organic species from aqueous solutions. This work lays the groundwork for the general design of polymers for the separation of a wide range of inhibitory compounds in biomass pretreatment streams.

Introduction

Pretreatment of biomass is a critical step in the biochemical route to low-carbon liquid transportation fuels. Steam explosion, ammonia fiber expansion (AFEX), and dilute acid treatments are frequently used to breakdown and/or reorganize lignin structures in biomass, making sugars more accessible for downstream saccharification and fermentation processes [1-5]. The resulting pretreated biomass stream contains major components (sugar, lignin, and so on) and minor components like organic extractives, water-soluble low molecular weight compounds such as furfural, hydroxymethylfurfural (HMF), and vanillin, as well as other larger degradation products [2]. The water-soluble components are often fermentation inhibitors, so it is desirable to reduce their production or separate them before subsequent bio-processing steps [6].

Several methods are used to remove inhibitory compounds from fermentation broths. Chemical methods can be used to react inhibitory compounds into inactive forms [3,6]. Biological techniques can also be used to degrade furfural and HMF into more benign molecules [7]. Separations using polymers, ion exchange resins, and activated carbon are also an option for inhibitor removal from the pre-fermentation broth [8-12]. For example, the selective removal of furfural from batch solutions has been demonstrated using polystyrene and methacrylic ester resins (Amberlite-based polymers XAD-4, XAD-7) without affecting sugar yields, thereby improving ethanol production in downstream fermentation [8]. Although properties such as column void volume and flow velocity are known to affect the separation, polymer hydrophobicity is the primary factor affecting the performance of both polymer resins [8].

A key thermodynamic parameter that describes the potential of a polymer to separate a target molecule is the partition coefficient [13-18]. Tailoring the traits of the polymer to the trace aqueous component is a common strategy for environmental contaminants like toluene and benzene, where hydrophobic polymers are used to extract low-solubility non-polar molecules from the aqueous solvent [16,19,20]. In contrast, polar fermentation inhibitors

* Correspondence: dts@uw.edu
Department of Chemical Engineering, University of Washington, Box 351750, Seattle, WA 98195-1750, USA

like furfural in the pre-fermentation broths will require a more polar polymeric species for an effective separation scheme. The purpose of this paper is to convert chemical intuition-based arguments into a practical theory for guiding polymer selection for optimal absorption-based separation schemes.

The critical parameter that quantifies the absorption partitioning for a species (i) between a polymer (p) and a solvent (water = w) is the equilibrium partition coefficient ($K^i_{p/w}$), which can be expressed experimentally by [16];

$$K^i_{p/w} = \frac{C_p}{C_w} \tag{1}$$

where C_p is the equilibrium concentration of the target compound in the polymer phase and C_w is its equilibrium concentration in the water phase. The larger the $K^i_{p/w}$ value, the greater the affinity the target molecule has for the chosen polymer phase. Equilibrium partitioning experiments are routinely performed to determine $K^i_{p/w}$ values, but a thermodynamic approach suitable for prediction of trace aqueous components will be a key advancement in the selection of the optimal polymer for use in separations. Several approaches to predicting the partition coefficient can be found in the literature. These include molecular connectivity index [21], linear solvation energy relationships (LSERs) [22], and Flory-Huggins theory [23-25]. In our situation, using a Flory-Huggins theoretical approximation is appropriate because it accounts for size disparities between aqueous solute and polymer [24,26,27] and can be further simplified for dilute, low molecular weight contaminants partitioning into large polymer chains, resulting in the expression:

$$\log\left(K^i_{p/w}\right) = -\log(S^i_w \bar{V}_i) - \frac{[1+\chi]}{2.303} \tag{2}$$

where $K^i_{p/w}$ is the equilibrium partition coefficient of the target molecule (i) between polymer (p) and water (w) phases, S^i_w and $\bar{V}_i$ are the water solubility and molar volume of the molecule (i) being extracted. The Flory-Huggins interaction parameter χ can be further approximated by using the molecule and polymer solubility parameters, δ_i and δ_p, respectively:

$$\chi = \frac{(\delta_i - \delta_p)^2 \bar{V}_i}{RT} \tag{3}$$

where R is the universal gas constant and T is temperature (K). Key simplifying assumptions in Equation 2 are the following: 1) $\bar{V}_i/\bar{V}_p \rightarrow 0$ (molar volume of molecule is significantly smaller than that of the polymer), and the trace contaminant is dilute in both the solution and polymer. Here, we show the use of the simple set of Equations 2 and 3 as the basis for guiding the selection of polymer for the targeted inhibitor furfural found in most biomass pretreatment broths.

Materials and methods

Sample preparation

PDMS samples

Slygard 184 and catalyst for making PDMS were obtained from Dow Corning (Midland, MI, USA). Samples of PDMS used in the experiment were prepared as follows. Uncured PDMS was mixed using the standard 10:1 polymer to catalyst ratio in a poly(methyl-methacrylate) (PMMA) mold 1.5 mm deep. This was cured at 70°C overnight to make an approximately 1.5-mm-thick sheet of PDMS. Samples were punched out of the PDMS sheet using a 2-mm diameter punch. The resulting PDMS plugs were sealed and kept until used.

Epoxy samples

Fast-curing epoxy (bisphenol A diglycidyl ether resin) was obtained from ITW Devcon (Danvers, MA, USA). Epoxy resin and hardener were poured and mixed on a sheet of paper and spread into a thin layer approximately 1 to 2 mm thick. This was cured at room temperature for 4 h, allowing the polymer resin to harden completely. Samples approximately 2 × 4 mm samples were cut out of the slab using a razor blade. Epoxy slabs were sealed and kept until used.

Organic/aqueous solutions

Furfural was purchased from Sigma-Aldrich (St. Louis, MO, USA) and used as received. Experimental solutions were prepared by adding the appropriate amount of furfural into 20 mL of deionized (DI) water. The solutions were mixed, tightly capped, and left to sit until fully dissolved. These solutions were used within 15 min of preparation.

Polymer absorption procedure

Furfural solutions in 20-mL glass scintillation vials were used for the absorption/partitioning procedure. Three sample plugs (PDMS) or slabs (epoxy) were introduced to each vial, tightly capped, and left to equilibrate undisturbed at room temperature for more than 48 h. The final concentration in the aqueous solution after equilibration is

$$C_f = C_o \frac{1}{1 + K(V_p/V_s)} \tag{4}$$

where C_f and C_o are the final and initial concentrations in the liquid sample (respectively), V_p and V_s are the volumes of the polymer and liquid sample (respectively), and K is the partition coefficient. From this equation, we see, due to the small size of our polymer pieces and scale of K (determined later), that there is negligible change in concentration after equilibration with the polymer.

Raman spectra acquisition

Spectra were collected using a Renishaw inVia Raman micro-spectrometer (Renishaw, Wotton-under-Edge, UK) attached to a Leica DM IRBE upright optical microscope (Leica, Wetzlar, Germany). A 785-nm diode laser operated at full power (nominal 180 mW) was used to irradiate samples through a 50× (N.A. 0.8) objective lens. The spot area was approximately 50 μm [2]. Scattered light was acquired through the same objective lens and detected on a thermoelectrically cooled (−60°C) CCD. Spectra were typically acquired for 10 s at 100% laser power, except where fluorescence was an issue. For example, spectra for the equilibrated epoxy samples were collected using 100 acquisitions of 1-s duration to avoid saturation of the detector from fluorescence. The spectrum of furfural (as received) was acquired at 1% of the nominal laser power due to the fluorescence of the sample. All sample spectra were collected using a wet sample holder sealed with an optical coverslip. Laser stability was assured using the 520 cm^{-1} peak of silicon as a reference. The PDMS peak at 1,410 cm^{-1} and the epoxy peak at 1,610 cm^{-1} was acquired with all polymer phase spectra and served as an internal standard that helped normalize for sample to sample variations in optical focus.

Raman peak normalization and data analysis

Spectral peaks were analyzed using Wire 2.0 software. All spectra were baseline subtracted using second-order polynomial or cubic spline functions, producing very flat baselines for subsequent analysis. Intensities were normalized by total acquisition time. Spectral peaks where fit to standard Voigt distribution profiles. Curve fit parameters were used to calculate the reported integrated peak areas. Peak areas are reported in counts per second (cps). Normalized peak intensities were analyzed in the R Statistical Software package to for the linear regression analysis and determination of 95% confidence intervals for all parameters reported here [http://www.r-project.org/].

Results and discussion

To the best of our knowledge, the simplifications of Flory-Huggins theory that result in Equations 2 and 3 have not been evaluated for absorption partitioning. We assess the reasonableness of this Flory-Huggins approximation for predicting polymer/compound partition coefficients by using widely reported experimental data for the aqueous partition of toluene into various polymer phases. Figure 1 (I) shows the partition coefficient predictions, $\log\left(K^i_{p/w}\right)$, calculated from Equations 2 and 3 for toluene as a function of polymer solubility parameter δ_p over the range represented by the polymers listed in Table 1. Here, we use a toluene solubility parameter of $\delta_i = 8.9$, an aqueous solubility of $S_w^{\text{Toluene}} = 5.1$ mM at

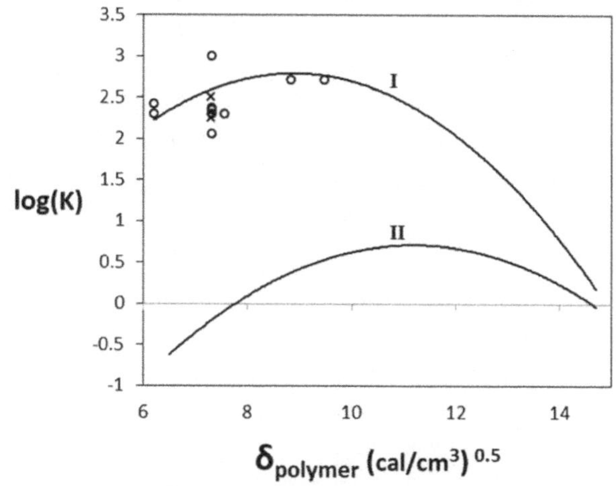

Figure 1 Flory-Huggins theoretical estimations. Flory-Huggins theoretical estimations of polymer/toluene (I) and polymer/furfural (II) partition coefficients (curves) for a realistic range of polymer solubility parameters ($6 \leq \delta_{polymer} \leq 14.5$). Experimental measurements for toluene partitioning into five different polymers is found in our prior published work (x) and results from other groups (o).

Table 1 Solubility parameters of different polymers we use to link theory to actual polymers of interest

Polymers	($\delta = (cal/cm^3)^{0.5}$
Poly(tetrafluoroethylene)	6.50
Poly(dimethylsiloxane)	7.30
Poly(butadiene)	7.57
Poly(ethylene)	8.00
Poly(propylene)	8.20
Poly(n-butyl methacrylate)	8.70
Poly(ethyl methacrylate)	9.04
Poly(n-butyl acrylate)	9.04
Poly(chloroprene)	9.04
Poly(styrene)	9.13
Poly(phenyloxide)	9.15
Poly(methyl methacrylate)	9.30
Acrylic	9.40
Poly(vinyl chloride)	9.50
Poly(vinyl acetate)	9.80
Poly(methyl acrylate)	10.02
Poly(ethylene terephthalate)	10.10
Bisphenol A epoxy resin	10.70
Cellulose acetate	11.70
Nylon 6,6	13.70
Poly(ethylene oxide)	14.70

Data in table extracted from [21].

25°C, and a molar volume of $\bar{V}_{\text{Toluene}} = 106.3$ mL/mol. The data in Table 1 shows that polymers such as PDMS ($\delta_p = 7.3$), polystyrene-co-butadiene (PSB, $\delta_p = 8.84$), and polyacrylonitrile-co-butadiene (PAB, $\delta_p = 9.48$) are close to the theoretical maximum log ($K^i_{p/w}$) for toluene.

Also included in Figure 1 are experimental values of $\log\left(K^{\text{Toluene}}_{p/w}\right)$ found in the literature for the partitioning of toluene into five different polymer phases. Symbols *x* and *o* represent experimental values from our work [16] and literature sources [28-31], respectively. From these data, we can see that the Flory-Huggins theoretical predictions adequately estimate the partition coefficient of toluene across several different polymer extraction phases. There is significant variation in the experimentally derived PDMS/toluene partition coefficients (Figure 1). We can only hypothesize why this is. The measurements come from a wide range of different experimental methods, including MIMS (membrane inlet mass spectrometry), Raman spectroscopy, UV-Vis spectroscopy, and IR spectroscopy. Nominally, to be reported as a partition coefficient, each method assumed equilibrium partitioning had been achieved between the PDMS phase and the aqueous toluene sample. Moreover, when we place each point at identical x-axis locations, we assume all PDMS polymers were identical. It is challenging to know exactly how these fundamental assumptions may interact to drive the wide variation in reported values for the partition coefficient for toluene in PDMS.

In the case of the fermentation inhibitor furfural, we also generate a similar theoretically curve to optimize partitioning and absorption using a furfural's solubility parameter $\delta i = 11.2$ [24], an aqueous solubility of $S^{\text{Furfural}}_{w} = 865$ mM, and a molar volume of $\bar{V}_{\text{Furfural}} = 83.2$ mL/mol. Figure 1 (II) shows the predicted furfural partition coefficient between the polymer and water phases as log ($K^{\text{Furfural}}_{p/w}$) versus polymer solubility parameter. Compared to toluene, the theoretical maximum log ($K^i_{p/w}$) for furfural is significantly lower due to its greater water solubility. Figure 1 (II) also displays a diverse range of possible log ($K^{\text{Furfural}}_{p/w}$) values from negative to positive, which shows the optimum range of polymer solubility parameters is $10 \le \delta_p \le 12.5$. Using this plot and Table 1, we can find the polymer phases that likely give the maximum furfural partitioning.

To experimentally demonstrate the robustness of our approach, we select a high-performing and low-performing polymer absorption phase using the data in Figure 1 (II). Fast-curing epoxy and PDMS are both cross-linking polymer networks with significantly different solubility parameters. PDMS, a conventional absorption polymer used often in analytical chemistry techniques, has a solubility parameter $\delta_p = 7.3$. Epoxy, a polymeric resin most widely used as an adhesive, has a solubility parameter $\delta_p = 10.7$. The Flory-Huggins approximation shows that for furfural, PDMS polymer should display less favorable partitioning with a negative log ($K^i_{p/w}$) value, while epoxy should have enhanced partitioning with a positive log ($K^i_{p/w}$) that is close to the theoretical maximum.

The two polymers were equilibrated in a 50-mM solution of furfural and water. Figure 2 is a visual illustration of the equilibrium partitioning of furfural from solution into the two polymer phases. Figure 2A shows pieces of epoxy resin (one before (left) and the other (right) after being introduce to 50-mM solution of furfural), while Figure 2B shows pieces of PDMS before and after the same equilibration. Figure 2C shows the color of 50-mM furfural solution (top) and furfural as received (bottom) contained in 2-ml glass vials. The as-received furfural (99% pure) was dark brown in color, even though pure furfural should be clear. The brown color in stored furfural is caused by acidic impurities and resins produced when furfural autoxidizes [32]. For example, 5-methylfurfural is a brown- colored impurity. Figure 2A shows that these impurities seem to partition strongly into the epoxy phase causing the color change from yellow (before) to brown (after) equilibration. However, the change seen in PDMS, before and after equilibration (B), is not as drastic. The strongly colored epoxy sample produced strong fluorescence during Raman spectral acquisition.

Raman fingerprinting scans show that the 1,372 cm^{-1} peak for the fermentation inhibitor furfural can be clearly discerned in all experimental polymer and solvent phases, even with a fluorescence impurity in the as received furfural. Figure 3 shows characteristic Raman spectra of furfural in four different solvent/polymer environments within the 1,300 to 1,750 cm^{-1} wavenumber window. Spectrum A shows the characteristic peaks of furfural (as-received) while spectrum B shows it dissolved in water. Spectra C and D are from the epoxy and PDMS polymers, respectively, equilibrated in aqueous furfural. The fingerprint scans show that the 1,372 cm^{-1} furfural peak is distinct from the background polymer and water phases, and each polymer phase has a strong background signal that can be used as an internal standard for peak intensity normalization.

To quantify partitioning, we do more careful fitting of the normalized furfural peak intensities as a function of aqueous furfural concentration in each phase being studied. Figure 4A shows the aqueous 1,372 cm^{-1} furfural peak, as dissolved in water at 50-, 35-, and 20-mM concentrations (curves I, II, and III, respectively). Figure 4B shows the same furfural peak in PDMS equilibrated at the same aqueous concentrations, again denoted, I, II, and III. The furfural peak location is slightly shifted due to the change in chemical environment. Figure 4C shows the furfural peak taken in the epoxy equilibrated with

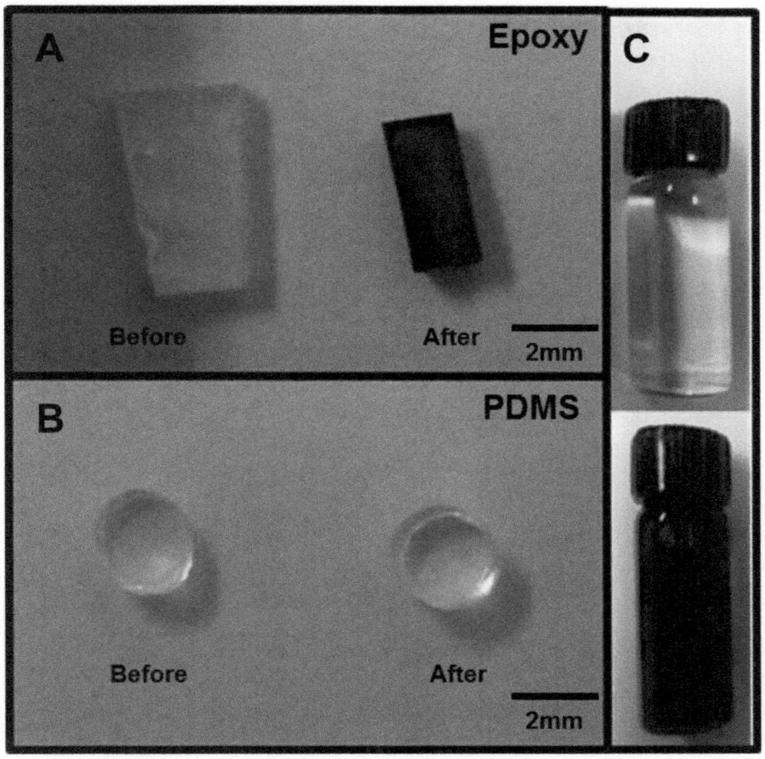

Figure 2 Two slabs of epoxy, two plugs of PDMS, and two vials of furfural. **(A)** Two slabs of epoxy before and after equilibration in 50 mM Furfural. **(B)** Two plugs of PDMS before and after equilibration in 50 mM Furfural. **(C)** Top 2-ml vial is filled with 50 mM furfural and the bottom is filled with furfural (as received).

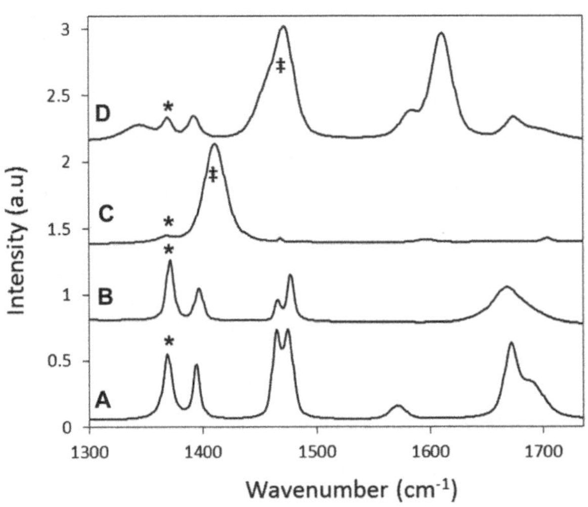

Figure 3 Fingerprinting Raman spectra of furfural in different phases. **(A)** Furfural (as received), **(B)** dissolved in water, **(C)** partitioned from water into PDMS, **(D)** partitioned from water into Epoxy. Asterisks (*) indicate furfural spectral peak used for analysis. (‡) indicate internal standards.

the aqueous furfural (also at identical concentrations, I, II, and III). The baseline-subtracted spectral data points are presented along with corresponding curve fits, and scale bars show the intensity of the Raman signal in counts per second (cps). Firstly, we see that the signal acquired in epoxy is enhanced compared to the water phase. We also see that the signal in PDMS is significantly attenuated compared to the water phase. It is evident (qualitatively) that aqueous furfural partitions preferentially in epoxy compared to PDMS. The range of furfural concentrations used here represents a balance between maximizing Raman signal (which favors high concentrations) and a desire to meet the constraints of our polymer absorption theory, namely, dilute solute in the polymer phases. Reported furfural concentrations in fermentation broths can be in the range 2 to 5 g/L which is exactly the range used here [5,33].

We quantify aqueous partitioning of furfural into both PDMS and epoxy phases by using the method detailed in prior work, though here we go more deeply into sampling methods that let us better estimate measurement uncertainty [16,20]. Figure 5 is a plot of the entire normalized dataset, with the linear regression fits (solid line) and corresponding 95% confidence (dashed) intervals (dashes) in epoxy (triangles), water (squares), and PDMS

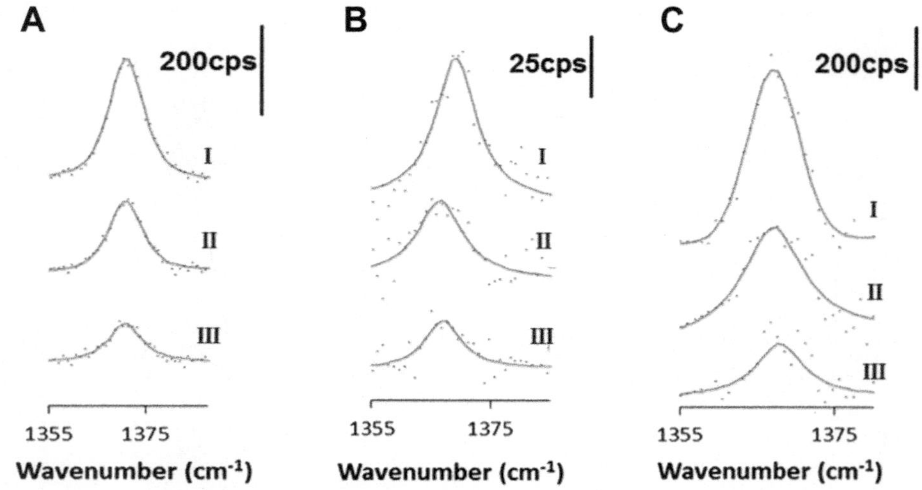

Figure 4 Raman spectra of the 1,372 cm^{-1} furfural peak. Panels **(A)**, **(B)**, and **(C)** are Raman spectra of the 1,372 cm^{-1} furfural peak acquired in the aqueous phase and the equilibrated PDMS and epoxy phases, respectively. Three different aqueous concentrations are shown: (I) 50 mM, (II) 35 mM, and (III) 20 mM. Data points are raw data, and lines are the Voigt curve fits. The normalize signal intensity is shown as counts per second (cps).

(circles) phases. The confidence intervals were generated using the R statistical software package. Replicate data points at each concentration are from six different spectra acquired at two different microscopic locations (per sample) on three equilibrated samples (two different trials). Peak areas were normalized to a PDMS or epoxy internal standard peaks at 1,410 or 1,610 cm^{-1}, respectively (Figure 3). This reduces variation based on solid phase sample preparation so that peaks can be directly related to furfural concentration (assuming negligible polymer swelling by the solvent or solute). We estimate the furfural partition coefficient between water and polymer $K^{\text{Furfural}}_{\text{Polymer/Water}}$ from the data in Figure 5 as:

$$K^{\text{Furfural}}_{p/w} = \frac{(\text{Slope})_{p/w}}{(\text{Slope})_w} \tag{5}$$

where $(\text{Slope})_{p/w}$ is the slope of the fit for furfural measured directly in either epoxy or PDMS phase in contact with the aqueous solution, and $(\text{Slope})_w$ is the slope of the best fit line measured directly in the aqueous phase. Based on the best fit slopes, the mean value of aqueous furfural partitioning into PDMS is $\log\left(K^{\text{Furfural}}_{\text{PDMS}/w}\right) = -0.47$ and into epoxy is $\log\left(K^{\text{Furfural}}_{\text{Epoxy}/w}\right) = 0.59$. The 95% confidence intervals for the furfural partition coefficient between PDMS and water is $-0.62 \leq log\left(K^{\text{Furfural}}_{\text{PDMS}/w}\right) \leq -0.24$ and for epoxy and water it is $0.41 \leq \log\left(K^{\text{Furfural}}_{\text{Epoxy}/w}\right) \leq 0.81$. We estimated these 95% confidence intervals by random sampling with replacement using the boot strapping method implemented in the R statistical software package. For the results presented here, the data was randomly sampled 1,000 times. Table 2 shows the experimentally derived and

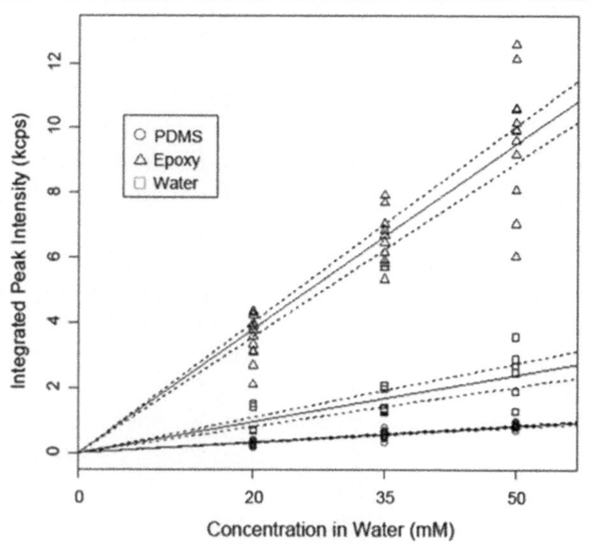

Figure 5 Integrated peak intensity vs. water concentration plot. Replicate measurements for the integrated 1,372 cm^{-1} furfural Raman peak intensities are shown as a function of the concentrations in the water phase. Linear fits (solid lines) and 95% confidence intervals (dashed lines) are presented for the Raman signals acquired in the equilibrated epoxy phase (triangle), equilibrated PDMS phase (circle), and the water phase (square).

Table 2 Experimental values of log(K_f)

SPME polymer phases	Log(K_f) $-\lambda \leq$ Mean $\leq +\lambda$	Log(K_f) (Flory-Huggins)
Epoxy	0.41 ≤ 0.59 ≤ 0.81	0.70
PDMS	−0.62 ≤ −0.47 ≤ −0.24	−0.21

Experimental values of log(Kf), including the mean and 95% confidence range (±λ), for partitioning from water into epoxy and PDMS are compared to the Flory-Huggins approximation.

predicted values of log K of both polymers in water. We see that our values match the predicted trend, in that PDMS gives a negative value of $\log\left(K^{\text{Furfural}}_{\text{PDMS}/w}\right)$, while fast-curing epoxy gives a positive $\log\left(K^{\text{Furfural}}_{\text{Epoxy}/w}\right)$ value as predicted by the Flory-Huggins approximations.

Conclusions and implications

We show the general effectiveness of using Flory-Huggins theory for screening polymer materials to use as solid phase absorbents for dilute aqueous solutes. Experimentally and theoretically, we show that PDMS has unfavorable partitioning of furfural from water, whereas epoxy has favorable partitioning. Flory-Huggins theory shows that, in the simple dilute limit we explored, the two main factors affecting the separation are the solubility of the solute in water and how closely matched the polymer-furfural solubility parameters were. Using data in the literature for toluene, we showed that this approach is reasonable for other solutes than furfural, and we believe it is a general approach that is applicable to other compounds present in pre-fermentation broths.

Our simplified Flory-Huggins approach has implications for bioprocess design. First, in a dilute mixture with multiple inhibitors, the partitioning calculation, to first approximation, is independent for each inhibitor species. That means optimal materials can be selected fairly easily. At the same time, selectivity (that is, desire to separate inhibitory solutes and not sugars) can also be assessed based on this theory, by looking for polymers that maximize inhibitor partitioning and minimize sugar partitioning. Secondarily, this work has implications for process design. Equation 4 describes the reduction in inhibitor compound concentration when a batch of pre-fermentation broth is contacted with a fresh absorbent polymer. Using furfural absorption into an epoxy resin as an example, we see that Equation 4 predicts a 50% reduction in furfural concentration from a single equilibrium stage when the polymer-to-solution volume ratio is approximately 1:4, since $K \approx 4$. Of course, one can design packed beds with multiple equilibrium stages to get higher separation factors. Bed regeneration can be carried out using clean water. Separation design based on cyclic operation that alternates between a regeneration loop and absorption loop can be conceived, though it is beyond the scope of this paper. Of course, accurate equilibrium stage calculations may require a more comprehensive thermodynamic model for partitioning into the solid phase, especially to accommodate the high concentrated sugar species, as our main purpose was to guide materials selection.

Abbreviations

AFEX: ammonia fiber expansion; cps: counts per second; DI: deionized; Epoxy/fast-curing epoxy: 5 bisphenol A diglycidyl ether; HMF: hydroxymethylfurfural; LSERs: linear solvation energy relationships; PAB: polyacrylonitrile-co-butadiene; PDMS: poly(dimethylsiloxane); PMMA: poly(methyl-methacrylate); PSB: polystyrene-co-butadiene.

Competing interests

The authors declare they have no competing interests in this work.

Authors' contributions

ICN carried out all experimental and theoretical work and was primary in writing, editing, and finalizing the manuscript. DTS advised in all experimental and theoretical works as well as helped write and revise the manuscript. All authors read and approved the final manuscript.

Acknowledgements

We gratefully acknowledge Dr. Daisuke Sasatani for providing guidance on the statistical analysis using the R statistical package. We also acknowledge major funding provided by the Northwest Advanced Renewables Alliance via National Institute for Food and Agriculture, U.S. Department of Agriculture grant 2011-68005-30416, the Boeing Sutter Endowment (DTS), and the Gates Foundation Millennium Scholarship program (ICN).

References

1. Eggeman T, Elander RT. Process and economic analysis of pretreatment technologies. Bioresour Technol. 2005;96:2019–25.
2. Lloyd TA, Wyman CE. Combined sugar yields for dilute sulfuric acid pretreatment of corn stover followed by enzymatic hydrolysis of the remaining solids. Bioresour Technol. 2005;96:1967–77.
3. Persson P, Andersson J, Gorton L, Larsson S, Nilvebrant NO, Jonsson LJ. Effect of different forms of alkali treatment on specific fermentation inhibitors and on the fermentability of lignocellulose hydrolysates for production of fuel ethanol. J Agric Food Chem. 2002;50:5318–25.
4. Ewanick S, Schmitt E, Gustafson R, Bura R. Use of Raman spectroscopy for continuous monitoring and control of lignocellulosic biorefinery processes. Pure Appl Chem. 2014;86:867–79.
5. Ewanick SM, Thompson WJ, Marquardt BJ, Bura R. Real-time understanding of lignocellulosic bioethanol fermentation by Raman spectroscopy. Biotechnol Biofuels. 2013;6:28.
6. Larsson S, Reimann A, Nilvebrant NO, Jonsson LJ. Comparison of different methods for the detoxification of lignocellulose hydrolyzates of spruce. Appl Biochem Biotechnol. 1999;77–9:91–103.
7. Ran H, Zhang J, Gao QQ, Lin ZL, Bao J. Analysis of biodegradation performance of furfural and 5-hydroxymethylfurfural by Amorphotheca resinae ZN1. Biotechnol Biofuels. 2014;7:51.
8. Weil JR, Dien B, Bothast R, Hendrickson R, Mosier NS, Ladisch MR. Removal of fermentation inhibitors formed during pretreatment of biomass by polymeric adsorbents. Ind Eng Chem Res. 2002;41:6132–8.
9. Kim Y, Kreke T, Hendrickson R, Parenti J, Ladisch MR. Fractionation of cellulase and fermentation inhibitors from steam pretreated mixed hardwood. Bioresour Technol. 2013;135:30–8.
10. Lee JW, Trinh LTP, Lee HJ. Removal of inhibitors from a hydrolysate of lignocellulosic biomass using electrodialysis. Sep Purif Technol. 2014;122:242–7.
11. Carter B, Squillace P, Gilcrease PC, Menkhaus TJ. Detoxification of a lignocellulosic biomass slurry by soluble polyelectrolyte adsorption for improved fermentation efficiency. Biotechnol Bioeng. 2011;108:2053–60.
12. Gurram RN, Datta S, Lin YJ, Snyder SW, Menkhaus TJ. Removal of enzymatic and fermentation inhibitory compounds from biomass slurries for enhanced biorefinery process efficiencies. Bioresour Technol. 2011;102:7850–9.
13. Lord H, Pawliszyn J. Introduction to SPME for in vivo analysis. Lc Gc Europe. 2012;25:180-+.
14. Difilippo EL, Eganhouse RP. Assessment of PDMS-water partition coefficients: implications for passive environmental sampling of hydrophobic organic compounds. Environ Sci Technol. 2010;44:6917–25.
15. Heglund DL, Tilotta DC. Determination of volatile organic compounds in water by solid phase microextraction and infrared spectroscopy. Environ Sci Technol. 1996;30:1212–9.
16. Nwaneshiudu IC, Yu QM, Schwartz DT. Quantitative solid-phase microextraction (SPME)-Raman spectroscopy for the detection of trace organics in water. Appl Spectrosc. 2012;66:1487–91.

17. Stahl DC, Tilotta DC. Screening method for nitroaromatic compounds in water based on solid-phase microextraction and infrared spectroscopy. Environ Sci Technol. 2001;35:3507–12.
18. Vuckovic D, Zhang X, Cudjoe E, Pawliszyn J. Solid-phase microextraction in bioanalysis: New devices and directions. J Chromatogr A. 2010;1217:4041–60.
19. Gorecki T, Yu XM, Pawliszyn J. Theory of analyte extraction by selected porous polymer SPME fibres. Analyst. 1999;124:643–9.
20. Nwaneshiudu IC, Nwaneshiudu CA, Schwartz DT. Separation and enhanced detection of anesthetic compounds using solid phase micro-extraction (SPME)-Raman spectroscopy. Appl Spectrosc. 2014;68:1254–9.
21. Brusseau ML. USING QSAR to evaluate phenomenological models for sorption of organic-compounds by soil. Environ Toxicol Chem. 1993;12:1835–46.
22. Poole SK, Poole CF. Chromatographic models for the sorption of neutral organic compounds by soil from water and air. J Chromatogr A. 1999;845:381–400.
23. Flory PI. Thermodynamics of high polymer solutions. J Chem Phys. 1942;10:51–61.
24. Georgi A, Kopinke FD. Validation of a modified Flory-Huggins concept for description of hydrophobic organic compound sorption on dissolved humic substances. Environ Toxicol Chem. 2002;21:1766–74.
25. Han SJ, Puech L, Law RV, Steinke JHG, Livingston A. Selection of elastomeric membranes for the separation of organic compounds in acidic media. J Membr Sci. 2002;199:1–11.
26. Chiou CT, Kile DE, Brinton TI, Malcolm RL, Leenheer JA, Maccarthy P. A comparison of water solubility enhancements of organic solutes by aquatic humic materials and commercial humic acids. Environ Sci Technol. 1987;21:1231–4.
27. Chiou CT. Linear and nonlinear sorption of nonpolar and polar organic compounds from water to soil. Abstr Pap Am Chem Soc. 2000;220:U317.
28. Flavin K, Hughes H, Dobbyn V, Kirwan P, Murphy K, Steiner H, et al. A comparison of polymeric materials as pre-concentrating media for use with ATR/FTIR sensing. Int J Environ Anal Chem. 2006;86:401–15.
29. Boscaini E, Alexander ML, Prazeller P, Mark TD. Membrane inlet proton transfer reaction mass spectrometry (MI-PTRMS) for direct measurements of VOCs in water. Int J Mass Spectrom. 2004;239:171–7.
30. Dorea HS, Bispo JRL, Aragao KAS, Cunha BB, Navickiene S, Alves JPH, et al. Analysis of BTEX, PAHs and metals in the oilfield produced water in the State of Sergipe, Brazil. Microchem J. 2007;85:234–8.
31. Wittkamp BL, Hawthorne SB, Tilotta DC. Determination of aromatic compounds in water by solid phase microextraction and ultraviolet absorption spectroscopy.1. Methodology. Anal Chem. 1997;69:1197–203.
32. Dunlop AP. Furfural formation and behavior. Ind Eng Chem. 1948;40:204–9.
33. Allen SA, Clark W, McCaffery JM, Cai Z, Lanctot A, Slininger PJ, et al. Furfural induces reactive oxygen species accumulation and cellular damage in Saccharomyces cerevisiae. Biotechnol Biofuels. 2010;3:2.

Phaeodactylum tricornutum photorespiration takes part in glycerol metabolism and is important for nitrogen-limited response

Aiyou Huang[1], Lixia Liu[2], Chen Yang[2] and Guangce Wang[1*]

Abstract

Background: Microalgae are potential sources of biofuels and high-value compounds. Mixotrophic conditions usually promote growth of microalgae. The pennate diatom *Phaeodactylum tricornutum*, with its short life cycle, completely sequenced genome, and ease of transformation, can be used as a model for studying carbon metabolism in microalgae.

Results: We compared the growth rate of *P. tricornutum* (IOCAS-001) under different conditions and labeled the cells using [^{13}C]glycerol (GL). The results revealed GL promoted the growth of *P. tricornutum*. Ser and Gly were synthesized via photorespiration. The ^{13}C enrichment of Ser and Gly under nitrogen-limited conditions was much higher compared to other amino acids, indicating the enhancement of photorespiration. Addition of sodium acetate decreased the growth rate of *P. tricornutum* under nitrogen-limited conditions. Our results indicated that the GL carbon backbone enters the Calvin cycle in the form of dihydroxyacetone phosphate (DHAP), producing xylulose 5-phosphate (X5P) with a GL2_3-generated carbon backbone distributed at X5P1_2 and ribose 5-phosphate (R5P) with GL1-derived carbon atoms at R5P1 and R5P2. Both R5P and X5P can be converted into ribulose-1,5-bisphosphate (RuBP). By oxygenation of RuBP carboxylase/oxygenase (Rubisco) and metabolism through photorespiration, these RuBPs generate Ser and Gly with GL1 or GL2-derived carbon atoms at position 1 and GL1 or GL3-derived carbon atoms at other positions, resulting in a low level of ^{13}C enrichment of Gly1 and Ser1.

Conclusion: Our results indicated different strains of *P. tricornutum* have different mechanisms for organic carbon metabolism. Photorespiration is involved in GL metabolism and is important for the nitrogen-limited response in *P. tricornutum*.

Classification: Metabolic flux analysis, microalgae

Keywords: *Phaeodactylum tricornutum*, Photorespiration, Glycerol metabolism, Nitrogen-limited response

Introduction

Microalgae have received increasing attention as potential sources of biofuels, high-value chemicals, pharmaceuticals, bioactive compounds, and so on. Variations of culture conditions, including carbon source, have been investigated in attempts to promote the production of high-value compounds. Microalgae cultivated under mixotrophic conditions usually have higher growth rate and accumulate greater biomass as well as specific high-value compounds at the plateau phase compared to photoautotrophic conditions [1-3]. Acetate, for example, enhanced carotenoid biosynthesis in the late-exponential growth phase in *Haematococcus pluvialis* [4], and glucose, the common carbohydrate in bacterial culture, can be used by *Chlorella* and promotes the growth rate significantly [5]. A clear illustration of carbon metabolism, especially utilization of organic substrates and the mechanism underlying accumulation of high-value compounds, is important for industrialization of microalgae production. Related studies should be conducted, especially in model organisms.

Diatoms, which are believed to produce around one-fifth of the primary productivity on earth [6,7], can produce approximately 46 tons of organic oil $ha^{-1} \cdot year^{-1}$

* Correspondence: gcwang@qdio.ac.cn
[1]Institute of Oceanology, Chinese Academy of Sciences, Qingdao 266071, China
Full list of author information is available at the end of the article

and are, therefore, potential sources of biodiesel fuel [8]. The pennate diatom *Phaeodactylum tricornutum* is an atypical diatom with a weakly silicified outer shell [9-12] that is ruptured easily by sonication or high-pressure homogenization, facilitating extraction of intracellular metabolites. *P. tricornutum* is a potential source of polyunsaturated fatty acids (PUFA) [13] and fucoxanthin [14], the marine carotenoid reported to exert anti-carcinogenic effects [15] as well as radical scavenging [16] and, most excitingly, to have anti-obesity properties [17,18]. The *P. tricornutum* (CCMP 2561) genome has been sequenced completely, providing a clear genetic background and revealing *P. tricornutum* adopts the metabolic pathways of both plants and animals [19]. *P. tricornutum* is easily transformed and can be regulated genetically [20-23]. Because of the characteristics mentioned above, *P. tricornutum* is considered to be a potential source for biodiesel and the production of high-value compounds, as well as a model for studying carbon metabolism in microalgae.

P. tricornutum is a photoautotrophic organism capable of photosynthesis using light energy and assimilating inorganic carbon sources, like plant cells. Also, there are reports of *P. tricornutum* using organic substrates as carbon and energy sources under mixotrophic conditions. The idea that *P. tricornutum* can use glycerol is quite widely known [1-3,24], whereas reports of its ability to use glucose were controversial [25,26]. To demonstrate a clear pathway for organic carbon metabolism in *P. tricornutum*, methods such as metabolic flux analysis can be used. Stationary metabolic flux analysis used ^{13}C-labeled carbon source as substrate. During metabolism, the carbon backbone of the substrate is broken and recombined in different pathways, producing metabolites with different labeling patterns that can be detected through gas chromatography mass spectrometry (GC-MS) or nuclear magnetic resonance (NMR). These data can be used to calculate the contributions of different pathways to the synthesis of a defined metabolite, that is, flux ratios. Combined with a stoichiometric reaction model and extracellular consumption and secretion rates, these flux ratios can be used to quantify the net carbon fluxes in central carbon metabolism [27]. Based on ^{13}C labeling experiment using glucose and glycerol as carbon sources, Zheng *et al.* found unusual pathways were active in *P. tricornutum* (CCMP 632). Ser and Gly were largely synthesized via the glyoxylate (GOC) cycle followed by photorespiratory reactions [26]. This provided important information about glucose metabolism in *P. tricornutum* (CCMP 632). Nevertheless, some previous studies had reported that *P. tricornutum* cannot utilize glucose at all [25], indicting that different strains of *P. tricornutum* have different efficiencies of glucose metabolism. Strains that lack a glucose-utilizing ability might have a different mechanism for organic carbon metabolism from strains that can use glucose.

In this study, we compared the growth rates of *P. tricornutum* (IOCAS-001) under different conditions and labeled cells with [^{13}C]glycerol (GL). The results revealed the strain in this study cannot use glucose. Ser and Gly were synthesized from GOC that origin from photorespiration but not the GOC cycle. Photorespiration was indeed involved in GL metabolism and was important for the nitrogen-limited response in *P. tricornutum*.

Results

Growth of *P. tricornutum* under different culture conditions

Influence of carbon source

We compared the growth rate of *P. tricornutum* (IOCAS-001) under conditions of no additional carbon source, 0.174 g·L^{-1} $NaHCO_3$, 0.02 M glucose, 0.02 M GL, and 0.02 M glycine. The growth rate of *P. tricornutum* on 0.02 M glucose or 0.02 M glycine was significantly lower (P <0.01) compared to 0.174 g·L^{-1} $NaHCO_3$ and even lower compared to that with no additional carbon source (Figure 1A,G). The growth rate of *P. tricornutum* with 0.02 M GL was higher (P <0.05) compared to no additional carbon source and even higher (P <0.05) than 0.174 g·L^{-1} $NaHCO_3$ after 10 days of culturing (Figure 1A,G), indicating addition of GL promoted the growth of *P. tricornutum*, whereas glucose or glycine might not be utilized as carbon sources by *P. tricornutum*.

Influence of GL concentration

Because ^{13}C-labeling GL is expensive, we optimized the concentration of GL to save substrate without affecting the growth rate of *P. tricornutum*. We compared the growth rate of *P. tricornutum* under conditions of no additional carbon source, 0.174 g·L^{-1} $NaHCO_3$, 0.005 M, 0.02 M, and 0.1 M GL. The results showed the growth rate of *P. tricornutum* on 0.005 M GL and with 0.174 g·L^{-1} $NaHCO_3$ were similar, whereas the growth rate of *P. tricornutum* on 0.02 M and 0.1 M GL were similar and higher (P <0.01) compared to 0.174 g·L^{-1} $NaHCO_3$ (Figure 1B,H), indicating 0.02 M GL was sufficient for the growth of *P. tricornutum*. We chose 0.02 M GL as the substrate for mixotrophic growth of *P. tricornutum*.

Influence of light

Using 0.02 M GL as the carbon source, we compared the growth rate of *P. tricornutum* under mixotrophic conditions (that is, GL assimilated simultaneously with CO_2 fixation under cool white fluorescent lights with a 12 h dark/12 h light cycle) and heterotrophic conditions (no light during culture). The results showed *P. tricornutum* can used glycerol to sustain growth under heterotrophic conditions, though the growth rate was much lower (P <0.01) compared to under mixotrophic conditions (Figure 1C,I), indicating light was important for

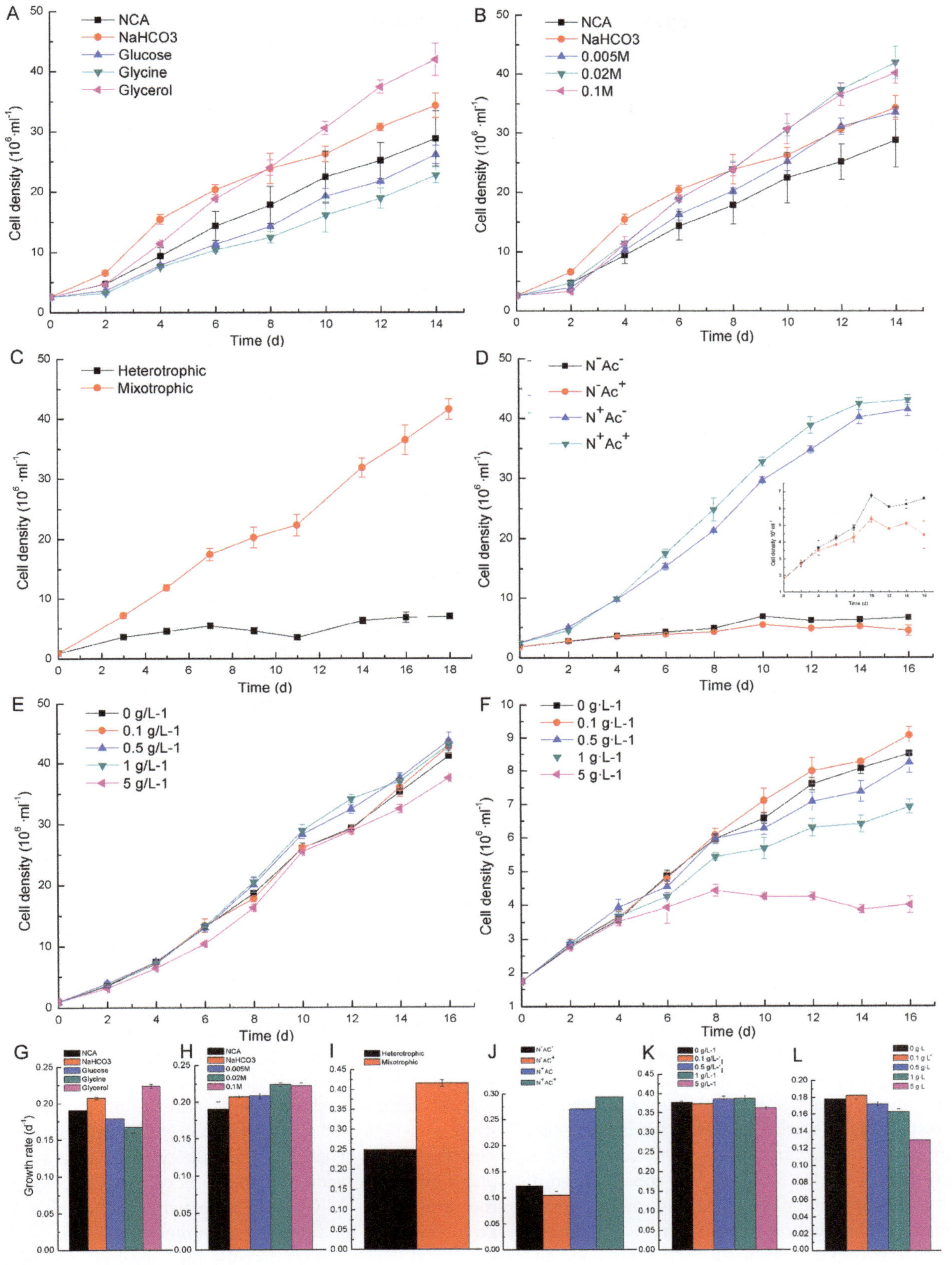

Figure 1 (See legend on next page.)

(See figure on previous page.)

Figure 1 Growth of *P. tricornutum* on different conditions. Cell densities and growth rates of *P. tricornutum* grown on different carbon sources **(A, G)**, under different GL concentrations **(B, H)**, under mixotrophic and heterotrophic conditions **(C, I)**, in the presence or absence of sodium acetate **(D, J)**, under different sodium acetate concentrations in the presence of nitrogen **(E, K)**, and under different sodium acetate concentrations in the absence of nitrogen **(F, L)**. A-F *Y* axis, cell density ($10^6 \cdot mL^{-1}$), *X* axis, culture time; G-L *Y* axis, growth rate (day^{-1}); NAC, no additional carbon source; N^+Ac^+, both sodium nitrate and sodium acetate plus; N^+Ac^-, sodium nitrate alone; N^-Ac^+, sodium acetate alone; N^-Ac^-, both sodium nitrate and sodium acetate minus. Error bars are the SD of three replicates.

growth of *P. tricornutum* even in the presence of GL as carbon source.

Influence of sodium acetate

To investigate the likely influence of sodium acetate on the growth rate of *P. tricornutum,* we compared the growth of *P. tricornutum* under different conditions (both sodium nitrate and sodium acetate plus, N^+Ac^+; sodium nitrate alone, N^+Ac^-; sodium acetate alone, N^-Ac^+; both sodium nitrate and sodium acetate minus, N^-Ac^-). The results indicted that in the presence of sodium nitrate, addition of sodium acetate can promote (P <0.01) the growth of P. tricornutum (Figure 1D,J). While under sodium nitrate minus condition, addition of sodium acetate limited (P <0.01) the growth of P. tricornutum (Figure 1D,J).

To further investigate the dose-response effect of acetate on growth rate of *P. tricornutum,* sodium acetate was added to different concentrations (0, 0.1, 0.5, 1, and 5 $g \cdot L^{-1}$) under sodium nitrate plus or minus conditions. The results indicted that, compared to acetate minus conditions, 0.1 $g \cdot L^{-1}$ sodium acetate had a negligible influence on growth of *P. tricornutum* under both sodium nitrate plus and minus conditions, 0.5 and 1 $g \cdot L^{-1}$ sodium acetate had a positive influence (P <0.05) under nitrate plus conditions and a negative effect (P <0.05) under nitrate minus conditions, while 5 $g \cdot L^{-1}$ sodium acetate limited (P <0.01) the growth of *P. tricornutum* under both nitrate plus and minus conditions (Figure 1E,F,K,L).

Change in glycerol and acetate concentration

To verify the consumption of glycerol and acetate, the supernatants from cultures under N^+Ac^+, N^+Ac^-, N^-Ac^+, and N^-Ac^- conditions were collected, and glycerol and acetate concentrations were determined. There was a significant decrease in glycerol concentration under N^+Ac^+ and N^+Ac^- conditions (from approximately 2000 to approximately 1600 mg · L^{-1}, Figure 2A,B), while the consumption of glycerol under N^-Ac^+ and N^-Ac^- conditions was not significant (Figure 2C,D). Similarly, the consumption of acetate was significant under N^+Ac^+ conditions (from approximately 960 to approximately 870 mg · L^{-1} sodium acetate, Figure 2E) but not obvious under N^-Ac^+ conditions (Figure 2F). It was likely that the decrease in glycerol and acetate concentrations under N^- condition was too slight to be detected. As the biomass did not increase much under N^- condition, possibly only a little glycerol and acetate were consumed.

Total lipid content

Total lipid content was determined to verify if consumption of acetate influenced synthesis of total lipid. Though acetate was consumed under N^+ Ac^+ conditions, there was no significant difference between total lipid contents of N^+Ac^+ and N^+Ac^- (Figure 3). There was no significant difference between total lipid contents of N^-Ac^+ and N^-Ac^-, either (Figure 3). Total lipid contents range from approximately 38% to approximately 46% cell dry weight (CDW).

^{13}C-labeling experiment

Acclimation of P. tricornutum to an organic carbon source under mixotrophic conditions

To investigate the metabolism of GL in *P. tricornutum,* we cultured *P. tricornutum* using [^{13}C]GL and detected the distribution of ^{13}C enrichments in amino acids. At first, the cultures used as inocula were grown on f/2 medium with $NaHCO_3$ as carbon source, and the level of [U-^{13}C]GL was 20%. Table 1 gives the corrected abundance of the mass isotopomers for hydrolyzed amino acids in 20% [U-^{13}C]GL labeling experiments; the mean fractional labeling (FL) was approximately 0.047. As the ^{13}C isotope is stable and the natural abundance is 0.011 [28], the experimental error of some fragments, which were slightly higher than the natural ^{13}C abundance, might be large. Thus, these data were not analyzed further. We suggested the reason might be because the cultures used as inocula were cultivated with $NaHCO_3$ as carbon source, and the cell might need time for acclimation before it can utilize GL when transferred to medium with GL as carbon source.

Next, cultures used for inocula were acclimated to organic carbon source for approximately 3 months by repeated transfer of cells in the mid-exponential growth phase into fresh medium with GL as carbon source, and the level of [U-^{13}C]GL was increased to 30%. Table 1 gives the FL for hydrolyzed amino acids in 30% [U-^{13}C]GL labeling experiments; the mean FL was approximately 0.150. Fragments from Phe and Tyr, which are generally synthesized from phosphoenolpyruvate (PEP) and erythrose-4-P (E4P), had similar values of mass

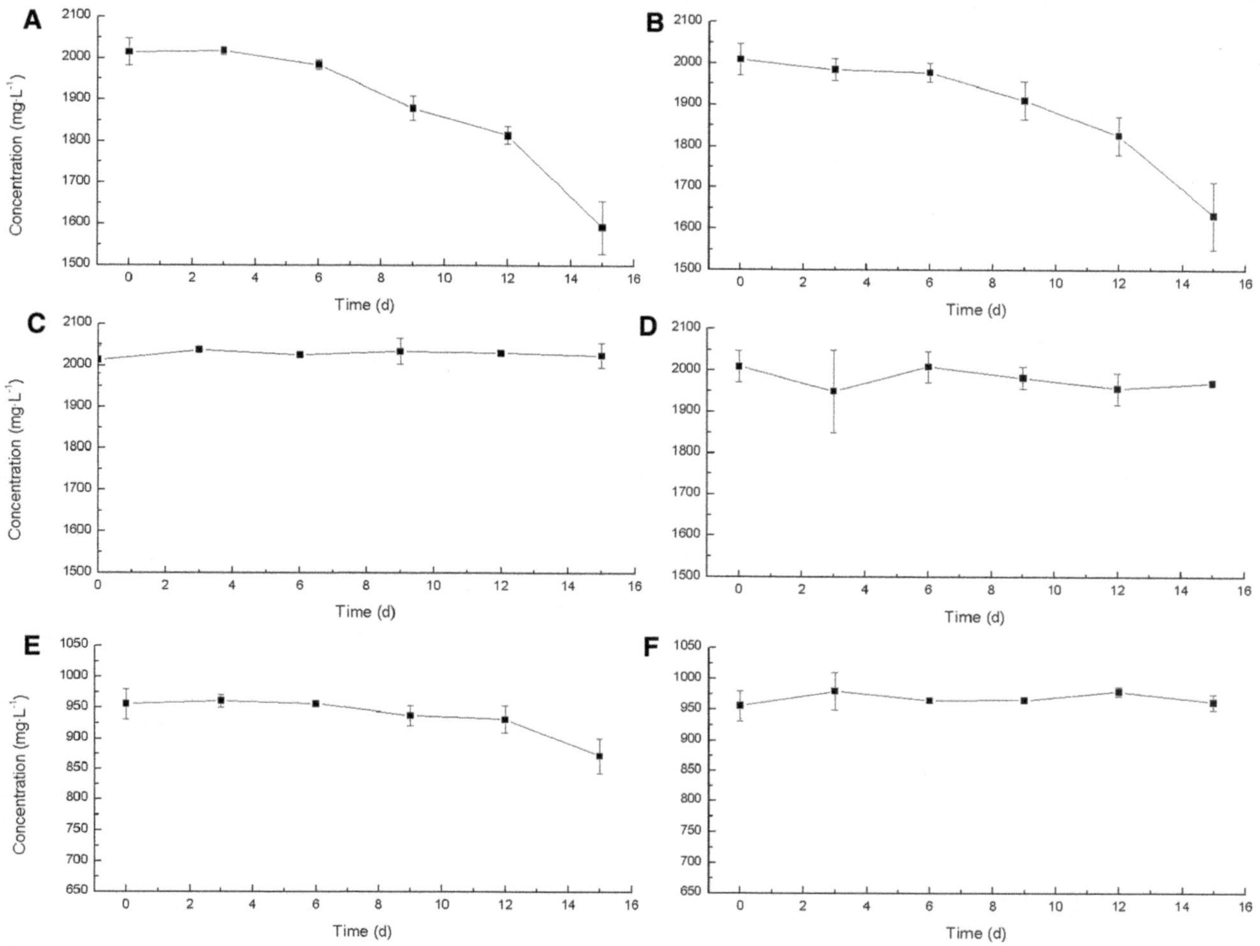

Figure 2 Change in glycerol and acetate concentrations under different conditions. Change of glycerol concentration under N^+Ac^+ **(A)**, N^+Ac^- **(B)**, N^-Ac^+ **(C)**, and N^-Ac^- **(D)** conditions and change of sodium acetate concentration under N^+Ac^+ **(E)** and N^-Ac^+ **(F)** conditions.

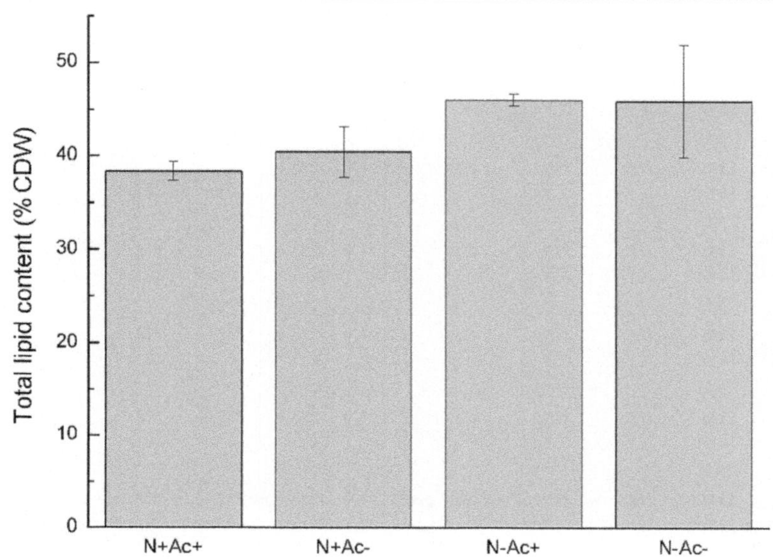

Figure 3 Total lipid contents under N^+Ac^+, N^+Ac^-, N^-Ac^+, and N^-Ac^- conditions.

Table 1 FLs for amino acids under different labeling experiments

Fragments	20% U-^{13}C[a]	30% U-^{13}C N+[b]	30% U-^{13}C N-[c]	1,3-^{13}C N+[d]	1,3-^{13}C N-[e]
M_ala_057	0.057 ± 0.003	0.180 ± 0.015	0.122 ± 0.006	0.190 ± 0.012	0.161 ± 0.005
M_ala_085	0.076 ± 0.001	0.195 ± 0.013	0.143 ± 0.005	0.202 ± 0.009	0.180 ± 0.004
M_asx_057	0.041 ± 0.001	0.146 ± 0.011	0.094 ± 0.003	0.151 ± 0.016	0.119 ± 0.003
M_asx_085	0.045 ± 0.001	0.149 ± 0.012	0.097 ± 0.003	0.159 ± 0.016	0.125 ± 0.004
M_asx_302	0.042 ± 0.001	0.149 ± 0.010	0.097 ± 0.003	0.148 ± 0.016	0.115 ± 0.004
M_glx_057	0.046 ± 0.001	0.168 ± 0.015	0.140 ± 0.001	0.175 ± 0.013	0.155 ± 0.005
M_glx_085	0.047 ± 0.002	0.170 ± 0.017	0.144 ± 0.004	0.183 ± 0.013	0.165 ± 0.003
M_glx_302	0.068 ± 0.004	0.170 ± 0.014	0.154 ± 0.022	0.157 ± 0.007	0.157 ± 0.015
M_gly_057	0.049 ± 0.002	0.172 ± 0.013	0.294 ± 0.025	0.190 ± 0.008	0.219 ± 0.002
M_gly_085	0.050 ± 0.003	0.173 ± 0.014	0.292 ± 0.025	0.254 ± 0.006	0.362 ± 0.004
M_his_057	-	0.127 ± 0.006	-	0.107 ± 0.029	0.103 ± 0.036
M_his_159	-	0.144 ± 0.026	-	0.127 ± 0.013	0.109 ± 0.008
M_his_302	-	0.134 ± 0.010	-	0.121 ± 0.015	0.096 ± 0.021
M_ile_015	0.041 ± 0.001	0.128 ± 0.006	0.047 ± 0.026	0.124 ± 0.012	0.072 ± 0.043
M_ile_085	0.048 ± 0.002	0.133 ± 0.005	0.069 ± 0.001	0.137 ± 0.014	0.109 ± 0.001
M_leu_015	0.041 ± 0.001	0.135 ± 0.005	0.060 ± 0.013	0.136 ± 0.013	0.109 ± 0.004
M_leu_085	0.047 ± 0.002	0.141 ± 0.005	0.069 ± 0.003	0.150 ± 0.014	0.120 ± 0.002
M_lys_057	0.042 ± 0.002	0.117 ± 0.006	0.073 ± 0.024	0.117 ± 0.012	0.090 ± 0.000
M_lys_159	0.051 ± 0.004	0.119 ± 0.006	0.089 ± 0.046	0.121 ± 0.016	0.095 ± 0.001
M_lys_302	0.064 ± 0.010	0.129 ± 0.005	0.078 ± 0.072	0.092 ± 0.009	0.087 ± 0.002
M_met_057	0.000 ± 0.000	0.148 ± 0.011	0.106 ± 0.031	0.163 ± 0.014	0.161 ± 0.003
M_met_085	0.000 ± 0.000	0.150 ± 0.011	0.142 ± 0.009	0.180 ± 0.017	0.179 ± 0.010
M_phe_057	0.050 ± 0.001	0.142 ± 0.006	0.069 ± 0.006	0.140 ± 0.013	0.118 ± 0.026
M_phe_085	0.048 ± 0.002	0.143 ± 0.006	0.073 ± 0.015	0.139 ± 0.014	0.101 ± 0.013
M_phe_302	0.045 ± 0.003	0.137 ± 0.009	0.056 ± 0.003	0.111 ± 0.012	0.073 ± 0.004
M_phe_sc	0.098 ± 0.002	0.196 ± 0.008	0.366 ± 0.165	0.198 ± 0.014	0.339 ± 0.191
M_pro_057	0.044 ± 0.001	0.140 ± 0.011	0.119 ± 0.018	0.158 ± 0.017	0.121 ± 0.008
M_pro_085	0.042 ± 0.001	0.148 ± 0.012	0.106 ± 0.005	0.162 ± 0.020	0.119 ± 0.001
M_ser_057	0.057 ± 0.001	0.183 ± 0.012	0.313 ± 0.011	0.225 ± 0.009	0.257 ± 0.001
M_ser_085	0.058 ± 0.001	0.184 ± 0.012	0.303 ± 0.014	0.269 ± 0.008	0.346 ± 0.005
M_ser_302	0.057 ± 0.001	0.185 ± 0.012	0.323 ± 0.018	0.199 ± 0.008	0.211 ± 0.001
M_thr_057	0.041 ± 0.001	0.124 ± 0.008	0.070 ± 0.002	0.128 ± 0.017	0.095 ± 0.001
M_thr_085	0.040 ± 0.005	0.125 ± 0.008	0.077 ± 0.006	0.132 ± 0.014	0.100 ± 0.001
M_thr_sc	0.117 ± 0.003	0.180 ± 0.008	0.155 ± 0.030	0.188 ± 0.014	0.167 ± 0.014
M_val_057	0.047 ± 0.002	0.145 ± 0.006	0.073 ± 0.004	0.150 ± 0.013	0.109 ± 0.003
M_val_085	0.054 ± 0.002	0.152 ± 0.005	0.087 ± 0.007	0.152 ± 0.013	0.118 ± 0.002
M_val_302	0.049 ± 0.001	0.152 ± 0.005	0.080 ± 0.005	0.138 ± 0.013	0.101 ± 0.004
M_tyr_057	0.048 ± 0.002	0.140 ± 0.005	0.063 ± 0.005	0.140 ± 0.013	0.103 ± 0.019
M_tyr_085	0.053 ± 0.001	0.143 ± 0.005	0.096 ± 0.041	0.140 ± 0.013	0.150 ± 0.098
M_tyr_302	0.047 ± 0.003	0.140 ± 0.010	0.064 ± 0.005	0.112 ± 0.011	0.078 ± 0.006
average FL	0.047 ± 0.005	0.150 ± 0.008	0.121 ± 0.005	0.157 ± 0.014	0.141 ± 0.007

[a]FLs for amino acids under 20% [U-^{13}C]glycerol labeling experiments, cultures with $NaHCO_3$ as carbon source was used as inocula; [b]FLs for amino acids under 30% [U-^{13}C]glycerol labeling experiments, nitrogen plus; [c]FLs for amino acids under 30% [U-^{13}C]glycerol labeling experiments, nitrogen minus; [d]FLs for amino acids under 50% [1,3-^{13}C]glycerol labeling experiments, nitrogen plus; [e]FLs for amino acids under 50% [1,3-^{13}C]glycerol labeling experiments, nitrogen minus.

distribution vector (MDV) (M_Phe_057, M_Phe_085, and M_Phe_302 *versus* M_Tyr_057, M_Tyr_085, and M_Tyr_302), confirming the consistency of GC-MS results.

The FL values of amino acids synthesized in different cell partitions were diluted to different degrees

Under mixotrophic conditions, different amino acid fragments had different FL values (Table 1) regardless of the type of labeling substrates ([U-^{13}C]GL or [1,3-^{13}C]GL). Eight amino acids (Lys, Thr, Ile, Leu, Tyr, His, Phe, and Val), which were reported to be synthesized in chloroplasts in higher plants, mainly had mostly lower average FL values, whereas average *FL* values for Ala, Ser, Gly, and Glx were higher (Table 2). This indicated GL was assimilated simultaneously with CO_2 fixation in chloroplasts, resulting in different dilutions of FL values between amino acids synthesized in different cell partitions.

^{13}C enrichment of Ser and Gly in the absence of nitrogen

As shown in Table 1, under nitrogen plus conditions, Ala had the highest average FL value followed by Ser, Gly, and Glx. In the absence of nitrogen, FL values of Ser and Gly were increased significantly with levels close to the ^{13}C enrichment of substrate, indicating the Ser and Gly metabolic pathway was important for the nitrogen-limitation response in *P. tricornutum*.

Positional labeling of Ser and Gly based on [U-^{13}C]GL or [1,3-^{13}C]GL

We calculated the positional labeling of Ser and Gly to investigate the likely pathway for Ser and Gly metabolism in *P. tricornutum*. When [U-^{13}C]GL was used as substrate, both Ser and Gly had similar ^{13}C enrichment in all positions, regardless of the nitrogen conditions (Table 3). When [1,3-^{13}C]GL was used as substrate in the presence of nitrogen, the ^{13}C enrichment of Ser1 was significantly lower (approximately half) compared to Ser2 and Ser3, and the ^{13}C enrichment of Gly1 was also approximately half compared to Gly2 (Table 3). Moreover, the ^{13}C enrichment of Ser1 and Gly1 was decreased significantly in the absence of nitrogen, suggesting the majority of Ser1 and Gly1 were derived from GL2, consistent with the report of Zheng *et al.* [26].

Table 2 Average FL values for hydrolyzed amino acids in nitrogen plus (U-^{13}C) labeling experiment

AA	FL ± SD
Lys	0.118 ± 0.006
Thr	0.124 ± 0.008
Ile	0.130 ± 0.005
Leu	0.138 ± 0.005
Tyr	0.141 ± 0.006
His	0.135 ± 0.015
Phe	0.140 ± 0.007
Val	0.149 ± 0.005
Asx	0.148 ± 0.011
Met	0.149 ± 0.011
Pro	0.148 ± 0.012
Gly	0.172 ± 0.014
Glx	0.169 ± 0.015
Ser	0.184 ± 0.012
Ala	0.188 ± 0.014

FL, fractional labeling; SD, standard deviation.

Change in intracellular metabolite pools in response to nitrate limitation and acetate addition

Table 4 shows the ratios of abundances of key cellular metabolites under different culture conditions (N^-Ac^+, N^-Ac^-, N^+Ac^+, and N^+Ac^-). Gly, ethylene glycol, and formate increased under N^- conditions (N^-Ac^+/N^+Ac^+ and N^-Ac^-/N^+Ac^-) and were reduced when acetate was added (N^-Ac^+/N^-Ac^- and N^+Ac^+/N^+Ac^-). Glu and Gln were reduced under N^- conditions (N^-Ac^+/N^+Ac^+ and N^-Ac^-/N^+Ac^-) and increased when acetate was added (N^-Ac^+/N^-Ac^- and N^+Ac^+/N^+Ac^-). Accordingly, alpha ketoglutarate acid (AKG), the product of deamidization of Glu, increased under N^- conditions (N^-Ac^+/N^+Ac^+ and N^-Ac^-/N^+Ac^-) and was reduced when acetate was added (N^-Ac^+/N^-Ac^- and N^+Ac^+/N^+Ac^-). This was in accord with the previous reports [29]. All amino acids, except Gly and *N*-Isovaleroylglycine, were reduced under N^- conditions (N^-Ac^+/N^+Ac^+ and N^-Ac^-/N^+Ac^-). Besides, intermediate products of orrnithine-urea cycle, including Asp, fumarate, and orrnithine, were reduced under N^- conditions (N^-Ac^+/N^+Ac^+ and N^-Ac^-/N^+Ac^-), in accord with the previous reports [30].

Discussion

P. tricornutum use of organic carbon

In this study, the growth rates of *P. tricornutum* with 0.02 M glucose, 0.02 M GL, and 0.02 M glycine were compared to those with 0.174 g·L^{-1} $NaHCO_3$ and no additional carbon source. The growth rate of *P. tricornutum* on 0.02 M glucose or 0.02 M glycine was significantly lower (P <0.01) compared to 0.174 g·L^{-1} $NaHCO_3$ and even lower compared to no additional carbon source (Figure 1G), but the growth rate of *P. tricornutum* on 0.02 M GL was higher (P <0.01) compared to 0.174 g·L^{-1} $NaHCO_3$ after 10 days of culture (Figure 1A). These results indicated GL can promote the growth of *P. tricornutum* significantly, whereas glucose and glycine might not be utilized by the *P. tricornutum* strain used in this study. The complete genome sequence revealed the *P. tricornutum* (CCMP 2561) genome lacks glucose permease (protein-Npi-phospho-L-histidine-

Table 3 Positional labeling of Ser and Gly

	U-^{13}C, N plus	U-^{13}C, N minus	1,3-^{13}C, N plus	1,3-13C, N minus
Ser1	0.175 ± 0.016	0.332 ± 0.006	0.137 ± 0.012	0.080 ± 0.010
Ser2	0.182 ± 0.010	0.313 ± 0.033	0.261 ± 0.005	0.341 ± 0.010
Ser3	0.175 ± 0.017	0.293 ± 0.010	0.276 ± 0.011	0.351 ± 0.001
Gly1	0.167 ± 0.011	0.295 ± 0.025	0.126 ± 0.010	0.076 ± 0.003
Gly2	0.169 ± 0.014	0.292 ± 0.025	0.254 ± 0.006	0.362 ± 0.004
Average FLs of AA	0.150 ± 0.008	0.121 ± 0.005	0.157 ± 0.014	0.141 ± 0.007
Substance FLs	0.3	0.3	0.33	0.33

FL, fractional labeling.

sugar Npi-phosphotransferase, EC:2.7.1.69) [19], a class of enzymes that catalyze the phosphorylation of glucose and is important for utilization of extracellular glucose by cells. It is likely that the absence of glucose permease results in the poor glucose utilization ability of *P. tricornutum* (IOCAS-001). For the slow growth rate on glycine, we proposed the ammonium produced during the metabolism of glycine might be toxic to *P. tricornutum* cells because *P. tricornutum* growing on NH_4Cl also has a slow growth rate (Additional file 1: Figure S1) in this study. Yet there are reports, including the report of Zheng *et al.*, that the mixotrophic growth rate of *P. tricornutum* on glucose or glycine is significantly higher compared to photoautotrophic conditions and accumulated a greater biomass at the plateau phase [1-3,26]. The inconsistent results about glucose utilization of *P. tricornutum* might be due to the use of different trains and different culture conditions. The strain fully sequenced was the *P. tricornutum* accession Pt1 8.6 deposited as CCMP2561 in the Provasoli-Guillard National Center for Culture of Marine Phytoplankton. The strain used by Zheng *et al.* was the *P. tricornutum* strain CCMP 632 obtained from the Provasoli-Guillard National Center for Marine Algae and Microbiota (NCMA) (East Boothbay, ME). The strain used in our study was screened from the East China Sea and maintained in our laboratory for decades and identified as *P. tricornutum* (IOCAS-001) according to 18s ribosomal RNA sequence. Thus, these strains are of different origins. It is probable that these strains are different in genome structure or have mutated (the uptake of a mobile genetic element that carries the genes necessary for glucose utilization or strain mutation over many sub-cultures) and resulted in different glucose utilization abilities. Further, experimental conditions might affect the results. Associated bacteria, for example, might help in the phosphorylation of glucose and thus help in the use of glucose by *P. tricornutum*. The *P. tricornutum* cultures used in this study were purified by growth on agar plates containing multi-antibiotics and shown to be bacteria-free by RT-PCR and sequencing (data not shown) using primers described previously [31,32]. It is likely the absence of associated bacteria explains why *P. tricornutum* has poor ability to use glucose or glycine.

GL metabolism in *P. tricornutum*

Our results showed the growth rate under heterotrophic conditions was much lower compared to mixotrophic conditions (Figure 1C,I), indicating light was important for GL metabolism in *P. tricornutum*.

In vivo, GL can be phosphorylated and converted into dihydroxyacetone phosphate (DHAP) and glyceraldehyde 3-phosphate (GAP) through the following steps:

1. ATP + glycerol → ADP + sn-glycerol 3-phosphate
2. sn-glycerol-3-phosphate + NAD^+ → DHAP + NADH + H^+
3. DHAP → GAP
4. GAP → 3-P-glycerate (3PG)

and, finally, DHAP and 3PG can be metabolized through the glucolysis pathway. The complete genome sequence revealed *P. tricornutum* (CCMP 2561) contains all the enzymes required for the reactions mentioned above, that is, (1) glycerol kinase [gi: 7203857], (2) glycerol-3-phosphate dehydrogenase [gi: 7205122], (3) triosephosphate isomerase [gi: 7202457], and (4) glyceraldehyde-3-phosphate dehydrogenase [gi: 7203381] and phosphoglycerate kinase [gi: 7195255] [19], indicating *P. tricornutum* (CCMP 2561) can use GL as a carbon source.

The average FL value of protein amino acids was approximately 0.150 for the 30% (U-^{13}C) labeling experiment (Table 1), indicating nearly half of the carbon came from CO_2 fixation and inocula biomass. This suggested GL was assimilated simultaneously with CO_2 fixation under mixotrophic conditions, which was likely why *P. tricornutum* under mixotrophic conditions had a higher growth rate compared to heterotrophic conditions. Nevertheless, since at least half of the carbon came from GL, the growth rate under heterotrophic conditions should be approximately half compared to mixotrophic conditions if the GL assimilated was independent of CO_2 fixation and light. This was not the case, as suggested by Figure 1C,I, indicating light influences the assimilation of GL. Following the steps described above, metabolism of GL needs large amounts of ATP, which can be generated through photophosphorylation. Besides, DHAP and GAP generated from GL

Table 4 Ratios of abundances of key cellular metabolites under different culture conditions determined by NMR

Group	Metabolite	N-AC+ / N-AC-	N-AC+ / N + AC+	N + AC+ / N + AC-	N-AC- / N + AC-
Alcohols	Ethanol	1.049*	1.067*	1.063	1.081
	Ethylene glycol	0.878	2.005*	1.293*	2.952**
Amines	Dimethylamine	1.153*	1.032	1.252*	1.120*
	Methylamine	1.143*	1.048	0.730**	0.670**
	Trimethylamine	0.804*	0.573**	1.140	0.812*
Amino acids	Alanine	1.176**	0.045**	0.965	0.037**
	Asparagine	-	0.000**	0.569**	0.000**
	Aspartate	-	0.000**	1.236**	0.000**
	Betaine	0.485**	0.385**	0.768**	0.608**
	Carnitine	1.079*	0.931	0.997	0.860*
	Glutamate	1.147*	0.257**	1.069	0.240**
	Glutamine	-	0.000**	1.220**	0.000**
	Glycine	0.657	1.241*	0.956	1.805*
	Isoleucine	0.964	0.168**	1.108	0.193**
	Leucine	0.876**	0.278**	0.880	0.279**
	Methionine	1.278*	0.067**	1.067	0.056**
	N-Isovaleroylglycine	1.333**	2.952**	1.448*	3.207**
	O-Acetylcarnitine	0.871**	0.853	1.107	1.083*
	Ornithine	-	0.000**	0.993	0.000**
	Phenylalanine	0.882*	0.462**	1.073	0.561**
	Proline	-	0.000**	1.177	0.000**
	Threonine	-	0.000**	1.516**	0.000**
	Tyrosine	0.959	0.678**	1.040	0.735**
	Valine	1.283**	0.154**	1.001	0.120**
Amino acid derivatives	Creatine	1.296**	1.658**	1.192**	1.526**
	Guanidoacetate	1.198*	0.330**	0.833	0.229**
Ammoniums Compounds	Choline	0.849*	0.028**	2.379**	0.080**
	O-Phosphocholine	0.790**	1.985**	0.712**	1.789**
	sn-Glycero-3-phosphocholine	0.945	1.180	0.975	1.218
Ketones	Acetone	0.578**	1.381**	0.647	1.547*
Methanol	Methanol	0.944	1.052	1.023	1.142*
Nucleic acid components	ATP	0.000**	0.000**	0.797	0.335**
	NAD+	1.002	-**	0.000	0.960
Organic acids	2-Oxoglutarate	0.874	22.509**	0.849	21.861**
	Acetate	1.077*	1.355*	1.079	1.357**
	Ascorbate	0.587**	0.327**	1.465*	0.817
	Formate	0.654**	1.099*	0.901	1.515**
	Fumarate	0.778**	0.462**	1.161	0.690*
	Isocitrate	-	-	0.000**	0.000**
	Lactate	0.586**	6.185**	2.303**	24.313**
	Pyruvate	0.938	0.455**	1.414	0.685**
	Succinate	0.694**	0.514**	0.991	0.734**

Table 4 Ratios of abundances of key cellular metabolites under different culture conditions determined by NMR *(Continued)*

Sugars	Glucose	1.203	1.303*	1.341**	1.453**
	Glucose-1-phosphate	0.916	-**	-	-**
	Glucose-6-phosphate	1.210*	2.031**	1.200**	2.014**
	Glycerol	0.823**	0.595**	1.022	0.739**
Others	Trigonelline	1.098*	0.859**	1.122	0.878

**Student *t* test *P* value <0.01, *Student *t* test *P* value <0.05, ⁻Divide by zero.

metabolism were intermediate metabolites of the Calvin cycle. It was likely that consumption of these metabolites through the Calvin cycle promotes GL metabolism in *P. tricornutum*. We focused on Ser, which had an unexpectedly low level of ^{13}C enrichment at position 1 and was reported to be synthesized via the GOC cycle followed by photorespiratory reactions [26], to investigate other likely pathways for GL metabolism in *P. tricornutum*.

As was reported, Ser can be synthesized directly from 3PG through a phosphorylated route [33] or from glycine by the glycolate pathway [34]. In *P. tricornutum*, Ser could be synthesized from GL by four routes (Figure 4): (1) Ser was synthesized directly from GL-derived 3PG through a phosphorylated route, in which case the carbon backbone in GL was not disrupted and Ser should have the same MDV as GL. (2) Ser was synthesized

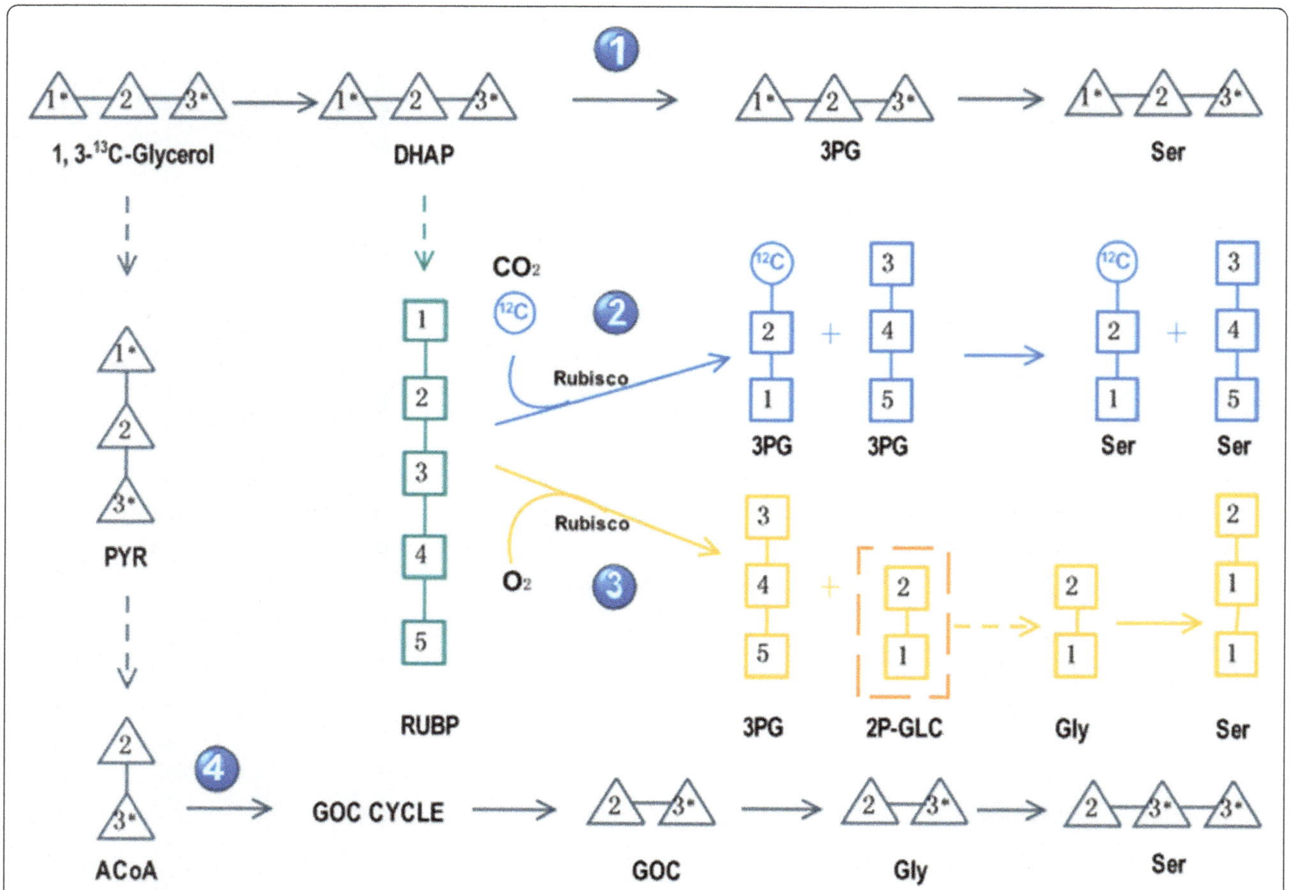

Figure 4 Scheme of four possible pathways for Ser arising from GL. (1) Ser was synthesized directly from GL-derived 3PG. The carbon backbone in GL was not disrupted and Ser had the same MDV as GL. (2) Ser was synthesized directly from Calvin cycle-derived 3PG, half of which would have an unlabeled carbon atom at position 1 deriving from atmospheric CO_2. The ^{13}C enrichment of Ser1 should be diluted regardless of the labeling pattern of GL. (3) Ser was synthesized from GOC through photorespiration. The carbon backbone in GL was recombined during the Calvin cycle and Ser1 was derived from RuBP2. (4) Ser was synthesized from GOC via the GOC cycle, as reported by Zheng *et al.* In this case, Ser1 originated solely from GL2 and the ^{13}C enrichment should be zero when [1, 3-^{13}C]GL was used as substrate.

directly from the Calvin cycle-derived 3PG, half of which would have an unlabeled carbon atom derived from atmospheric CO_2 in position 1. In this case, the ^{13}C enrichment of Ser1 should be diluted by half regardless of the labeling pattern of GL. (3) Ser was synthesized from GOC through photorespiration (oxygenation of ribulose-1,5-bisphosphate, RuBP). (4) Ser was synthesized from GOC via the GOC cycle, as reported by Zheng *et al.* [26]; in this case, Ser1 origin solely from GL2, whose ^{13}C enrichment should be zero using [1, 3-^{13}C]GL as substrate. Our results showed the ^{13}C enrichment of Ser1 was significantly lower (ca. half under normal condition and much lower under nitrogen-limited condition) compared to Ser2 and Ser3 when [1, 3-^{13}C]GL was used as substrate and so was the ^{13}C enrichment of Gly1 relative to Gly2 (Table 3). Nevertheless, both Ser and Gly had similar ^{13}C enrichment in all positions when [U-^{13}C]GL was used as substrate. Thus, it is unlikely Ser was synthesized solely from route 1, route 2 or route 4. It is more likely that Ser was largely synthesized, as in higher plants [34], from Gly via the photorespiratory pathway (route 3). In other words, the Calvin cycle and photorespiration took part in GL metabolism in *P. tricornutum*.

The likely response of photorespiration to nitrogen limitation

It was reported that photorespiration was important for nitrogen assimilation [35]. To investigate the probable relationship between photorespiration and nitrogen assimilation in *P. tricornutum*, we labeled *P. tricornutum* in the absence of nitrogen using [U-^{13}C]GL or [1,3-^{13}C]GL as substrate. As Table 1 shows, the ^{13}C enrichment of Ser and Gly were increased significantly when those of other amino acids decreased in the absence of nitrogen, regardless of the type of substrate. Synthesis of protein was reduced under nitrogen-limited conditions [36], meaning the *de novo* synthesis of amino acids would be reduced also. Thus, it was anticipated the ^{13}C enrichment of most amino acids would be reduced. The high FL values of Ser and Gly under nitrogen-limited conditions suggested most Ser and Gly were synthesized *de novo* using carbon backbones obtained through GL metabolism. As mentioned above, Ser and Gly were synthesized through photorespiration (Figure 4, route 3); thus, the increase of Ser and Gly synthesis indicated the increase of photorespiration under nitrogen-limited conditions, consistent with the suggestion that photorespiration was important for nitrogen assimilation [35]. In addition, the response of diatom to nitrogen limitation was different from that of higher plants and green algae. Instead of photosynthesis, the breakdown of intracellular stores responded for carbon backbone demand for reassimilation of nitrogen [36]. We proposed that photorespiration might be important for redistribution of nitrogen under nitrogen-limited condition.

Influence of sodium acetate on growth of *P. tricornutum* with or without nitrogen

It was reported photorespiration bypass can improve photosynthesis rates and biomass production in *Arabidopsis thaliana* [37,38]. One bypass was suggested by converting one molecule of GOC with one molecule of acetyl coenzyme A (ACoA) into malate by malate synthase [37-39]. Malate can be metabolized through other pathway (TCA cycle, GOC cycle, and so on) and thus reduce carbon loss through the photorespiratory pathway. To investigate likely photorespiration bypass and its influence on the growth of *P. tricornutum*, we compared the growth of *P. tricornutum* with or without sodium acetate, which can be converted into ACoA by ACoA synthase. The results indicated sodium acetate promoted the growth of *P. tricornutum* in the presence of nitrogen, but growth was decreased in the absence of nitrogen. We proposed that when sodium acetate was added, ACoA was synthesized by ACoA synthase. Excess ACoA promoted the condensation reaction between ACoA and GOC and resulted in the photorespiration bypass in *P. tricornutum*, which improved biomass production in the present of nitrogen. When nitrogen was limited, *P. tricornutum* had a slower growth rate in the presence of sodium acetate, suggesting photorespiration was important for the nitrogen-limited response in *P. tricornutum*, consistent with reports that photorespiration was important for nitrogen assimilation [35]. Together with the report that genes encoding GOC cycle-related enzymes were down-regulated significantly under nitrogen deprivation [40], we proposed down-regulation of the GOC cycle allows more GOC catalysis through photorespiration. The enhancement of photorespiration might facilitate the redistribution of nitrogen and thus helped *P. tricornutum* to survive under nitrogen-limited conditions. These results also provided evidences for GL metabolism through route 3 but not route 4 (Figure 4). In route 3, GOC was generated from oxygenation of RuBP. Excess ACoA would transfer GOC into malate, thus reduced carbon loss through following photorespiratory reactions, which would benefit growth under nitrogen-replete conditions but be unfavorable under nitrogen-limited conditions. While in route 4, glycerol was converted into ACoA and then GOC through the GOC cycle. GOC was metabolized through photorespiratory reactions. Excess ACoA would enhance route 4 (photorespiratory reactions), which should be favorable for growth under nitrogen-limited condition. This was not the case in our results; thus, we suggested that glycerol was largely metabolized through route 3 (Figure 4) in the strain used in this study.

Metabolic analysis provided further evidences for the importance of photorespiration in nitrogen-limited response

To obtain further insight into the importance of photorespiration in nitrogen-limited response, the composition of the intracellular metabolite pool under different culture conditions (N^-Ac^+, N^-Ac^-, N^+Ac^+, and N^+Ac^-) was measured by NMR and the ratios of abundances of key cellular metabolites were calculated (Table 4). Gly increased under N^- conditions (N^-Ac^+/N^+Ac^+ and N^-Ac^-/N^+Ac^-) and was reduced when acetate was added (N^-Ac^+/N^-Ac^- and N^+Ac^+/N^+Ac^-). In addition, ethylene glycol and formate, which can be produced from intermediate metabolites of photorespiration (glycolate → glycolaldehyde → ethylene glycol and glyoxylate → oxalate → formate), had the same expression pattern as Gly (Table 4). These results supported our hypotheses that photorespiration was up-regulated under N^- condition, whereas addition of acetate would result in photorespiration bypass and suppress the following steps of photorespiration. Besides, Glu and Gln, which participated in photorespiration as nitrogen carriers, were reduced under N^- conditions (N^-Ac^+/N^+Ac^+ and N^-Ac^-/N^+Ac^-) and increased when acetate was added (N^-Ac^+/N^-Ac^- and N^+Ac^+/N^+Ac^-). Accordingly, AKG (the precursor of Glu) increased under N^- conditions (N^-Ac^+/N^+Ac^+ and N^-Ac^-/N^+Ac^-) and was reduced when acetate was added (N^-Ac^+/N^-Ac^- and N^+Ac^+/N^+Ac^-). In theory, Glu and Gln might have the same expression pattern as Gly (increased under N^- conditions). Yet Glu and Gln also played as nitrogen carriers for biosynthesis. Since cell growth and most amino acids were reduced under nitrogen-limited condition (Figure 1D and Table 4), it was probable that an approximately 0.25-fold amount of Glu was sufficient to support photorespiration.

Likely GL metabolism pathway in *P. tricornutum*

Notably, the FL values of Ser and Gly in the absence of nitrogen based on [U-^{13}C]GL were approximately 0.3, close to the level of substrate. During photorespiration, 2P-glycolate is generated from oxygenation of RuBP, whose ^{13}C enrichment was inevitably diluted by CO_2 during the Calvin cycle. Since Ser and Gly were synthesized from GOC, which was generated from 2P-glycolate, why were they not diluted by CO_2? Obviously, only when the carbon atoms in position 1 and position 2 of pentose-5-phosphate (P5P1 and P5P2) were derived solely from GL could GOC retain a high level of ^{13}C enrichment. This cannot be the case when GL entered the Calvin cycle as GAP, as GAP can also be derived from carboxylation of RuBP whose ^{13}C enrichment would be diluted. Thus, GL likely entered the Calvin cycle in another form, for example, DHAP. Supposing all DHAP was generated from GL (Figure 5), RuBP regeneration started with the condensation reaction of one molecule of GAP and one molecule of DHAP, generating one molecule of fructoce-6-P (F6P) with GL3_1-generated carbon atoms at F6P1_3. Next, F6P1_2 was transferred to a second molecule of GAP, generating one molecule of xylulose-5-P (X5P) with the GL 3_2-generated carbon backbone at X5P1_2 and one molecule of erythrose-4-P (E4P) with a GL1-derived carbon atom distributed at E4P1. Next, E4P was condensed with a second molecule of DHAP, generating one molecule of sedoheptulose-7-P (S7P) with a GL1-derived carbon atom distributed at S7P4 and the GL3_1-derived carbon backbone distributed at S7P1_3. Sequentially, S7P1_2 was transferred to a third molecule of GAP, generating one molecule of X5P with a GL3_2-generated carbon backbone distributed at X5P1_2 and one molecule of ribose-5-P (R5P) with GL1-derived carbon atoms distributed at R5P1 and R5P2. Both R5P and X5P can be converted into ribulose-5-P (Ru5P), which can finally be converted into RuBP. By oxygenation by Rubisco and metabolism through photorespiration, these RuBPs can generate three molecules of Gly: one had carbon atoms generated from GL1 at both positions and the other two had GL2 at Gly1 and GL3 at Gly2. Generated from these Gly, Ser would have a GL1 or GL2-derived carbon atom at Ser1 and GL1 or GL3 at Ser2 and Ser3. In this pathway, Ser and Gly would have a low level of ^{13}C enrichment at position 1 and similar high level of ^{13}C enrichment at other positions. As GL1-derived Ser1 and Gly1 are basically derived from R5P, the ^{13}C enrichment of Gly1 and Ser1 would depend on the ratio of R5P to X5P used in the generation of RuBP. When more R5P is used, a higher ^{13}C enrichment at Gly1 and Ser1 would occur. This might account for the different ^{13}C enrichment in Gly1 and Ser1 under different nitrogen conditions, providing further evidence for GL metabolism through route 3. Ser1 and Gly1 were significantly lower under nitrogen-limited conditions compared to when ample nitrogen was present, indicating the proportion of RuBP generated from R5P decreased. It was probable that R5P was used for synthesis of other compounds, for example, mRNA. *P. tricornutum* might reserve mRNA synthesis under nitrogen-limited conditions to prepare for recovery under optimized conditions, consistent with our results for transcriptome and proteome of *P. tricornutum* under nitrogen-limited conditions (unpublished data).

Conclusion

Our results indicated the strain used in this study (IOCAS-001) was different from that in the study of Zheng *et al.* [26]. Strains that lack a glucose-utilizing ability have a different mechanism for organic carbon metabolism from strains that can use glucose. Photorespiration is involved in GL metabolism and is important for the nitrogen-limited response in *P. tricornutum*.

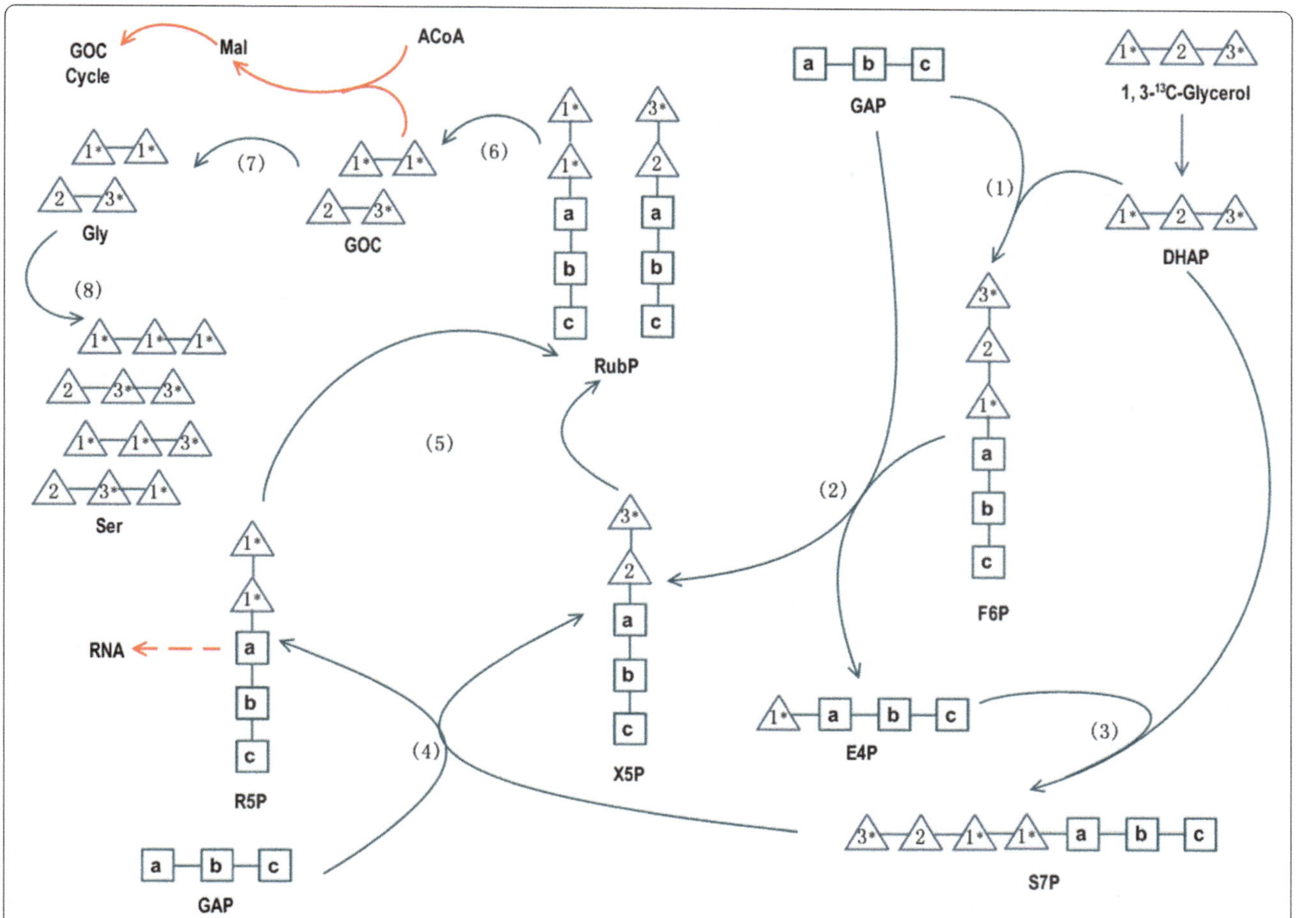

Figure 5 Scheme of possible pathways for GL metabolism in *P. tricornutum*. (1) One molecule of F6P was derived through a condensation reaction between one molecule of GAP and one molecule of DHAP. Carbon backbone from GL3_1 was distributed at F6P1_3. (2) F6P1_2 was transferred to a second molecule of GAP, generating one molecule of X5P with GL3_2-generated carbon backbone distributed at X5P1_2 and one molecule of E4P with GL1-derived carbon atom distributed at E4P1. (3) E4P condensate with a second molecule of DHAP, generating one molecule of S7P with GL1-derived carbon atom distributed at S7P4 and GL3_1-derived carbon backbone distributed at S7P1_3. (4) S7P1_2 were transferred to a third molecule of GAP, generating one molecule of X5P with GL3_2-generated carbon backbone distributed at X5P1_2 and one molecule of R5P with GL1-derived carbon atoms distributed at R5P1 and R5P2. (5) RuBP generated from R5P and X5P. (6) Generation of GOC (GOC) from oxygenation of RuBP. (7) Generation of Gly from GOC. A third of Gly had carbon atoms generated from GL1 at both positions, and others had GL2 at Gly1 and GL3 at Gly2. (8) Generation of Ser from Gly. Ser had GL1 or GL2-derived carbon atom at Ser1 and GL1 or GL3 at Ser2 and Ser3.

Materials and methods

Strains and culture conditions

Axenic cultures of *P. tricornutum* were screened from the East China Sea and maintained in our laboratory for decades and identified as *P. tricornutum* (IOCAS-001) according to morphological observation and the 18s ribosomal RNA sequence. The cultures were grown in f/2 medium [41] made with steam-sterilized $NaHCO_3$-free artificial seawater [42] supplemented with f/2 inorganic nutrients, trace elements and vitamins (sterilized by filtration). One volume of cells in mid-exponential growth phase was harvested by centrifugation for 10 min at 5000×*g*, washed with sterilized $NaHCO_3$-free artificial seawater, and inoculated into nine volumes of fresh medium. Then, 100 mg · L^{-1} each of ampicillin, kanamycin, and streptomycin were added to avoid bacterial contamination. $NaHCO_3$ (0.174 g·L^{-1}) or a source of organic carbon (see below) was added. For nitrogen-free experiments, sodium nitrate was omitted from f/2 inorganic nutrients. For experiments testing the influence of sodium acetate, it was added to a final concentration of 1 g·L^{-1}. To further test the influence of sodium acetate concentration, it was added to final concentrations of 0, 0.1, 0.5, 1, and 5 g·L^{-1}, respectively. Each treatment was conducted in triplicate. Cultures were grown at 20°C under cool white fluorescent lights at approximately 100 μmol m^{-2}·s^{-1} with a 12 h dark/12 h light cycle for mixotrophic conditions and completely dark for heterotrophic conditions. Cell growth was monitored by measurement of absorbance at 730 nm (A_{730nm}) using a UV/visible spectrophotometer (Uv-1800; Shimadzu Scientific Instruments, Japan). Cell number was counted using a haemacytometer, and their relationship with the value of A_{730nm} was determined.

Selection of carbon source and optimization of GL concentration

Cells were cultured with 0.02 M GL, glucose, or glycine as carbon source. To determine the optimal concentration of GL, cells were cultured with 0.005, 0.02, and 0.1 M GL as carbon source. A_{730nm} was measured and compared to growth with no additional carbon source and with $NaHCO_3$ at a final concentration of 0.174 g·L^{-1} as carbon source.

Determination of GL and acetate concentrations

The culture was centrifuged at 8,000×*g* for 10 min at room temperature. The supernatant was used to detect extracellular metabolites by LC as described previously [43]. Basically, 10 μL sample were loaded onto a Bio-Rad HPX-87H column (Bio-Rad; Hercules, CA, USA) connected to an HPLC system (Agilent 1200) (GMI; Ramsey, MN, USA), eluted with 5 mM H_2SO_4 at a flow rate of 0.5 mL·min^{-1}, and then detected with a differential refraction detector.

Total lipid analysis

Cells were harvested and quickly froze in liquid nitrogen and then freeze-dried. Total lipids were extracted as described by Bligh and Dyer [44] with minor modifications. Briefly, approximately 20 mg dried cells were mixed with 1 mL of chloroform/methanol (1:1). The mixture was stirred vigorously for 5 min, after which 0.3 mL of 0.2 M H_3PO_4 (containing 1 M KCl) was added and mixed. After centrifugation at 5000 × g for 5 min, the solvent phase was recovered. This process was repeated three times. The solvent phases were combined, washed three times with distilled water, and then evaporated at room temperature under a ventilated fume hood. The total lipid content was obtained by weighing and expressed as percent cell dry weight (% CDW).

^{13}C labeling experiment

GL (0.02 M) was used as the ^{13}C-labeling substrate. Mid-exponential growth phase cells (with 0.174 g·L^{-1} $NaHCO_3$ or 0.02 M GL as carbon source) were harvested by centrifugation for 10 min at 5000×*g*, washed with sterilized $NaHCO_3$-free artificial seawater, and inoculated into nine volumes of fresh medium containing a 20%:80% (or 30%:70%) *w/w* mixture of [U-^{13}C]GL (99%; Cambridge isotope laboratories)/unlabeled GL or 50%:50% w/w mixture of [1,3-^{13}C]GL (99%; Cambridge isotope laboratories)/unlabeled GL. For labeling under N-limited conditions, $NaNO_3$ was omitted from the medium. The labeling experiment lasted for 10 days under normal conditions and a week under N-limited conditions when the cells grew to mid-exponential growth phase. Each treatment was conducted in triplicate.

Sample preparation and gas chromatography-mass spectrometry analysis

Samples for GC-MS analysis were prepared as described previously [27]. After labeling, cells were harvested by centrifugation for 10 min at 5000×*g*, washed with distilled water, and placed into 1.5 mL Eppendorf tubes and centrifuged for 5 min at 10,000×*g*. The supernatants were discarded, and cell pellets were hydrolyzed in 6 M HCl at approximately 110°C for 12 h then dried overnight at approximately 80°C. Subsequently, each dried sample was dissolved in 100 μL of water-free pyridine then derivatized by the addition of 50 μL of *N*-tert-butyldimethylsilyl-*N*-methyltrifluoroacetamide (MTBSTFA) and heated at 85°C for 60 min. The sample was centrifuged for 5 min at 10,000×*g*, and the supernatant was filtered through 0.22-μm pore-size filters and then loaded onto a GC-MS apparatus for analysis as previously described [45]. GC-MS was performed with an Agilent 6890-5973 GC-MS (Conquer Scientific; San Diego, CA, USA) system equipped with an Agilent HP-5MS column (30 m × 0.25 mm × 0.25 μm) with helium as the carrier gas. The oven temperature was programmed at 100°C for 2 min then increased to 260°C at a rate of 5°C min^{-1} and kept at 260°C for 10 min. The injection volume was 2 mL, and the MS device was operated in scan mode.

GC-MS data processing

The GC-MS data were processed as previously described [27]. The mass distribution vector (MDV) for each fragment of 13 amino acid residues (Ala, Gly, Val, Asx, Glx, Pro, His, Phe, Tyr, Ser, Thr, Lys, and Met) and GL were assigned according to Eq. (1), where m_0 is the fractional abundance of molecules with monoisotopic mass and $m_{i>0}$ is the abundance of fragments with greater mass [46] and corrected for the natural abundance of stable isotopes of O, N, Si, S, H, and C [27].

$$MDV_a = \begin{bmatrix} (m_0) \\ (m_1) \\ \vdots \\ (m_n) \end{bmatrix} \text{with} \sum\nolimits_{i=0}^{n} m_i = 1 \qquad (1)$$

The fractional labeling (FL) of each fragment was calculated according to Eq. (2), where *mi* has the same meaning as Eq. (1) [45].

$$FL = \frac{\sum_{i=0}^{n} i \cdot m_i}{n \cdot \sum_{i=0}^{n} m_i} \quad (2)$$

Metabolite analysis

Cells cultured under N^-Ac^+, N^-Ac^-, N^+Ac^+, and N^+Ac^- conditions were harvested and quickly frozen in liquid nitrogen, then freeze-dried. Metabolites were extracted and analysis by NMR as described previously [47]. Student *t* test was used to detect significance of differences between samples.

Additional file

Additional file 1: Figure S1. Growth of *P. tricornutum* on NH_4Cl and $NaNO_3$ as nitrogen sources.

Abbreviations

PUFA: polyunsaturated fatty acids; GC-MS: gas chromatography mass spectrometry; NMR: nuclear magnetic resonance; GL: glycerol; CDW: cell dry weight; FL: fractional labeling; PEP: phosphoenolpyruvate; E4P: erythrose-4-P; MDV: mass distribution vector; DHAP: dihydroxyacetone phosphate; GAP: glyceraldehyde 3-phosphate; 3PG: 3-P-glycerate; RuBP: ribulose-1,5-bisphosphate; ACoA: acetyl coenzyme A; P5P: pentose-5-phosphate; F6P: fructoce-6-P; X5P: xylulose-5-P; E4P: erythrose-4-P; S7P: sedoheptulose-7-P; R5P: ribose-5-P; Ru5P: ribulose-5-P; MTBSTFA: *N*-tert-butyldimethylsilyl-*N*-methyltrifluoroacetamide.

Competing interests

The authors declare that they have no competing interests.

Authors' contributions

AYH carried out the experiments, performed the data analysis, and drafted the manuscript. LXL cultured the *P. tricornutum*, prepared the samples, and participated in data analysis. CY conceived of the study and performed the data analysis. GCW conceived of the study and drafted the manuscript. All authors read and approved the manuscript.

Acknowledgements

This work was supported by the National Natural Science Foundation of China (41406169), International Science & Technology Cooperation Program of China (ISTCP, 2015DFG32160), Ministry of Science and Technology of the PRC fundamental research work (NO. 2012FY112900-01), Science and Technology Strategic Pilot of the Chinese Academy of Sciences (XDA05030401), and the General Financial Chinese Grant from the China Postdoctoral Science Foundation (2014 M551970).

Author details

[1]Institute of Oceanology, Chinese Academy of Sciences, Qingdao 266071, China. [2]Institute of Plant Physiology and Ecology, Shanghai Institute for Biological Sciences, Chinese Academy of Sciences, Shanghai 200032, China.

References

1. Garci MCC, Sevilla JMF, Fernandez FGA, Grima EM, Camacho FG. Mixotrophic growth of *Phaeodactylum tricornutum* on glycerol: growth rate and fatty acid profile. J Appl Phycol. 2000;12:239–48.
2. Garcia MCC, Miron AS, Sevilla JMF, Grima EM, Camacho FG. Mixotrophic growth of the microalga *Phaeodactylum tricornutum* - influence of different nitrogen and organic carbon sources on productivity and biomass composition. Process Biochem. 2005;40:297–305.
3. Ceron Garcia MC, Camacho FG, Miron AS, Sevilla JMF, Chisti Y, Grima EM. Mixotrophic production of marine microalga *Phaeodactylum tricornutum* on various carbon sources. J Microbiol Biotechnol. 2006;16:689–94.
4. Kobayashi M, Kakizono T, Nagai S. Enhanced carotenoid biosynthesis by oxidative stress in acetate-induced cyst cells of a green unicellular alga, *Haematococcus pluvialis*. Appl Environ Microbiol. 1993;59:867–73.
5. Lee YK, Ding SY, Hoe CH, Low CS. Mixotrophic growth of *Chlorella sorokiniana* in outdoor enclosed photobioreactor. J Appl Phycol. 1996;8:163–9.
6. Field C, Behrenfeld M, Randerson J, Falkowski P. Primary production of the biosphere: integrating terrestrial and oceanic components. Sci. 1998;281:237–40.
7. Falkowski P, Barber R, Smetacek V. Biogeochemical controls and feedbacks on ocean primary production. Sci. 1998;281:200–6.
8. Demirbas A. Biodiesel: a realistic fuel alternative for diesel engines. London: Springer Verlag; 2008.
9. Francius G, Tesson B, Dague E, Martin-Jezequel V, Dufrene YF. Nanostructure and nanomechanics of live *Phaeodactylum tricornutum* morphotypes. Environ Microbiol. 2008;10:1344–56.
10. De Martino A, Meichenin A, Shi J, Pan KH, Bowler C. Genetic and phenotypic characterization of *Phaeodactylum tricornutum* (Bacillariophyceae) accessions. J Phycol. 2007;43:992–1009.
11. Borowitzka M, Volcani B. The polymorphic diatom *Phaeodactylum tricornutum*: ultrastructure of its morphotypes. J Phycol. 1978;14:10–21.
12. Gutenbrunner S, Thalhamer J, Schid A. Mm (1994) proteinaceaous and immunochemical distinctions between the oval and fusirom morphotypes of *Phaeodactylum tricornutum* (Bacillariophyceae). J Phycol. 1994;30:129–36.
13. Patil V, Reitan KI, Knutsen G, Mortensen LM, Källqvist T, Olsen E, et al. Microalgae as source of polyunsaturated fatty acids for aquaculture. Curr Top Plant Biol. 2005;6:57–65.
14. Owens TG, Wold ER. Light-harvesting function in the diatom *Phaeodactylum tricornutum* I. Isolation and characterization of pigment-protein complexes. Plant Physiol. 1986;80:732–8.
15. Kim JM, Araki S, Kim DJ, Park CB, Takasuka N, Baba-Toriyama H, et al. Chemopreventive effects of carotenoids and curcumins on mouse colon carcinogenesis after 1, 2-dimethylhydrazine initiation. Carcinogenesis. 1998;19:81–5.
16. Nomura T, Kikuchi M, Kubodera A, Kawakami Y. Proton-donative antioxidant activity of fucoxanthin with 1, 1-diphenyl-2-picrylhydrazyl (DPPH). IUBMB Life. 1997;42:361–70.
17. Maeda H, Hosokawa M, Sashima T, Funayama K, Miyashita K. Fucoxanthin from edible seaweed, *Undaria pinnatifida*, shows antiobesity effect through ucp1 expression in white adipose tissues. BBRC. 2005;332:392–7.
18. Matsumoto M, Hosokawa M, Matsukawa N, Hagio M, Shinoki A, Nishimukai M, et al. Suppressive effects of the marine carotenoids, fucoxanthin and fucoxanthinol on triglyceride absorption in lymph duct-cannulated rats. Eur J Nutr. 2010;49:243–9.
19. Bowler C, Allen AE, Badger JH, Grimwood J, Jabbari K, Kuo A, et al. The *Phaeodactylum* genome reveals the evolutionary history of diatom genomes. Nature. 2008;456:239–44.
20. Apt KE, Grossman A, Kroth-Pancic P. Stable nuclear transformation of the diatom *Phaeodactylum tricornutum*. Mol Gen Genet MGG. 1996;252:572–9.
21. De Risco V, Raniello R, Maumus F, Rogato A, Bowler C, Falciatore A. Gene silencing in the marine diatom *Phaeodactylum tricornutum*. NAR. 2009;37:e96–6.
22. Siaut M, Heijde M, Mangogna M, Montsant A, Coesel S, Allen A, et al. Molecular toolbox for studying diatom biology in *Phaeodactylum tricornutum*. Gene. 2007;406:23–35.
23. Zaslavskaia L, Lippmeier J, Kroth P, Grossman A, Apt K. Transformation of the diatom *Phaeodactylum tricornutum* (Bacillariophyceae) with a variety of selectable marker and reporter genes. J Phycol. 2000;36:379–86.
24. Ceron-Garcia MC, Fernandez-Sevilla JM, Sanchez-Miron A, Garcia-Camacho F, Contreras-Gomez A, Molina-Grima E. Mixotrophic growth of *Phaeodactylum tricornutum* on fructose and glycerol in fed-batch and semi-continuous modes. Bioresour Technol. 2013;147:569–76.
25. Ukeles R, Rose WE. Observations on organic-carbon utilization by photosynthetic marine microalgae. Mar Biol. 1976;37:11–28.
26. Zheng YT, Quinn AH, Sriram G. Experimental evidence and isotopomer analysis of mixotrophic glucose metabolism in the marine diatom *Phaeodactylum tricornutum*. Microb Cell Fact. 2013;12:16.
27. Nanchen A, Fuhrer T, Sauer U. Determination of metabolic flux ratios from ^{13}C-experiments and gas chromatography–mass spectrometry data. Methods Mol Biol. 2007;358:92.

28. Rosman K, Taylor P. Isotopic compositions of the elements 1997. JPCRD. 1998;27:1275.
29. Guerra LT, Levitan O, Frada MJ, Sun JS, Falkowski PG, Dismukes GC. Regulatory branch points affecting protein and lipid biosynthesis in the diatom *Phaeodactylum tricornutum*. Biomass & Bioenergy. 2013;59:306–15.
30. Allen AE, Dupont CL, Obornik M, Horak A, Nunes-Nesi A, McCrow JP, et al. Evolution and metabolic significance of the urea cycle in photosynthetic diatoms. Nature. 2011;473:203–9.
31. Edwards U, Rogall T, Blocker H, Emde M, Bottger EC. Isolation and direct complete nucleotide determination of entire genes. Characterization of a gene coding for 16S-ribosomal RNA. NAR. 1989;17:7843–53.
32. Weisburg WG, Barns SM, Pelletier DA, Lane DJ. 16 s ribosomal DNA amplification for phylogenetic study. J Bacteriol. 1991;173:697–703.
33. Colman B, Norman EG. Serine synthesis in cyanobacteria by a nonphotorespiratory pathway. Physiol Plant. 1997;100:133–6.
34. Husic DW, Husic HD, Tolbert NE, Black Jr CC. The oxidative photosynthetic carbon cycle or C_2 cycle. Crit Rev Plant Sci. 1987;5:45–100.
35. Rachmilevitch S, Cousins AB, Bloom AJ. Nitrate assimilation in plant shoots depends on photorespiration. Proc Natl Acad Sci U S A. 2004;101:11506–10.
36. Hockin NL, Mock T, Mulholland F, Kopriva S, Malin G. The response of diatom central carbon metabolism to nitrogen starvation is different from that of green algae and higher plants. Plant Physiol. 2012;158:299–312.
37. Peterhansel C, Blume C, Offermann S. Photorespiratory bypasses: how can they work? JExB. 2013;64:709–15.
38. Maier A, Fahnenstich H, Von Caemmerer S, Engqvist M, Weber A, Flugge U, et al. Glycolate oxidation in A. *thaliana* chloroplasts improves biomass production. Frontiers in Plant Science. 2012;3:38.
39. Fahnenstich H, Scarpeci TE, Valle EM, Flügge U-I, Maurino VG. Generation of hydrogen peroxide in chloroplasts of *Arabidopsis* overexpressing glycolate oxidase as an inducible system to study oxidative stress. Plant Physiol. 2008;148:719–29.
40. Yang Z-K, Niu Y-F, Ma Y-H, Xue J, Zhang M-H, Yang W-D, et al. Molecular and cellular mechanisms of neutral lipid accumulation in diatom following nitrogen deprivation. Biotechnology for Biofuels. 2013;6:67.
41. Guillard, Robert RL. Culture of phytoplankton for feeding marine invertebrates. In: Smith WL, Chanley MH, editors. Culture of marine invertebrate animals. New York: Plenum Press; 1975. p. 29–60.
42. Harrison P, Waters R, Taylor F. A broad spectrum artificial sea water medium for coastal and open ocean phytoplankton. J Phycol. 1980;16:28–35.
43. Sun L, Huang A, Gu W, Ma Y, Zhu D, Wang G. Hydrogen production by *Enterobacter cloacae* isolated from sugar refinery sludge. IJHE. 2015;40:1402–7.
44. Bligh EG, Dyer WJ. A rapid method of total lipid extraction and purification. Can J Biochem Physiol. 1959;37:911–7.
45. Xiong W, Liu L, Wu C, Yang C, Wu Q. ^{13}C-tracer and gas chromatography-mass spectrometry analyses reveal metabolic flux distribution in the oleaginous microalga *Chlorella protothecoides*. Plant Physiol. 2010;154:1001–11.
46. Fischer E, Sauer U. Metabolic flux profiling of *Escherichic coli* mutants in central carbon metabolism using GC-MS. EJBio. 2003;270:880–91.
47. Ge F, Huang WC, Chen Z, Zhang CY, Xiong Q, Bowler C. et al. Methylcrotonyl-coa carboxylase regulates triacylglycerol accumulation in the model diatom *Phaeodactylum tricornutum*. Plant Cell. 2014;26:1681–97.

pH-induced change in cell susceptibility to butanol in a high butanol-tolerant bacterium, *Enterococcus faecalis* strain CM4A

Manabu Kanno, Hideyuki Tamaki*, Yasuo Mitani, Nobutada Kimura, Satoshi Hanada and Yoichi Kamagata

Abstract

Background: Though butanol is considered as a potential biofuel, its toxicity toward microorganisms is the main bottleneck for the biological butanol production. Recently, butanol-tolerant bacteria have been proposed as alternative butanol production hosts overcoming the end product inhibition. One remaining key issue to be addressed is how physicochemical properties such as pH and temperature affect microbial butanol tolerance during cultivation and fermentation.

Results: We investigated the pH effect on butanol tolerance of a high butanol-tolerant bacterium, *Enterococcus faecalis* strain CM4A. The strain grew over a broad pH range (pH 4.0 to 12.0) and preferred alkaline pH (pH 8.0 and 10.0) in the absence of butanol. However, in the presence of butanol, strain CM4A grew better under acidic and neutral pH conditions (pH 6.0 and 6.8). Membrane fatty acid analysis revealed that the cells exposed to butanol exhibited increased cyclopropane and saturated fatty acids, which contribute to butanol tolerance of the strain by decreasing membrane fluidity, more evidently at acidic and neutral pH than at alkaline pH. Meanwhile, the strain grown under alkaline pH without butanol increased short chain fatty acids, which is involved in increasing membrane fluidity for alkaline adaptation. Such a change was not observed in the cells grown under alkaline pH with butanol. These results suggested that strain CM4A simultaneously exposed to butanol and alkali stresses was not likely able to properly adjust membrane fluidity due to the opposite response to each stress and thereby showed low butanol tolerance under alkaline pH. Indeed, the cells exposed to butanol at alkaline pH showed an irregular shape with disrupted membrane structure under transmission electron microscopy observation, which also indicated the impact of butanol and alkali stresses on functioning of cellular membrane.

Conclusion: The study clearly demonstrated the alkaline pH-induced increase of cell susceptibility to butanol in the tested strain. Our findings indicate the non-negligible impact of pH on microbial butanol tolerance, providing a new insight into efficient butanol production.

Keywords: Butanol- and alkali-tolerant microorganism, Physiological response, Butanol and pH stresses, Biofuel

Background

Microbial fuel production from renewable resources has gained increased attention in view of energy security and environmental concerns. Although *n*-butanol (referred to as butanol) is considered to be a promising biofuel due to its desirable chemical properties such as low volatility and miscibility with gasoline [1], butanol fermentation process is still economically unfavorable. Major limitation in biological production of butanol is its high toxicity toward microorganisms [2]. Butanol is known to intercalate into the cell membrane and break the lipid hydrogen bonds, resulting in disruption of membrane structure and cell death [3].

Heterologous butanol production has become realized by the development of genetic and metabolic engineering [4,5], and thus, butanol-tolerant bacteria have been regarded as alternative hosts overcoming growth interruption problem during butanol production. Recently, a number of butanol-tolerant bacteria able to grow in the presence of greater than 2.0% (vol/vol) butanol have

* Correspondence: tamaki-hideyuki@aist.go.jp
Bioproduction Research Institute, National Institute of Advanced Industrial Science and Technology (AIST), Tsukuba, Ibaraki 305-8566, Japan

been reported [6-11]. We isolated an aerobic butanol-tolerant strain CM4A from grease-contaminated soil, which was most closely related to *Enterococcus faecalis* with high 16S rRNA gene sequence similarity (99.6%) [11]. Strain CM4A exhibited superior butanol tolerance with the ability to grow in the presence of up to 3.5% butanol without assimilation and degradation.

Microbial butanol tolerance is possibly influenced by physicochemical properties such as temperature and pH for cultivation and fermentation. However, very little is known about the influence of those culture conditions to butanol tolerance, since the well-studied bacterial species grew under the limited temperature and pH conditions. Indeed, the effect of culture pH on the butanol tolerance has not been investigated, although only a few studies on the effect of temperature were reported [6,12].

In this study, we aimed to investigate how culture pH affects microbial butanol tolerance using *E. faecalis* strain CM4A, which can grow over a broad pH range (pH 4.0 to 12.0). This study further characterized the response of cell membrane to both butanol and pH stresses.

Results and discussion

Alkali tolerance of *E. faecalis* strain CM4A

E. faecalis strain CM4A was cultured aerobically in a glucose-rich medium. The strain grew over a broad pH range (pH 4.0 to 12.0), with the highest growth rate at pH 8.0, but the highest maximum optical density at 600 nm (OD_{600}) was obtained at pH 10.0 (Figure 1). The alkali tolerance of this strain was comparable to that of obligate alkaliphiles, such as *Alkaliphilus transvaalensis* and *Bacillus marmarensis* [13,14]. Strain CM4A is thus considered as an alkali-tolerant microorganism. The ability to withstand alkaline pH up to 11.9 has been reported in the type strain of *E. faecalis* [15], which is consistent with the results of strain CM4A.

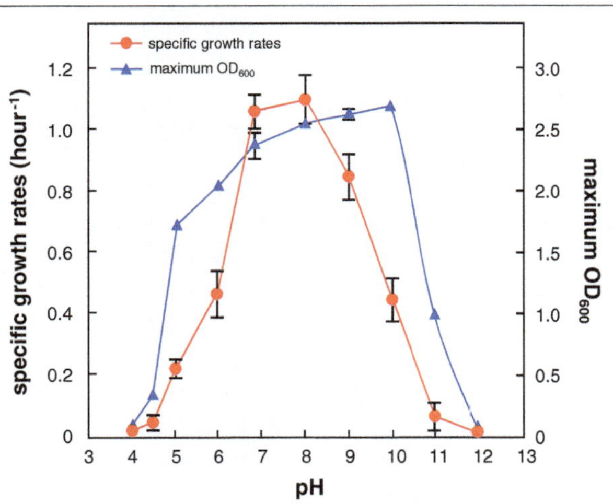

Figure 1 Growth characteristics of *E. faecalis* strain CM4A at various pHs. Effect of pH on growth as determined by specific growth rates (red circle) and maximum optical density at 600 nm (OD_{600}) (blue triangle) in the absence of butanol. The values and error bars represent the mean and SD of triplicate experiments.

pH effect on butanol tolerance of *E. faecalis* strain CM4A

To elucidate an effect of pH on butanol tolerance, strain CM4A was tested for butanol tolerance under various pHs. Since cells did not grow sufficiently at pH 5.0 or pH 11.0 in the presence of 2.0% butanol (data not shown), the relative growth rates were compared under four different pH conditions (pH 6.0, 6.8, 8.0, 10.0) in the presence and absence of butanol. Interestingly, the strain exhibited the highest growth rate at neutral pH (pH 6.8) in the presence of 2.0% butanol, although its

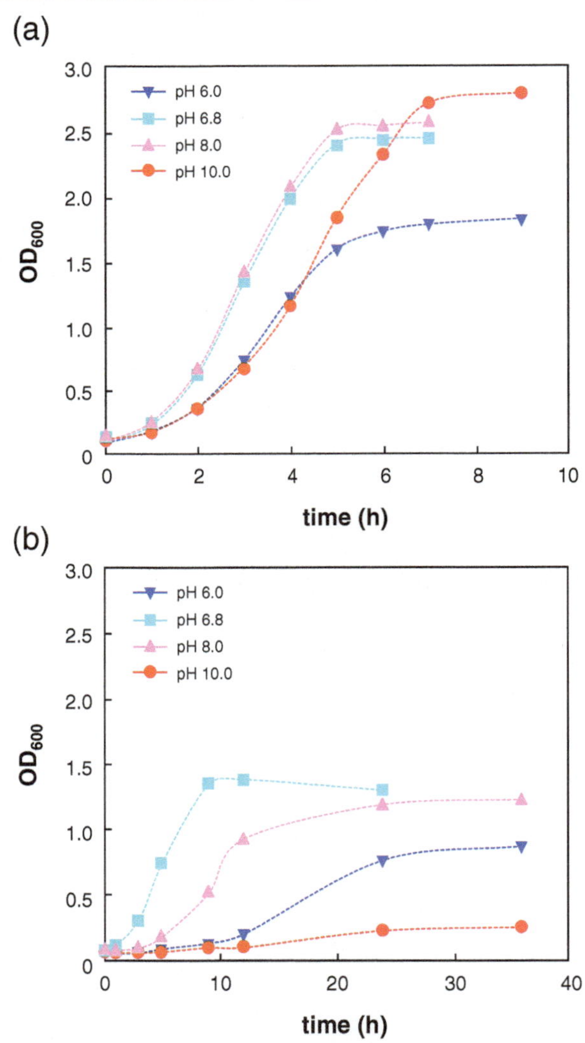

Figure 2 Butanol tolerance assay of *E. faecalis* strain CM4A. Growth curves of strain CM4A in the absence **(a)** and presence **(b)** of 2.0% butanol under different pH conditions. The values represent the mean of triplicate experiments. OD_{600}, optical density at 600 nm.

optimum growth was observed at alkaline pH (pH 8.0) in the absence of butanol (Figure 2). Note that this alkali-tolerant strain also showed significantly higher growth rate at pH 6.0 than at pH 10.0 in the presence of butanol ($p < 0.01$ by *t*-test). Likewise, other butanol-tolerant strains, *Bacillus amyloliquefaciens* strain FW5A and *Lysinibacillus xylanilyticus* strain SK7A in the phylum *Firmicutes* [11], also showed relatively high butanol tolerance at acidic pH conditions (see Additional file 1 and Additional file 2), suggesting that this phenomenon might be a common trait in butanol-tolerant strains within the *Firmicutes*.

Changes in the fatty acid compositions of *E. faecalis* strain CM4A in the absence of butanol at alkaline pH

To elucidate the mechanism of alkali tolerance, the response of the cell membrane to alkali challenge was further investigated. In particular, changes in the fatty acid composition of strain CM4A were determined under different pH conditions (pH 6.0, 6.8, 8.0, 10.0) in the absence of butanol. The cellular fatty acids of strain CM4A were mainly composed of cyclopropane and normal saturated or unsaturated chains in the range from C14 to C20, with a high abundance of palmitic acid (C16:0) and *cis*-vaccenic acid (C18:1ω7c). No significant change was observed between acidic and neutral pH ($p > 0.01$ by *t*-test). In contrast, the proportion of C14 chain length fatty acids (myristic acid (C14:0) and *cis*-tetradecenoic acid (C14:1ω7c)) tended to increase with increasing pH (Table 1). Such increase in the proportion of short chain fatty acids are known to increase membrane fluidity [16] that is a key to microbial alkaline adaptation. It has been reported, indeed, that *E. coli*, *Listeria monocytogenes*, and alkaliphilic *Bacillus* spp. also enhance membrane fluidity under alkaline conditions by increasing unsaturated or branched-chain fatty acids [17-20]. Membrane fluidity is known to affect the configuration and activity of membrane proteins such as ATP synthase and various transporters (for example, cation/proton antiporters, and so on), which are regulating proton entry and cytoplasmic retention that serve to prevent intracellular alkalinization [21-23]. Taken together, our findings revealed that strain CM4A altered membrane fatty acid compositions for increasing its membrane fluidity and subsequently adapted to alkaline conditions.

Changes in the fatty acid compositions of *E. faecalis* strain CM4A in the presence of butanol at different pHs

The fatty acid composition of strain CM4A was also investigated at different pHs in the presence of 2% butanol. The increase in the proportion of both cyclopropane fatty acid (*cis*-11,12-methylene octadecanoic acid (cyclo-C19:0)) and total saturated fatty acids in response to butanol was found regardless of the culture pH (Table 1). These alterations causing a decrease in membrane fluidity likely contribute to maintain cellular integrity upon exposure to butanol as suggested in the previous studies [3,11,24]. It is noteworthy that the responses of the cell membrane to butanol exposure were more evident at acidic pH (Table 1). For example, the cells grown at pH 6.0 contained higher proportions of cyclopropane fatty acid (3.5 ± 1.1%) and total saturated fatty acids (57.9 ± 4.7%) than the cells grown at other pH conditions. In contrast, these changes were little observed in the cells grown at pH 8.0 and 10.0, although the proportion of saturated fatty acids was slightly high

Table 1 Changes in the fatty acid composition of strain CM4A*

	No solvents				2.0% butanol			
	pH 6.0	pH 6.8	pH 8.0	pH 10.0	pH 6.0	pH 6.8	pH 8.0	pH 10.0
C14:0	4.2 ± 0.5	4.6 ± 1.3	6.5 ± 1.9	6.9 ± 0.7	5.8 ± 0.3	5.0 ± 1.9	5.8 ± 0.1	4.6 ± 1.9
C14:1ω7c	0.1 ± 0.2	0.2 ± 0.2	0.5 ± 0.2	0.6 ± 0.1	0.3 ± 0.3	0.3 ± 0.3	0.6 ± 0.1	0.2 ± 0.2
C16:0	41.5 ± 1.1	41.6 ± 1.0	39.6 ± 3.3	39.2 ± 2.9	49.4 ± 4.1	44.8 ± 1.8	41.0 ± 1.9	43.5 ± 1.7
C16:1	0.8 ± 0.7	1.2 ± 1.1	0.9 ± 0.9	0.9 ± 0.9	1.0 ± 1.0	1.9 ± 2.0	1.3 ± 1.5	0.8 ± 0.7
C16:1ω7c	4.4 ± 1.3	6.1 ± 2.0	6.5 ± 0.8	7.5 ± 1.1	6.4 ± 1.2	7.2 ± 2.6	7.5 ± 1.8	6.9 ± 4.4
C18:0	2.2 ± 0.6	1.5 ± 0.2	1.3 ± 0.5	0.9 ± 0.2	2.7 ± 1.2	1.6 ± 0.6	1.2 ± 0.6	5.1 ± 3.6
C18:1ω7c	46.3 ± 2.4	44.9 ± 4.8	44.3 ± 3.0	43.8 ± 1.7	30.4 ± 0.9	37.6 ± 4.9	42.0 ± 4.1	38.6 ± 2.8
cyclo-C19:0	0.6 ± 0.8	ND	0.4 ± 0.7	ND	*3.5 ± 1.1*	*1.3 ± 0.6*	*0.7 ± 0.2*	*0.3 ± 0.3*
C20:1	0.0 ± 0.1	ND	0.1 ± 0.1	ND	0.5 ± 0.9	0.1 ± 0.2	ND	ND
Total C14	*4.3 ± 0.5*	*4.8 ± 1.5*	*6.9 ± 2.1*	*7.5 ± 0.8*	*6.1 ± 0.6*	*5.4 ± 2.3*	*6.4 ± 0.1*	*4.8 ± 1.8*
Total saturated fatty acids	47.9 ± 0.6	47.6 ± 1.6	47.4 ± 4.7	47.1 ± 3.3	*57.9 ± 4.7*	*51.5 ± 1.6*	*47.9 ± 1.4*	*53.2 ± 1.8*

*Each fatty acid composition is described as a percentage of the whole-cell lipids. The values are the means ± standard deviations of three independent measurements. The double-bond positions of C16:1 and C20:1 were not identified. 'Total saturated fatty acids' is the sum of myristic acid (C14:0), palmitic acid (C16:0), and stearic acid (C18:0). Abbreviations: C14:1ω7c, *cis*-tetradecenoic acid; C16:1ω7c, palmitoleic acid; C18:1ω7c, *cis*-vaccenic acid; cyclo-C19:0, *cis*-11,12-methylene octadecanoic acid; ND, not detected. The italicized values are discussed in the text.

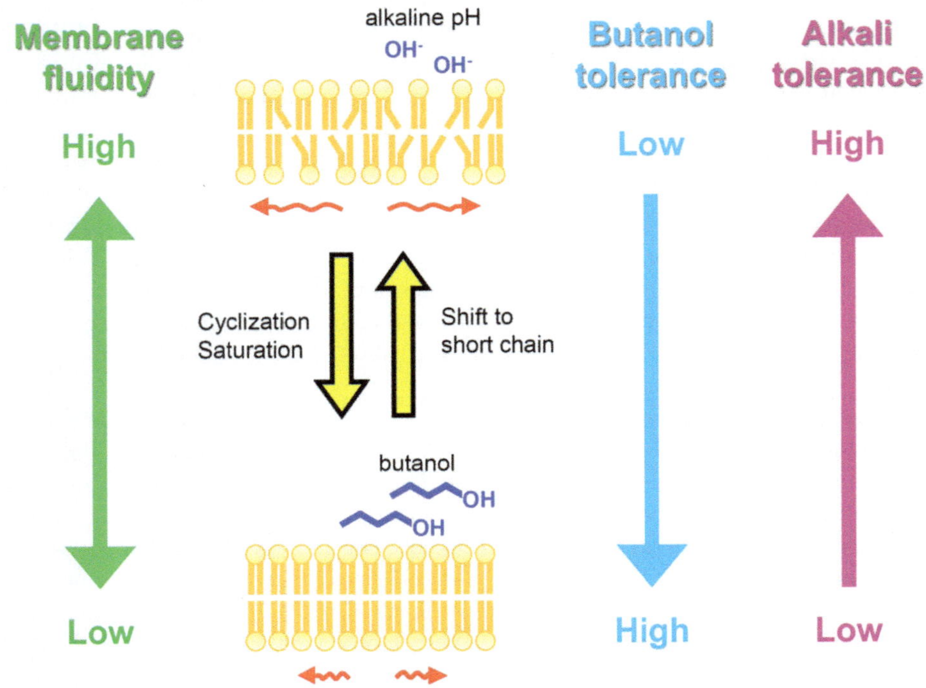

Figure 3 Schematic representation of membrane fluidity adjustment in *E. faecalis* strain CM4A. The opposite mechanisms involved in membrane fluidity adjustment behind butanol and alkali tolerance. The strain alters the proportion of membrane fatty acid components (that is cyclization, saturation, or shift to short chain) in response to butanol exposure or alkaline pH.

at pH 10.0 (53.2 ± 1.8%). The results suggested that the cells failed to decrease membrane fluidity at alkaline pHs in the presence of butanol, which appeared to be a main cause of lowered butanol tolerance under alkaline conditions.

Furthermore, the increase in the proportion of C14 chain length fatty acids with increasing pH, which occurred in the absence of butanol, was not observed in the cells grown with butanol (Table 1). This indicated that the adjustment of membrane fluidity in response to alkali challenge did not work in the presence of butanol. Thus, whereas strain CM4A would tolerate butanol by decreasing membrane fluidity, it would adapt to alkaline pH by increasing membrane fluidity (Figure 3). Due to

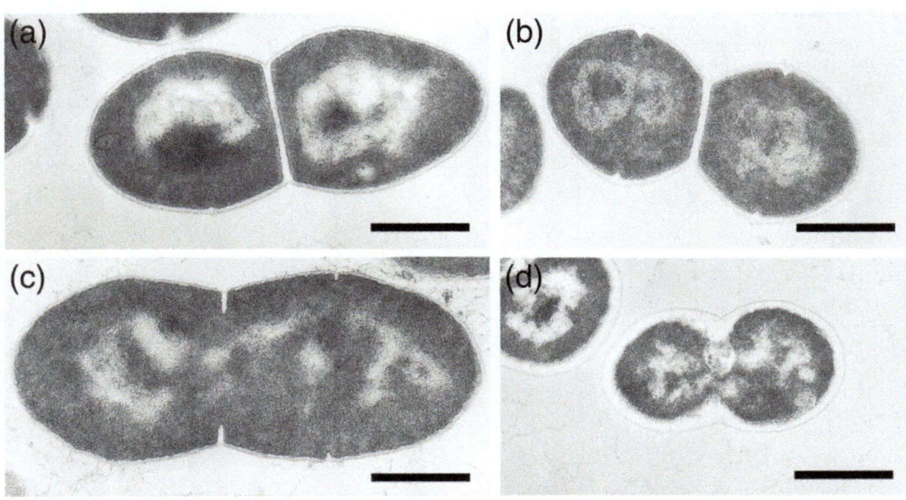

Figure 4 Cell morphology of *E. faecalis* strain CM4A. Transmission electron micrographs of the cell grown at pH 6.8 **(a, b)** or pH 10.0 **(c, d)** in the absence **(a, c)** and presence **(b, d)** of 2.0% butanol. Bars, 0.5 μm.

these two mechanisms working oppositely, the strain might lose the ability to adjust membrane fluidity under simultaneous exposure to both butanol and alkali stresses; hence, the strain might increase cell susceptibility to butanol under alkaline conditions. Indeed, by transmission electron microscopy, the cells grown with 2.0% butanol at pH 10.0 showed irregular shape with disrupted membrane structure (Figure 4). This is contrary to the cells grown in other conditions, which showed well-preserved cell structure (Figure 4).

Conclusions

In this paper, we present a conceptual study of combined effects of butanol and pH stresses on bacterial cells. It is convincingly concluded that a high butanol-tolerant bacterium, *E. faecalis* strain CM4A, shows lower ability to tolerate butanol under alkaline conditions due to the interactive effects of butanol and alkali stresses. To our knowledge, this is the first report describing the effect of pH on microbial butanol tolerance. High butanol-tolerant bacteria can be considered as an alternative butanol production host. Our findings indicate the non-negligible impact of pH on butanol tolerance of bacterial species, which provide a new insight into an efficient heterologous butanol production using alternative bacterial host.

Methods

Strains and culture conditions

Previously isolated butanol-tolerant strains able to grow in greater than 2.0% (vol/vol) butanol at pH 6.8, *E. faecalis* CM4A, *Bacillus* sp. FW5A, and *Lysinibacillus* sp. SK7A in the phylum *Firmicutes* [11], were used in this study. Cells were grown aerobically with shaking at 30°C in tryptone-glucose-yeast (TGY) medium consisting of (per liter) 20 g tryptone, 5 g glucose, 5 g yeast extract, 7 ml basal salt solution, 1 ml vitamin solution [25], and 25 mM buffer as described below.

Growth test under different pH conditions

To assess the effect of pH, strains were cultured under various pH conditions (pH 3.0 to 12.5). Growth was monitored by measuring the optical density (OD_{600}) as previously described [6,8-11]. The pH value of culture medium was adjusted by using the following appropriate buffer: sodium citrate buffer (pH 3.0), sodium acetate buffer (pH 4.0 to 5.0), potassium phosphate buffer (pH 6.0 to 9.0), sodium carbonate buffer (pH 10.0), and disodium hydrogen phosphate buffer plus NaOH (pH 11.0 to 12.5). After autoclaving, the pH value was measured by a pH meter and readjusted by adding NaOH or HCl. The pH value did not change by more than 0.3 from the initial pH during the period of exponential growth, in which physiological and morphological characterization was conducted.

Butanol tolerance assay

Growth in the presence of *n*-butanol was evaluated in the same way as mentioned above. Specific growth rate was calculated from the linear range of exponential growth. Relative growth rate was defined as the specific growth rate in the presence of butanol relative to that without butanol.

Physiological and morphological characterization

Membrane fatty acid composition and cell morphology of strain CM4A in the late exponential growth phase were investigated as previously described [11]. Briefly, fatty acid methyl esters were extracted from whole-cell methanolyzed products into *n*-hexane and then analyzed by gas chromatography-mass spectrometry system. Cell morphology was observed by using transmission electron microscopy as previously described [26].

Additional files

Additional file 1: Figure S1. Butanol tolerance assay of *Bacillus amyloliquefaciens* strain FW5A. Growth curves of strain FW5A in the absence (a) and presence (b) of 2.0% butanol under different pH conditions. The values represent the mean of triplicate experiments.

Additional file 2: Figure S2. Butanol tolerance assay of *Lysinibacillus xylanilyticus* strain SK7A. Growth curves of strain SK7A in the absence (a) and presence (b) of 2.0% butanol under different pH conditions. The values represent the mean of triplicate experiments.

Abbreviations

C14:0: myristic acid; C14:1ω7c: *cis*-tetradecenoic acid; C16:0: palmitic acid; C16:1ω7c: palmitoleic acid; C18:0: stearic acid; C18:1ω7c: *cis*-vaccenic acid; cyclo-C19:0: *cis*-11,12-methylene octadecanoic acid; OD_{600}: optical density at 600 nm; TGY: tryptone-glucose-yeast.

Competing interests

The authors declare that they have no competing interests.

Authors' contributions

MK performed all the experiments, analyzed the data, and drafted the manuscript. HT coordinated the study. HT participated in planning, writing, and data interpretation. YM, NK, SH, and YK contributed to data interpretation and revised the manuscript. All the authors read and approved the final manuscript.

Acknowledgements

This work was supported by the New Energy and Industrial Technology Development Organization (NEDO) program in Japan. We thank Xian-Ying Meng at AIST for electron microscopy and Mizuho Muramatsu and Aya Akiba at AIST for fatty acid analysis.

References

1. Schwarz WH, Gapes JR. Butanol-rediscovering a renewable fuel. BioWorld Eur. 2006;1:16–9.

2. Liu S, Qureshi N. How microbes tolerate ethanol and butanol. N Biotechnol. 2009;26:117–21.
3. Huffer S, Clark ME, Ning JC, Blanch HW, Clark DS. The role of alcohols in growth, lipid composition, and membrane fluidity of yeast, bacteria, and archaea. Appl Environ Microbiol. 2011;77:6400–8.
4. Atsumi S, Hanai T, Liao JC. Non-fermentative pathways for synthesis of branched-chain higher alcohols as biofuels. Nature. 2008;451:86–9.
5. Nielsen DR, Leonard E, Yoon SH, Tseng HC, Yuan C, Prather KLJ. Engineering alternative butanol production platforms in heterologous bacteria. Metab Eng. 2009;11:262–73.
6. Knoshaug EP, Zhang M. Butanol tolerance in a selection of microorganisms. Appl Biochem Biotechnol. 2009;153:13–20.
7. Ruhl J, Schmid A, Blank LM. Selected *Pseudomonas putida* strains can grow in the presence of high butanol concentrations. Appl Environ Microbiol. 2009;75:4653–6.
8. Li J, Zhao JB, Zhao M, Yang YL, Jiang WH, Yang S. Screening and characterization of butanol-tolerant micro-organisms. Lett Appl Microbiol. 2010;50:373–9.
9. Kataoka N, Tajima T, Kato J, Rachadech W, Vangnai A. Development of butanol-tolerant *Bacillus subtilis* strain GRSW2-B1 as a potential bioproduction host. AMB Express. 2011;1:1–11.
10. Ting CNW, Wu J, Takahashi K, Endo A, Zhao H. Screened butanol-tolerant *Enterococcus faecium* capable of butanol production. Appl Biochem Biotechnol. 2012;168:1672–80.
11. Kanno M, Katayama T, Tamaki H, Mitani Y, Meng XY, Hori T, et al. Isolation of butanol- and isobutanol-tolerant bacteria and physiological characterization of their butanol tolerance. Appl Environ Microbiol. 2013;79:6998–7005.
12. Baer SH, Blaschek HP, Smith TL. Effect of butanol challenge and temperature on lipid composition and membrane fluidity of butanol-tolerant *Clostridium acetobutylicum*. Appl Environ Microbiol. 1987;53:2854–61.
13. Takai K, Moser DP, Onstott TC, Spoelstra N, Pfiffner SM, Dohnalkova A, et al. *Alkaliphilus transvaalensis* gen. nov., sp. nov., an extremely alkaliphilic bacterium isolated from a deep South African gold mine. Int J Syst Evol Microbiol. 2001;51:1245–56.
14. Denizci AA, Kazan D, Erarslan A. *Bacillus marmarensis* sp. nov., an alkaliphilic, protease-producing bacterium isolated from mushroom compost. Int J Syst Evol Microbiol. 2010;60:1590–4.
15. Flahaut S, Hartke A, Giard JC, Auffray Y. Alkaline stress response in *Enterococcus faecalis*: adaptation, cross-protection, and changes in protein synthesis. Appl Environ Microbiol. 1997;63:812–4.
16. Mykytczuk NCS, Trevors JT, Leduc LG, Ferroni GD. Fluorescence polarization in studies of bacterial cytoplasmic membrane fluidity under environmental stress. Prog Biophys Mol Biol. 2007;95:60–82.
17. Clejan S, Krulwich TA, Mondrus KR, Setoyoung D. Membrane lipid composition of obligately and facultatively alkalophilic strains of *Bacillus* spp. J Bacteriol. 1986;168:334–40.
18. Yeo IH, Han SK, Yu JH, Bai DH. Isolation of novel alkalophilic *Bacillus alcalophilus* subsp. YB380 and the characteristics of its yeast cell wall hydrolase. J Microbiol Biotechnol. 1998;8:501–8.
19. Yuk HG, Marshall DL. Adaptation of *Escherichia coli* O157:H7 to pH alters membrane lipid composition, verotoxin secretion and resistance to simulated gastric fluid acid. Appl Environ Microbiol. 2004;70:3500–5.
20. Giotis ES, McDowell DA, Blair IS, Wilkinson BJ. Role of branched-chain fatty acids in pH stress tolerance in *Listeria monocytogenes*. App Environ Microbiol. 2007;73:997–1001.
21. Borochov H, Shinitzky M. Vertical displacement of membrane proteins mediated by changes in microviscosity. Proc Natl Acad Sci U S A. 1976;73:4526–30.
22. Shinitzky M, Borochov H, Wilbrandt W. Lipid fluidity as a physiological regulator of membrane transport and enzyme activities. In: Lassen UV, With HH, Wiesing IO, editors. Membrane Transport in Erythrocytes. Copenhagen: Munksgaard; 1980. p. 91–102.
23. Padan E, Bibi E, Ito M, Krulwich TA. Alkaline pH homeostasis in bacteria: new insights. Biochim Biophys Acta. 2005;1717:67–88.
24. Pini CV, Bernal P, Godoy P, Ramos JL, Segura A. Cyclopropane fatty acids are involved in organic solvent tolerance but not in acid stress resistance in *Pseudomonas putida* DOT-T1E. Microb Biotechnol. 2009;2:253–61.
25. Hanada S, Kawase Y, Hiraishi A, Takaichi S, Matsuura K, Shimada K, et al. *Porphyrobacter tepidarius* sp. nov., a moderately thermophilic aerobic photosynthetic bacterium isolated from a hot spring. Int J Syst Evol Microbiol. 1997;47:408–13.
26. Mitani Y, Meng XY, Kamagata Y, Tamura T. Characterizationn of LtsA from *Rhodococcus erythropolis*, an enzyme with glutamine amidotransferase activity. J Bacteriol. 2005;187:2582–91.

Permissions

All chapters in this book were first published in BFB, by BioMed Central; hereby published with permission under the Creative Commons Attribution License or equivalent. Every chapter published in this book has been scrutinized by our experts. Their significance has been extensively debated. The topics covered herein carry significant findings which will fuel the growth of the discipline. They may even be implemented as practical applications or may be referred to as a beginning point for another development.

The contributors of this book come from diverse backgrounds, making this book a truly international effort. This book will bring forth new frontiers with its revolutionizing research information and detailed analysis of the nascent developments around the world.

We would like to thank all the contributing authors for lending their expertise to make the book truly unique. They have played a crucial role in the development of this book. Without their invaluable contributions this book wouldn't have been possible. They have made vital efforts to compile up to date information on the varied aspects of this subject to make this book a valuable addition to the collection of many professionals and students.

This book was conceptualized with the vision of imparting up-to-date information and advanced data in this field. To ensure the same, a matchless editorial board was set up. Every individual on the board went through rigorous rounds of assessment to prove their worth. After which they invested a large part of their time researching and compiling the most relevant data for our readers.

The editorial board has been involved in producing this book since its inception. They have spent rigorous hours researching and exploring the diverse topics which have resulted in the successful publishing of this book. They have passed on their knowledge of decades through this book. To expedite this challenging task, the publisher supported the team at every step. A small team of assistant editors was also appointed to further simplify the editing procedure and attain best results for the readers.

Apart from the editorial board, the designing team has also invested a significant amount of their time in understanding the subject and creating the most relevant covers. They scrutinized every image to scout for the most suitable representation of the subject and create an appropriate cover for the book.

The publishing team has been an ardent support to the editorial, designing and production team. Their endless efforts to recruit the best for this project, has resulted in the accomplishment of this book. They are a veteran in the field of academics and their pool of knowledge is as vast as their experience in printing. Their expertise and guidance has proved useful at every step. Their uncompromising quality standards have made this book an exceptional effort. Their encouragement from time to time has been an inspiration for everyone.

The publisher and the editorial board hope that this book will prove to be a valuable piece of knowledge for researchers, students, practitioners and scholars across the globe.

List of Contributors

Yvonne Stolze
Institute for Genome Research and Systems Biology, CeBiTec, Bielefeld University, Universitätsstraße 25, D-33615 Bielefeld, Germany

Martha Zakrzewski
QIMR Berghofer Medical Research Institute Herston, 300 Herston Road, Brisbane QLD 4006, Australia

Irena Maus
Institute for Genome Research and Systems Biology, CeBiTec, Bielefeld University, Universitätsstraße 25, D-33615 Bielefeld, Germany

Felix Eikmeyer
Institute for Genome Research and Systems Biology, CeBiTec, Bielefeld University, Universitätsstraße 25, D-33615 Bielefeld, Germany

Sebastian Jaenicke
Bioinformatics Resource Facility, CeBiTec, Bielefeld University, Universitätsstraße 25, D-33615 Bielefeld, Germany

Nils Rottmann
NORTH-TEC Maschinenbau GmbH, Oldenhörn 1, 25821 Bredstedt, Germany

Clemens Siebner
Institute for Genome Research and Systems Biology, CeBiTec, Bielefeld University, Universitätsstraße 25, D-33615 Bielefeld, Germany

Alfred Pühler
Institute for Genome Research and Systems Biology, CeBiTec, Bielefeld University, Universitätsstraße 25, D-33615 Bielefeld, Germany

Andreas Schlüter
Institute for Genome Research and Systems Biology, CeBiTec, Bielefeld University, Universitätsstraße 25, D-33615 Bielefeld, Germany

Kazuhiko Kurosawa
Department of Biology, Massachusetts Institute of Technology, 77 Massachusetts Avenue, Cambridge, MA 02139, USA

Andreas Radek
Department of Biology, Massachusetts Institute of Technology, 77 Massachusetts Avenue, Cambridge, MA 02139, USA
Institute of Bio- and Geosciences, IBG-1: Biotechnology, Systems Biotechnology, Forschungszentrum Juelich, 52425 Juelich, Germany

Jens K Plassmeier
Department of Biology, Massachusetts Institute of Technology, 77 Massachusetts Avenue, Cambridge, MA 02139, USA

Anthony J Sinskey
Department of Biology, Massachusetts Institute of Technology, 77 Massachusetts Avenue, Cambridge, MA 02139, USA
Engineering Systems Division, Massachusetts Institute of Technology, 77 Massachusetts Avenue, Cambridge, MA 02139, USA

Tim Snoek
Laboratory for Genetics and Genomics, Centre of Microbial and Plant Genetics (CMPG), KU Leuven, Kasteelpark Arenberg 22, 3001 Leuven, Belgium
Laboratory for Systems Biology, VIB, Bio-Incubator, Gaston Geenslaan 1, 3001 Leuven, Belgium

Martina Picca Nicolino
Laboratory for Genetics and Genomics, Centre of Microbial and Plant Genetics (CMPG), KU Leuven, Kasteelpark Arenberg 22, 3001 Leuven, Belgium
Laboratory for Systems Biology, VIB, Bio-Incubator, Gaston Geenslaan 1, 3001 Leuven, Belgium

Stefanie Van den Bremt
Laboratory of Enzyme, Fermentation and Brewing Technology, KU Leuven technologiecampus Ghent, Gebroeders
De Smetstraat 1, 9000 Ghent, Belgium

Stijn Mertens
Laboratory for Genetics and Genomics, Centre of Microbial and Plant Genetics (CMPG), KU Leuven, Kasteelpark Arenberg 22, 3001 Leuven, Belgium
Laboratory for Systems Biology, VIB, Bio-Incubator, Gaston Geenslaan 1, 3001 Leuven, Belgium

Veerle Saels
Laboratory for Genetics and Genomics, Centre of Microbial and Plant Genetics (CMPG), KU Leuven, Kasteelpark Arenberg 22, 3001 Leuven, Belgium
Laboratory for Systems Biology, VIB, Bio-Incubator, Gaston Geenslaan 1, 3001 Leuven, Belgium

Alex Verplaetse
Laboratory of Enzyme, Fermentation and Brewing Technology, KU Leuven technologiecampus Ghent, Gebroeders
De Smetstraat 1, 9000 Ghent, Belgium

Jan Steensels
Laboratory for Genetics and Genomics, Centre of Microbial and Plant Genetics (CMPG), KU Leuven, Kasteelpark Arenberg 22, 3001 Leuven, Belgium
Laboratory for Systems Biology, VIB, Bio-Incubator, Gaston Geenslaan 1, 3001 Leuven, Belgium

Kevin J Verstrepen
Laboratory for Genetics and Genomics, Centre of Microbial and Plant Genetics (CMPG), KU Leuven, Kasteelpark Arenberg 22, 3001 Leuven, Belgium
Laboratory for Systems Biology, VIB, Bio-Incubator, Gaston Geenslaan 1, 3001 Leuven, Belgium

Zara Merali
The Biorefinery Centre, Institute of Food Research, Norwich Research Park, Colney, Norwich NR4 7UA, UK

Samuel R A Collins
The Biorefinery Centre, Institute of Food Research, Norwich Research Park, Colney, Norwich NR4 7UA, UK

Adam Elliston
The Biorefinery Centre, Institute of Food Research, Norwich Research Park, Colney, Norwich NR4 7UA, UK

David R Wilson
The Biorefinery Centre, Institute of Food Research, Norwich Research Park, Colney, Norwich NR4 7UA, UK

Andres Käsper
Biogold OÜ, Lossi 19A, 12616 Tallinn, Estonia

Keith W Waldron
The Biorefinery Centre, Institute of Food Research, Norwich Research Park, Colney, Norwich NR4 7UA, UK

Navya Menon
BBSRC/EPSRC Centre for Synthetic Biology of Fine and Speciality Chemicals, Manchester Institute of Biotechnology, Faculty of Life Sciences, 131 Princess Street, The University of Manchester, Manchester M1 7DN, UK

András Pásztor
Molecular Plant Biology, Department of Biochemistry, Tykistökatu 6A 6krs, University of Turku, FI 20014 TURUN YLIOPISTO, Turku, Finland

Binuraj RK Menon
BBSRC/EPSRC Centre for Synthetic Biology of Fine and Speciality Chemicals, Manchester Institute of Biotechnology, Faculty of Life Sciences, 131 Princess Street, The University of Manchester, Manchester M1 7DN, UK

Pauli Kallio
Molecular Plant Biology, Department of Biochemistry, Tykistökatu 6A 6krs, University of Turku, FI 20014 TURUN YLIOPISTO, Turku, Finland

Karl Fisher
BBSRC/EPSRC Centre for Synthetic Biology of Fine and Speciality Chemicals, Manchester Institute of Biotechnology, Faculty of Life Sciences, 131 Princess Street, The University of Manchester, Manchester M1 7DN, UK

M Kalim Akhtar
Molecular Plant Biology, Department of Biochemistry, Tykistökatu 6A 6krs, University of Turku, FI 20014 TURUN YLIOPISTO, Turku, Finland

David Leys
BBSRC/EPSRC Centre for Synthetic Biology of Fine and Speciality Chemicals, Manchester Institute of Biotechnology, Faculty of Life Sciences, 131 Princess Street, The University of Manchester, Manchester M1 7DN, UK

Patrik R Jones
Molecular Plant Biology, Department of Biochemistry, Tykistökatu 6A 6krs, University of Turku, FI 20014 TURUN YLIOPISTO, Turku, Finland
Department of Life Sciences, Imperial College London, Sir Alexander

Nigel S Scrutton
BBSRC/EPSRC Centre for Synthetic Biology of Fine and Speciality Chemicals, Manchester Institute of Biotechnology, Faculty of Life Sciences, 131 Princess Street, The University of Manchester, Manchester M1 7DN, UK

Shih-Hsin Ho
Organization of Advanced Science and Technology, Kobe University, 1-1 Rokkodai, Nada-ku, Kobe 657-8501, Japan

Akihito Nakanishi
Organization of Advanced Science and Technology, Kobe University, 1-1 Rokkodai, Nada-ku, Kobe 657-8501, Japan

Xiaoting Ye
Organization of Advanced Science and Technology, Kobe University, 1-1 Rokkodai, Nada-ku, Kobe 657-8501, Japan

Jo-Shu Chang
Department of Chemical Engineering, National Cheng Kung University, Tainan 701, Taiwan
Research Center for Energy Technology and Strategy, National Cheng Kung University, Tainan 701, Taiwan
Center for Bioscience and Biotechnology, National Cheng Kung University, Tainan 701, Taiwan

Chun-Yen Chen
Center for Bioscience and Biotechnology, National Cheng Kung University, Tainan 701, Taiwan

Tomohisa Hasunuma
Organization of Advanced Science and Technology, Kobe University, 1-1 Rokkodai, Nada-ku, Kobe 657-8501, Japan

Akihiko Kondo
Department of Chemical Science and Engineering, Graduate School of Engineering, Kobe University, 1-1 Rokkodai, Nada-ku, Kobe 657-8501, Japan
Biomass Engineering Program, RIKEN, 1-7-22 Suehiro-cho, Tsurumi-ku, Yokohama, Kanagawa 230-0045, Japan

Shaolong Sun
No. 35 Tsing Hua East Road, Haidian District, Beijing Key Laboratory of Lignocellulosic Chemistry, Beijing Forestry University, Beijing 100083, China

Jialong Wen
No. 35 Tsing Hua East Road, Haidian District, Beijing Key Laboratory of Lignocellulosic Chemistry, Beijing Forestry University, Beijing 100083, China

Shaoni Sun
No. 35 Tsing Hua East Road, Haidian District, Beijing Key Laboratory of Lignocellulosic Chemistry, Beijing Forestry University, Beijing 100083, China

Run-Cang Sun
No. 35 Tsing Hua East Road, Haidian District, Beijing Key Laboratory of Lignocellulosic Chemistry, Beijing Forestry University, Beijing 100083, China

Jinyong Yan
Key Laboratory of Biofuels, and Shandong Provincial Key Laboratory of Energy Genetics, Qingdao Institute of Bioenergy and Bioprocess Technology, Chinese Academy of Sciences, No. 189 Songling Road, 266101 Qingdao, Shandong, China

Yi Liu
Key Laboratory of Biofuels, and Shandong Provincial Key Laboratory of Energy Genetics, Qingdao Institute of Bioenergy and Bioprocess Technology, Chinese Academy of Sciences, No. 189 Songling Road, 266101 Qingdao, Shandong, China
University of Chinese Academy of Sciences, No. 19A Yuquan Road, 100049 Beijing, China

Cong Wang
Key Laboratory of Biofuels, and Shandong Provincial Key Laboratory of Energy Genetics, Qingdao Institute of Bioenergy and Bioprocess Technology, Chinese Academy of Sciences, No. 189 Songling Road, 266101 Qingdao, Shandong, China

Bingnan Han
Key Laboratory for Marine Drugs, Department of Pharmacy, Renji Hospital, Shanghai Jiao Tong University School of Medicine, No. 145 Shandongzhong Road, 200127 Shanghai, China

Shengying Li
Key Laboratory of Biofuels, and Shandong Provincial Key Laboratory of Energy Genetics, Qingdao Institute of Bioenergy and Bioprocess Technology, Chinese Academy of Sciences, No. 189 Songling Road, 266101 Qingdao, Shandong, China

Vera Novy
Institute of Biotechnology and Biochemical Engineering, Graz University of Technology, NAWI Graz, Petersgasse 12/I, 8010 Graz, Austria

Karin Longus
Institute of Biotechnology and Biochemical Engineering, Graz University of Technology, NAWI Graz, Petersgasse 12/I, 8010 Graz, Austria

Bernd Nidetzky
Institute of Biotechnology and Biochemical Engineering, Graz University of Technology, NAWI Graz, Petersgasse 12/I, 8010 Graz, Austria
Austrian Centre of Industrial Biotechnology, Petersgasse 14, A-8010 Graz, Austria

Jelmer Tamis
Department of Biotechnology, Delft University of Technology, Julianalaan 67, 2628 BC Delft, The Netherlands

Dimitry Y Sorokin
Department of Biotechnology, Delft University of Technology, Julianalaan 67, 2628 BC Delft, The Netherlands
Winogradsky Institute of Microbiology, RAS, Leninskii avenue, 14, Leninskii avenue, 32a, Moscow 119991, Russia

Yang Jiang
Department of Biotechnology, Delft University of Technology, Julianalaan 67, 2628 BC Delft, The Netherlands

Mark C M van Loosdrecht
Department of Biotechnology, Delft University of Technology, Julianalaan 67, 2628 BC Delft, The Netherlands

Robbert Kleerebezem
Department of Biotechnology, Delft University of Technology, Julianalaan 67, 2628 BC Delft, The Netherlands

Jie Zhang
Shandong Provincial Key Laboratory of Energy Genetics, Qingdao Institute of Bioenergy and Bioprocess Technology, Chinese Academy of Sciences, 189, Songling Road, Qingdao 266101, People's Republic of China
Qingdao Engineering Laboratory of Single Cell Oil, Qingdao Institute of Bioenergy and Bioprocess Technology, Chinese Academy of Sciences, 189, Songling Road, Qingdao 266101, People's Republic of China
University of Chinese Academy of Sciences, Chinese Academy of Sciences, 19, Yuquan Road, Beijing 100049, People's Republic of China

Ya-Jun Liu
Shandong Provincial Key Laboratory of Energy Genetics, Qingdao Institute of Bioenergy and Bioprocess Technology, Chinese Academy of Sciences, 189, Songling Road, Qingdao 266101, People's Republic of China
Qingdao Engineering Laboratory of Single Cell Oil, Qingdao Institute of Bioenergy and Bioprocess Technology, Chinese Academy of Sciences, 189, Songling Road, Qingdao 266101, People's Republic of China

Gu-Zhen Cui
Shandong Provincial Key Laboratory of Energy Genetics, Qingdao Institute of Bioenergy and Bioprocess Technology, Chinese Academy of Sciences, 189, Songling Road, Qingdao 266101, People's Republic of China
Qingdao Engineering Laboratory of Single Cell Oil, Qingdao Institute of Bioenergy and Bioprocess Technology, Chinese Academy of Sciences, 189, Songling Road, Qingdao 266101, People's Republic of China

Qiu Cui
Shandong Provincial Key Laboratory of Energy Genetics, Qingdao Institute of Bioenergy and Bioprocess Technology, Chinese Academy of Sciences, 189, Songling Road, Qingdao 266101, People's Republic of China
Key Laboratory of Biofuels, Qingdao Institute of Bioenergy and Bioprocess Technology, Chinese Academy of Sciences, 189, Songling Road, Qingdao 266101, People's Republic of China
Qingdao Engineering Laboratory of Single Cell Oil, Qingdao Institute of Bioenergy and Bioprocess Technology, Chinese Academy of Sciences, 189, Songling Road, Qingdao 266101, People's Republic of China

Xinxing Lv
State Key Laboratory of Microbial Technology, School of Life Science, Shandong University, No.27 Shanda South Road, Jinan 250100, Shandong, People's Republic of China

Fanglin Zheng
State Key Laboratory of Microbial Technology, School of Life Science, Shandong University, No.27 Shanda South Road, Jinan 250100, Shandong, People's Republic of China

Chunyan Li
State Key Laboratory of Microbial Technology, School of Life Science, Shandong University, No.27 Shanda South Road, Jinan 250100, Shandong, People's Republic of China

Weixin Zhang
State Key Laboratory of Microbial Technology, School of Life Science, Shandong University, No.27 Shanda South Road, Jinan 250100, Shandong, People's Republic of China

Guanjun Chen
State Key Laboratory of Microbial Technology, School of Life Science, Shandong University, No.27 Shanda South Road, Jinan 250100, Shandong, People's Republic of China

Weifeng Liu
State Key Laboratory of Microbial Technology, School of Life Science, Shandong University, No.27 Shanda South Road, Jinan 250100, Shandong, People's Republic of China

Chia-wen C Hsieh
Department of Geosciences and Natural Resource Management, Faculty of Science, University of Copenhagen, Rolighedsvej 23, DK-1958 Frederiksberg C, Denmark

David Cannella
Department of Geosciences and Natural Resource Management, Faculty of Science, University of Copenhagen, Rolighedsvej 23, DK-1958 Frederiksberg C, Denmark

Henning Jørgensen
Present address: Center for Bioprocess Engineering, Department of Chemical and Biochemical Engineering, Technical University of Denmark, Søltofts Plads, Building 229, DK-2800 Kgs. Lyngby, Denmark

Claus Felby
Department of Geosciences and Natural Resource Management, Faculty of Science, University of Copenhagen, Rolighedsvej 23, DK-1958 Frederiksberg C, Denmark

Lisbeth G Thygesen
Department of Geosciences and Natural Resource Management, Faculty of Science, University of Copenhagen, Rolighedsvej 23, DK-1958 Frederiksberg C, Denmark

Junfeng Wang
Key Laboratory of Biofuels, Qingdao Institute of Bioenergy and Bioprocess Technology, Chinese Academy of Sciences, Qingdao, Shandong 266101, People's Republic of China

Jinli Liu
Low-Carbon Energy Conversion Science and Engineering Center, Shanghai Advanced Research Institute, Chinese Academy of Sciences, Shanghai 201203, People's Republic of China

Tianzhong Liu
Key Laboratory of Biofuels, Qingdao Institute of Bioenergy and Bioprocess Technology, Chinese Academy of Sciences, Qingdao, Shandong 266101, People's Republic of China

Yang Liu
Chengdu Institute of Biology, Chinese Academy of Sciences, No.9 Section 4, Renmin Nan Road, 610041 Chengdu, China
University of Chinese Academy of Sciences, No.19A Yuquan Road, 100049 Beijing, China
Key Laboratory of Environmental and Applied Microbiology, Chinese Academy of Sciences, No.9 Section 4, Renmin Nan Road, 610041 Chengdu, China
Environmental Microbiology Key Laboratory of Sichuan Province, No.9 Section 4, Renmin Nan Road, 610041 Chengdu, China

Yang Fang
Chengdu Institute of Biology, Chinese Academy of Sciences, No.9 Section 4, Renmin Nan Road, 610041 Chengdu, China
Key Laboratory of Environmental and Applied Microbiology, Chinese Academy of Sciences, No.9 Section 4, Renmin Nan Road, 610041 Chengdu, China
Environmental Microbiology Key Laboratory of Sichuan Province, No.9 Section 4, Renmin Nan Road, 610041 Chengdu, China

Mengjun Huang
Chengdu Institute of Biology, Chinese Academy of Sciences, No.9 Section 4, Renmin Nan Road, 610041 Chengdu, China
University of Chinese Academy of Sciences, No.19A Yuquan Road, 100049 Beijing, China
Key Laboratory of Environmental and Applied Microbiology, Chinese Academy of Sciences, No.9 Section 4, Renmin Nan Road, 610041 Chengdu, China
Environmental Microbiology Key Laboratory of Sichuan Province, No.9 Section 4, Renmin Nan Road, 610041 Chengdu, China

Yanling Jin
Chengdu Institute of Biology, Chinese Academy of Sciences, No.9 Section 4, Renmin Nan Road, 610041 Chengdu, China
Key Laboratory of Environmental and Applied Microbiology, Chinese Academy of Sciences, No.9 Section 4, Renmin Nan Road, 610041 Chengdu, China
Environmental Microbiology Key Laboratory of Sichuan Province, No.9 Section 4, Renmin Nan Road, 610041 Chengdu, China

Jiaolong Sun
Chengdu Institute of Biology, Chinese Academy of Sciences, No.9 Section 4, Renmin Nan Road, 610041 Chengdu, China
Key Laboratory of Environmental and Applied Microbiology, Chinese Academy of Sciences, No.9 Section 4, Renmin Nan Road, 610041 Chengdu, China
Environmental Microbiology Key Laboratory of Sichuan Province, No.9 Section 4, Renmin Nan Road, 610041 Chengdu, China

Xiang Tao
Chengdu Institute of Biology, Chinese Academy of Sciences, No.9 Section 4, Renmin Nan Road, 610041 Chengdu, China
Key Laboratory of Environmental and Applied Microbiology, Chinese Academy of Sciences, No.9 Section 4, Renmin Nan Road, 610041 Chengdu, China
Environmental Microbiology Key Laboratory of Sichuan Province, No.9 Section 4, Renmin Nan Road, 610041 Chengdu, China

Guohua Zhang
Chengdu Institute of Biology, Chinese Academy of Sciences, No.9 Section 4, Renmin Nan Road, 610041 Chengdu, China
Key Laboratory of Environmental and Applied Microbiology, Chinese Academy of Sciences, No.9 Section 4, Renmin Nan Road, 610041 Chengdu, China
Environmental Microbiology Key Laboratory of Sichuan Province, No.9 Section 4, Renmin Nan Road, 610041 Chengdu, China

Kaize He
Chengdu Institute of Biology, Chinese Academy of Sciences, No.9 Section 4, Renmin Nan Road, 610041 Chengdu, China
Key Laboratory of Environmental and Applied Microbiology, Chinese Academy of Sciences, No.9 Section 4, Renmin Nan Road, 610041 Chengdu, China
Environmental Microbiology Key Laboratory of Sichuan Province, No.9 Section 4, Renmin Nan Road, 610041 Chengdu, China

Yun Zhao
Key Laboratory of Bio-Resources and Eco-Environment, Ministry of Education, College of Life Sciences, Sichuan University, N0.24 South Section 1, Yihuan Road, 610064 Chengdu, China

Hai Zhao
Chengdu Institute of Biology, Chinese Academy of Sciences, No.9 Section 4, Renmin Nan Road, 610041 Chengdu, China
Key Laboratory of Environmental and Applied Microbiology, Chinese Academy of Sciences, No.9 Section 4, Renmin Nan Road, 610041 Chengdu, China

Environmental Microbiology Key Laboratory of Sichuan Province, No.9 Section 4, Renmin Nan Road, 610041 Chengdu, China

Haijia Su
Beijing Key Laboratory of Bioprocess, Beijing University of Chemical Technology, No.15, Beisanhuan East Road, Beijing 100029, People's Republic of China

Luo Liu
Beijing Key Laboratory of Bioprocess, Beijing University of Chemical Technology, No.15, Beisanhuan East Road, Beijing 100029, People's Republic of China

Shaojie Wang
Beijing Key Laboratory of Bioprocess, Beijing University of Chemical Technology, No.15, Beisanhuan East Road, Beijing 100029, People's Republic of China

Qingfeng Wang
Beijing Key Laboratory of Bioprocess, Beijing University of Chemical Technology, No.15, Beisanhuan East Road, Beijing 100029, People's Republic of China

Yixin Jiang
Beijing Key Laboratory of Bioprocess, Beijing University of Chemical Technology, No.15, Beisanhuan East Road, Beijing 100029, People's Republic of China

Xiaocong Hou
Beijing Key Laboratory of Bioprocess, Beijing University of Chemical Technology, No.15, Beisanhuan East Road, Beijing 100029, People's Republic of China

Tianwei Tan
Beijing Key Laboratory of Bioprocess, Beijing University of Chemical Technology, No.15, Beisanhuan East Road, Beijing 100029, People's Republic of China

A Joe Shaw
Mascoma Corporation, Lebanon, NH 03766, USA
Novogy Inc., 85 Bolton St, Cambridge, MA 02140, USA

Bethany B Miller
Mascoma Corporation, Lebanon, NH 03766, USA

Stephen R Rogers
Mascoma Corporation, Lebanon, NH 03766, USA

William R Kenealy
Mascoma Corporation, Lebanon, NH 03766, USA
Verdezyne Inc., 2715 Loker Avenue West, Carlsbad, CA 92010, USA

Alex Meola
Mascoma Corporation, Lebanon, NH 03766, USA

Ashwini Bhandiwad
Thayer School of Engineering, Dartmouth College, Hanover, NH 03755, USA
Energy Biosciences Institute, 2151 Berkeley Way, Berkeley, CA 94704, USA

W Ryan Sillers
Mascoma Corporation, Lebanon, NH 03766, USA
Myriant Corporation, 66 Cummings Park, Woburn, MA 01801, USA

Indraneel Shikhare
Mascoma Corporation, Lebanon, NH 03766, USA

David A Hogsett
Mascoma Corporation, Lebanon, NH 03766, USA
OPX Biotechnologies Inc., 2425 55th Street, Boulder, CO 80301, USA

Christopher D Herring
Mascoma Corporation, Lebanon, NH 03766, USA
Thayer School of Engineering, Dartmouth College, Hanover, NH 03755, USA

Ikechukwu C Nwaneshiudu
Department of Chemical Engineering, University of Washington, Box 351750, Seattle, WA 98195-1750, USA

Daniel T Schwartz
Department of Chemical Engineering, University of Washington, Box 351750, Seattle, WA 98195-1750, USA

Aiyou Huang
Institute of Oceanology, Chinese Academy of Sciences, Qingdao 266071, China

Lixia Liu
Institute of Plant Physiology and Ecology, Shanghai Institute for Biological Sciences, Chinese Academy of Sciences, Shanghai 200032, China

Chen Yang
Institute of Plant Physiology and Ecology, Shanghai Institute for Biological Sciences, Chinese Academy of Sciences, Shanghai 200032, China

Guangce Wang
Institute of Oceanology, Chinese Academy of Sciences, Qingdao 266071, China

Manabu Kanno
Bioproduction Research Institute, National Institute of Advanced Industrial Science and Technology (AIST), Tsukuba, Ibaraki 305-8566, Japan

Hideyuki Tamaki
Bioproduction Research Institute, National Institute of Advanced Industrial Science and Technology (AIST), Tsukuba, Ibaraki 305-8566, Japan

Yasuo Mitani
Bioproduction Research Institute, National Institute of Advanced Industrial Science and Technology (AIST), Tsukuba, Ibaraki 305-8566, Japan

Nobutada Kimura
Bioproduction Research Institute, National Institute of Advanced Industrial Science and Technology (AIST), Tsukuba, Ibaraki 305-8566, Japan

Satoshi Hanada
Bioproduction Research Institute, National Institute of Advanced Industrial Science and Technology (AIST), Tsukuba, Ibaraki 305-8566, Japan

Yoichi Kamagata
Bioproduction Research Institute, National Institute of Advanced Industrial Science and Technology (AIST), Tsukuba, Ibaraki 305-8566, Japan